Ergebnisse der Mathematik und ihrer Grenzgebiete

Band 16

Herausgegeben von
P. R. Halmos · P. J. Hilton · R. Remmert · B. Szőkefalvi-Nagy

Unter Mitwirkung von
L. V. Ahlfors · R. Baer · F. L. Bauer · R. Courant
A. Dold · J. L. Doob · S. Eilenberg · M. Kneser · G. H. Müller
M. M. Postnikov · B. Segre · E. Sperner

Geschäftsführender Herausgeber: P. J. Hilton

Lamberto Cesari

Asymptotic Behavior and Stability Problems in Ordinary Differential Equations

Third Edition

With 37 Figures

Springer-Verlag New York Heidelberg Berlin 1971

Lamberto Cesari
University of Michigan
Ann Arbor, MI 48104, U.S.A.

AMS Subject Classifications (1970)
Primary 34A15, 34A30, 34C05, 34C15, 34C25, 34D05, 34D10, 34D20
Secondary 34E25, 34E15, 34E20

ISBN-13: 978-3-642-85673-0 e-ISBN-13: 978-3-642-85671-6

DOI: 10.1007/978-3-642-85671-6

Softcover reprint of the hardcover 1st edition 1971.

Preface to the First Edition

In the last few decades the theory of ordinary differential equations has grown rapidly under the action of forces which have been working both from within and without: from within, as a development and deepening of the concepts and of the topological and analytical methods brought about by LYAPUNOV, POINCARÉ, BENDIXSON, and a few others at the turn of the century; from without, in the wake of the technological development, particularly in communications, servomechanisms, automatic controls, and electronics. The early research of the authors just mentioned lay in challenging problems of astronomy, but the line of thought thus produced found the most impressive applications in the new fields.

The body of research now accumulated is overwhelming, and many books and reports have appeared on one or another of the multiple aspects of the new line of research which some authors call "qualitative theory of differential equations".

The purpose of the present volume is to present many of the viewpoints and questions in a readable short report for which completeness is not claimed. The bibliographical notes in each section are intended to be a guide to more detailed expositions and to the original papers.

Some traditional topics such as the Sturm comparison theory have been omitted. Also excluded were all those papers, dealing with special differential equations motivated by and intended for the applications. General theorems have been emphasized wherever possible. Not all proofs are given but only the typical ones for each section and some are just outlined.

I wish to thank the colleagues who have read parts of the manuscript and have made suggestions: W.R. FULLER, R.A. GAMBILL, M. GOLOMB, J.K. HALE, N.D. KAZARINOFF, C.R. PUTNAM, and E. SILVERMAN. I am indebted to A.W. RANSOM and W.E. THOMPSON for helping with the proofs. Finally, I want to express my appreciation to the Springer Verlag for its accomplished and discerning handling of the manuscript.

June 1958

LAMBERTO CESARI
Lafayette, Ind.

Preface to the Second Edition

This second edition, which has become necessary within so short a time, presents no major changes.

However new results in the line of work of the author and of J. K. Hale have made it advisable to rewrite section (8.5). Also, some references to most recent work have been added.

June 1962

Lamberto Cesari
University of Michigan
Ann Arbor

Preface to the Third Edition

Some minor changes have made in the text with respect to the second edition, particularly at the pages 86 and 95.

In 1964 a Russian edition of this book appeared (MIR, Moscow), cared for by V. V. Nemyckii, who added a few appendices and updated the bibliography to the year 1961. It has been impossible to include these additions in the present edition.

November 1970

Lamberto Cesari
University of Michigan
Ann Arbor

Contents

Chapter I

The concept of stability and systems with constant coefficients

§ 1. Some remarks on the concept of stability

1.1. Existence, uniqueness, continuity. We shall often consider systems of n first order normal differential equations

$$x_i' = f_i(t, x_1, \ldots, x_n), \qquad i = 1, 2, \ldots, n, \tag{1.1.1}$$

where $x_i' = dx_i/dt$ and where t is real and f_i, x_i real or complex. We shall often denote $x = (x_1, \ldots, x_n)$ as a "point" or a "vector" and t as the "time", and E_n shall denote the space of the points x. Often (1.1.1) is obtained by a transformation of the m second order Lagrange equations relative to a mechanical system with m degrees of freedom, and thus $n = 2m$.

Sometimes we shall denote also the $(n+1)$-tuple $(t, x_1, \ldots, x_n)$ as a "point", and E_{n+1} shall denote the space of the points $(t, x_1, \ldots, x_n)$, or (t, x). We shall be concerned with the behavior of the solutions of (1.1.1) for $t \geq t_0$ and $t \to \infty$ (at the right of t_0), or for $t \leq t_0$ and $t \to -\infty$ (at the left of t_0) for some given t_0. The functions f_i are supposed to be defined in convenient sets S of "points" $(t, x_1, \ldots, x_n)$. Expressions like open, or closed sets S, are considered as self-explanatory. We shall say often that a set S is open at the right [left] of t_0 if S is open when we restrict ourselves to points with $t \geq t_0$ [$t \leq t_0$]. Finally, expressions like continuity of the functions f_i at a point $(t, x_1, \ldots, x_n)$, or of the functions $x_i(t)$ at a time t do not need explanations. By using vector notations, the system (1.1.1) can be written in the form

$$x' = f(t, x), \tag{1.1.2}$$

where $x' = dx/dt$, and where x, f are the vectors

$$x = (x_1, \ldots, x_n), \qquad x_i = x_i(t), \qquad f = (f_1, \ldots, f_n), \qquad f_i = f_i(t, x).$$

If f does not depend on t, then system (1.1.2) is called *autonomous*. If f is periodic in t, of some period T, i.e., $f(t+T, x) = f(t, x)$ for all t and x, then (1.1.2) is called *periodic*.

We shall denote by $\|u\| = |u_1| + \cdots + |u_n|$ the norm of any vector u. If t_0, $x_0 = (x_{10}, \ldots, x_{n0})$, $b > 0$, $a > 0$, are given, we shall often consider sets S (tubes) defined by $S = [t_0 \leq t \leq t_0 + a, \|x - x_0\| \leq b]$, or $S = [t_0 \leq t < +\infty, \|x - x_0\| \leq b]$. It is only for the sake of simplicity that we have used the norm $\|u\|$ mentioned above in defining a tube S instead of

$$\|u\| = \max\,[|u_1|, \ldots, |u_n|], \quad \text{or} \quad \|u\| = (|u_1|^2 + \cdots + |u_n|^2)^{\frac{1}{2}},$$

for instance. The last one, or Euclidean norm, will be denoted by $\|u\|_e$.

Well known theorems of existence and uniqueness hold for system (1.1.2). We mention here only a few (see E. KAMKE [1]; S. LEFSCHETZ [2]).

(1.1. i) If $f(t, x)$ is continuous in a set $S = [t_0 \leq t \leq t_0 + a, \|x - x_0\| \leq b]$, then there is a vector function

$$x = x(t) = x(t; t_0, x_0), \; t_0 \leq t \leq t_0 + a_1, \; \|x(t) - x_0\| \leq b, \qquad (1.1.3)$$

which is a solution of (1.1.2) in $[t_0, t_0 + a_1]$ and satisfies the initial condition $x(t_0; t_0, x_0) = x_0$, where a_1 is some number $0 < a_1 \leq a$ (it may well occur that the solution $x(t; t_0, x_0)$ has actually a maximal interval of existence $[t_0, t_0 + \alpha]$ with $0 < \alpha < a$).

The same condition assures also the existence of $x(t; t_0, x_1)$ in $t_0 \leq t \leq t_0 + a_1$, for all x_1 with $\|x_1 - x_0\| \leq b_1$ for some $a_1, b_1, 0 < a_1 \leq a$, $0 < b_1 \leq b$. It may be pointed out that, if a_1 is the maximum number $0 < a_1 \leq a$, for which $\|x(t) - x_0\| \leq b$ for all $t_0 \leq t \leq t_0 + a_1$, then either $\|x(a_1) - x_0\| = b$, or $a_1 = a$. If $f(t, x)$ is continuous for all $t \geqq t_0$, $\|x - x_0\| \leq b$, then $x(t; t_0, x_0)$ may exist in $[t_0, +\infty)$, or in some maximal interval $[t_0, t_0 + a_1]$, $a_1 < +\infty$, and then $\|x(t_0 + a_1) - x_0\| = b$.

A Lipschitz condition of the type $\|f(t, x_2) - f(t, x_1)\| \leq K\|x_2 - x_1\|$ for all $(t, x_1), (t, x_2) \in S$ and some constant $K > 0$ assures the uniqueness of the solution $x(t; t_0, x_0)$ and also its continuity with respect to the initial vector x_0; that is, given $\varepsilon > 0$ there exists $\delta = \delta(\varepsilon; f, x_0, a_1) > 0$, $0 < \delta \leq b_1$, such that $\|x_1 - x_0\| \leq \delta$ implies $\|x(t; t_0, x_1) - x(t; t_0, x_0)\| \leq \varepsilon$ for all $t_0 \leq t \leq t_0 + a_1$. Thus, $x(t; t_0, x_0)$ can be said to be uniformly continuous in $[t_0, t_0 + a_1]$ with respect to the initial vector x_0.

The same condition assures also the continuity of $x(t; t_0, x_0)$ as a function of t, x_0 (and even of t_0) with analogous restrictions as to t, t_0, x_0 (see E. KAMKE [1]; S. LEFSCHETZ [2]).

More precisely let us mention the following statements.

(1.1. ii) If $f(t, x)$ is continuous in a set $S = [t_0 \leq t \leq t_0 + a, \|x - x_0\| \leq b]$, and $\|f(t, x_1) - f(t, x_2)\| \leq K\|x_1 - x_2\|$ for all $(t, x_1), (t, x_2) \in S$, and some constant K, then $x(t; t_0, x_0)$ is uniquely determined by the initial condition $x(t_0) = x_0$. In addition, if $x(t; t_0, x_0)$ exists in $[t_0, t_0 + a_1]$, $0 < a_1 \leq a$, $a_1 < +\infty$, and $\|x(t) - x_0\| < b$ in $[t_0, t_0 + a_1]$, then there is a b_1, $0 < b_1 \leq b$, such that $x(t; t_0, u)$, $u = (u_1, \ldots, u_n)$, exists in the same interval $[t_0, t_0 + a_1]$ for all $\|u - x_0\| \leq b_1$, and $x(t; t_0, u)$ is a continuous function of t and u for $t_0 \leq t \leq t_0 + a_1$ and $\|u - x_0\| \leq b_1$.

(1.1. iii) Under the conditions of (1.1. ii), if the components f_i of f have continuous partial derivatives $f_{ij} = \partial f_i / \partial x_j$ in S, $i, j = 1, \ldots, n$, then the components x_i of $x(t; t_0, u)$ have continuous partial derivatives $x_{ij} = \partial x_i/\partial u_j$ with respect to $u_j, i, j = 1, \ldots, n$, for $t_0 \leq t \leq t_0 + a_1$, $\|u - x_0\| \leq b_1$. If $f(t, x, \mu)$ is a continuous function of t, x, and a parameter μ, $\mu_1 \leq \mu \leq \mu_2$, then $x(t; t_0, u, \mu)$ is a continuous function of t, u, μ. Analogous statements hold for the continuity of the partial derivatives of higher order of x provided the components f_i of f have corresponding continuous partial derivatives (cf. E. KAMKE [1], E. A. CODDINGTON and N. LEVINSON [3]).

The Lipschitz condition in (1.1. ii) may be replaced by $\|f(t, x_1) - f(t, x_2)\| \leq L(\|x_1 - x_2\|)$ for all $(t, x_1), (t, x_2) \in S$, where $L(r)$, $0 \leq r \leq 2b$, denotes a continuous function with $L(0) = 0$, $L(r) > 0$ for $0 < r \leq 2b$, $\int_{0+}^{2b} dr/L(r) = +\infty$ (W. F. OSGOOD [1]).

It may be interesting to notice that any condition of uniqueness is also a condition of continuity with respect to the initial values. Also, it is noteworthy that a lower bound for the number a_1 in statement (1.1. i) can be given: if $\|f(t, x)\| < M$ for all $0 \leq t \leq a$, $\|x - x_0\| \leq b$, then, one may choose $a_1 \geq \min[a, M b^{-1}]$ (E. KAMKE [1]; S. LEFSCHETZ [2]). If $f(t, x)$ is continuous for all $t \geq t_0$ and real vectors x, and a solution $x(t; t_0, x_0)$ exists only in a maximal interval $[t_0, \bar{t})$, $t_0 < \bar{t} < +\infty$, then $\|x(t, t_0; x_0)\| \to +\infty$ as $t \to \bar{t} - 0$. This fact is a consequence of (1.1. i) and of the evaluation of a_1 given above (E. KAMKE [1]; or A. WINTNER [4]). An example of a system presenting this behavior is given below in (1.3), example 1.

We shall be interested in the existence of solutions $x(t; t_0, x_0)$ in the whole interval $[t_0, +\infty)$. Linear systems, i.e., systems of the form

$$x_i' = \sum_{j=1}^{n} a_{ij}(t)\, x_j + f_i(t), \qquad i = 1, 2, \ldots, n, \tag{1.1.4}$$

where $a_{ij}(t)$, $f_i(t)$ are continuous functions of t for $t \geq t_0$, have such a property, i.e., every solution $x(t; t_0, x_0)$ exists in $[t_0, +\infty)$ (E. KAMKE [1]) and, by (1.1. ii), is unique.

For systems (1.1.2) conditions for the existence of $x(t; t_0, x_0)$ in the whole interval $[t_0, +\infty)$ have been given by A. WINTNER as follows:

(1.1. iv) If $f(x)$ is a continuous function of x for all vectors x, and $\|f(x)\| \leq K\|x\|$ for some constant K and all x, then $x(t; t_0, x_0)$ exists in $[t_0, +\infty)$ for all x_0 (A. WINTNER [4]).

(1.1. v) If $f(t, x)$ is a continuous function of t and x for all $t \geq t_0$ and x, if there is a function $L(r) > 0$, $r \geq 0$, with $\int^{+\infty} dr/L(r) = +\infty$, and $\|f(t, x)\|_e \leq L(\|x\|_e)$ for all $t \geq t_0$ and x, then $x(t; t_0, x_0)$ exists in $[t_0, +\infty)$ for all x_0 (A. WINTNER [4]).

Proof of (1.1. v). Let $r = (x_1^2 + \cdots + x_n^2)^{\frac{1}{2}}$. Then, for any solution $x = x(t)$, $t_0 \leq t < \bar{t}$, we have, by differentiation and manipulation, $(r r')^2 \leq (x_1^2 + \cdots + x_n^2)(x_1'^2 + \cdots + x_n'^2)$, and hence, by (1.1. i), also $(r r')^2 \leq r^2(f_1^2 + \cdots + f_n^2)$. Thus $(r r')^2 \leq r^2\|f\|_e^2$, and $dt \geq |dr|/L(r)$. If $x(t; t_0, x_0)$ exists only in $[t_0, \bar{t})$, $\bar{t} < +\infty$, then $\|x(t)\|_e$

$\to +\infty$, hence $r = r(t) \to +\infty$, and, by integration, $+\infty > \bar{t} - t_0 \geq \int_0^{+\infty} dr/L(r) = +\infty$, a contradiction. (1.1. v) is thereby proved, hence, also (1.1. iv) is proved.

Condition (1.1. v) can be replaced by the more general one $\|f(t, x)\|_e \leq m(t)\, L(\|x\|_e)$, where $m(t)$, $t \geq t_0$, is any continuous function and $L(r)$ is as in (1.1. v) (R. Conti [7]). An analogous condition holds for f complex.

In this book we will encounter a number of conditions assuring the existence of $x(t; t_0, x_0)$ in $[t_0, +\infty)$. Let us mention that N. P. Erugin [7, 8], M. A. Krasnoselskii and M. G. Krein [2], T. Yoshizawa [1], have given other conditions to the same purpose. Obviously, the considerations above concerning $t \geq t_0$ can be repeated for $t \leq t_0$ and are omitted. This will be the rule in the present book.

According to C. Carathéodory [I, pp. 665–688] the concept of a solution of a system of ordinary differential equations can be generalized by postulating that (1.1.1), or (1.1.2), or (1.1.4) by satisfied only for almost all t, provided we restrict ourselves to solutions $x(t)$ which are absolutely continuous (AC). This is essentially due to the fact that $x(t; t_0, x_0)$ is actually a solution of the integral equation

$$x(t) = x_0 + \int_{t_0}^{t} f[u, x(u)]\, du.$$

The existence, uniqueness, and continuity theorems given above hold also in this new setting, which may allow some relaxation in the conditions required on f. A condition of uniqueness replacing condition (1.1. ii) as well as the Osgood condition mentioned above is the following one due to L. Tonelli [5]: $\|(f(t, x_1) - f(t, x_2)\| \leq \varphi(t)\, L(\|x_1 - x_2\|)$, where $\varphi(t)$ is an L-integrable function in $[t_0, t_0 + a]$ and $L(r)$, $0 \leq r \leq 2b$, is a continuous function with $L(0) = 0$, $L(r) > 0$ for $0 < r \leq 2b$, and $\int_{0+}^{2b} dr/L(r) = +\infty$.

Remark. If $f(t, x)$ is a continuous function of t and x for $a \leq t \leq b$ and all x, if $x_n(t)$, $a \leq t \leq b$, $n = 1, 2, \ldots$, are solutions of (1.1.1) in $[a, b]$ with $\|x_n(t)\| \leq M$, and $x_n(t) \to x(t)$ as $n \to \infty$ for all t, $a \leq t \leq b$, then $x(t)$, $a \leq t \leq b$, is a solution of (1.1.1). Indeed, $f(t, x)$ is continuous in $a \leq t \leq b$, $\|x\| \leq M$, and hence $\|f(t, x)\| \leq N$ for some N. Since $x_n' = f(t, x_n)$, we conclude that $\|x_n'(t)\| \leq N$ for all t and n, and this implies that the functions $x_n(t)$, $a \leq t \leq b$, $n = 1, 2, \ldots$, are equicontinuous. Thus, the convergence $x_n(t) \to x(t)$ is uniform in $[a, b]$. By $x_n(t) = x_n(a) + \int_a^t f[u, x_n(u)]\, du$, as $n \to \infty$, we deduce $x(t) = x(a) + \int_a^t f[u, x(u)]\, du$, and finally $x'(t) = f[t, x(t)]$, $a \leq t \leq b$.

1.2. Stability in the sense of Lyapunov. We shall now be concerned with the existence of a solution $x(t; t_0, x_0)$ of (1.1.2) for all $t \geq t_0$ and its behavior as $t \to +\infty$ (asymptotic behavior). There are many properties which may be used in the characterization of such a behavior. We shall mention the Lyapunov stability in the present article, the boundedness in (1.4), and a score of other properties in (1.5).

If $f(t, x)$ is continuous in a set S of points (t, x) and S is open at the right of t_0, if a solution $x(t) = x(t; t_0, x_0)$ of (1.1.2) exists in the infinite interval $t_0 \leq t < +\infty$, and $[t, x(t)] \in S$ for all $t \geq t_0$, then $x(t; t_0, x_0)$ is said to be *stable (at the right) in the sense of* Lyapunov if (α) there exists a $b_1 > 0$, such that every solution $x(t; t_0, x_1)$ exists in $t_0 \leq t < +\infty$ and

$[t, x(t)] \in S$ for all $t \geq t_0$ whenever the initial vector x_1 satisfies $\|x_1 - x_0\| \leq b_1$; (β) given $\varepsilon > 0$, there is a $\delta = \delta(\varepsilon; f, x_0)$, $0 < \delta \leq b_1$, such that $\|x_1 - x_0\| \leq \delta$ implies $\|x(t; t_0, x_1) - x(t; t_0, x_0)\| \leq \varepsilon$ for all $t_0 \leq t < +\infty$. Thus stability in the sense of Lyapunov turns out to be the same as uniform continuity of $x(t; t_0, x_0)$ in $[t_0, +\infty)$ with respect to the initial vector x_0. The same solution $x(t; t_0, x_0)$ is said to be *asymptotically stable (at the right)* if (α), (β) hold, and (β') there exists a $\delta = \delta(f, x_0)$, $0 < \delta \leq b_1$, such that $\|x_1 - x_0\| \leq \delta$ implies $\|x(t; t_0, x_1) - x(t; t_0, x_0)\| \to 0$ as $t \to +\infty$.

If f satisfies a Lipschitz condition as in (1.1) in each set $S = [t_0 \leq t \leq t_1, \|x - x_0\| \leq b, t_1 > t_0]$, with b sufficiently large, then the solution $x(t; t_0, x_0)$ is uniquely determined not only by the point (t_0, x_0) but also by any point (t_2, x_2) with $x_2 = x(t_2)$, $t_0 \leq t_2 \leq t_1$. The continuity theorem we have mentioned in (1.1) assures then that the Lyapunov (asymptotic) stability (or instability) of the solution $x(t; t_0; x_0)$ is not affected by replacing the couple (t_0, x_0) by any other couple (t_2, x_2), $x_2 = x(t_2)$, $t_0 \leq t_2 \leq t_1$.

If we replace the interval $[t_0, +\infty)$ by the interval $(-\infty, t_0]$ in all previous considerations, then we will define Lyapunov and asymptotic stability at the left. Lyapunov and asymptotic stability at both sides are then self-explanatory. To simplify the notations in the present book, by *stability* and *asymptotic stability* we will always mean stability and asymptotic stability in the sense of LYAPUNOV. Also, unless indicated otherwise, we will always understand stability at the right.

If the conditions (α), (β), or (β') above are satisfied only by the solutions $x(t; t_0, x_0)$ of a given manifold M of solutions of (1.1.2), then we have *Lyapunov* or *asymptotic conditional stabilities.*

For linear systems

$$x_i' = \sum_{h=1}^{n} a_{ih}(t)\, x_h + f_i(t), \qquad i = 1, 2, \ldots, n, \tag{1.2.1}$$

$t_0 \leq t < +\infty$, where $a_{ih}(t)$, $f_i(t)$ are continuous functions in $[t_0, +\infty)$, condition (α) is always satisfied. In addition, if one solution $x(t)$ is stable, then all solutions are stable, and we may say that the system (1.2.1) is stable. If one solution is asymptotically stable, then all solutions are asymptotically stable, and we may say that the system (1.2.1) is asymptotically stable.

Indeed, let x_0, x be two solutions of (1.2.1) with initial values u_0, u at $t = t_0$, and $x_0 + \Delta x$, $x + \Delta x$ the solutions with initial values $u_0 + \Delta u$, $u + \Delta u$. Since x_0 is stable, given $\varepsilon > 0$, there is $\delta > 0$ such that $\|\Delta x\| < \varepsilon$ for all $\|\Delta u\| < \delta$, $t \geq t_0$, and this in turns implies the stability of x. The same holds for asymptotic stability.

According to the definitions above, asymptotic stability implies Lyapunov stability (also, see remark at the end of 6.2).

A normal differential equation of the *n-th* order

$$x^{(n)} = f(t, x, \ldots, x^{(n-1)}), \tag{1.2.2}$$

where $x^{(h)} = d x^h / d t^h$, $h = 0, 1, \ldots, n$, can be reduced to a system (1.1.1), or (1.1.2), by the standard transformation $x = x_1$, $x' = x_2, \ldots, x^{(n-1)} = x_n$. Hence a solution $x(t) = x(t; t_0, \eta_0, \ldots, \eta_{n-1})$, $t_0 \leq t < +\infty$, of (1.2.2), satisfying the initial condition $x^{(h)}(t_0) = \eta_h$, $h = 0, 1, \ldots, n-1$, will be said to be *stable at the right in the sense of* LYAPUNOV if (α) there is a $b_1 > 0$ such that every solution $\bar{x}(t) = x(t; t_0, \bar{\eta}_0, \ldots, \bar{\eta}_{n-1})$ exists in $[t_0, +\infty)$ for every $(\bar{\eta}_0, \ldots, \bar{\eta}_{n-1})$ with $\Sigma |\bar{\eta}_h - \eta_h| < b_1$, where Σ is extended over all $h = 0, 1, \ldots, n-1$; (β) given $\varepsilon > 0$, there exists a δ, $0 < \delta \leq b_1$, such that $\Sigma |\bar{\eta}_h - \eta_h| \leq \delta$ implies $\Sigma |\bar{x}^{(h)}(t) - x^{(h)}(t)| \leq \varepsilon$ for all t, $t_0 \leq t < +\infty$. The same solution $x(t)$ shall be said to be asymptotically stable at the right if (α), (β) hold, and (β') there exists a δ_0, $0 \leq \delta_0 \leq b_1$, such that $\Sigma |\bar{\eta}_h - \eta_h| \leq \delta_0$ implies $\Sigma |\bar{x}^{(h)}(t) - x^{(h)}(t)| \to 0$ as $t \to +\infty$. Analogous definitions hold for stability at the left, or at both sides.

Finally, any normal differential system of order $n = n_1 + \cdots + n_k$,

$$x_i^{(n_i)} = f_i(x_1, \ldots, x_1^{(n_1-1)}, x_2, \ldots, x_2^{(n_2-1)}, \ldots, x_k, \ldots, x_k^{(n_k-1)}), \tag{1.2.3}$$

$i = 1, 2, \ldots, k$, can be reduced to a system (1.1.1) by a transformation analogous to the one used for (1.2.2), and the concept of stability in the sense of LYAPUNOV may be given in a similar way. For reference see A. LYAPUNOV [3].

For nonnormal systems of differential equations, the questions of existence, uniqueness, and continuity are much more difficult to answer. Nevertheless the definitions of stability given above for systems (1.1.1), (1.2.2), (1.2.3) apply formally also to nonnormal systems, as

$$F_i(t, x_1, \ldots, x_n, x_i', \ldots, x_n') = 0, \quad i = 1, 2, \ldots, n; \tag{1.2.4}$$

$$F(t, x, x', \ldots, x^{(n)}) = 0; \tag{1.2.5}$$

$$F_i(t, x_1, \ldots, x_1^{(n_1)}, x_2, \ldots, x_2^{(n_2)}, \ldots, x_k, \ldots, x_k^{(n_k)}) = 0, \quad i = 1, 2, \ldots, k. \tag{1.2.6}$$

1.3. Examples. 1. The solution $x = 0$ of the equation $x' = x^2$, x real, is stable neither at the right nor at the left since every real solution $x = x_0(1 - t x_0)^{-1}$ with $x_0 > 0$ ($x_0 < 0$) ceases to exist at $t = x_0^{-1}$. This shows that condition (α) does not hold in the present case (see illustration).

2. The solution $x = -1$ of the equation $x' = 1 - x^2$ is not stable at the right since all (real) solutions $x = \tanh(t - t_0 + k)$, $k = \operatorname{arctanh} x_0$, $-1 < x_0 < +1$, approach $+1$ as $t \to +\infty$. Every solution $x = \tanh(t - t_0 + k)$, as well as the solution $x = +1$, is asymptotically stable at the right (see illustration).

3. Every solution of the equation $x' = 0$ is stable at both sides, but asymptotically stable at neither side. Indeed, $x(t; t_0, x_0) = x_0 =$ constant for every t, and $|x(t; t_0, x_1) - x(t; t_0, x_0)| = |x_1 - x_0|$ for every t.

4. Every solution of the equation $x' + x = 0$ is asymptotically stable at the right and unstable at the left, since $x(t) = C e^{-t}$, C constant, and thus $|x(t; t_0, x_1) - x(t; t_0, x_0)| \to 0$ as $t \to +\infty$ and $\to +\infty$ as $t \to -\infty$.

5. The solution $x=0$ of the equation $x''-x=0$ is conditionally stable at the right with respect to the manifold M of the solutions of the form $x=Ce^{-t}$.

6. The solution $x=0$ of the equation $x'-|x|=0$ is (conditionally) asymptotically stable with respect to the manifold M of the solutions $x(t)\leq 0$.

1

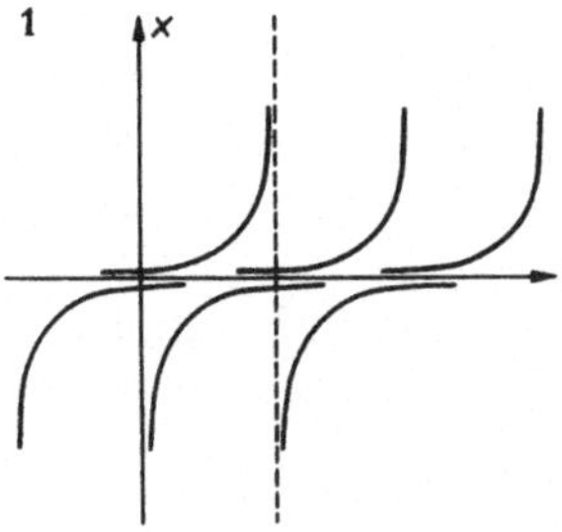

2

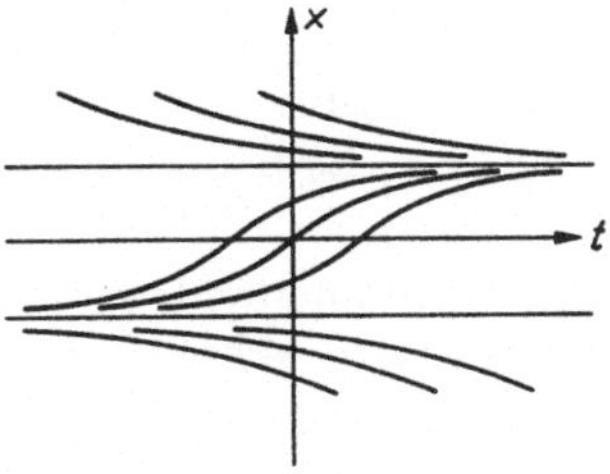

7. Every nonzero solution of the equation $x''=-2^{-1}(x^2+(x^4+4x'^2)^{\frac{1}{2}})x$ is stable neither at the right nor at the left. Indeed, every solution has the form $x=c\sin(ct+d)$, c, d constants, and thus for every two solutions $x_1=c_1\sin(c_1t+d_1)$, $x_2=c_2\sin(c_2t+d_2)$, with $c_1\neq 0$, $c_2\neq 0$, c_2/c_1 irrational, we have $\overline{\lim}\,|x_1-x_2|=|c_1|+|c_2|$ as $t\to+\infty$, as well as $t\to-\infty$. The zero solution is stable at both sides.

1.4. Boundedness. Given a system (1.1.2) of n first order differential equations, a solution $x=x(t;t_0,x_0)$, $t_0\leq t<+\infty$, is said to be *bounded at the right* if $\|x\|\leq M$ for all $t\geq t_0$ and some finite $M>0$; that is, if $\Sigma|x_i(t)|\leq M$ for all $t\geq t_0$, $i=1,2,\ldots,n$.

Given a differential equation (1.2.2) of order n, a solution $x(t)=x(t;t_0,\eta_0,\ldots,\eta_{n-1})$ will be called *bounded at the right up to the derivatives of order* k if $|x^{(h)}(t)|\leq M$ for all $t\geq t_0$ and $h=0,1,\ldots,k$. The same solution will be called *bounded at the right* if it is bounded at the right up to the derivatives of order $n-1$. Analogous definitions hold for boundedness at the left.

Boundedness and stability are independent concepts as the following examples show.

Example 8. Every solution of the equation $x'=1$ is of the form $x=C+t$ and thus is unbounded at both sides though stable at both sides.

Example 9. Every nonzero solution $x(t)$ of the equation considered in example 7 is bounded at both sides, though unstable at both sides.

Nevertheless, boundedness and stability are strictly connected for linear systems, as the following remarks show.

For homogeneous systems

$$x_i'=\sum_{h=1}^{n}a_{ih}(t)\,x_h\,,\qquad i=1,2,\ldots,n\,,\tag{1.4.1}$$

where $a_{ih}(t)$ are continuous functions in $[t_0,+\infty)$, the following elementary theorem holds: The solutions $x(t)$ of (1.4.1) are all stable at the right if and only if they are all bounded at the right (3.9. i).

For nonhomogeneous linear systems

$$x_i' = \sum_{h=1}^{n} a_{ih}(t)\, x_h + f_i(t), \qquad i = 1, 2, \ldots, n, \tag{1.4.2}$$

where $a_{ih}(t)$, $f_i(t)$ are continuous functions in $[t_0, +\infty)$, the following theorem holds: If the solutions $x(t)$ of (1.4.2) are all bounded at the right, they are all stable at the right; if they are all stable at the right and one is known to be bounded at the right, then they are all bounded at the right (3.9. i).

Analogous theorems hold for boundedness and stability at the left. Also, analogous theorems hold for n-th order linear equations

$$x^{(n)} + a_1(t)\, x^{(n-1)} + \cdots + a_n(t)\, x = 0, \qquad t_0 \leqq t < +\infty, \tag{1.4.3}$$

$$x^{(n)} + a_1(t)\, x^{(n-1)} + \cdots + a_n(t)\, x = f(t), \quad t_0 \leqq t < +\infty, \tag{1.4.4}$$

where $f(t)$, $a_i(t)$, $i = 1, 2, \ldots, n$, are continuous functions in $[t_0, +\infty)$ and where boundedness has to be understood as boundedness up to and including the derivatives of order $n-1$.

For equations (1.4.3), where the coefficients $a_{ih}(t)$ are continuous and bounded in $[t_0, +\infty)$, a nonelementary theorem of E. Esclangon [1] (see also E. Landau [1]) states that if $|x(t)| \leqq M$ for all $t_0 \leqq t < +\infty$ and some M, then there exists another constant M_1 such that $|x^{(h)}(t)| \leqq M_1$ for all $t_0 \leqq t < +\infty$ and $h = 0, 1, 2, \ldots, n-1$.

1.5. Other types of requirements and comments. Stability in the sense of Lyapunov and boundedness are only two of the possible requirements which can be demanded of a solution $x(t)$ of a system (1.1.1) existing in $[t_0, +\infty)$; that is, two of the relevant properties whose presence may or may not be used to characterize the behavior of a solution $x(t)$ of (1.1.1) as $t \to +\infty$.

First of all, we have already mentioned various modifications of the Lyapunov stability; asymptotic stability, conditional stability. If we suppose that all the functions $f_i(t; x_1, \ldots, x_n)$ in (1.1.1) are defined for $t \geqq t_0$ and all $(x_1, \ldots, x_n)$ of a given region R of points (vectors) $x = (x_1, \ldots, x_n)$ containing x_0, we may ask whether

$$\| x(t; t_0, x_1) - x(t; t_0, x_0) \| \to 0$$

as $t \to +\infty$ for every initial vector $x_1 \in R$ and then we will say that $x(t; t_0, x_0)$ is *asymptotically stable in the large relatively to R.* Of course R may be the whole $x_1, \ldots, x_n$-space (real or complex) and in any case we have to assure that $x(t; t_0, x_1)$ exists in $[t_0, +\infty)$ for every $x_1 \in R$. Stability in the large has been recently investigated by A. I. Lure [7], N. P. Erugin [6], M. A. Aizerman [1, 2, 3, 4], E. A. Barbasin and N. N. Krasovskii [1, 2].

We may require stability only for some components x_i of x, or for all. For instance a solution $x_0(t)$ of an n-th order differential equation, $n \geqq 2$, may be called stable, if given $\varepsilon > 0$ and t_0, there is a $\delta > 0$ such that $|\bar{x}(t_0) - x_0(t_0)| \leqq \delta$ implies $|\bar{x}(t) - x_0(t)| \leqq \varepsilon$ for all $t > t_0$, no matter whether an analogous fact holds when the derivatives up to the order $n-1$ are taken into consideration as usual.

The same modifications hold for boundedness. For instance, we may ask whether the solutions of a differential equation (1.2.1) are bounded, no matter if some of the derivatives are not (see e.g., A. Lyapunov [3]).

It may be mentioned here that E. ROUTH in a particular question proposed to denote as stable a solution $x(t)$ of a second order equation $x'' = F(x, x', t)$, x real, which (a) exists for all $t \geq t_0$, (b) $x(t) \to 0$ as $t \to +\infty$, and (c) $x(t) = 0$ at infinitely many points $t = t_n$ with $t_n \to +\infty$ as $n \to \infty$ (oscillatory solutions approaching zero as $t \to +\infty$).

Given a solution $x(t; t_0, x_0)$ of system (1.1.1), $t_0 \leq t < +\infty$, we may require that given $\varepsilon > 0$ there exists a $\delta > 0$ such that for all $\bar{t} \geq t_0$ and $\bar{x}$ with $\|\bar{x} - x(\bar{t}; t_0, x_0)\| < \delta$, the solution $x(t; \bar{t}, \bar{x})$ exists in $[\bar{t}, +\infty)$ and $\|x(t; t_0, x_0) - x(t; \bar{t}, \bar{x})\| < \varepsilon$ for all $t \geq \bar{t}$. Then $x(t; t_0, x_0)$ is said to be *uniformly stable in the sense of* LYAPUNOV at the right. Analogous definitions can be given at the left, or for uniform asymptotic stability. The concept of uniform stability was already considered by A. LYAPUNOV [3], and then discussed by K. PERSIDSKII [6], L. ERMOLAEV [1], and others [see (3.9) for uniform stability and (6.8), (7.3) for other concepts].

Let us consider the functions $D(t) = \|x(t; t_0, x_1) - x(t; t_0, x_0)\|$, $t_0 \leq t < +\infty$, and $\varepsilon(\delta) = \operatorname{Sup} D(t)$ where the supremum is taken for all $t_0 \leq t < +\infty$ and $\|x_1 - x_0\| \leq \delta$. In case of Lyapunov stability we have $\varepsilon(\delta) \to 0$ as $\delta \to 0$. We may ask whether $\varepsilon(\delta) \leq M\delta$, or $\varepsilon(\delta) \leq M\delta^\alpha$ for some $M \geq 1$ and $0 < \alpha \leq 1$, and all δ sufficiently small.

In the case of asymptotic stability we have $D(t) \to 0$ as $t \to +\infty$ for $\|x_1 - x_0\|$ sufficiently small, and we may ask whether $D(t)$ is an infinitesimal of some prescribed type, e.g., $D(t) \leq M e^{-\alpha t}$, or $D(t) < M t^{-\alpha}$ for some M, $\alpha > 0$ and all $t_0 \leq t < +\infty$ (exponential, harmonic stability, etc.). We may inquire whether $D(t)$ is integrable in $[t_0, +\infty)$, or more generally whether $D(t) \in L^\alpha$ in $[t_0, +\infty)$ for some $\alpha > 0$; i.e., $D^\alpha(t)$ is integrable in $[t_0, +\infty)$. Even in the case where $D(t)$ does not approach zero as $t \to +\infty$, it may occur that $D(t) \in L^\alpha$ for some $\alpha > 0$. In addition an evaluation of the magnitude or growth of $D(t)$ in $[t_0, t]$ as $t \to +\infty$, or of its integral in $[t_0, t]$, may be of interest.

When only real solutions are considered, we may ask whether the components $x_i(t)$ of $x(t)$ remain of constant sign for t large, or if they are oscillatory in caracter, and, if they approach zero, whether they approach zero monotonically or not. In this sense the classical and recent oscillation theorems for linear and nonlinear differential equations fall in the frame of the present discussion. Since the word "stability" is often misused, the expression "qualitative theory of differential equations" may be preferred.

The considerations above refer mainly to $t \to +\infty$, or $t \to -\infty$ on the real axis, and we will suppose that this is the case most of the times. Nevertheless, t could approach ∞ in the complex field, either in any neighborhood of ∞, or in some sector, or some other set of the complex plane, and then the question of the behavior of the solutions as $t \to \infty$ could be discussed analogously.

Finally it should be pointed out that the transformation $t = 1/z$ (or others) transforms a neighborhood (real, or complex) of $t = \infty$ into a neighborhood (real, or complex) of $z = 0$ for the transformed equation. Should $z = 0$ be an "ordinary" or a "regular singular" point for the transformed equation, then Cauchy or Fuchs theories would yield complete information on the behavior of the solutions (cf. § 3 and § 10).

1.6. Stability of equilibrium. The concept of LYAPUNOV's stability as given in (1.2) was considered long before LYAPUNOV in connection with the question of the "mechanical stability" of a position of equilibrium of a conservative system Σ with constraints independent of t. If Σ has m degrees of freedom, if $q_1, q_2, \ldots, q_m$ is any system of Lagrangian coordinates and $V = V(q_1, \ldots, q_m)$ the potential energy of Σ, then the behavior of Σ is described by the system of m second order Lagrangian equations

$$\partial L/\partial q_i - (d/dt)(\partial L/\partial q_i') = 0, \qquad i = 1, 2, \ldots, m, \tag{1.6.1}$$

where $L = T - V$, and $T = T(q_1, \ldots, q_m, q_1', \ldots, q_m')$ is the kinetic energy of Σ. If (1.6.1) has a constant solution $q_i = q_{i0}, q_i' = 0, i = 1, 2, \ldots, m$, then the solution represents a possible stationary state E of Σ, or a position of equilibrium. The mechanical stability of E is generally considered as expressed by the Lyapunov stability of the above constant solution of system (1.6.1). By a displacement it is always possible to transfer the equilibrium point to the origin.

It may be mentioned here that by introducing the generalized momenta $p_i = \partial T/\partial q_i'$ and the Hamiltonian function $H = H(q_1, \ldots, q_m, p_1, \ldots, p_m)$ defined by $H = T + V = 2T - L = \sum_{i=1}^{m} p_i q_i - L$, equations (1.6.1) are reduced to the $2m$ Hamilton equations

$$dq_i/dt = \partial H/\partial p_i, \quad dp_i/dt = -\partial H/\partial q_i, \quad i = 1, \ldots, m. \tag{1.6.2}$$

A theorem of J. L. LAGRANGE [1] assures that a position of equilibrium of a conservative system is stable if V has a minimum there. Conversely, A. LYAPUNOV [3] has proved under restrictions that the same position is unstable if V has no minimum there. All this is connected with the "second method" of A. LYAPUNOV and we will refer briefly to it in § 7.

1.7. Variational systems. Given a system (1.1.1) and a solution $x(t; t_0, x_0)$, $t_0 \leq t < +\infty$, contained in a region S of points (t, x) open at the right of t_0, then the question of the stability of $x(t; t_0, x_0)$ at the right can always be reduced, at least formally, to the question of the stability at the right of the solution $u = 0$ of some new system (1.1.1). Indeed, by putting

$$x = x(t; t_0, x_0) + u, \tag{1.7.1}$$

we transform the solution $x = x(t; t_0, x_0)$ into a solution $u = 0$ and the system (1.1.1) into the new system

$$u' = f[t, x(t) + u] - f[t, x(t)] = F(t, u),$$

where $F(t, 0) = 0$ for every $t \geq t_0$. If we suppose that the components f_i of f have partial derivatives $f_{ij} = \partial f_i/\partial x_j$ continuous in S and we put

$$\left.\begin{aligned} a_{ij}(t) &= f_{ij}[t, x(t)], \\ F_i(t, u) &= f_i[t, x(t) + u] - f_i[t, x(t)] \\ &= a_{i1}(t) u_1 + a_{i2}(t) u_2 + \cdots + a_{in}(t) u_n + X_i(t, u), \\ i = 1, \ldots, n, &\quad u = (u_1, \ldots, u_n), \quad t_0 \leq t < +\infty, \end{aligned}\right\} \tag{1.7.2}$$

then the vector $X(t, u)$ of components $X_i(t, u)$ is a continuous function of t and u in a set S of the $(t, u_1, \ldots, u_n)$-space, open at the right of

$t=t_0$ and containing the half straight-line $[t \geq t_0, u=0]$. In addition, from (1.7.2) we have $X(t, 0)=0$ for every t, and, from TAYLOR'S formula, we deduce $X(t, u) \to 0$ as $\|u\| \to 0$ for every $t \geq t_0$ [uniformly in every $t_0 \leq t \leq t_1$, $t_1 < +\infty$]. System (1.1.1) is transformed into the system

$$u_i' = \sum_{j=1}^{n} a_{ij}(t)\, u_j + X_i(t, u), \qquad i=1, \ldots, n, \tag{1.7.3}$$

which admits of the trivial solution $u=0$. If we suppose that the components f_i of f have both first and second partial derivatives continuous in S then, by TAYLOR'S formula, we have

$$X(t, u)/\|u\| \to 0 \quad \text{as} \quad \|u\| \to 0,$$

for every $t \geq 0$ [uniformly in every $t_0 \leq t \leq t_1$, $t_1 < +\infty$].

If the functions f_i are analytic in S with respect to $x_1, \ldots, x_n$ for every $t \geq t_0$, then we have

$$X_i(t; u_1, \ldots, u_n) = \sum_{h=2}^{\infty} \sum_{i_1+\cdots+i_n=h} p_{i_1 i_2 \ldots i_n}(t)\, u_1^{i_1} u_2^{i_2} \ldots u_n^{i_n},$$

where $i_1, i_2, \ldots, i_n \geq 0$ are all integers, the coefficients $p_{i_1 i_2 \ldots i_n}(t)$ are continuous functions of t, and the series above converge absolutely and uniformly in a neighborhood U of $(0, 0, \ldots, 0)$ for every $t \geq t_0$.

In any case the stability of the solution $x(t; t_0, x_0)$ with respect to system (1.1.1) is reduced to the stability of $u=0$ with respect to system (1.7.3). The system (1.7.3) is called the *variational system* relative to the solution $x(t; t_0, x_0)$ of system (1.1.1). Obviously, if system (1.1.1) is linear, then also (1.7.3) is linear $[X \equiv 0]$ and the transformation is trivial. If (1.1.1) is nonlinear, then X is not zero, and the system (1.7.3), is nonlinear.

It is natural to associate the linear system

$$u_i' = \sum_{j=1}^{n} a_{ij}(t)\, u_j \tag{1.7.4}$$

with system (1.7.3) and this one is said to be the *linear variational system* relative to $x(t; t_0, x_0)$ and system (1.1.1). It is natural also to ask whether the stability or instability of the solution $u=0$ for system (1.7.4) implies stability or instability of the solution $u=0$ for system (1.7.3). In general there is no such close connection between systems (1.7.3) and (1.7.4) as the following examples show.

Examples: 1. The solution $x=0$ is stable at either side for the linear equation $x'=0$, but is unstable at either side for the nonlinear equation $x'=x^2$ (cf. 1.3, no. 1).

2. The solution $x=0$ is unstable at the right for the linear equation $x'=x$ since the solutions have all the form $x=Ce^t$. On the other hand, the solution $x=0$ is asymptotically stable at the right for the nonlinear system $x'=x-e^t x^3$ since all solutions of this equation have the form $x=C e^t[1+(\frac{2}{3})C^2(e^{3t}-1)]^{-\frac{1}{2}}$, ≥ 0, and hence $x \to 0$ as $t \to +\infty$.

Nevertheless, under mild conditions, the stability and instability of the solution $u=0$ with respect to the linear system (1.7.4) imply stability and instability respectively of the solution $u=0$ with respect to the nonlinear system (1.7.3), as A. LYAPUNOV essentially discovered [3]. We shall give precise theorems along this line in § 6 and § 8.

The generality of the just mentioned statement has led to a definition of stability which is widely used in applied mathematics. A solution $x=x_0(t)$, for which the solution $u=0$ of the corresponding linear variational system (1.7.4) is stable, is said to be *infinitesimally stable*.

1.8. Orbital stability. For the sake of simplicity we shall suppose now that the system (1.1.1) is autonomous, i.e., of the form $x'=f(x)$, $x=(x_1, \ldots, x_n)$, and that a solution $x=X(t)$, $-\infty<t<+\infty$, of it is known which is periodic of some period T, i.e., $X(t+T)=X(t)$ for all t. Since the system is autonomous, also $x=X(t-\gamma)$ is a solution for every constant γ (phase). In this condition $x=X(t)$ defines a closed path C (orbit) in the x-space E_n. We shall denote by $\{x, C\}$ the usual distance of the point x from C (i.e., the infimum of all the distances from x to any point p of C). The orbit C is said to be *stable* provided, given any $\varepsilon>0$, there is a $\delta>0$ such that every solution $x(t)$ which passes at some time $t=t_0$ at a point $x_0\in E_n$ with $\{x_0, C\}<\delta$ has the property $\{x(t), C\}<\varepsilon$ for all $t\geq t_0$. The same orbit C is said to be *asymptotically stable* provided it is stable and there is an $\varepsilon_0>0$ such that every solution $x(t)$ which passes at some time $t=t_0$ at a point $x_0\in E_n$ with $\{x_0, C\}<\varepsilon_0$ has the property $\{x(t), C\}\to 0$ as $t\to+\infty$. A further more particular case is the *asymptotic stability with asymptotic phase* for which the additional requirement is made that $x(t)-X(t-\gamma)\to 0$ as $t\to+\infty$ for a convenient constant γ (phase). For instance, the system of the example 1 of 1.9 presents orbits which are all stable, but not asymptotically stable, according to the above definition. For these definitions see, for instance, E. ROUTH [3], A. LYAPUNOV [3], E. CODDINGTON and N. LEVINSON [3].

1.9. Stability and change of coordinates. Neither Lyapunov stability, nor boundedness have invariant characters with respect to a general change of coordinates (unknown functions) $x_1, x_2, \ldots, x_n$, or of independent variable t. Thus for a mechanical system it may happen that a solution is stable with respect to a given system of Lagrangian coordinates and is unstable with respect to another system.

Examples. 1. The system

$$x' = -y(x^2+y^2)^{\frac{1}{2}} \qquad y' = x(x^2+y^2)^{\frac{1}{2}}$$

has the 2-parameter family of solutions

$$x = c\cos(ct+d), \qquad y = c\sin(ct+d),$$

(c, d arbitrary constants). The solution $x=0$, $y=0$ is stable, all other solutions are unstable. Nevertheless, if new unknowns r, d are introduced by means of the relations

$$x = r\cos\theta, \qquad y = r\sin\theta, \qquad \theta = rt+d,$$

then the given system is transformed into the system

$$r' = 0, \quad d' = 0,$$

whose solutions $r = c_1$, $d = c_2$, (c_1, c_2 constants) are all stable.

2. The pendulum equation $x'' + \sin x = 0$ has a family of solutions of the form $x = c \sin[\varphi(c)t + d]$, ($c$, d constants), where $\varphi(c)$ denotes a function of c which can be expressed (though not easily) in terms of elliptic functions. The solution $x = 0$ is stable while all other solutions of this family are unstable. The transformation

$$x = r \sin[\varphi(r)t + d], \quad y = r \cos[\varphi(r)t + d]$$

transforms the given differential equation into the same system $r' = 0$, $d' = 0$ above whose solutions are all stable.

3. If a material point P in a given plane α is attracted by a fixed center Q in α proportionally to the inverse of the square of its distance from this center, then we may consider the elliptic trajectories described by P which are all unstable (except the equilibrium solution $P = Q$ which is stable). Such instability is due to the fact that the times of revolutions are functions of the axes of the ellipses and thus points P, P' very close to each other at the time $t = 0$ but traveling on slightly different ellipses, may find themselves at opposition and thus at great distance from each other in due course of time. Such instability occurs both in cartesian and polar coordinates. Nevertheless the movement of P is stable with respect to the variable

$$u = r - p(1 + e \cos\theta)$$

where p and e are the parameter and the eccentricity of the ellipse described by P (A. LYAPUNOV [3]).

4. Any system (1.1.2) which has a system of n first integrals $\Phi_i(x, t) = c_i$, $i = 1, \ldots, n$, that can be solved for $x_1, \ldots, x_n$, $x_i = \varphi_i(t, c)$ for $t_0 \leq t < +\infty$, where $c = (c_1, \ldots, c_n)$, can be reduced to equilibrium $y_i' = 0$, $i = 1, \ldots, n$, by the coordinate transformation $y_i = \Phi_i(x, t)$. All the solutions $y_i = c_i$, $i = 1, \ldots, n$, of the transformed system are stable.

1.10. Stability of the *m*-th order in the sense of G. D. BIRKHOFF. Motivated by astronomical research of the nineteenth century, and following H. POINCARÉ [6, 7], G. D. BIRKHOFF has deeply investigated a concept of stability [20, 23] which has some interest in general dynamics and astronomy (e.g., the three body problem). We shall give here only the formal definition and mention a few properties (see G. D. BIRKHOFF [20]). Let $x' = f(x)$, $x = (x_1, \ldots, x_n)$, be an autonomous differential system and let $(\bar{x}_1, \ldots, \bar{x}_n)$ be a point of equilibrium, i.e., $f(\bar{x}) = 0$. Such a point is said to be *stable of the m-th order* if it possesses the following properties:

(A) There exist two constants K, L such that every solution $x = x(t; x_0, t_0)$ with $\|x_0 - \bar{x}\| \leq \varepsilon$ satisfies the relation $\|x(t; x_0, t_0) - \bar{x}\| \leq K\varepsilon$ for all t with $|t - t_0| \leq L\varepsilon^{-m}$.

(B) For all solutions $x(t; x_0, t_0)$ as in (A) and all fixed T and all polynomials $P(x)$ in $x_1, \ldots, x_n$, whose terms have degrees $\geq s$, the function $P[x(t); t_0, x_0]$, [in particular, the components $x_i(t; t_0, x_0)$ of x] can be represented in $[t_0 - T, t_0 + T]$ by means of trigonometrical sums

$$A_0 + \Sigma(A \cos\lambda_i t + B_i \sin\lambda_i t), \quad |\lambda_i - \lambda_j| \geq \lambda > 0,$$

of not more than $N + 1$ terms with an error $\leq Q\varepsilon^{m+s}$ in $[t_0 - T, t_0 + T]$, (Q, λ, N constants, $s = 1, 2, \ldots$). Stability of order m implies stability of every order $1, \ldots, m - 1$. Property (A) alone is denoted by BIRKHOFF as "perturbative stability" (of order m); property (B) alone is denoted as "trigonometric stability" (of order m).

Stability of all orders is denoted as *complete stability*. Perturbative stability (of first, or higher order) is similar, but independent from LYAPUNOV stability at both sides. Nevertheless, perturbative stability of a linear system with constant coefficients, $x_i' = \sum_j^n a_{ij} x_j$, is equivalent to the boundedness in $(-\infty, +\infty)$ of all solutions x, and, hence (1.4), to the LYAPUNOV stability at both sides.

If we suppose that the components of $f(x)$, say $f_i(x_1, \ldots, x_n)$, $i = 1, \ldots, n$, are analytic functions of $x_1, \ldots, x_n$, around $(\bar{x}_1, \ldots, \bar{x}_n)$, and we perform an analytic transformation $x = \Phi(y)$, where Φ is also analytic around the transformed point of equilibrium $\bar{y}$, then system $x' = f(x)$ is transformed into an analogous system $y' = g(y)$. Both perturbative and trigonometrical stabilities are invariant with respect to analytic transformations Φ. On these questions see G. D. BIRKHOFF [20, 23, 25].

1.11. A general remark and bibliographical notes. There are many different definitions of "stability" which have little to do with one another, though there are intricate connections between them. Each represents a desirable property of a given solution. On the other hand, the property indicated by LYAPUNOV's definition which seems the most natural of all, is not present in many important cases. No known definition is acceptable in all cases. For an analysis of the concept see A. WINTNER [35], J. J. STOKER [1, 4], G. D. BIRKHOFF [20], and other expositions.

For collateral reading on the general questions dealt with in this book the reader is referred to the following well known text books and reports: A. ANDRONOV and C. E. CHAIKIN [1], R. BELLMAN [12], N. BOGOLIUBOV and N. M. KRYLOV (S. LEFSCHETZ, ed.) [34], B. V. BULGAKOV [10]; E. A. CODDINGTON and N. LEVINSON [3], D. GRAFFI [16], A. L. LURE [8], E. KAMKE [1], S. LEFSCHETZ [2, 3], I. G. MALKIN [22], L. A. MACCOLL [1], N. MINORSKY [6], V. V. NEMYCKII [5], V. V. NEMYCKII and V. STEPANOV [1], G. SANSONE [16], G. SANSONE and R. CONTI [2], V. M. STARZINSKII [5], J. J. STOKER [1], S. P. STRELKOV [1], A. WINTNER [35], A. M. LETOV [7], YU. A. MITROPOLSKII [8], I. M. RAPOPORT [4], N. BOGOLIUBOV and YU. A. MITROPOLSKII [1], M. A. AIZERMAN [7], G. N. DUBOSIN [2], N. G. CETAEV [5], I. G. MALKIN [9], W. KAPLAN [3], N. W. MCLACHLAN [5].

§ 2. Linear systems with constant coefficients

2.1. Matrix notations. By $A, B, \ldots$ we shall denote $m \times n$ matrices $[a_{ij}]$, $[b_{ij}], \ldots$, i.e. matrices with m rows and n columns, whose elements a_{ij}, b_{ij}, $i = 1, \ldots, m$, $j = 1, \ldots, n$, are real or complex numbers. By $A = B$ we mean that the matrices are equal, or identical, i.e. they both have m rows and n columns and $a_{ij} = b_{ij}$, $i = 1, \ldots, m$, $j = 1, \ldots, n$. By the zero matrix, or 0, we denote the matrix whose elements are all zero. If $m = 1$, $n = 1$, the matrix is a number; if $m = 1$, $n > 1$, or $m > 1$, $n = 1$, the matrix is called a prime, or a vector. If not indicated otherwise we think of a vector as an $n \times 1$ matrix and denote it by a small letter. If $m = n$ the matrix is said to be a square matrix of order n and the terms a_{ii}, $i = 1, \ldots, n$, are said to form its main diagonal; their sum, $tr A$, is called the trace of A. If $m = n$ and $a_{ij} = 0$ for $i \neq j$, then A is said to be a diagonal matrix; if $m = n$, $a_{ij} = 0$ for $i \neq j$, $a_{ij} = 1$ for $i = j$, than the matrix is said to be the identy matrix or unit matrix and denoted by $A = I = [\delta_{ij}]$. By $C = A + B$, $D = \alpha A = A\alpha$ (α real or complex, both A and B $m \times n$ matrices) we denote the matrices $[c_{ij}]$, $[d_{ij}]$ defined by $c_{ij} = a_{ij} + b_{ij}$, $d_{ij} = \alpha a_{ij} = a_{ij}\alpha$. Finally if A is an $m \times p$ matrix, and B a $p \times n$ matrix, by $E = AB$ we denote the $m \times n$ matrix $[e_{ij}]$ defined by $e_{ij} = a_{i1} b_{1j} + \cdots + a_{ip} b_{pj}$. Equality, addition, and multiplication, as defined above, have the well known formal properties sketched below:

(2.1. i) (a) $A = A$; (b) $A = B$ implies $B = A$; (c) $A = B$, $B = C$ implies $A = C$.

(2.1. ii) (a) $A + B = B + A$; (b) $(A + B) + C = A + (B + C)$;

(2.1. iii) (a) $\alpha(A + B) = \alpha A + \alpha B$; (b) $(\alpha + \beta) A = \alpha A + \beta A$;

(2.1. iv) (a) $(AB) C = A(BC)$; (b) $(A + B) C = AC + BC$, $C(A + B) = CA + CB$.

Given any $m \times n$ matrix $A = [a_{ij}]$, then the complex conjugate matrix $\bar{A} = [\bar{a}_{ij}]$ of A is the $m \times n$ matrix whose elements $\bar{a}_{ij}$ are the complex conjugates of a_{ij}, and the transpose $A_{-1} = [b_{ij}]$ of A is the $n \times m$ matrix defined by $b_{ij} = a_{ji}$. If $m = n$, and $A = A_{-1}$, then A is said to be symmetric, if $A = \bar{A}_{-1}$ then A is said to be Hermitian. From these definitions it follows that $\overline{AB} = \bar{A}\bar{B}$, and $(AB)_{-1} = B_{-1}A_{-1}$.

Any $m' \times n'$ matrix $B = [a_{i_r j_s}]$, $1 \leq m' \leq m$, $1 \leq n' \leq n$, obtained by extracting from an $m \times n$ matrix $A = [a_{ij}]$ the m' rows of indices $i_1, \ldots, i_{m'}$, and the n' columns of indices $j_1, \ldots, j_{n'}$, is said to be a minor of the matrix A. If $m' = n'$, $m = n$, $i_r = j_r$, $r = 1, \ldots, n'$, then B is said to be a principal minor of A. If $m' = n' = n - 1$, $m = n$, then B can be thought of as obtained from A by suppressing the i-th row and the j-th column of A and is then denoted by A_{ij}, or minor of the element a_{ij} of A.

By $\det A = \det [a_{ij}]$ we shall denote as usual the determinant of the $n \times n$ matrix $A = [a_{ij}]$. If A, B, C, are $n \times n$ matrices and $C = AB$, then $\det C = \det A \cdot \det B$; and thus $\det C = 0$ if and only if $\det A = 0$, or $\det B = 0$. The number $\alpha_{ij} = (-1)^{i+j} \det A_{ij}$ is said to be the cofactor of a_{ij} in A. An $n \times n$ matrix $B = [b_{ij}] = A^{-1}$ is said to be the inverse of the $n \times n$ matrix $A = [a_{ij}]$ if $BA = I$. It is well known that a matrix A has an inverse A^{-1} if and only if $\det A \neq 0$ and then $A^{-1} = [b_{ij}]$ is uniquely defined by $b_{ji} = \alpha_{ij}/\det A$ and $A^{-1} A = A A^{-1} = I$. Also $\det A^{-1} = (\det A)^{-1}$; more generally $\det A^m = (\det A)^m$ for all integers $m \geq 0$ and even for all integers $m \gtreqless 0$ if $\det A \neq 0$. Finally $(AB)^{-1} = B^{-1}A^{-1}$ for all $n \times n$ matrices A, B with $\det A$, $\det B \neq 0$, and $(A^{-1})_{-1} = (A_{-1})^{-1}$. Thus the symbol A_{-1}^{-1} is unambiguous.

If A is an $n \times n$ matrix, by the rank h of A, $0 \leq h \leq n$, we denote the maximum order of its square minors whose determinant is $\neq 0$. By the nullity ν of A we mean $\nu = n - h$, $0 \leq \nu \leq n$. Thus $\nu > 0$ if and only if $\det A = 0$.

The polynomial of degree n, $f(\varrho) = \det(\varrho I - A) = \det[\delta_{ij}\varrho - a_{ij}]$ is said to be the characteristic polynomial of A, and

$$f(\varrho) = \varrho^n - S_1 \varrho^{n-1} + S_2 \varrho^{n-2} - \cdots \pm S_n,$$

where S_r denotes the sum of all principal minors of A of order r, and thus $S_n = \det A$. The equation $f(\varrho) = 0$ is then the characteristic equation of A and its distinct roots $\varrho_1, \ldots, \varrho_m$, $1 \leq m \leq n$, the characteristic roots of A. For each root ϱ_r we shall consider the multiplicity μ_r, $1 \leq \mu_r \leq n$, of ϱ_r for the equation $f(\varrho) = 0$, and the nullity ν_r, $1 \leq \nu_r \leq n$, of the matrix $\varrho_r I - A$, $r = 1, 2, \ldots, m$. Obviously $\mu_1 + \cdots + \mu_m = n$.

If $A = [a_{ij}]$ is an $n \times n$ matrix and $P = [p_{ij}]$ any $n \times n$ matrix with $\det P \neq 0$, then the $n \times n$ matrix $B = P^{-1} A P$ is said to be obtained from A by a similarity transformation and this relation is denoted by $B \sim A$. Any matrix B obtained from a matrix A by performing a given permutation of its n rows and the same permutation of its columns is certainly similar to A. The following statements are immediately proved:

(2.1. v) (a) $A \sim A$; (b) $A \sim B$ implies $B \sim A$; (c) $A \sim B$, $B \sim C$ implies $A \sim C$;

(2.1. vi) (a) $A \sim B$ implies $\det A = \det B$ and A and B have the same characteristic polynomial and the same characteristic roots with the same multiplicities and nullities.

If $A = [a_{ij}]$ is an $n \times n$ matrix and $B_s = [b_{uv}^{(s)}]$, $s = 1, \ldots, N$, are $n_s \times n_s$ matrices with $n_1 + \cdots + n_N = n$, and $a_{ij} = b_{uv}^{(s)}$ whenever $i = n_1 + \cdots + n_{s-1} + u$, $j = n_1 + \cdots + n_{s-1} + v$, $u, v = 1, \ldots, n_s$, $s = 1, \ldots, N$, and $a_{ij} = 0$ otherwise, then we say that A is the direct sum (cf. S. PERLIS [1]) of the matrices B_s and we write $A = \operatorname{diag}(B_1, B_2, \ldots, B_N)$. In other words, A is made up by the "boxes" $B_1, \ldots, B_N$, adjacent to one another along the main diagonal, while all other elements of A are zero. If A is the direct sum of the matrices $B_1, \ldots, B_N$, then we have $\det A = \det B_1 \cdot \det B_2 \cdot \ldots \cdot \det B_N$. Since $\varrho I - A$ is also the direct sum of the matrices $\varrho I - B_s$, $s = 1, \ldots, N$, we also have

$$\det(\varrho I - A) = \det(\varrho I - B_1) \cdot \ldots \cdot \det(\varrho I - B_N);$$

in other words, the characteristic polynomial of A is the product of the characteristic polynomials of the matrices B_s.

In matrix theory the following theorem is proved concerning the reduction of a matrix to one of its canonical forms (JORDAN's) in the complex field by means of similarity transformations.

(2.1. viii) Given an $n \times n$ matrix A of characteristic roots ϱ_r, $r = 1, \ldots, m$, with multiplicities μ_r and nullities ν_r, there are complex matrices P with $\det P \neq 0$ such that $J = P^{-1} A P$ is the direct sum of N matrices $C_s = [c_{ij}^{(s)}]$ of orders n_s with $c_{ii} = \varrho_s'$, $i = 1, \ldots, n_s$, $c_{i,i+1} = 1$, $i = 1, 2, \ldots, n_s - 1$, $c_{ij} = 0$ otherwise, where $\varrho_s' = \varrho_r$ for some $r = 1, \ldots, m$. The N matrices C_s are uniquely determined up to an arbitrary permutation.

There are in the theory of matrices various proofs of (2.1. viii). Most of them construct the matrix P in steps, as a product of matrices $P_1, P_2, \ldots$, so chosen that the matrices $P_1^{-1} A P_1$, $P_2^{-1} P_1^{-1} A P_1 P_2, \ldots$, assume more and more the diagonal aspect in the sense of the theorem (processes of diagonalization by similarity transformation). It is possible to proceed in such a way that at each step the elements of the matrices $P_1, P_2, \ldots$, are obtained by solving algebraic linear systems whose determinants are always $\neq 0$.

The matrices C_s are diagonal only if $n_s = 1$. Also, we have in any case $\det(\varrho I - C_s) = (\varrho - \varrho_s)^{n_s}$, and since the characteristic polynomial of A is the product of the characteristic polynomials of the C_s (2.1. vi), we conclude that for every characteristic root ϱ_r, of A there must be a number of matrices C_s having $\varrho_s' = \varrho_r$ in the main diagonal with $\Sigma^{(r)} n_s = \mu_r$ where $\Sigma^{(r)}$ ranges over all s with $\varrho_s' = \varrho_r$. These matrices C_s are called the companion matrices of ϱ_r (S. PERLIS [1]). On the other hand, for each C_s the matrix $\varrho_s' I - C_s$ has determinant zero but by suppressing the first row and the last column of $\varrho_s' I - C_s$, (whose elements are all zeros) we obtain a diagonal matrix whose diagonal elements are all -1. Thus $\varrho_s' I - C_s$ has rank $n_s - 1$ and nullity 1. Analogously the matrix $\varrho_r I - J$ has determinant zero, but by suppressing the rows and columns mentioned above corresponding to the companion matrices of ϱ_r we get a minor of maximal order $n - \Sigma^{(r)}(1)$ and determinant $\neq 0$. Thus the nullity ν_r of $\varrho_r I - A$ is given by $\nu_r = \Sigma^{(r)}(1)$. From the two equalities

$$\mu_r = \Sigma^{(r)} n_s, \qquad \nu_r = \Sigma^{(r)}(1)$$

where we always have $n_s \geq 1$, we conclude as follows:

(2.1. ix) For each characteristic root ϱ_r of an $n \times n$ matrix A we have $\nu_r \leq \mu_r$, i.e., the nullity ν_r of the matrix $\varrho_r I - A$ is always $\leq$ the multiplicity μ_r of ϱ_r in the characteristic polynomial $f(\varrho) = \det(\varrho I - A)$.

(2.1. x) The matrix A has a diagonal canonical form if and only if $\nu_r = \mu_r$, $r = 1, \ldots, m$; i.e., if and only if for each characteristic root ϱ_r the nullity ν_r is equal to the multiplicity μ_r.

(2.1. xi) For each characteristic root ϱ_r the companion matrices C_s (with $\varrho_s' = \varrho_r$) have all orders $n_s = 1$ if and only if $\nu_r = \mu_r$.

Given a matrix $A(t) = [a_{ij}(t)]$ whose elements are all differentiable functions of t, by derivative $A'(t) = DA(t) = dA/dt$ is meant the matrix $[a_{ij}'(t)]$.

Similarly, we may define $\int_a^b A(t)\,dt$. The formal theorems on derivatives and integrals hold as usual, say, e.g., $(A+B)' = A' + B'$, $(AB)' = A'B + AB'$, $(aA)' = a'A + aA'$. If $A(t)$ is an $n \times n$ matrix as above with $\det A \neq 0$, then from $AA^{-1} = I$, by differentiation and manipulation, we may deduce $(A^{-1})' = -A^{-1}A'A^{-1}$.

The formula for the derivative of a determinant is better deduced directly from the definition of determinant as usual, and reads $(\det X(t))' = \sum_{i=1}^{n} \det X_i(t)$, where $X_i(t)$ is the matrix obtained from X by replacing the elements of its i-th row (column) by their derivatives.

We shall denote by the norm $\|A\|$ of a $n \times n$ matrix $A = [a_{ij}]$ the sum $\|A\| = \Sigma |a_{ij}|$ of the absolute values of all its elements. Thus, if $C = AB$, it follows immediately $\|C\| \leq \|A\|\,\|B\|$. In particular, if $y = Ax$, where x, y are n-vectors ($n \times 1$ matrices) and A an $n \times n$ matrix, we have $\|y\| \leq \|A\|\,\|x\|$ where $\|x\|$, $\|y\|$ are the norms of the vectors x, y (1.1). Also $\|cA\| = |c|\,\|A\|$, $\|A+B\| \leq \|A\| + \|B\|$, and, if $A(t)$ is function of t, $a \leq t \leq b$, then $\left\| \int_a^b A(t)\,dt \right\| \leq \int_a^b \|A(t)\|\,dt$.

If A_h, $h = 0, 1, 2, \ldots$, denote real or complex $m \times n$ matrices the concepts of limit and series

$$A = \lim_{h \to \infty} A_h, \qquad S = \sum_{h=0}^{\infty} A_h,$$

are selfexplanatory since they reduce to the $m \times n$ limits or series of corresponding elements.

Given a real or complex $n \times n$ matrix A, by the matrix e^A (exponential matrix of A) is denoted the sum of the series

$$e^A = I + (1/1!)\,A + (1/2!)\,A^2 + (1/3!)\,A^3 + \cdots,$$

and this series is certainly convergent since $\|A^h\| \leq \|A\|^h$ for all $h \leq 0$ and, therefore, each of the n^2 series components is minorant of the convergent series whose elements are $(1/h!)\,\|A\|^h$. It is immediately proved that $e^A e^B = e^{A+B}$ for any two $n \times n$ matrices provided $AB = BA$. Note that, if $A \sim B$ then $e^A \sim e^B$. Indeed, if $B = P^{-1}AP$, then we have $B^h = P^{-1}A^hP$ for all $h \geq 0$ and finally $e^B = P^{-1}e^AP$. If $A = \operatorname{diag}(a_1, \ldots, a_n)$, then $e^A = \operatorname{diag}(e^{a_1}, \ldots, e^{a_n})$. If A is the direct sum of matrices $B_1, \ldots, B_m$, then e^A is the direct sum of the matrices $e^{B_1}, \ldots, e^{B_m}$. Thus if $A = J$ is the Jordan canonical form of (2.1. viii) then J is the direct sum of the matrices C_s of orders n_s discussed there, $n_1 + \cdots + n_m = n$, and e^J is the direct sum of the matrices e^{C_s}. Suppose $C_s = [c_{jk}]$ is one of these matrices, say $c_{jj} = r$, $c_{j,j+1} = 1$, $c_{jk} = 0$ otherwise, and put $C_s = rI + Z$, where $Z = [z_{jk}]$, $z_{jj} = 0$, $z_{j,j+1} = 1$, $z_{jk} = 0$ otherwise. By direct computation we see that $Z^h = [z_{jk}^{(h)}]$ is the matrix with $z_{j,j+h}^{(h)} = 1$, $z_{jk}^{(h)} = 0$ otherwise, if $1 \leq h \leq n_s - 1$. For $h \geq n_s$ we have $Z^h = 0$, i.e., all large powers of Z are zero. Thus the series for e^Z reduces to a polynomial expression in Z and we have $e^Z = [\xi_{jk}]$ with $\xi_{jj} = 1$, $\xi_{j,j+1} = 1/1!$, $\xi_{j,j+2} = 1/2!, \ldots$, and $\xi_{jk} = 0$ if $j > k$. Thus e^Z is completely defined and we have $e^{C_s} = e^r e^Z$.

Finally e^J is the direct sum of the matrices e^{C_s}. Note that, if t is any real or complex number, then $Ct = rtI + Zt$, $e^{Ct} = e^{rt}e^{Zt}$, and $e^{Zt} = [\zeta_{jk}]$ with $\zeta_{jj} = 1$, $\zeta_{j,j+1} = t/1!$, $\zeta_{j,j+2} = t^2/2!$, ..., and $\zeta_{jk} = 0$ for $j > k$. It is important for our purpose to observe inally that if A is any $n \times n$ matrix and t real or complex, we have

$$(d/dt)\, e^{At} = A\, e^{At}.$$

This identity can be proved by direct differentiation of the series for e^{At},

$$e^{At} = I + (t/1!)\, A + (t/2!)\, A^2 + (t/3!)\, A^3 + \cdots,$$

or by reduction of A to canonical form and the consideration of the matrices $e^{C_s t}$ above. Finally we will need the following theorem:

(2.1. xii) For any real or complex $n \times n$ matrix A with $\det A \neq 0$ there are (infinetely many) complex matrices B with $e^B = A$, and we will denote them as $B = \ln A$.

Proof. Suppose first $A = J$ have the canonic form of (2.1. viii). If all $n_s = 1$, then J is diagonal, $A = \operatorname{diag}(\lambda_1, \ldots, \lambda_n)$, $\lambda_j \neq 0$, $j = 1, \ldots, n$, and we have $B = \operatorname{diag}(\ln \lambda_1, \ldots, \ln \lambda_n)$. Otherwise we may consider J as the direct sum of the matrices C_s of orders n_s, and determine B as the direct sum of matrices B_s of the same orders n_s. For each matrix C_s put $C_s = rI + Z$. Let us observe that for any complex number y, $|y| < 1$, we have $1 + y = \exp \ln(1 + y)$, and hence

$$1 + y = \sum_{h=0}^{\infty} (1/h!) \left(\sum_{k=0}^{\infty} (-1)^k k^{-1} y^k \right)^h,$$

and this identity could be verified by actual computations. Thus the same identity holds when y is replaced by the matrix $r^{-1}Z$, that is, we have

$$I + r^{-1}Z = \sum_{h=0}^{\infty} (1/h!) \left(\sum_{k=0}^{\infty} (-1)^k k^{-1} r^{-k} Z^k \right)^h,$$

where $Z^k = 0$ for all $k \geq n_s$. Consequentely, we may assume

$$\ln(I + r^{-1}Z) = \sum_{k=0}^{\infty} (-1)^k k^{-1} r^{-k} Z^k,$$

and finally

$$\ln C_s = \ln(rI + Z) = (\ln r)\, I + \sum_{k=0}^{n-1} (-1)^k k^{-1} r^{-k} Z^k.$$

Thus $\ln J$ is defined as the direct sum of the matrices $\ln C_s$, $s = 1, \ldots, m$. For any $n \times n$ matrix A we have $A = PJP^{-1}$ for some matrix P with $\det P \neq 0$, and we may assume $B = \ln A = P \ln J\, P^{-1}$.

2.2. First applications to differential systems. If $A = [a_{ij}(t)]$ denotes an $n \times n$ continuous matrix function of t, $t \geq t_0$, and $x(t) = [x_i(t)]$ a vector function of t, we shall consider the homogeneous linear system

$$x_i' = \sum_{j=1}^{n} a_{ij}(t)\, x_j, \qquad i = 1, \ldots, n, \quad \text{or} \quad x' = Ax. \tag{2.2.1}$$

By a fundamental system of solutions $X(t) = [x_{ij}(t)]$, of (2.2.1) we shall denote an $n \times n$ matrix $X(t)$ whose n columns are independent solutions of (2.2.1). Sometimes we may suppose that these n solutions are determined by the initial conditions

$$x_{ij}(t_0) = \delta_{ij}, \qquad i = 1, \ldots, n, \qquad (j = 1, \ldots, n),$$

where $\delta_{ij} = 1$, or 0, according as $i = j$, or $i \neq j$; i.e., $X(t_0) = I$ where I is the unit matrix. Then we have $x(t) = X(t)\, x(t_0)$ for every solution $x(t)$ of (2.2.1). Indeed $X(t)\, x(t_0)$, as a linear combination of solutions of (2.2.1), is a solution of (2.2.1), and since $X(t_0)\, x(t_0) = x(t_0)$, the product $X(t)\, x(t_0)$, by the uniqueness theorem (1.1. ii), coincides with $x(t)$. From the formula for the derivative of a determinant we obtain also, as usual, that $\det X(t)$ satisfies the first order equation

$$(d/dt)(\det X) = (\operatorname{tr} A) \det X,$$

and hence we have the Jacobi-Liouville formula

$$\det X(t) = \det X(t_0) \exp \int_{t_0}^{t} (\operatorname{tr} A)\, dt. \tag{2.2.2}$$

If $f(t) = [f_i(t),\ i = 1, \ldots, n]$ denotes any n-vector, we shall consider also the nonhomogeneous linear system

$$x_i' = \sum_{j=1}^{n} a_{ij}(t)\, x_j + f_i(t), \qquad i = 1, \ldots, n, \quad \text{or} \quad x' = A x + f. \tag{2.2.3}$$

Then if $x(t)$ is any solution of (2.2.3), $y(t)$ the solution of the homogeneous system $y' = A y$ determined by the same initial conditions $y(t_0) = x(t_0)$, if $Y(t)$ is the fundamental system of solutions of $y' = A y$ with $Y(t_0) = I$, then the following relation holds:

$$x(t) = y(t) + \int_{t_0}^{t} Y(t)\, Y^{-1}(\alpha)\, f(\alpha)\, d\alpha. \tag{2.2.4}$$

Indeed, the second member verifies (2.2.3), satisfies the same initial conditions as $x(t)$, and thus coincides with $x(t)$ by the uniqueness theorem (1.1. ii).

If A is a constant matrix, and we assume $t_0 = 0$, then $Y(t)\, Y^{-1}(\alpha)$ is the fundamental system of solutions of (2.2.1) determined by the initial conditions $Y(t)\, Y^{-1}(\alpha) = I$ at $t = \alpha$, and the same for $Y(t - \alpha)$; hence $Y(t)\, Y^{-1}(\alpha) = Y(t - \alpha)$ and finally (2.2.4) becomes

$$x(t) = y(t) + \int_{t_0}^{t} Y(t - \alpha)\, f(\alpha)\, d\alpha,$$

for A a constant matrix and $Y(0) = I$.

2.3. Systems with constant coefficients. The system of first order homogeneous linear differential equations

$$x_i' = \sum_{j=1}^{n} a_{ij}\, x_j, \qquad i = 1, \ldots, n, \tag{2.3.1}$$

can be written in the form $x' = A x$ where $A = [a_{ij}]$ is a constant $n \times n$ matrix, $x = (x_1, \ldots, x_n)$ is an $n \times 1$ matrix, or n-vector, function of t, and $x'=dx/dt$. If P is any constant $n\times n$ matrix with $\det P\neq 0$ and $y=(y_1,\dots,y_n)$ is an $n\times 1$ matrix, or n-vector, function of t, related to x by the formula $x=Py$, or $y=P^{-1}x$, then (2.3.1) is transformed into the system $y'=By$, where $B=P^{-1}AP$. Thus, by (2.1), there are matrices P which transform A into its canonical form J discussed in (2.1), $J=\operatorname{diag}[C_1, C_2, \dots, C_N]$, where each matrix C_s of order n_s is defined as in (2.1) and $n_1+\dots+n_N=n$. For every s let $h=n_1+\dots+n_{s-1}$. Then, if $n_s=1$, the $(h+1)$-th equation of the system $y'=Jy$ has the form

$$y'_{h+1}=\varrho'_s\, y_{h+1}. \tag{2.3.2}$$

If $n_s>1$, then the equations of indices $h+1,\dots,h+n_s$ of the same system have the form

$$\left.\begin{aligned} &y'_{h+1}=\varrho'_s\, y_{h+1}+y_{h+2},\\ &y'_{h+2}=\varrho'_s\, y_{h+2}+y_{h+3},\ \dots,\ y'_{h+n_s}=\varrho'_s\, y_{h+n_s}, \end{aligned}\right\} \tag{2.3.3}$$

where $s=1, 2, \dots, N$. Each system (2.3.2) has the solution

$$y_{h+1}=e^{\varrho'_s t}.$$

Thus a corresponding solution of the system $y'=Jy$ is obtained by putting $y_j=0$ for all $1\leq j\leq h$, $h+2\leq j\leq n$. Each system (2.3.3) has n_s independent solutions of the form

$$\left.\begin{aligned} &y_{h+1}=\left(t^{\alpha-1}/(\alpha-1)!\right)e^{\varrho'_s t},\quad y_{h+2}=\left(t^{\alpha-2}/(\alpha-2)!\right)e^{\varrho'_s t},\ \dots,\\ &y_{h+\alpha}=e^{\varrho'_s t},\quad y_{h+\alpha+1}=\dots=y_{n_s}=0, \end{aligned}\right\} \tag{2.3.4}$$

where α is one of the integers $\alpha=1, 2, \dots, n_s$. The corresponding n_s solutions of the system $y'=By$ are then obtained by putting $y_j=0$ for all $1\leq j\leq h$, $h+n_s+1\leq j\leq n$. If we denote by $Y(t)$ the matrix of all n solutions of system $y'=Jy$ defined above, we have $Y(0)=I$, and thus $Y(t)$ is certainly a fundamental system. According to (2.1) system $x'=Ax$ has the system of solutions $X=e^{At}$, and since $X(0)=I$ obviously $X(t)$ is a fundamental system of solutions of (2.1). For $A=J$, $Y=e^{Jt}$ is a fundamental system of solutions of $y'=Jy$, where e^{Jt} is the direct sum of the matrices $e^{C_s t}$. By comparison with (2.1) it is easy to recognize that (2.3.4) is exactly $e^{C_s t}$. Since the $x_i\,[y_i]$ are linear combinations of the $y_s\,[x_s]$ with constant coefficients, we conclude that system (2.3.1) has all solutions x_i bounded in $[0, +\infty)$ if and only if the same occurs for the solutions y of $y'=Jy$, and this occurs if and only if $R(\varrho_r)\leq 0$, $r=1,\dots,m$, and if, for those roots (if any) with $R(\varrho_r)=0$ all companion matrices C_s have orders $n_s=1$. By (2.1) we know that this occurs if and only if $\mu_r=\nu_r$. Thus we conclude as follows:

(2.3. i) The system $x'=Ax$ has solutions all bounded in $[0, +\infty)$ if and only if $R(\varrho_r)\leq 0$, $r=1,\dots,m$, and if, for those roots (if any) with $R(\varrho_r)=0$, we have $\mu_r=\nu_r$. Also, system $x'=Ax$ has solutions all approaching zero as $t\to+\infty$ if and only if $R(\varrho_r)<0$, $r=1,\dots,m$.

An n-th order linear homogeneous differential equation with constant coefficients

$$y^{(n)} + a_1 y^{(n-1)} + \cdots + a_n y = 0 \qquad (2.3.5)$$

can be written in the form (2.3.1) by the substitution $y = x_1$, $y' = x_2, \ldots,$ $y^{(n-1)} = x_n$ and then it yields the linear system

$$x_1' = x_2, \quad x_2' = x_3, \ldots, x_{n-1}' = x_n, \quad x_n' = -a_n x_1 - \cdots - a_1 x_n,$$

whose matrix $A = [a_{ij}]$ has a quite typical form. It is an elementary exercise to prove that $\det(\varrho I - A) = \varrho^n + a_1 \varrho^{n-1} + \cdots + a_n$ and that for every characteristic root ϱ_r we have $\nu_r = 1$. A consequence of (2.3. i) is then

(2.3. ii) Equation (2.3.5) has all solutions bounded in $[0, +\infty)$ together with all their derivatives if and only if $R(\varrho_r) \leq 0$, $r = 1, 2, \ldots, m$, and if, for all roots ϱ_r (if any) with $R(\varrho_r) = 0$, we have $\mu_r = 1$. Also, equation (2.3.5) has all solutions approaching zero (with all their derivatives) if and only if $R(\varrho_r) < 0$, $r = 1, 2, \ldots, m$.

We add here the following simple remark concerning system $x' = Ax$.

(2.3. iii) If a is any real number $a > R(\varrho_r)$, $r = 1, \ldots, m$, then there is a constant $C > 0$, $(C = C[A, a])$, such that $\|x(t)\| \leq C\|x(0)\| \exp(at)$ for every solution $x(t)$ of $x' = Ax$.

Proof. Let $\alpha = \max R(\varrho_s)$, $\nu = \max n_s$, and let $c > 0$, be a constant such that $1, t, \ldots, t^\nu \leq c \exp(a - \alpha)t$ for all $t \geq 0$. Such a constant certainly exists since $t^s / \exp(a - \alpha)t \to 0$ as $t \to +\infty$, $s = 0, 1, \ldots, \nu$. Then, by $|\exp(\varrho_s t)| \leq \exp(\alpha t)$, we deduce $|y_i(t)| < c \exp(at)$ for each element of the matrix $Y(t)$ above. Hence, $\|Y(t)\| \leq n^2 c \exp(at)$ for all $t \geq 0$. Now for every solution x of $x' = Ax$ we have $x = Py$, $y = P^{-1}x$, and $y(t) = Y(t)y(0)$ since $Y(0) = I$. Thus $x(t) = Py(t) = PY(t)y(0) = PY(t)P^{-1}x(0)$ and $\|x(t)\| \leq \|P\| \cdot \|Y(t)\| \cdot \|P^{-1}\| \|x(0)\| \leq C\|x(0)\| \exp(at)$ for some constant C.

2.4. The Routh-Hurwitz and other criteria. The considerations above show that the question of the boundedness in $[0, +\infty]$ of all solutions of a system (2.3.1) or an equation (2.3.5) is reduced to a question of algebra. Thus, any condition assuring that the characteristic roots ϱ_i have the properties above may be of interest for the problem under discussion. One of the best known conditions is due to E. J. Routh [1] and A. Hurwitz [1].

(2.4. i) If $F(z) = z^n + a_1 z^{n-1} + \cdots + a_n$ is a polynomial with real coefficients, let $D_1 = a_1$,

$$D_k = \det \begin{bmatrix} a_1 & a_3 & a_5 & \cdots & a_{2k-1} \\ 1 & a_2 & a_4 & \cdots & a_{2k-2} \\ 0 & a_1 & a_3 & \cdots & a_{2k-3} \\ 0 & 1 & a_2 & \cdots & a_{2k-4} \\ \cdot & \cdot & \cdot & \cdots & \cdot \\ 0 & 0 & 0 & \cdots & a_k \end{bmatrix}, \qquad k = 2, 3, \ldots, n,$$

with $a_j = 0$ for $j > n$. If all determinants D_k are positive, $k = 1, 2, \ldots, n$, then all zeros of $F(z)$ have negative real parts.

For instance, if $F(z) = z^3 + 11z^2 + 6z + 6$, we have $D_1 = 11$, $D_2 = 60$, $D_3 = 360$, and $F(z)$ has all roots with negative real parts.

See for references M. MARDEN [1, p. 141]. The same book refers also to analogous conditions for polynomials $F(z)$ with complex coefficients. A more involved condition assuring that the roots of $F(z)$ either have real parts negative, or have real parts zero and are simple (as required by 2.3. ii), has been given by T. VIOLA [1]. The Hurwitz criterion is also a particular case of more comprehensive statements concerning the number of zeros of $F(z)$ whose real parts are above, or below a given number, or between two given numbers. Either theory of residues, or Sturm sequences, are used in the proofs of these statements (M. MARDEN [1]).

By the remarks of (1.4) conditions (2.3 i) or (2.3. ii) are also sufficient conditions for the stability in the sense of LYAPUNOV (resp. asymptotic stability) at the right of all solutions of system (2.3.1) [or differential equation (2.3.5)].

For n large the use of the Routh-Hurwitz criterion is impractical, and other equivalent processes replace it quite well, namely the very same processes by means of which that criterion is usually proved. We mention here briefly some pertinent statements.

(2.4. ii) A necessary condition in order that the real polynomial $F(z) = z^n + a_1 z^{n-1} + \cdots + a_n$ have all its roots with negative real parts, is that all (real) coefficients $a_1, \ldots, a_n$ are positive.

Proof. Indeed, the roots $z_1, \ldots, z_n$ (each repeated as many times as its multiplicity) are real, or in complex conjugate pairs. Hence the polynomial $F(z)$ is the product of factors either of the form $(z - \alpha - i\beta)(z - \alpha + i\beta) = z^2 - 2\alpha z + (\alpha^2 + \beta^2) = z^2 + az + b$ with $a > 0$, $b > 0$, or of the form $z - \alpha = z + a$, with $a > 0$. By successive multiplications we necessarily obtain a polynomial $F(z)$ which has its coefficients all $\neq 0$ and positive.

Now let us consider together with $F(z)$, the polynomial $G(z) = a_1 z^{n-1} + a_3 z^{n-3} \ldots$, whose last terms is $a_{n-1} z$, or a_n according as n is even or odd. We may well suppose now $a_1, a_2, \ldots, a_n$ all real and positive. Let us perform on $F(z)$, $G(z)$ the usual finite process for the determination of their highest common factor, i.e., determine the Sturmian finite sequence

$$F = G d_1 + f_2, \quad G = f_2 d_2 + f_3, \quad f_2 = f_3 d_3 + f_4, \ldots$$

where $d_1 = b_1 z + 1$, $d_2 = b_2 z$, $d_3 = b_3 z, \ldots$, are all polynomials of the first degree.

(2.4. iii) A necessary and sufficient condition in order that all roots of the real polynomial $F(z)$ have negative real parts is that the numbers $b_1, b_2, \ldots$ all be positive.

For instance, if $F(z) = z^3 + 11z^2 + 6z + 6$, we have $G(z) = 11z^2 + 6$, and, by successive divisions, we have $d_1 = (1/11)z + 1$, $d_2 = (121/60)z$, $d_3 = (10/11)z$, and hence b_1, b_2, b_3 are positive, and $F(z)$ has all roots with negative real parts.

Proof of (2.4. iii*).* Indeed, the numbers b_i and HURWITZ' determinants D_i are related by formulas usually proved in algebra, namely

$$b_1 = D_1^{-1}, \quad b_2 = D_1^2 D_2^{-1}, \quad b_2 = D_2^2 D_1^{-1} D_3^{-1}, \quad b_3 = D_3^2 D_2^{-1} D_4^{-1}, \ldots.$$

In general $b_i = D_i^2 D_{i-1}^{-1} D_{i+1}^{-1}$ (see e.g. D. F. LAWDEN [1]; M. MARDEN [1]). Thus the numbers b_i are all positive if and only if the numbers D_i are all positive, and thus (2.4. iii) follows from (2.4. i).

Let us observe finally that the study of $F(z)$ on the imaginary axis of the complex z-plane can be done easily by putting $z = iy$. Then

$$F(iy) = A(y) + iB(y) = (a_n - a_{n-2}y^2 + \cdots) + i(a_{n-1}y - a_{n-3}y^3 + \cdots)$$

and the zeros of F on the imaginary axis are the common real roots of the two real polypomials A and B. An important theorem of the

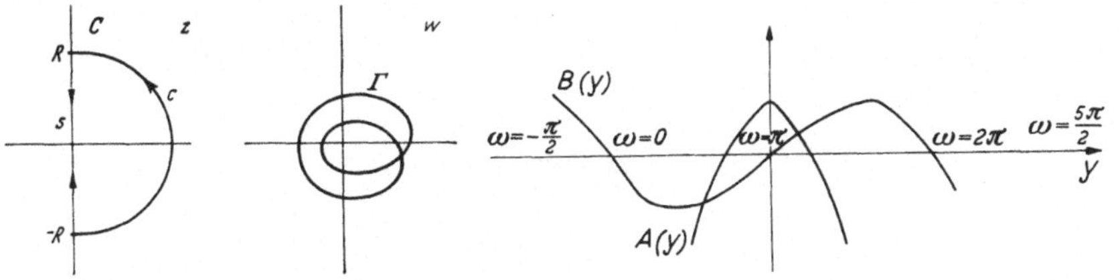

theory of complex functions states that the number of the zeros of an analytic regular function $F(z)$ within a closed path C ($F \neq 0$ on C) is given by $\Omega/2\pi$ where Ω is the variation of the argument of $F(z)$ along C. In other words, as z describes C, the complex variable $w = F(z)$ describes a closed path Γ in the w-plane and $|\Omega/2\pi|$ is the number of times by which Γ encircles the origin $w = 0$. (In Topology this number is called the topological index of Γ with respect to the origin $w = 0$ (see P. ALEXANDROV and H. HOPF [1], p. 462). If C is the path which is the composite of the half circumference c $\left[c = R\exp(i\theta),\ -\frac{\pi}{2} \leq \theta \leq \frac{\pi}{2}\right]$, and the segment s between the points Ri and $-Ri$ then C for large R will contain all roots with positive real parts. If $F(z)$ has no imaginary root, then $F(z) \neq 0$ on C for large R. On c the term z^n of $F(z)$ is predominant and hence the variation of the argument along c is $n\pi$ $(1 + O(R^{-1})$. Along s we have $\arg F(z) = \arctan[B(y)/A(y)]$, and thus a detailed study of this real function of y in $(-\infty, +\infty)$ yields the variation of the argument of w along s, and finally Ω. Thus we have the following:

(2.4. iv) A necessary and sufficient condition for asymptotic stability is that $F(z)$ have no pure imaginary roots and that $\Omega = 0$.

For instance if $F(z) = z^3 + 11z^2 + 6z + 6$, we have $n = 3$, $A(y) = 6 - 11y^2$, $B(y) = 6y - y^3$, and the graphs of $A(y)$, $B(y)$ show that (1) A and B have no common root, hence $F(z)$ has no purely imaginary root; (2) if $\omega(y) = \arctan(B/A)$, and we assume $\omega(0) = \pi$, then $\omega(-\infty) = -\frac{\pi}{2}$, $\omega(+\infty) = \frac{5\pi}{2}$. Thus $\arg F(z)$ has variations -3π on s, and $+3\pi$ on C; i.e., $\Omega = 0$, and $F(z)$ has all its roots with negative real part.

For a great number of applications of the methods discussed above (2.4. i, ii, iii, iv) see the recent book by D. F. LAWDEN [1]. The method which has lead to (2.4. iv) is closely related to the Nyquist diagram (2.8. vii).

2.5. Systems of order 2. The considerations of (2.3), (2.4) may be usefully exemplified by the following examples.

$$\text{I.}\quad x' = a x_1 + b x_2, \qquad x_2' = c x_1 + d x_2, \qquad a, b, c, d \text{ real constants.} \tag{2.5.1}$$

The characteristic equation is $\varrho^2 - (a+d)\varrho + (ad - bc) = 0$ and (2.5.1), for $t \to +\infty$ presents the following cases: 1. $a + d < 0$, $ad - bc > 0$; roots with nega-

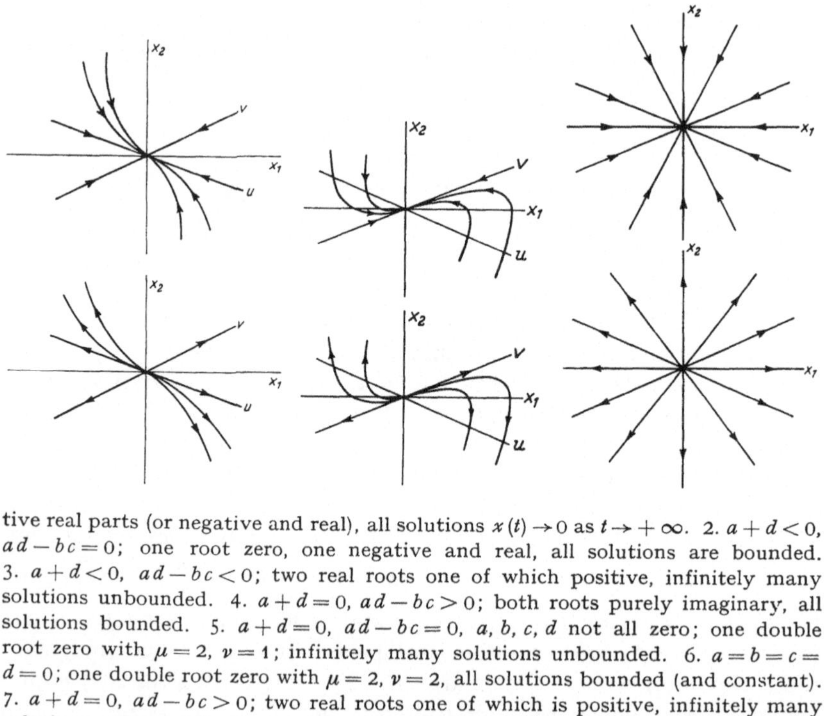

tive real parts (or negative and real), all solutions $x(t) \to 0$ as $t \to +\infty$. 2. $a + d < 0$, $ad - bc = 0$; one root zero, one negative and real, all solutions are bounded. 3. $a + d < 0$, $ad - bc < 0$; two real roots one of which positive, infinitely many solutions unbounded. 4. $a + d = 0$, $ad - bc > 0$; both roots purely imaginary, all solutions bounded. 5. $a + d = 0$, $ad - bc = 0$, a, b, c, d not all zero; one double root zero with $\mu = 2$, $\nu = 1$; infinitely many solutions unbounded. 6. $a = b = c = d = 0$; one double root zero with $\mu = 2$, $\nu = 2$, all solutions bounded (and constant). 7. $a + d = 0$, $ad - bc > 0$; two real roots one of which is positive, infinitely many solutions unbounded. 8. $a + d > 0$; at least one root with real positive part (or real and positive), infinitely many solutions unbounded.

Another viewpoint in the analysis of the solutions of system (2.5.1) is the following one which has far reaching consequences in the discussion of nonlinear systems. We shall consider the solutions of (2.5.1) as trajectories in the $x_1 x_2$ plane, and study their behavior as $t \to +\infty$. We shall suppose $ad - bc \neq 0$ which excludes zero roots for the characteristic equation. The following cases shall be taken into consideration.

(a) Two real distinct roots of the same sign, say $\varrho_2 < \varrho_1 < 0$, or $0 < \varrho_1 < \varrho_2$; i.e. $(a+d)^2 - 4(ad - bc) > 0$, and $a + d < 0$, or $a + d > 0$. Then system (2.5.1) is transformable by means of a linear real transformation to the canonical form $u' = \varrho_1 u$, $v' = \varrho_2 v$, whose solutions are $u = A\exp(\varrho_1 t)$, $v = B\exp(\varrho_2 t)$, A, B arbitrary constants. These solutions represent the u-axis ($A \neq 0$, $B = 0$), the v-axis ($A = 0$, $B \neq 0$) and the curves $v/B = (u/A)^{\varrho_2/\varrho_1}$ ($A \neq 0$, $B \neq 0$). If $\varrho_2 < \varrho_1 < 0$, then $u \to 0, v \to 0$, $v/u = (B/A)\exp(\varrho_2 - \varrho_1)t \to 0$ as $t \to +\infty$; if $0 < \varrho_1 < \varrho_2$ the same occurs as $t \to -\infty$. The trajectories are represented in the illustrations where the arrows correspond to increasing values of t. The point $(0, 0)$ is said to be a *stable*, or *unstable node*, according as $\varrho_1, \varrho_2 < 0$, or $\varrho_1, \varrho_2 > 0$.

(b) One double root ϱ (necessarily real) with $\mu = 2$, $\nu = 1$; i.e. $(a+d)^2 - 4(ad - bc) = 0$, (and thus $\varrho = (a+d)/2$) and $a - d, b, c$ not all zero. Then system (2.5.1) is transformable by means of a real linear transformation to the canonical system $u' = \varrho u$, $v' = \varrho v + u$ whose solutions are $u = A\exp(\varrho t)$, $v = (At + B)\exp(\varrho t)$. Thus

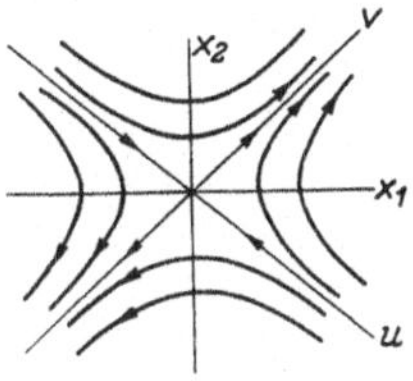

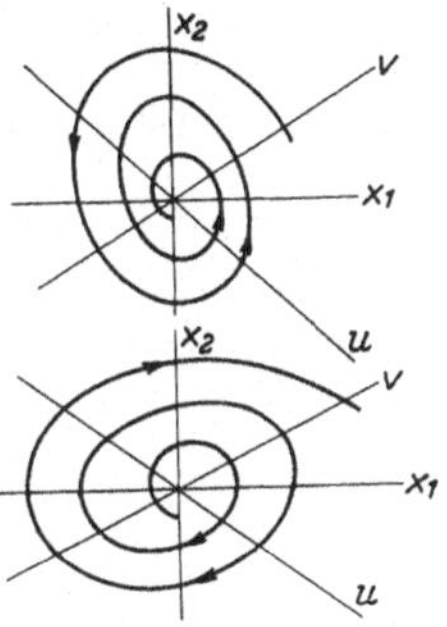

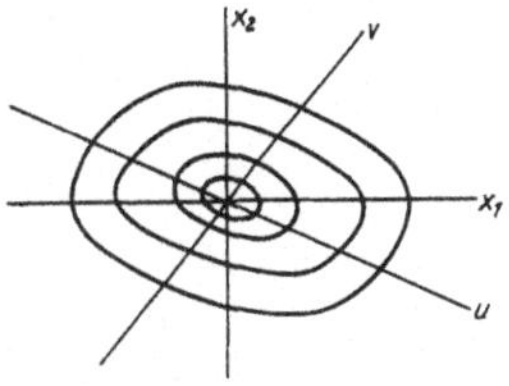

for $A = 0$, the corresponding solutions form the v-axis. If $\varrho < 0$, i.e. $a + d < 0$, then $u \to 0$, $v \to 0$ $v/u \to \infty$ as $t \to +\infty$; if $\varrho > 0$, i.e., $a + d > 0$, the same occurs as $t \to -\infty$. The trajectories (for $A \neq 0$, $B \neq 0$) cross the u-axis at $t = -B/A$. The trajectories are represented in the illustrations. The point $(0, 0)$ is said to be a *stable*, or *unstable node* according as $\varrho < 0$, or $\varrho > 0$.

(c) One double root ϱ (necessarily real) with $\mu = 2$, $\nu = 2$; i.e., $a = d = \varrho$, $b = c = 0$. Then system (2.5.1) has the form $x_1 = \varrho x_1$, $x_2' = \varrho x_2$ and its solutions are $x_1 = A\exp(\varrho t)$, $x_2 = B\exp(\varrho t)$, which form the x_1-axis, the x_2-axis, and the straight lines $x_2/x_1 = B/A$, ($A \neq 0$, $B \neq 0$). If $\varrho = a = d < 0$, then $x_1 \to 0$, $x_2 \to 0$ as $t \to +\infty$. If $\varrho = a = d > 0$, the same occurs as $t \to -\infty$. The trajectories are represented in the illustrations, and the point $(0, 0)$ is said to be a *stable*, or *unstable node* according as $\varrho < 0$, or $\varrho > 0$.

(d) Two real distinct roots of different signs, say $\varrho_1 < 0 < \varrho_2$; i.e., $(a+d)^2 - 4(ad - bc) > 0$, $(a+d)(ad - bc) < 0$. Then, the discussion proceeds as in (a) only now for $A \neq 0$, $B \neq 0$, the curves have the equations $(u/A)^{\varrho_1}(v/B)^{\varrho_2} = 1$ and $u \to 0$, $v \to \infty$ as $t \to +\infty$. The trajectories are represented in the illustrations, and the point $(0, 0)$ is said to be a *saddle point*.

(e) Two complex conjugate roots $\varrho, \bar{\varrho} = \alpha \pm i\beta$, $\alpha \neq 0$; i.e., $(a+d)^2 - 4(ad - bc) < 0$, $a + d \neq 0$, and $\alpha = (a+d)/2$. Then, as we have noticed in (2.1), system (2.5.1) is transformable, by means of a linear transformation, to the system $u' = \varrho u$, $v' = \bar{\varrho} v$, and each real solution (x_1, x_2) of (2.5.1) is a linear combination with conjugate coefficients of complex conjugate solutions u of $u' = \varrho u$, and $v = \bar{u}$ of $v' = \bar{\varrho} v$. Now in the complex u-plane the first equation has solutions of the form $u = A e^{\varrho t} = m e^{in} e^{\alpha t} e^{i\beta t} = m e^{\alpha t} e^{i(n + \beta t)}$. Thus if $u = r e^{i\theta}$, $r \neq 0$, we have $r = m e^{\alpha t}$, $\theta = n + \beta t + 2k\pi$. If $\alpha < 0$, then $r \to 0$ as $t \to +\infty$; if $\alpha > 0$, the same occurs as

$t \to -\infty$. In any case $\theta \to \infty$ as $t \to \infty$. The trajectories are represented in the illustrations. The point (0, 0) is said to be a *stable* or *unstable spiral point* according as $\alpha < 0$, or $\alpha > 0$.

(f) Two purely imaginary complex roots $\varrho, \bar{\varrho} = \pm i\beta$, $\beta \neq 0$; i.e., $a+d=0$, $\beta^2 = ad-bc > 0$. The discussion proceeds as in (e) only that here $r =$ constant, i.e., the trajectories are circles in the complex u-plane, and are ellipses in the $x_1 x_2$-plane (see illustration). The point (0, 0) is said to be a *center*.

II. $$x'' + 2gx' + fx = 0, \; f, g \text{ real constants}, \tag{2.5.2}$$

This equation, by putting $x_1 = x$, $x_2 = x'$, is reduced to the system $x_1' = x_1$, $x_2' = -fx_1 - 2gx_2$; hence $a=0$, $b=1$, $c=-f$, $d=-2g$, $a+d=-2g$, $ad-bc=f$.

The discussion is analogous to the one above, only the cases (b) and (c) are excluded. The $x_1 x_2$-plane, now that $x_1 = x$, $x_2 = x'$, is said to be the *phase plane.* The point (0, 0) is a center if $g=0$, $f>0$, is a stable (unstable) spiral point if $f>g^2$, $g>0\,[g<0]$, is a saddle point if $f<g^2$, $f<0$, is a stable (unstable) node if $f \leq g^2$, $g>0\,[g<0]$. The solutions of (2.5.2) are of the form

$$x = A\,e^{\alpha t} \sin(\gamma t - \lambda) \quad \text{if} \quad f > g^2,$$

$$x = A\,e^{r_1 t} + B\,e^{r_2 t} \quad \text{if} \quad f < g^2,$$

$$x = (A\,t + B)\,e^{\alpha t} \quad \text{if} \quad f = g^2,$$

where A, B, λ are arbitrary constants. If $f > g^2$ then $\alpha = -g$, $\gamma = (f-g^2)^{\frac{1}{2}} > 0$; and, as usual, the following terms are used: A amplitude, λ phase, $\delta = -2\pi\alpha/\gamma$ logarithmic decrement, $T = 2\pi/\gamma$ period, $\nu = T^{-1} = \gamma/2\pi$ frequency, the nonzero solutions being all oscillatory (5.1). If $f<g^2$, then $r_1, r_2 = -g \pm (g^2-f)^{\frac{1}{2}}$; if $f=g^2$, $\alpha = -g$, and the solutions are not oscillatory in either case. All solutions $x(t)$ together with $x'(t)$ approach zero as $t \to +\infty$ if and only if $g>0$, $f>0$; are all bounded if and only if $g \geq 0$, $f>0$, or $g>0$, $f \geq 0$.

If $x(t)$ is the displacement of a physical system at the time t, and $g \geq 0$, $f>0$, then its motion is said to be aperiodic and overcritically damped if $g>0$, $f<g^2$; aperiodic and critically damped if $g>0$, $f=g^2$; oscillatory and undercritically damped if $g>0$, $f>g^2$; simply harmonic if $g=0$, $f>0$.

It is typical of all linear oscillations that the frequency ($\nu = \gamma/2\pi$ above) is a constant, independent of the amplitude A. This independence ceases in general with nonlinear systems.

2.6. Nonhomogeneous systems. If A is a constant matrix and $f(t)$ a vector function, then we shall consider the nonhomogeneous system

$$x' = A\,x + f. \tag{2.6.1}$$

(2.6. i) If $\|f(t)\| \leq c\,e^{bt}$, $R(\varrho_i) < b$, $i=1,\ldots,n$, then also $\|x(t)\| < N e^{bt}$, $t \geq t_0$, for every solution $x(t)$ of (2.6.1) and some constants N and t_0.

In particular if $f(t)$ is bounded, and $R(\varrho_i) < 0$, $i=1,\ldots,n$, then all $x(t)$ are bounded ($t \geq t_0$). Indeed, if $y(t)$ is the solution of the system $y' = A\,y$ having the same initial conditions at $t=t_0=0$, if $Y(t)$ is the fundamental solution of $y' = A\,y$ with $Y(0) = I$, then

$$x(t) = y(t) + \int_0^t Y(t-\alpha)\,f(\alpha)\,d\alpha,$$

$$\|x(t)\| \leq \|y(t)\| + \int_0^t \|Y(t-\alpha)\|\,\|f(\alpha)\|\,d\alpha.$$

From $\|f(t)\| \leq C e^{bt}$, $R(\varrho_i) \leq a < b$, $i = 1, \ldots, n$, it follows that $\|y(t)\|$, $\|Y(t)\| < M e^{at}$ for some constant M and $t \geq 0$, and hence

$$\|x(t)\| \leq M e^{at} + M C \int_0^t e^{a(t-\alpha)} e^{b\alpha} d\alpha$$
$$= M e^{at} + M C (b-a)^{-1} (e^{bt} - e^{at}) \leq M [1 + C(b-a)^{-1}] e^{bt}.$$

The conclusion of (2.6. i) does not hold necessarily if $R(\varrho_i) = b$ for some i, even if for the corresponding roots ϱ_i we have $\mu_i = \nu_i$, as the following example shows: $x_1' = x_2$, $x_2' = -x_1 + 2\cos t$ (i.e., $y'' + y = 2\cos t$) whose solutions $x_1 = t \sin t + C \sin(t-\gamma)$, $x_2 = t\cos t + \sin t + C\cos(t-\gamma)$ are all unbounded. For the case now excluded we mention the statement: (2.6. ii) If the system $x' = Ax$ has all its solutions bounded in $[0, +\infty)$, if $\int^{+\infty} \|f(t)\| dt < +\infty$, then also the solutions of the system $x' = Ax + f$ are all bounded in $[0, +\infty)$. The proof is a modification of the previous one. (See § 3.) Analogous statements hold for a nonhomogeneous differential equation

$$x^{(n)} + a_1 x^{(n-1)} + \cdots + a_n x = f(t)$$

with constant coefficients.

2.7. Linear resonance. We have already mentioned that the equation

$$x'' + k x' + \omega^2 x = 0, \qquad k \geq 0, \ \omega > 0, \tag{2.7.1}$$

has solutions all bounded and oscillatory if $k^2 < 4\omega^2$, i.e., $k < 2\omega$. We shall now consider the nonhomogeneous equation

$$x'' + k x' + \omega^2 x = A \cos m t, \qquad k \geq 0, \qquad \omega > 0, \qquad m \geq 0, \qquad m \neq \omega, \tag{2.7.2}$$

and its limiting case

$$x'' + k x' + \omega^2 x = A \cos \omega t, \qquad k \geq 0, \qquad \omega > 0. \tag{2.7.3}$$

As usual, $k x'$ is said to be the damping force, $\omega^2 x$ the restoring spring force, $A \cos mt$ an external sinusoidal force (input). If x_1, x_2, x_3 are the solutions of the equations (2.7.1), (2.7.2), (2.7.3) respectively satisfying the initial conditions $x(0) = \eta_0$, $x'(0) = \eta_1$, then

$$x_1(t) = [\eta_0 \cos\gamma t + \gamma^{-1}(\eta_1 - \eta_0 \alpha) \sin \gamma t] e^{\alpha t},$$

$$x_2(t) = [(\eta_0 - A\Delta\cos\lambda)\cos\gamma t + \gamma^{-1}(\eta_1 - A\Delta m \sin\lambda -$$
$$- \alpha\eta_0 + \alpha A \Delta \cos\lambda)\sin\gamma t] e^{\alpha t} + A\Delta\cos(mt - \lambda),$$

$$x_3(t) = [\eta_0\cos\gamma t + \gamma^{-1}(\eta_1 - \alpha\eta_0 - A k^{-1})\sin\gamma t] e^{\alpha t} + A k^{-1}\omega^{-1}\sin\omega t,$$

where

$$\alpha = -2^{-1} k, \qquad \gamma = (\omega^2 - 4^{-1} k^2)^{\frac{1}{2}} > 0,$$

$$\cos\lambda = (\omega^2 - m^2)\Delta, \qquad \sin\lambda = k m \Delta, \qquad \Delta = [(\omega^2 - m^2)^2 + k^2 m^2]^{-\frac{1}{2}}.$$

For $m = 0$, $\omega > 0$, we have $\Delta = \omega^{-2}$, $\cos\lambda = 1$, $\sin\lambda = 0$, and

$$x_2(t) = [(\eta_0 - A\omega^{-2})\cos\gamma t + \gamma^{-1}(\eta_1 - \alpha\eta_0 + \alpha A \omega^{-2})\sin\gamma t] e^{\alpha t} + A\omega^{-2}.$$

For $k = 0$, $\omega > 0$, we have $\alpha = 0$, $\gamma = \omega$, and

$$x_3(t) = \eta_0 \cos\omega t + \omega^{-1}\eta_1 \sin\omega t + (2\omega)^{-1} A t \sin\omega t.$$

Obviously $\gamma \to \omega$ as $k \to 0$ and $\omega/2\pi$ is said to be the "natural frequency" for (2.7.2). Now, if $k > 0$, then $\alpha < 0$, and the terms containing $e^{\alpha t}$ above approach zero as $t \to +\infty$ (transient), so that, for large t, we have

$$x_1(t) \approx 0, \quad x_2(t) \approx A\,\Delta \cos(mt - \lambda), \quad x_3(t) \approx A\,k^{-1}\omega^{-1} \sin \omega t$$

(steady state). In other words, the "input" $A \cos mt$ generates through (2.7.2) an analogous "output" $A\Delta \cos(mt - \lambda)$ of the same frequency, different phase, and amplitude $A\Delta$, while the constant input A generates a constant output $A\omega^{-2}$. The ratio

$$\mu = A\,\Delta : A\,\omega^{-2} = \Delta\,\omega^2 = \left[\left(1 - \left(\frac{m}{\omega}\right)^2\right)^2 + \left(\frac{k}{\omega}\right)^2 \left(\frac{m}{\omega}\right)^2\right]^{-\frac{1}{2}}$$

is called the (dimensionless) amplification factor. If $\xi = m/\omega$, $\eta = k/\omega$, we have $\mu = \mu(\xi) = [(1 - \xi^2)^2 + \eta^2 \xi^2]^{-\frac{1}{2}}$ and $\mu(0) = 1$, $\mu(+\infty) = 0$ for every η. We have now $d\mu/d\xi = -\xi\mu^3(2\xi^2 + \eta^2 - 2)$; hence, if $\eta \geqq \sqrt{2}$, i.e., $k \geqq \sqrt{2}\omega$, then $\mu(\xi)$ is a decreasing function of ξ and $1 > \mu > 0$ for all $\xi > 0$. If $\eta < \sqrt{2}$, i.e., $k < \sqrt{2}\omega$, then $\mu(\xi)$ has a maximum at $\xi = (1 - 2^{-1}\eta^2)^{\frac{1}{2}}$ given by $\mu = \eta^{-1}(1 - \eta^2/4)^{-\frac{1}{2}}$. It is usually said that for $k \geqq \sqrt{2}\omega$ there is no resonance while for $k < \sqrt{2}\omega$ there is resonance around the resonance frequency $m/2\pi = \omega(1 - k^2/2\omega^2)^{\frac{1}{2}}/2\pi$ and that, at this frequency, the amplification factor, or resonance factor, is $\mu_{res} = \omega k^{-1}(1 - k^2/4\omega^2)^{-\frac{1}{2}}$. Obviously $m_{res} \to \omega$, $\mu_{res} \to +\infty$ as $k \to 0$. In other words, the resonance is most remarkable for small damping coefficient k and $m/2\pi$ close to the resonance frequency $m_{res}/2\pi$, where $m_{res}/2\pi \to \omega/2\pi$ (natural frequency) as $k \to 0$.

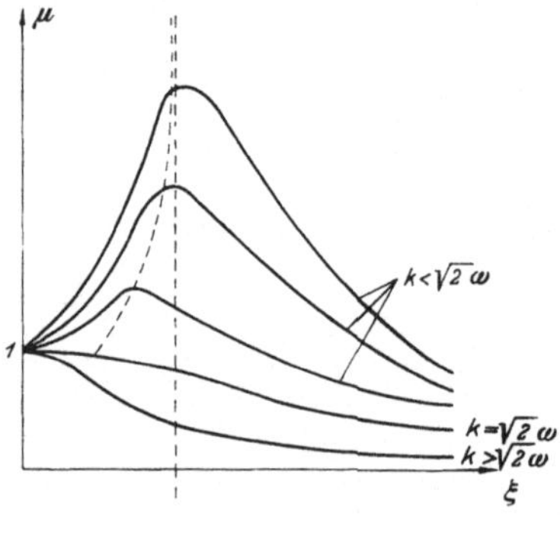

The phenomenon considered above, though quite significant, is rarely completely materialized in any application. In all practical cases the resonance phenomenon is less crude. The deep reason is that all mechanical, or physical system are not linear, but they are more and more similar to linear systems the smaller are their deplacements from their position of equilibrium (see pendulum, elastic spring, elastic string). As soon as the oscillations build up, the system becomes essentially nonlinear, and then the natural frequency is not a constant [end of (2.4)], but a function of the amplitude. Generally the natural frequency changes with the amplitude making the resonance less stringent. See this book in (§ 8) and E. W. Brown [1].

2.8. Servomechanisms. (a) *General considerations.* Let $u(t)$ be a (known) function and $x(t)$ another function (unknown) related by the differential equation

$$a_0 x^{(n)} + \cdots + a_n x = b_0 u^{(m)} + \cdots + b_m u, \quad a_0 \neq 0, \tag{2.8.1}$$

where $a_0, \ldots, a_n, b_0, \ldots, b_m$ denote constants. By using as usual the operator D we have

$$P(D)\,x = Q(D)\,u, \tag{2.8.2}$$

where P, Q denote polynomials in D. If $f(t) = Q(D)\,u$, then any solution x of (2.8.1) is related to the solution y of

$$P(D)\,y = 0, \tag{2.8.3}$$

which has the same initial conditions as x at $t=0$, by the formula

$$x(t) = y(t) + \int_0^t Y(t-\alpha) f(\alpha)\, d\alpha,$$

where $Y(t)$ is the solution of (2.8.3) with $Y(0) = \cdots = Y^{(n-2)}(0) = 0$, $Y^{(n-1)}(0) = 1$ (2.2). An analogous formula in terms of u can be obtained as follows for $0 \le m \le n-1$. Let $W_0(t)$ be the particular solution of (2.8.3) which satisfies the conditions $W_0(0) = \eta_0, \ldots, W_0^{(n-1)}(0) = \eta_{n-1}$, defined by

$$a_0\eta_0 = b_{m-n+1}, \quad a_0\eta_1 + a_1\eta_0 = b_{m-n+2}, \ldots, a_0\eta_{n-1} + \cdots + a_{n-1}\eta_0 = b_m,$$

where $b_j = 0$ if $j < 0$. Then we have

$$x(t) = y(t) + \int_0^t W_0(t-\alpha)\, u(\alpha)\, d\alpha \tag{2.8.4}$$

as is easy to verify.

A great number of physical systems are regulated by (2.8.1) and they are often called (linear) "servomechanisms" (in a general sense), where some "input" $u(t)$ (signal, flow, power, etc.) generates, through a device V, an "output" $x(t)$ of the same or of different kind, and $u(t)$, $x(t)$, functions of time, are connected by relation (2.8.1), though sometimes, in the applications, the coefficients a_i, b_i are never actually determined. The device V may be as simple as the one-mesh electric circuit mentioned below, or as vast and complicated as mentioned in the remark at the beginning of (2.9).

The statements below are well-known for linear problems. The first one is usually called the principle of superposition. The remaining ones are really corollaries of it.

(2.8. i) $P(D)\, x_i = Q(D)\, u_i$, $i = 1, \ldots, N$, implies $P(D)\, (\Sigma x_i) = Q(D)\, (\Sigma u_i)$.

(2.8. ii) $P(D)\, x_0 = 0$, $P(D)\, x_1 = Q(D)\, u$ implies $P(D)\, (x_0 + x_1) = Q(D)\, u$.

(2.8. iii) $P(D)\, x_i = Q(D)\, u$, $i = 1, 2$, implies $P(D)\, (x_1 - x_2) = 0$.

The wide use of (2.8. i) in all kind of applications (since it allows the separation of various kinds of "inputs") need not be mentioned here. The following statement is also a trivial consequence of (2.8. ii, iii):

(2.8. iv) If u is a given function and x a solution of (2.8.1), then x is stable (asymptotically stable) for equation (2.8.1) if and only if the solution $y = 0$ is stable (asymptotically stable) for equation (2.8.3).

Therefore, the stability of the solutions x of (2.8.1) is reduced to the question of the stability of the solution $y = 0$ of (2.8.3) and this in turn to the questions of algebra discussed in (2.4). In a wide range of applications only asymptotic stability is of interest. Indeed plain stability is very labile and may change into instability by the slightest variation of the physical system of which the differential problem under discussion is, often, a very approximate representation. (Some purely imaginary roots may move to the right half of the complex plane and generate instability.) Thus asymptotic stability is essential. We have already seen in (2.4) some basic tools related to the Routh-Hurwitz criterion by means of which the question of stability can be answered. (For applications see, e.g., D. F. Lawden [1].) We shall suppose, in the following lines that all roots of the equation $a_0\varrho^n + \cdots + a_n = 0$ have negative real parts, and hence all coefficients a_i are different from zero and have the same sign.

If u is a given function (input) then the output x depends upon the initial conditions. If x_1, x_2 are two different solutions of (2.8.1) corresponding to different and in reality, unpredictable initial conditions, then $x_1 - x_2$ is a solution of (2.8.3) (by 2.8. iii), and then $x_1 - x_2 \to 0$ as $t \to +\infty$ because of the asymptotic stability (and even $|x_1 - x_2| < c e^{-at}$ for some constants $a > 0$, $c > 0$, because of 2.3. iii). Thus x_1 and x_2 (if they do not approach zero themselves) have the "same behavior" as $t \to +\infty$ (steady state). Thus any particular solution x_1 of (2.8.1) is valid for the description of the steady state as $t \to +\infty$.

Now we have to describe this steady state (response) in terms of the input $u(t)$. A solution $x(t)$ of (2.8.1) for $m < n$ is given by

$$x(t) = \int_{-\infty}^{t} W_0(t-\alpha)\, u(\alpha)\, d\alpha, \tag{2.8.5}$$

and this integral is absolutely convergent if u is bounded. If u is a constant $A \neq 0$, then a particular solution x_0 of (2.8.1) is the constant $B = A\, b_m/a_n$ (which may be zero), and we may denote b_m/a_n as an amplification factor (for a constant input). If u is a harmonic oscillation, say $u = A \sin \omega t$, of frequency $\omega/2\pi$, then a solution x of (2.8.1) and (2.8.2) can be easily determined by the use of complex variable theory. Indeed, $u = A \exp(i\omega t)$, $x = C \exp(i\omega t)$ verify (2.8.2) provided $C\,P(i\omega) = A\,Q(i\omega)$. Hence, if μ and Φ denote the modulus and the argument of $Q(i\omega)/P(i\omega)$, we have

$$x = I(A\,\mu e^{i\Phi} e^{i\omega t}) = A\,\mu \sin(\omega t + \Phi).$$

In other words, the response x (steady state) of (2.8.2) to the input $A \sin \omega t$ is a harmonic oscillation of the same frequency, amplitude $A\mu$, and phase Φ. The number μ is then the amplification factor. Both $\mu = \mu(\omega)$, $\Phi = \Phi(\omega)$ depend upon ω. By interpreting μ and Φ as the polar coordinates of an auxiliary $\xi\eta$-plane, we have the harmonic response diagram (see the examples below). Frequencies ω_0, $0 < \omega_0 < +\infty$, at which $\mu(\omega)$ has a maximum (if any) are called resonance frequencies.

The complex function $Y(i\omega) = Q(i\omega)/P(i\omega)$ is called the frequency response function relative to equation (2.8.1) i.e., to the corresponding physical system Σ (a servomechanism) which, under this aspect is called, sometimes, a filter.

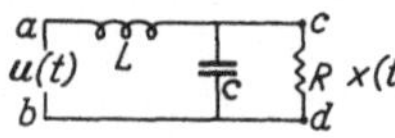

For instance, if $u(t)$, $x(t)$ are voltages between the terminals ab, cd of the electric circuit of the illustration, then $LCx'' + LR^{-1}x' + x = u$, or

$$(D^2 + 2g\,D + f^2)\,x = f^2 u, \tag{2.8.6}$$

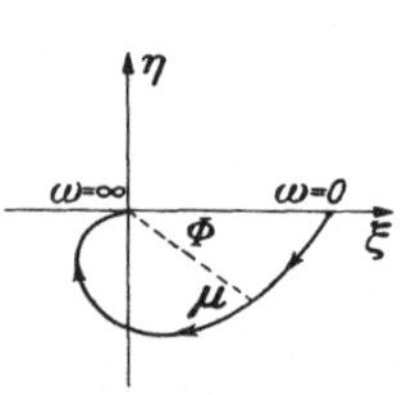

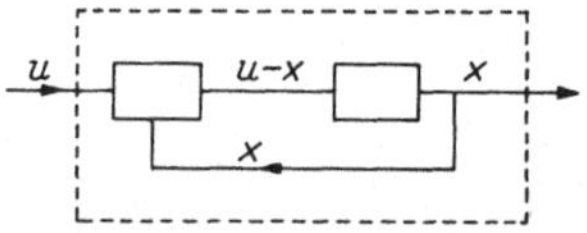

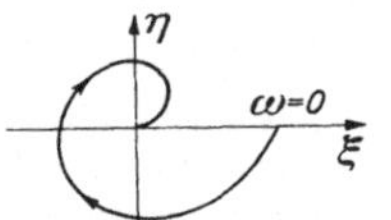

where g, f are positive constants. Then the frequency response function is $Y(i\omega) = f^2[f^2 - \omega^2 + 2ig\omega]^{-1}$. A harmonic oscillation $u = A \sin \omega t$ generates the output $x = A\,\mu \sin(\omega t + \Phi)$ (steady state) with $\mu = f^2[(f^2 - \omega^2)^2 + 4g^2\omega^2]^{-\frac{1}{2}}$, $\tan \Phi = -2g\omega(f^2 - \omega^2)^{-1}$. The harmonic response diagram is given in the illustration.

As another example suppose that a quantity, say x, which is being controlled by some regulator is continuously confronted with the controlling quantity u and the difference $u - x$ influences the output x through some device so that $(a\,D^2 + b\,D)\,x = c(u - x)$, a, b, c positive constants. Then, through simplifications, we again have the same equation (2.8.6).

For the equation $(D + a)^4 x = a^4 u$ we have $\mu = a^4(a^2 + \omega^2)^{-2}$, $\Phi = -4 \arctan(\omega/a)$; hence, as ω varies from 0 to $+\infty$, the point of polar coordinates (μ, Φ) describes the curve shown in the illustration (harmonic response diagram).

(b) *The weighting function.* It is convenient to define the function $W(t)$, $-\infty < t < +\infty$, by $W(t) = W_0(t)$ if $t \geq 0$, $W(t) = 0$ if $t < 0$. Thus we have, from (2.8.5),

$$x(t) = \int_{-\infty}^{t} W_0(t-\alpha)\, u(\alpha)\, d\alpha = \int_{0}^{+\infty} u(t-\alpha)\, W_0(\alpha)\, d\alpha,$$

and finally

$$x(t) = \int_{-\infty}^{+\infty} u(t-\alpha)\, W(\alpha)\, d\alpha. \tag{2.8.7}$$

The function $W(t)$ is called the weighting function of the filter. It is particularly important to consider the rectangular input $u(t) = 0$ for $t < 0$ and $t > 1/m$, $u(t) = m$ for $0 \leq t \leq 1/m$, $m > 0$. Then we have $x(t) = m \int_{0}^{1/m} W(t-\alpha)\, d\alpha$. A graph of this function gives an idea how the filter behaves under a sudden impulse, shows the time lag (b) of the filter, and yields information on the transient.

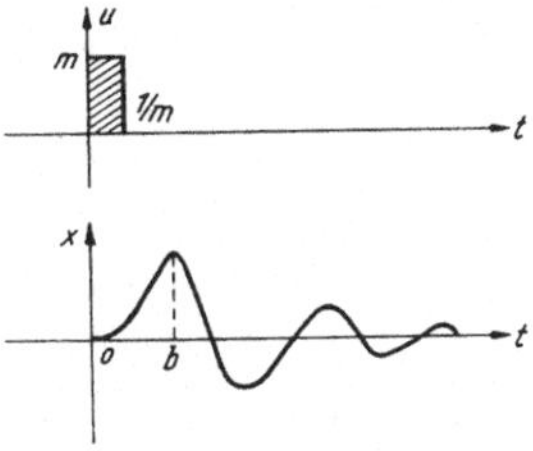

Finally, we have $x(t) \to W(t)$ as $m \to +\infty$. It is usual to consider the Dirac delta function or impulse function, $\delta(t) = 0$ for $t \neq 0$, with $\int_{-\infty}^{+\infty} \delta(t)\, dt = 1$, and to conclude that the weighting function $W(t)$ is the response of the filter to the impulse function. In the theory of distributions of L. Schwartz this important point has received rigorous mathematical formulation.

Also, of particular interest, is the case where $u(t)$ is the unit-step function $u(t) = 0$ if $t < 0$, $u(t) = 1$ if $t \geq 0$.

(c) *The Fourier transform.* Let us observe that if $u(t)$, $-\infty < t < +\infty$, is any periodic function of period T, then we may write $u(t)$ as a Fourier series, say $u(t) \sim \sum_{n=-\infty}^{+\infty} a_n e^{in\omega t}$, where $a_n = (1/T) \int_{-T/2}^{T/2} u(\alpha)\, e^{-in\omega\alpha} d\alpha$. We shall expect the output to be given by

$$x(t) \sim \sum_{n=-\infty}^{+\infty} a_n\, Y(i\, n\, \omega)\, e^{in\omega t}.$$

This is actually the case under the usual condition that $u(t)$ is L-integrable in $[0, T]$ and $m \leq n - 2$, and then the series for $x(t)$ above converges uniformly.

If $u(t)$, $-\infty < t < +\infty$, is any nonperiodic function, we may write $u(t)$ as a Fourier integral

$$u(t) \sim (1/2\pi) \int_{-\infty}^{+\infty} U(i\omega)\, e^{i\omega t} d\omega, \quad \text{where } U(i\omega) = \int_{-\infty}^{+\infty} u(t)\, e^{-i\omega t} dt.$$

Then we may expect the output to be given by

$$X(t) \sim (1/2\pi) \int_{-\infty}^{+\infty} U(i\omega)\, Y(i\omega)\, e^{i\omega t} d\omega$$

and this is really the case under usual conditions on $u(t)$, say $\int_{-\infty}^{+\infty} |u(t)|\, dt < +\infty$ and $m \leq n - 2$. Hence, if $X(w)$ is the Fourier transform of $x(t)$, we have

$$X(i\omega) = U(i\omega)\, Y(i\omega), \quad \text{or:}$$

(2.8. v) The Fourier transform of the filter output is equal to the Fourier transform of the input multiplied by the frequency response function.

Finally by putting $u(t) = e^{i\omega t}$ in (2.8.7), we have

$$x(t) = e^{-i\omega t} \int_{-\infty}^{+\infty} W(\alpha)\, e^{-i\omega\alpha}\, d\alpha = e^{i\omega t}\, Y(i\omega), \quad \text{or:}$$

(2.8. vi) The frequency response of a stable filter is the Fourier transform of the weighting function

$$Y(i\omega) = \int_{-\infty}^{+\infty} W(\alpha)\, e^{-i\omega\alpha}\, d\alpha .$$

(d) *The Nyquist stability criterion.* It should be mentioned first that the frequency response function of a given filter, or servomechanism Σ, can be determined experimentally, and this is often done without any knowledge of the differential equation (2.8.1) regulating Σ.

The combination of known systems Σ_1, Σ_2 and a comparator Δ according to the scheme of the illustration is called a feedback, or *closed loop system* of which

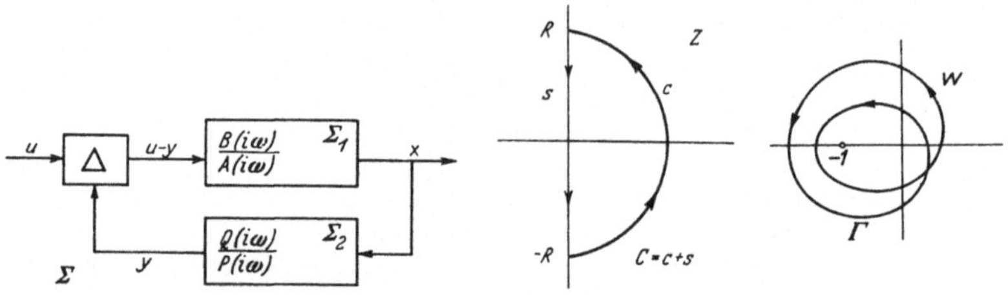

an example had been already given under (a). A controlled quantity $x = x(t)$ is confronted with the controlling quantity u through a device Σ_2 which transforms $x(t)$ into $y(t)$, and the difference $u - y$ produced by the comparator Δ generates the output x through a device Σ_1. Thus we will have

$$A(D)\, x = B(D)\,(u - y), \qquad P(D)\, y = Q(D)\, x,$$

where A, B, P, Q are polynomials in D, and, by elimination of y,

$$[A(D)\, P(D) + B(D)\, Q(D)]\, x = B(D)\, P(D)\, u .$$

The frequency response function of the closed loop system is

$$Z(i\omega) = B(i\omega)\, P(i\omega)/[A(i\omega)\, P(i\omega) + B(i\omega)\, Q(i\omega)]$$

and the stability of the closed loop system is assured if all the roots of the equation

$$A(\lambda)\, P(\lambda) + B(\lambda)\, Q(\lambda) = 0$$

have negative real parts, i.e., if the same can be said of the zeros of the function

$$S(z) = \frac{B(z)\, Q(z)}{A(z)\, P(z)} + 1 .$$

Now, by suppressing in Σ the comparator Δ and the connection between Δ and Σ_2, we have an "open loop" system Σ_0 regulated by $A(D)\, x = B(D)\, u$, and $P(D)\, y = Q(D)\, x$; hence by

$$A(D)\, P(D)\, y = B(D)\, Q(D)\, u .$$

The response function of the open loop system Σ_0 is $Y_0(i\omega) = B(i\omega)\,Q(i\omega)/A(i\omega)\,P(i\omega) = S(i\omega) - 1$ and the stability of Σ_0 depends on the equation $A(\lambda)\,P(\lambda) = 0$.

If Σ_0 is known to be stable, then $A(\lambda)\,P(\lambda) = 0$ has all its roots with negative real parts, hence $S(\lambda)$ has no poles on or at the right of the imaginary axis. If we consider again the closed path C of (2.4) in the complex z-plane (see illustration) we have another application of the theorem of the theory of complex functions mentioned there. Let Γ be the image of C in the complex w-plane, where $w = Y_0(z)$, i.e., the image of C under the open loop response function Z_0. Then the number of roots of the equation $S(z) = 0$ within C is equal to the number of times Γ encircles the point $w = -1$ [provided no root of $S(z)$ is on C]. For large R this number is equal to the number of roots of $S(z)$ to the right of the imaginary axis. The following conclusion can be drawn:

(2.8. vii) (NYQUIST's stability criterion.) A sufficient condition in order that the closed loop Σ is stable is that the open loop Σ_0 is stable and the image Γ of the path C under the open loop response function does not encircle $w = -1$ for large R.

2.9. Bibliographical notes. The tremendous growth of the general theory of servomechanisms is typical of the last twenty years. The regulators of the movements of large telescopes controlled by the feeble light of a distant star, the regulators of an automatically run chemical plant, the feedback electric systems containing tubes, the nerve cell, are only a few examples of the applications. The considerations of (2.8) intend only to give a glimpse of the first elements of the theory of servomechanisms and to establish a bridge. For the theory we refer to the excellent books: L. A. MACCOLL [1], H. M. JAMES, N. B. NICHOLS, and R. S. PHILLIPS [1], and to G. S. BROWN and D. P. CAMPBELL [1], D. F. LAWDEN [1], R. OLDENBURG [1].

For extension of the theory of servomechanisms to the case of more than one controlling and controlled quantity, see the quoted books, and, e.g., the papers M. GOLOMB and E. USDIN [1], F. H. RAYMOND [1], A. M. POPOVSKII [1]. We should also mention, in this respect, the research of B. V. BULGAKOV [7, 9] concerning consistency and bounds for the solutions of linear differential systems of the form

$$\sum_{j=1}^{n} f_{ij}(D)\,x_j = \sum_{p=1}^{n} g_{ip}(D)\,u_p, \quad i = 1, \ldots, n.$$

Differential-difference equations and systems (systems with time lags) have a very large literature and we mention here the classical papers of O. HILB [1], C. LOVE [1], and the more recent ones D. R. HARTREE [1], A. D. MYSKIS [1–9], N. D. HAYES [1], L. A. PIPES [1], and H. SPÄTH [1], R. BELLMAN [13], E. PINNEY [3]. The question of the stability of the solutions, in particular, is discussed in TA LI [1], A. ANDRONOV and A. MAYER [3]. Extensions to the nonlinear case will be mentioned in § 8.

From the very large amount of recent literature on linear servomechanisms we mention A. ANDRONOV and A. MAYER [1], B. FOGAGNOLO [1], E. GRÜNWALD [1], B. V. BULGAKOV [2, 3], N. F. BARBER and F. URSELL [1], Z. S. BLOH [1]. The last author has determined conditions in order that the response function to the unit step function be monotone.

In connection with stability of servomechanisms M. V. MEEROV [1] has discussed conditions under which the polynomial $P(z) + m\,Q(z) = 0$ has roots with negative real parts when both $P(z)$, $Q(z)$ are known to have this property.

Chapter II

General linear systems

§ 3. Linear systems with variable coefficients

3.1. A theorem of LYAPUNOV. We shall consider first linear differential systems

$$x_i' = \sum_{j=1}^{n} a_{ij}(t)\, x_j, \qquad i = 1, \ldots, n, \tag{3.1.1}$$

i.e., $x' = A(t)\,x$, $x = (x_1, \ldots, x_n)$, $A(t) = [a_{ij}(t)]$, and linear differential equations

$$x^{(n)} + a_1(t)\, x^{(n-1)} + \cdots + a_n(t)\, x = 0, \tag{3.1.2}$$

whose coefficients $a_{ij}(t)$, $a_i(t)$ are somehow "asymptotic" to constants a_{ij}, a_i [e.g., $a_{ij}(t) \to a_{ij}$, or $a_i(t) \to a_i$ as $t \to +\infty$]. A number of important results have been obtained in this line, to the effect that, under convenient hypotheses, more and more general, the solutions of (3.1.1) and (3.1.2) behave, as $t \to +\infty$ (or do not) as the solutions of the corresponding system, or equation, with constant coefficients

$$x_i' = \sum_{j=1}^{n} a_{ij}\, x_j, \qquad i = 1, 2, \ldots, n, \tag{3.1.3}$$

$$x^{(n)} + a_1\, x^{(n-1)} + \cdots + a_n\, x = 0. \tag{3.1.4}$$

A typical theorem of this sort is the following one which may be traced to LYAPUNOV [3], though it is given here in a more general form.

(3.1. i) If $A = [a_{ij}]$ is a constant matrix and $C = [f_{ij}(t)]$ is a matrix whose elements are continuous functions of t and $\|C\| < c$ for some sufficiently small $c > 0$ and all $t \geq t_0$, then, if all solutions of the system $x' = A\,x$ approach zero as $t \to +\infty$, also all solutions of the system $x' = [A + C(t)]\,x$ approach zero as $t \to +\infty$. The constant c may depend upon the matrix A.

The condition of the theorem relative to C is certainly satisfied if $C(t) \to 0$ as $t \to +\infty$; i.e., if $A + C(t)$ is asymptotic to A as $t \to +\infty$. An analogous theorem holds for differential equations. The condition that all solutions of $x' = A\,x$ approach zero as $t \to +\infty$ can be expressed by saying that all characteristic roots of A have negative real parts. The statement (3.1. i) does not hold any more if we replace, both in the hypothesis and conclusion, "approach zero as $t \to +\infty$", by "are bounded for large t". Indeed the differential equation $x'' + x = 0$ has all solutions bounded while the equation $x'' - (2/t)\,x' + x = 0$ has the fundamental system of solutions $\sin t - t\cos t$, $\cos t + t\sin t$, and thus all its nontrivial solutions are unbounded.

3.2. A proof of (3.1.i). We shall begin with the following statement which was given as an independent lemma first by R. BELLMAN [1, 12], and which can be partially traced in G. PEANO [1] and T. H. GRONWALL [1]. (On the same subject see also N. LEVINSON [6], H. WEYL [4], L. GIULIANO [1]). For an application of the lemma to uniqueness theorems see I. BIHARI [1].

(3.2. i) If $u(t) \geq 0$, $v(t) \geq 0$, $0 \leq t < +\infty$, are given functions, $u(t)$ continuous, $v(t)$ L-integrable in every finite interval, if for some non-negative constant C we have

$$u(t) \leq C + \int_0^t u(\alpha)\, v(\alpha)\, d\alpha, \qquad t \geq 0, \tag{3.2.1}$$

then we have also

$$u(t) \leq C \exp \int_0^t v(\alpha)\, d\alpha, \qquad t \geq 0.$$

Indeed, if $C > 0$, by algebraic manipulation of (3.2.1), we have

$$u\,v \Big/ \left(C + \int_0^t u\,v\,d\alpha\right) \leq v$$

and, by integration,

$$\log\left(C + \int_0^t u\,v\,d\alpha\right) - \log C \leq \int_0^t v(\alpha)\,d\alpha,$$

or

$$u \leq C + \int_0^t u\,v\,d\alpha \leq C \exp \int_0^t v(\alpha)\,d\alpha.$$

If $C = 0$, then (3.2.1) holds for every constant $C_1 > 0$ and then we have $0 \leq u(t) \leq C_1 \exp \int_0^t v(\alpha)\,d\alpha$, $t \geq 0$. This relation, as $C_1 \to 0$, implies $u(t) \equiv 0$. Thereby (3.2. i) is proved.

Now let us prove (3.1.i). By (2.2) we know that if $Y(t)$ is the fundamental system of solutions of $y' = Ay$ with $Y(0) = I$, then the solution x of $x' = (A + C)x$ having the same initial conditions as y can be written in the integral form

$$x = y + \int_0^t Y(t - \alpha)\, C(\alpha)\, x(\alpha)\, d\alpha. \tag{3.2.2}$$

We have now $\|y\| \leq c_1 e^{-at}$, $\|Y\| \leq c_2 e^{-at}$ for all $t \geq 0$ and some constants $c_1, c_2, a > 0$. Hence

$$\|x\| \leq \|y\| + \int_0^t \|Y(t-\alpha)\|\, \|C(\alpha)\|\, \|x(\alpha)\|\, d\alpha,$$

$$\|x\| \leq c_1 e^{-at} + \int_0^t c\, c_2\, e^{-a(t-\alpha)} \|x(\alpha)\|\, d\alpha,$$

$$\|x\|\, e^{at} \leq c_1 + \int_0^t c\, c_2\, e^{a\alpha} \|x(\alpha)\|\, d\alpha.$$

Finally by (3.2. i) we have

$$\|x\|\, e^{at} \leq c_1\, e^{c c_2 t},$$

and thus $\|x\| \to 0$ as $t \to +\infty$ provided $c c_2 < a$ (R. BELLMAN [12]).

3.3. Boundedness of the solutions. In a posthumous paper of *P.* FATOU [1] the erroneous statement is made that if $f(t)$, $0 \leq t < +\infty$, is a real function with $f(t) \to c$ as $t \to +\infty$ and $c > 0$, then all solutions of the equation

$$x'' + f(t)\, x = 0 \tag{3.3.1}$$

are bounded in $[0, +\infty)$. This statement was proved to be wrong by R. CACCIOPPOLI [1], O. PERRON [9] and A. WINTNER [6] by means of examples. The following theorem, proved by M. HUKUHARA and M. NAGUMO [1] as well as by R. CACCIOPPOLI [1], sets right FATOU's statement:

(3.3. i) If $f(t)$, $0 \leq t < +\infty$, is a real continuous function, if $f(t) \to c$ as $t \to +\infty$, with $c > 0$, and $\int^{+\infty} |f(t) - c|\, dt < +\infty$, then all solutions of (3.3.1) are bounded in $[0, +\infty)$.

This theorem was extended to n-th order differential equations (3.1.2) by M. HUKUHARA as we will mention below. If a differential equation (3.1.2) has coefficients not necessarily continuous but, say, L-integrable in any finite interval, then, according to C. CARATHÉODORY (1.1), the solutions $x(t)$ should be determined in the class of all functions $x(t)$ which are absolutely continuous (AC) in $[0, +\infty)$ together with $x'(t), \ldots, x^{(n-1)}(t)$. Thus $x^{(n)}(t)$ exists almost everywhere (a.e.), and satisfies (3.1.2) a.e. This assumption will be made everywhere in the following. With this assumption the condition that $f(t)$ is continuous in (3.3. i) can be removed. Also, the condition that $f(t) \to c$ is not essential. The limit of $f(t)$ as $t \to +\infty$ need not exist. It is enough that for some constant $c > 0$ we have $\int^{+\infty} |f(t) - c|\, dt < +\infty$. The generalization of (3.3. i) to equations (3.1.2) can now be formulated as follows:

(3.3. ii) If $f_i(t)$, $0 \leq t \leq +\infty$, are measurable functions and

$$\int_0^{+\infty} |f_i(t)|\, dt < +\infty, \qquad i = 1, \ldots, n,$$

if c_i are real numbers such that the equation

$$x^{(n)} + c_1\, x^{(n-1)} + \cdots + c_n\, x = 0 \tag{3.3.2}$$

has all solutions bounded in $[0, +\infty)$, then also the equation

$$x^{(n)} + [c_1 + f_1(t)]\, x^{(n-1)} + \cdots + [c_n + f_n(t)]\, x = 0 \tag{3.3.3}$$

has all solutions bounded (Dini-Hukuhara theorem).

This theorem which may be traced back to U. DINI [1] in more particular situations, was proved by H. SPÄTH [2], and M. HUKUHARA [2] for $f_i(t)$ continuous, by L. CESARI [2] for $f_i(t)\to 0$, and under the conditions above by D. CALIGO [3], R. BELLMAN [1], H. WEYL [4], and N. LEVINSON [6]. For systems the theorem above can be given as follows and was essentially proved by the same authors.

(3.3. iii) If $A=[a_{ij}]$ is a constant matrix, if $C(t)=[f_{ij}(t)]$ is a matrix whose coefficients are measurable functions of t and $\int_0^{+\infty}\|C(t)\|\,dt<+\infty$, if all solutions of the system $x'=Ax$ are bounded, then also all solutions of the system $x'=[A+C(t)]x$ are bounded in $[0,+\infty)$.

Proof of (3.3. ii*) and (3.3.* iii*).* Since (3.3. iii) includes (3.3. ii) it suffices to prove (3.3. iii). By (2.2) we have

$$x = y+\int_0^t Y(t-\alpha)\,C(\alpha)\,x(\alpha)\,d\alpha,$$

where now $\|y\|\le c_1$, $\|Y\|\le c_2$ for all $t\ge 0$. Hence

$$\|x\|\le\|y\|+\int_0^t\|Y(t-\alpha)\|\,\|C(\alpha)\|\,\|x(\alpha)\|\,d\alpha,$$

$$\|x\|\le c_1+\int_0^t c_2\|C(\alpha)\|\cdot\|x(\alpha)\|\,d\alpha,$$

and, by (3.2. i)

$$\|x\|\le M,\qquad\text{where}\quad M=c_1\exp\left(c_2\int_0^{+\infty}\|C(\alpha)\|\,d\alpha\right)$$

for all $t\ge 0$. This completes the proof of (3.3. iii).

3.4. Further conditions for boundedness. Another remark concerning equation (3.3.1) is due to A. WIMAN [1, 2], R. CACCIOPPOLI [1] and G. ASCOLI [1, 2], who proved the following statement:

(3.4. i) If $f(t)$, $0\le t<+\infty$, is a real continuous function, if $f(t)\to c$ as $t\to+\infty$ with $c>0$, and $f(t)$ is of bounded variation in $[0,+\infty)$ [in particular, monotone in $[0,+\infty)$], then all solutions of equation (3.3.1) are bounded in $[0,+\infty)$.

A. WINTNER [10] proved in addition that the solutions are asymptotic to the solutions of the limit equation $z''+cz=0$ [see (3.7)].

An immediate extension of statement (3.4. i) to n-th order differential equations, analogous to (3.3. ii), is not true as the first of the two following examples shows. Consider the two equations (L. CESARI [3]) [cf. (3.1)]:

$$\text{(a)}\quad x''-(2/t)\,x'+x=0,\qquad\text{(b)}\quad x''+(2/t)\,x'+x=0,$$

having monotone coefficients in $[1,+\infty)$ approaching, as $t\to+\infty$, the coefficients of the equation with constant coefficients (c) $x''+x=0$ whose solutions are all bounded in $[1,+\infty)$. Equation (a) has the fundamental system of solutions $\sin t-t\cos t$, $\cos t+t\sin t$, and thus all its nontrivial solutions are unbounded; equation (b) has the fundamental system $t^{-1}\sin t$, $t^{-1}\cos t$, and thus all its solutions are bounded in $[1, +\infty)$ (and even approach zero as $t \to +\infty$). The following theorem gives an extension of theorem (3.4. i) to n-th order differential equations and, besides, shows the reason for the different behavior of the solutions of the equations (a) and (b).

(3.4. ii) If the real-valued continuous functions $f_i(t)$, $t_0 \leq t < +\infty$, are of bounded variations in $[t_0, +\infty)$ and $f_i(t) \to 0$ as $t \to +\infty$, if c_i are real numbers such that the equation (3.3.2) has all solutions bounded in $[t_0, +\infty)$, if the roots $\varrho_i(t)$ [functions of t] of the algebraic equation

$$\varrho^n + [c_1 + f_1(t)]\,\varrho^{n-1} + \cdots + [c_n + f_n(t)] = 0 \qquad (3.4.1)$$

have real parts $R[\varrho_i(t)] \leq 0$ for all $t_0 \leq t < +\infty$, then all solutions of the equation

$$x^{(n)} + [c_1 + f_1(t)]\,x^{(n-1)} + \cdots + [c_n + f_n(t)]\,x = 0 \qquad (3.4.2)$$

are bounded in $[t_0, +\infty)$ (L. Cesari [3]).

An outline of the proof of this statement and of the following ones will be given in (3.5). As it is immediately seen, equation (b) verifies, and equation (a) does not, the latter condition.

Statements (3.3. ii) and (3.4. ii) can be included in the following unique statement.

(3.4. iii) If $f_i(t)$ are real functions of bounded variation in $[t_0, +\infty)$ and $f_i(t) \to 0$ as $t \to +\infty$, if $g_i(t)$ are L-integrable functions in $[t_0, +\infty)$, if c_i are real numbers such that equation (3.3.2) has all solutions bounded in $[t_0, +\infty)$, if the roots $\varrho_i(t)$ [functions of t] of equation (3.4.1) have real parts $R[\varrho_i(t)] \leq 0$ for all $t_0 \leq t < +\infty$, then the differential equation

$$x^{(n)} + [c_1 + f_1(t) + g_1(t)]\,x^{(n-1)} + \cdots + [c_n + f_n(t) + g_n(t)]\,x = 0 \qquad (3.4.3)$$

has all solutions bounded in $[t_0, +\infty)$ (L. Cesari [3]).

The roots $\varrho_i(t)$ of equation (3.4.1), which are functions of t, have been considered by L. Cesari. They have been called *generalized characteristic roots* (N. I. Gavrilov [1, 2]). The following statement extends (3.4. iii) to linear systems. An outline of the proof will be given in (3.5).

(3.4. iv) If the constant matrix $A = [a_{ij}]$ has all characteristic roots σ_i with real parts $R(\sigma_i) \leq 0$ and those roots σ_i with $R(\sigma_i) = 0$ are simple, if the matrix $B(t) = [b_{ij}(t)]$ has elements of bounded variation in $[t_0, +\infty)$ and $b_{ij}(t) \to 0$ as $t \to +\infty$ and the characteristic roots $\varrho_i(t)$ of the matrix $A + B(t)$ (generalized characteristic roots) have real parts $R[\varrho_i(t)] \leq 0$ for all $t_0 \leq t < +\infty$, if the matrix $C(t) = [C_{ij}(t)]$ has elements absolutely integrable in $[t_0, +\infty)$, i.e., $\int^{+\infty} \|C(t)\|\,dt < +\infty$, then the differential system $x' = [A + B(t) + C(t)]\,x$ has solutions all bounded in $[t_0, +\infty)$ (L. Cesari [3]).

Both statements (3.4. iii), (3.4. iv) assure, of course, LYAPUNOV's stability of every solution (1.2). A slightly stronger conclusion has been observed recently by R. CONTI [5] and will be mentioned in (3.9). If $B(t)$ has elements absolutely continuous in $[t_0, +\infty)$, then the condition that they be of bounded variation in $(t_0, +\infty)$ can be expressed by $\int^{+\infty} \|B'(t)\|\, dt < +\infty$.

Under the hypotheses of (3.4. iv) for any root σ_i we have $R(\sigma_i) \leqq 0$ and for those roots σ_i with $R(\sigma_i) = 0$ (if any) we have $\mu_i = \nu_i = 1$, where μ_i and ν_i are the multiplicity and the nullity of σ_i as defined in (2.1). Thus the condition above is more restrictive than the requirement that the solutions of the system $x_i' = \sum_{j=1}^{n} a_{ij} x_j$ are all bounded, which requires only $\mu_i = \nu_i$ for the same roots σ_i with $R(\sigma_i) = 0$. L. CESARI [3] has given an example showing that the restriction $\mu = \nu = 1$ cannot be removed. On the other hand (3.4. iv) is general enough to include (3.4. iii), since for the systems obtained by transformation of a differential equation (1.1; 1.2) we have necessarily $\nu_i = 1$ (2.1).

An extension of (3.4. iv) in connection with a remark of L. A. GUSAROV [1] has been given by U. BARBUTI [3]. For a nonlinear extension of (3.4. iv) see I. BIHARI [2]. The method of the proof is essentially the same as for (3.4. ii, iii, iv) and will be outlined below (3.5).

3.5. The reduction to *L*-diagonal form and an outline of the proofs of (3.4. iii) and (3.4. iv). The proofs of the theorems (3.4. ii, iii, iv) are based on a process (L. CESARI [3]) which has been used successively by various authors (N. Y. LYASCENKO [2], B. P. DEMIDOVIC [2], N. LEVINSON [6], I. M. RAPOPORT [4], R. CONTI [5], U. BARBUTI [3], I. BIHARI [2]) and which L. RAPOPORT refers to in his book as the reduction to L-diagonal form.

We will sketch below the process, after a few preliminary remarks.

Let us denote by $\varrho_i(t)$ the n roots of any equation $\varrho^n + F_1(t)\,\varrho^{n-1} + \cdots + F_n(t) = 0$, and by λ_i the n roots of the equation $\varrho^n + c_1 \varrho^{n-1} + \cdots + c_n = 0$, where $F_s(t) \to c_s$ as $t \to +\infty$, $s = 1, \ldots, n$. There is an enumeration of the roots λ_i and, for every t, an enumeration of the roots $\varrho_i(t)$ such that (a) $\varrho_i(t)$, $i = 1, 2, \ldots, n$, are continuous functions of t at every t where the F_s are continuous; (b) $\varrho_i(t) \to \lambda_i$ as $t \to +\infty$, $i = 1, \ldots, n$; (c) If λ_i is *simple*, then $\varrho_i(t)$ is simple in $[t_2, +\infty)$ for t_2 large enough. In addition, if all F_s are of bounded variation in $[t_0, +\infty)$, [absolutely continuous], then the same holds for the root $\varrho_i(t)$ in $[t_2, +\infty)$ (t_2 large enough).

These statements can be proved, for instance, by using the logarithmic indicator formula of the theory of complex functions (a proof was given by L. CESARI in [3]). In (c) the hypothesis that λ_i is simple is essential as the following example shows. The equation $\varrho^2 - (\sin^2 t/t^2) = 0$ has the roots $\varrho = \pm \sin t/t$, none of which is of bounded variation in any $[t_2, +\infty)$, while $\sin^2 t/t^2$ is of bounded variation and absolutely continuous in $[1, +\infty)$. The limiting equation $\lambda^2 = 0$ has the double root $\lambda = 0$.

Now let us proceed to outline the process of reduction to L-diagonal form of a given matrix $A + B(t)$, $t \geq t_0$, with $B(t) \to 0$ as $t \to +\infty$, under the following simplifying assumptions: 1. The elements of the matrix B are absolutely continuous and of bounded variation in $[t_0, +\infty)$; 2. The characteristic roots of the constant matrix A are all simple. We consider the constant matrix A and a process of diagonalization of A as mentioned in (2.1). This process leads to the building of certain matrices P with $\det P \neq 0$, by solving certain linear algebraic systems with determinants all $\neq 0$. Thus, if $2m > 0$ is the minimum of the absolute value of all these determinants, they will be bounded away from zero by $2m$. Let us now apply the *same* process of diagonalization to the matrix $A + B(t)$ for every $t \geq t_0$. We will determine certain analogous matrices $P(t)$ by solving certain linear systems whose determinants approach, as $t \to +\infty$, the same determinants above. Thus all new determinants, functions of t, will be bounded away from zero by at least $m > 0$ for all t large enough, say on $[t_2, +\infty)$ for some t_2. Thus we will have analogous bounds in $[t_2, +\infty)$ for both the matrices $P(t)$ and their inverses $P^{-1}(t)$. In the whole process we perform only operations of addition, subtraction, multiplication, and division with divisors bounded away from zero on functions which are: 1. elements $b_{ij}(t)$ of the matrix B; 2. generalized roots $\varrho_i(t)$. Since all these functions are absolutely continuous and of bounded variation on $[t_2, +\infty)$ for some t_2, the same will occur for the elements of the matrices $P(t)$. The same holds, therefore, for the final matrix $P(t)$ which is the product of all these matrices. Therefore, we will have $|\det P(t)| < M$, $|\det P^{-1}(t)| < M$, and $P^{-1}(A + B)P = \Lambda = \operatorname{diag}[\varrho_1(t), \ldots, \varrho_n(t)]$ for all $t \geq t_2$.

We shall now denote by L the class of the matrices $C(t)$ whose elements are measurable functions of t with $\int^{+\infty} \|C(t)\|\, dt < +\infty$. If $C(t)$ is any such matrix and we transform the system

$$x' = [A + B(t) + C(t)]\, x \tag{3.5.1}$$

by the transformation $x = Py$, or $y = P^{-1}x$, we obtain the system

$$y' = [\Lambda(t) + D(t)]\, y, \tag{3.5.2}$$

where $\Lambda = P^{-1}(A + B)P$ is the diagonal matrix $[\varrho_1(t), \ldots, \varrho_n(t)]$, and $D(t) = P^{-1}CP - P^{-1}P'$ is of class L. A system of the form (3.5.2) with $\Lambda(t)$ a diagonal matrix and $D(t)$ of class L is called "of L diagonal form" by I. M. Rapoport. It is clear that the solutions of system (3.5.1) are bounded for large t if and only if the same occurs for the solutions of system (3.5.2). If we now suppose $R[\varrho_i(t)] \leq 0$ for all $t \geq t_2$ and $i = 1, 2, \ldots, n$, the boundedness of the solutions of system (3.5.2) can be proved as follows. First, system (3.5.2) can be written in the form

$$y_i' = \varrho_i(t)\, y_i + \Phi_i(t), \qquad i = 1, \ldots, n, \tag{3.5.3}$$

where

$$\Phi_i(t) = \sum_{j=1}^{n} d_{ij}(t)\, y_i(t), \quad \text{and} \quad D(t) = [d_{ij}(t)].$$

Then (3.5.3) can be written in the integral form

$$y_i(t) = c_i \exp \int_{t_2}^{t} \varrho_i(\alpha)\, d\alpha + \int_{t_2}^{t} \Phi_i(\beta) \left[\exp \int_{\beta}^{t} \varrho_i(\alpha)\, d\alpha\right] d\beta, \qquad i = 1, \ldots, n,$$

and, since $R[\varrho_i(t)] \leq 0$, $i = 1, 2, \ldots, n$, also

$$|y_i(t)| \leq |c_i| + \int_{t_2}^{t} \sum_{j=1}^{n} |d_{ij}(\beta)|\, \|y(\beta)\|\, d\beta, \qquad i = 1, \ldots, n.$$

Moreover, if $C = |c_1| + \cdots + |c_n|$, also

$$\|y(t)\| \leq C + \int_{t_2}^{t} \|D(\beta)\| \cdot \|y(\beta)\|\, d\beta.$$

By (3.2. i) we have

$$\|y(t)\| \leq C \exp \int_{t_2}^{t} \|D(\beta)\| \, d\beta,$$

and since $D(t)$ is of class L, also $\|y(t)\| \leq M < +\infty$, where M is a convenient constant.

In the case that some of the roots λ_i, $i = 1, \ldots, n$, with real part negative, are not distinct, then the whole algebraic process described above can be modified, allowing diagonal matrices $\Lambda(t)$ with convenient "ones" above or below the main diagonal as for constant matrices. In the case where the elements of the matrix $B(t)$ are supposed to be of bounded variation in $[t_0, +\infty)$ (but not necessarily either continuous or absolutely continuous) and the roots λ_i are distinct, then the roots $\varrho_i(t)$ are distinct for $t \geq t_2$ and of bounded variation on $[t_2, +\infty)$ for some t_2 but not necessarily continuous nor absolutely continuous there. Thus the direct application of the whole process above must be modified so as to avoid the differentiation of singular functions (L. Cesari [3], I. M. Rapoport [4]).

3.6. Other conditions. Other conditions have been applied to the matrices A, B, C of the previous theorems to assure boundedness, and thus stability, of all solutions of a linear homogeneous system. We mention here the following two statements concerning systems of second and first order homogeneous differential equations respectively.

(3.6. i) If A is a constant, real, symmetric, negative definite matrix, if $B(t)$ is a symmetric real matrix whose elements are absolutely continuous functions such that $\int^{+\infty} \|B'(t)\| \, dt < +\infty$, and $(1+c)\,|x B x| \leq |x A x|$ for all real vectors x and some fixed constant $c > 0$, if $\int^{+\infty} \|C(t)\| \, dt < +\infty$, then all solutions of the system $x'' = [A + B(t) + C(t)]\, x$ are bounded for $t \to +\infty$ (R. Bellman [4]).

(3.6. ii) If A is a constant matrix, if $B(t)$ is a matrix whose elements are absolutely continuous functions bounded in $[t_0, +\infty)$, if all solutions of the system $y' = A y$ are bounded, if $\det\big(A + B(t)\big) \geq h > 0$ for all $t \geq t_0$, if $\int^{+\infty} \|B' + B(A+B)\| \, dt < +\infty$, then all solutions of the system $x' = [A + B(t)]\, x$ are bounded in $[t_0, +\infty)$ (R. Conti [3]).

For other conditions for boundedness of the solutions of a linear system $x' = A(t)\, x$, see N. I. Gavrilov [1, 2] (determinantal criteria), R. Vinograd [1, 2, 3], and V. I. Zubov [1].

3.7. Asymptotic behavior. A precise comparison between the solutions of a given system, say (3.1.1), and the solutions of the "limiting" system (3.1.3) has been the object of a great deal of research. We mention here O. Perron [6, 9], H. Späth [1, 2], N. Y. Lyascenko [2], R. Bellman [1, 4], S. Faedo [1], N. Levinson [6, 9], A. Wintner [1, 6, 10].

First we shall mention that if $y' = A y$ is a system with constant coefficients and solutions all bounded in $[0, +\infty)$, if $\int^{+\infty} \|C(t)\| \, dt < +\infty$, then not only are the solutions of the system $|x' = [A + C(t)]\, x|$ all bounded in $[0, +\infty)$ (3.3), but they are "asymptotically" equivalent to the solutions of the system $y' = A y$, at least in the sense that for every $x\,[y]$ there is a $y\,[x]$ such that $x - y \to 0$ as $t \to +\infty$ (N. Levinson [6, 9]). (See in (3.10) Wintner's extension of this statement.) I mention below some of the most recent results, as recently obtained by N. Levinson [6, 9] and R. Bellman [4] by means of variants of the process of reduction to L-diagonal form.

(3.7. i) If A is a constant matrix with real distinct characteristic roots $\lambda_1, \ldots, \lambda_n$, if $\int^{+\infty} \|C(t)\|\, dt < +\infty$, then there are n solutions $x^{(1)}, \ldots, x^{(n)}$ of the system $x' = [A + C(t)]\, x$ such that

$$x^{(k)} = e^{\lambda_k t}\,[c_k + o(1)] \quad \text{as} \quad t \to +\infty,$$

where the c_k are constants, $k = 1, \ldots, n$.

(3.7. ii) If A is a constant matrix with distinct characteristic roots $\lambda_1, \ldots, \lambda_n$, if $\|B(t)\| \to 0$ as $t \to +\infty$, then there are n solutions $x^{(1)}, \ldots, x^{(n)}$ of the system $x' = [A + B(t)]\, x$, such that

$$c_2 \exp\left[R(\lambda_k t) + d_2 \int_{t_0}^{t} \|B(\alpha)\|\, d\alpha\right] \leq \|x^{(k)}\| \leq C_1 \exp\left[R(\lambda_k t) + d_1 \int_{t_0} \|B(\alpha)\|\, d\alpha\right]$$

for all $t \geq t_0$, where c_1, c_2, d_1, d_2 denote convenient constants; in particular

$$\log\|x^{(k)}\|/t \to R(\lambda_k) \quad \text{as} \quad t \to +\infty, \quad k = 1, \ldots, n.$$

For more stringent results see N. Levinson [6, 9] and R. Bellman [4]. Other results have been given by S. Faedo [3].

The theorems above have analogous formulations for the solutions $x(t)$ of the differential equation

$$x^{(n)} + f_1(t)\, x^{(n-1)} + \cdots + f_n(t)\, x = 0 \tag{3.7.1}$$

where $f_i(t)$ are continuous functions of t in $[t_0, +\infty)$. We will not list such theorems. We prefer instead to recall here a result in the context of a quite different view point:

(3.7. iv) If $\int^{+\infty} t^{k-1} |f_k(t)|\, dt < +\infty$, $k = 1, 2, \ldots, n-1$, then $x^{(n-1)}(t)$ has a finite limit as $t \to +\infty$ (J. E. Wilkins [1]; R. Bellman [4]).

For $n = 2$ this result is a corollary of a theorem of O. Haupt [4]. A very elegant proof of it, for $n = 2$, has been given by M. Boas, R. P. Boas, N. Levinson [1]. Another proof for $n = 2$ under less general conditions has been given by D. Caligo [7].

Finally we recall here that U. Dini [2] and M. Bôcher [2] had already observed the possibility of extending the Fuchs theory concerning the behavoir around a regular singular point of the solutions of linear differential equations to the non-analytic case [cf. (1.5)]. The more detailed results of O. Haupt [1] and J. E. Wilkins [1], A. Wintner [3] on the subject cannot be discussed here. S. Faedo [1] has considered equations (3.7.1) where $f_i(t)$ has the form $f_i(t) = \mu_i t^{i-n} + q_i(t)$ (μ_i constant) $q_i(t)$ is a continuous function of t in some interval $[0, b]$ with $\int_0^b t^{n-i-1} |q_i(t)| (\log t)^{N-1}\, dt < +\infty$, where N denotes the maximum multiplicity of the roots of the indicial equation. Then under these hypotheses the integrals of (3.7.1) behave as $t \to 0$ as the corresponding integrals of the Fuchs-type equation. See also A. Ghizzetti [2], I. M. Sobol [1]. For a further investigation on the subject see A. Wintner [30].

3.8. Linear asymptotic equilibrium. We shall first state and prove the following statement whose interest goes beyond the application we will give below.

(3.8. i) If $A(t)$, $0 \leq t < +\infty$, is a matrix whose elements are integrable functions in every finite interval, then every AC solution $x(t)$ of the system $x' = A(t)\,x$ verifies the relation

$$\|x(t)\| \leq \|x(0)\| \exp \int_0^t \|A(\alpha)\|\, d\alpha. \tag{3.8.1}$$

Proof. From $x' = A\,x$ we deduce successively

$$x(t) = x(0) + \int_0^t A(\alpha)\, x(\alpha)\, d\alpha,$$

$$\|x(t)\| \leq \|x(0)\| + \int_0^t \|A(\alpha)\|\, \|x(\alpha)\|\, d\alpha,$$

and, by (3.2. i), we deduce (3.8.1).

Analogous results using different norms have been obtained by T. Kitamura [1], and H. Teyama [1]. A first application of (3.8. i) is the following one.

(3.8. ii) If $\int^{+\infty} \|A(t)\|\, dt < +\infty$, then every AC solution $x(t)$ of the system $x' = A(t)\,x$, has a finite limit $x(t) \to c$, c a constant vector, as $t \to +\infty$.

Proof. From (3.8. i) it follows immediately that $x(t)$ is bounded. On the other hand, to prove (3.8. ii), we have only to prove that $A\,x$ is integrable in $[0, +\infty)$. Indeed we have

$$\int_0^{+\infty} \|A\,x\|\, dt \leq \int_0^{+\infty} \|A\|\, \|x\|\, dt \leq \int_0^{+\infty} \|A\|\, \|x(0)\| \left(\exp \int_0^t \|A\|\, d\alpha\right) dt$$
$$= \|x(0)\| \left[\exp \int_0^{+\infty} \|A\|\, dt - 1\right] < +\infty.$$

(3.8. iii) Under the conditions of (3.8. ii) for every constant vector c there is a solution $x(t)$ of $x' = A(t)\,x$ such that $x(t) \to c$ as $t \to +\infty$.

Proof. If $X(t) = [x_{ij}(t)]$ is a fundamental system of solutions of the system and $x_{ij}(t) \to c_{ij}$ as $t \to +\infty$, we have

$$\det X(t) = \det X(0) \exp \int_0^t \operatorname{tr} A(t)\, dt,$$

hence

$$\det [c_{ij}] = \det X(+\infty) = \det X(0) \exp \int_0^{+\infty} \operatorname{tr} A(t)\, dt \neq 0.$$

Thus every vector c is a linear combination of the n columns of the matrix $[c_{ij}]$, and the same linear combination of the n columns of the matrix $X(t)$ approaches c as $t \to +\infty$.

It may be pointed out now that the system $x' = 0$, $x = (x_1, \ldots, x)$ has for solutions all the constant vectors $x = c$. If $x_1, \ldots, x_n$ are the Lagrangian coordinates of a mechanical system Σ, then it can be said that the system Σ is in equilibrium. Now a differential system $x' = A(t)\,x$ whose solutions $x(t)$ have a limit $x(t) \to c$ as $t \to +\infty$, and such that, given c, there exists $x(t)$ with $x(t)\to c$, may be said to have *linear asymptotic equilibrium*. Thus (3.8. ii, iii) can be restated by saying that if $\int^{+\infty} \|A(t)\|\, dt < +\infty$, the system $x' = A(t)\, x$ has linear asymptotic equilibrium.

Other conditions for linear asymptotic equilibrium have been given by W. J. TRJITZINSKY [1] and A. WINTNER [2, 7, 8, 23]. Recently A. WINTNER [30] has given the following statement:

(3.8. iv) If the limit $\int^{+\infty} A(t)\, dt$ exists and $B(t) = \int_t^{+\infty} A(\tau)\, d\tau$, if either

$$\int^{+\infty} \|B(t)\, A(t)\|\, dt < +\infty, \quad \text{or} \quad \int^{+\infty} \|A(t)\, B(t)\|\, dt < +\infty,$$

then the system $x' = A(t)\, x$ has linear asymptotic equilibrium.

The following theorem along the same line holds:

(3.8. v) If $A(t)$, $f(t)$ are $n\times n$ and $n\times 1$ continuous matrices, if $\|A(t)\| < M$ in $[0, +\infty)$, and $\int^{+\infty} A(t)\, dt$ exists; if $\int^{+\infty} dt \left\| \int_t^{+\infty} A(\alpha)\, d\alpha \right\| < +\infty$, and $\int^{+\infty} \left\| A(t) \int_0^t f(\alpha)\, d\alpha \right\| dt < +\infty$, then for every solution $x(t)$ of the system $x' = A(t)\, x + f(t)$ there is a solution $y(t)$ of the system $y' = f(t)$ with $x - y \to 0$ as $t \to +\infty$ (T. MANACORDA [5]).

3.9. Systems with variable coefficients. The theorems (3.3. iii, 3.4. ii, iii, iv) suggest possible extension to the case where A itself is a matrix function of t. The case where $A(t)$ is a given periodic matrix proved to be particularly fruitful. In this instance the theory is now quite advanced (A. LYAPUNOV [3], R. BELLMAN [4], G. ASCOLI [4], A. WINTNER [1, 3, 17], R. CONTI [5]). To show its deep lines we should first recall two refinements of the Lyapunov concept of stability (1.1) for linear systems, namely *uniform* and *restrictive* stability (see A. LYAPUNOV [3], L. ERMOLAEV [1], G. ASCOLI [4], R. CONTI [5]).

We have already mentioned in (1.2) that if a solution $x(t)$ of a linear system

$$x' = A(t)\, x \tag{3.9.1}$$

is stable (in the sense of LYAPUNOV) (say at the right), then all solutions of (3.9.1) are stable, and the system (3.9.1), therefore, can be called *stable*. The same holds for asymptotic stability. If we denote by $X(t)$ (an $n\times n$ matrix) a fundamental solution of (3.9.1), thus satisfying the relations $X' = A(t)\, X$, $X(t_0) = I$, I the unit matrix, then we have:

(3.9. i) System (3.9.1) is stable (at the right) if and only if $\|X(t)\| < M$ for some constant M and all $t \geq t_0$.

Proof. For every solution $x(t)$ of (3.9.1) and every other solution $\bar{x}(t)$ we have $\bar{x}(t) = x(t) + X(t)\,[\bar{x}(t_0) - x(t_0)]$. Indeed, the right hand member is certainly a solution of (3.9.1) and has the value $\bar{x}(t_0)$ at $t=t_0$. Thus by the uniqueness theorem [see (1.1)] it coincides with $\bar{x}(t)$ for all t. Suppose $\|X(t)\| \leq M$. Given $\varepsilon > 0$, we may take $\delta = M^{-1}\varepsilon$. Then $\|\bar{x}(t_0) - x(t_0)\| < \delta$ implies $\|\bar{x}(t) - x(t)\| \leq \|X(t)\| \cdot \|\bar{x}(t_0) - x(t_0)\| \leq M \cdot M^{-1}\varepsilon$. Vice versa, suppose that, given $\varepsilon > 0$, there is a $\delta > 0$ such that $\|\bar{x}(t_0) - x(t_0)\| < \delta$ implies $\|\bar{x}(t) - x(t)\| < \varepsilon$ for all $t \geq t_0$. By taking $\bar{x}_i(t_0) = x_i(t_0) + \delta$, $\bar{x}_j(t_0) = x_j(t_0)$ for $j \neq i$, we have $\varepsilon \geq \|\bar{x}(t) - x(t)\| = \|X_i(t)\|\delta$, where $X_i(t)$ denotes the i-th column of the matrix $X(t)$. Thus $\|X_i(t)\| \leq \varepsilon/\delta$ for all $t \geq t_0$, where i is any integer $i = 1, 2, \ldots, n$. Finally, $\|X(t)\| \leq n\varepsilon/\delta$. Thereby (*i*) is proved.

A solution $x(t)$ of system (3.9.1) is said to be *uniformly stable* (at the right) (cf. 1.5) if, given $\varepsilon > 0$, there is $\delta = \delta(\varepsilon)$ such that for every $t_0 \geq 0$ and every solution $\bar{x}(t)$ of (3.9.1) with $\|\bar{x}(t_0) - x(t_0)\| < \delta$ we have $\|\bar{x}(t) - x(t)\| < \varepsilon$ for all $t \geq t_0$. Obviously if a solution $x(t)$ of (3.9.1) is uniformly stable, every solution of (3.9.1) is also uniformly stable and the system (3.9.1) can be called uniformly stable. Denote by $X(t)$ the fundamental solution $X(t)$ of (3.9.1) with $X(0) = I$.

(3.9. ii) The system (3.9.1) is uniformly stable (at the right) if and only if $\|X(t)\, X^{-1}(\tau)\| \leq M$ for some constant M and all $0 \leq \tau \leq t$.

Proof. For every solution $x(t)$ of (3.9.1) and every other solution $\bar{x}(t)$ of (3.9.1) we have $\bar{x}(t) - x(t) = X(t)\, X^{-1}(t_0)\, [\bar{x}(t_0) - x(t_0)]$. The proof is then the same as for (3 9 i).

A system (3.9.1) is said to be *restrictively stable* if (3.9.1) is stable together with the adjoint system

$$y' = -\bar{A}_{-1}(t)\, y. \tag{3.9.2}$$

where $\bar{A}_{-1}$ denotes the conjugate of the transpose of the matrix A (2.1). We shall denote by $X(t)$ the same fundamental solution as in (ii). Then $\bar{X}^{-1}_{-1}(t)$ is a fundamental solution of (3.9.2) as is immediately proved by the relations discussed in (2.2). (See, also, E.A. Coddington and N. Levinson [3].)

(3.9. iii) System (3.9.1) is restrictively stable (at the right) if and only if $\|X(t)\| \leq M$, $\|X^{-1}(t)\| \leq M$ for some constant M, and $t \geq t_0$; i.e., if and only if $\|X(t)\, X^{-1}(\tau)\| \leq M$ for some M and all $t \geq 0$, $\tau \geq 0$. This theorem is a consequence of (3.9. i).

By the previous considerations it follows that restrictive stability implies uniform stability, and this, in turn, implies stability. The converses are not true as can be shown by examples and the use of the previous statements.

(3.9. iv) For systems with constant coefficients, stability and uniform stability are equivalent. For the same systems restrictive stability is equivalent to stability at both sides.

Indeed, $X(t)\, X^{-1}(\tau) = X(t-\tau)$ [see (2.2)] and the statement follows then immediately by (3.9. i, ii; iii).

R. Conti [5] has proved that under the same hypotheses as (3.4. iv) the systems considered there are not only stable but also uniformly stable.

We need now the following further concept. A system (3.9.1) is said to be *reducible* [reducible to zero] (A. LYAPUNOV [3], N.P. ERUGIN [1]) if there is a matrix $L(t)$ whose elements are absolutely continuous functions, which is bounded together with $L^{-1}(t)$ in $[t_0, +\infty)$ and such that $L^{-1}AL - L^{-1}L' =$ a constant matrix in $[t_0, +\infty)$ [= the zero matrix]. Then the transformation $x = L(t)\,y$ transforms (3.9.1) into a system with constant coefficients [to the system $y' = 0$] as it is immediately verified [cf. (3.5)].

(3.9. v) A system (3.9.1) is restrictively stable if and only if it is reducible to zero.

Proof. If (3.9.1) is restrictively stable and $X(t)$ is a fundamental solution, then $\|X(t)\|$, $\|X^{-1}(t)\| \leq M$ for all $t \geq t_0$, and the transformation $x = Xy$ transforms (3.9.1) into $X(t)\,y' = 0$ and hence into $y' = 0$ by left multiplication by $X^{-1}(t)$. Conversely, if $L(t)$ is a matrix as in the definition of reducibility to zero, then $L^{-1}AL - L^{-1}L' = 0$, and also, by left multiplication by L, $L' = AL$. Thus L is a fundamental system solution of (3.9.1), and $L(t)$ is bounded together with $L^{-1}(t)$ for $t \geq t_0$.

(3.9. vi) A system (3.9.1) is uniformly stable if it is stable and reducible.

Proof. The transformation $x = Ly$ transforms (3.9.1) into $y' = By$ with $B = L^{-1}AL - L^{-1}L'$. Thus, if $X = LY$, we have $Y = L^{-1}X$, $X^{-1} = Y^{-1}L^{-1}$, and the boundedness of X implies the boundedness of Y. Thus $y' = By$ is stable, and, also, uniformly stable, i.e., $\|Y(t)\,Y^{-1}(\tau)\| \leq M$ for some M and all $0 \leq \tau \leq t$. Finally $\|X(t)\,X^{-1}(\tau)\| = \|L(t)\,Y(t)\,Y^{-1}(\tau)\,L^{-1}(\tau)\| \leq \|L\|\,\|Y(t)\,Y^{-1}(\tau)\|\,\|L^{-1}\| \leq N$ for some N and $0 \leq \tau \leq t$. Thus (3.9.1) is uniformly stable.

A system (3.9.1) with periodic coefficients is reducible; hence, if it is stable, it is uniformly stable. Indeed, as we shall see in § 4, (3.9.1) admits of a fundamental system of the form $X(t) = P(t)\,M(t)$ where $P(t)$ is periodic, $\det P(t) \neq 0$ and $M(t)$ is the fundamental system of solutions of a system with constant coefficients. Thus $x = Py$ transforms the given system into $y' = By$ with the fundamental solution $Y = M(t)$, i.e., into a system with constant coefficients. For other conditions on reducibility see V.A. YAKUBOVIC [1], and A. WINTNER [3].

We have now the following criteria for restrictive stability.

(3.9. vii) If (3.9.1) is stable and $R\int_0^t \operatorname{tr} A(\alpha)\,d\alpha \geq -d > -\infty$ for all $t \geq 0$, then (3.9.1) is restrictively stable.

Indeed, $X(t)$ is bounded. On the other hand,

$$|\det X(t)| = \left|\exp\int_0^t \operatorname{tr} A(\alpha)\,d\alpha\right| = \exp R\int_0^t \operatorname{tr} A(\alpha)\,d\alpha \geq e^{-d} > 0,$$

where d is a convenient constant. Consequently, $X^{-1}(t)$ is bounded.

(3.9. viii) If (3.9.2) is stable and $R\int_0^t \operatorname{tr} A(\alpha)\,d\alpha \leq d < +\infty$ for all $t \geq 0$, then (3.9.2) is restrictively stable.

Indeed, $R \operatorname{tr}[-\bar{A}_{-1}(t)] = -R \operatorname{tr} A(t)$.

(3.9. ix) If for some constants $M > 0$, $0 < m < 1$, and all $t \geq 0$ we have

$$\left\| \int_0^t A(\alpha)\, d\alpha \right\| \leq M, \qquad \int_0^t \left\| \left\{ \int_\beta^t A(\alpha)\, d\alpha \right\} A(\beta)\, d\beta \right\| \leq m,$$

then (3.9.1) is restrictively stable (R. CONTI [5]).

In particular under the conditions of (3.8. iv) the system is restrictively stable.

We can now state the following generalization of the Dini-Hukuhara theorem (3.3. ii). A further generalization to nonlinear systems will be proved in (6.2).

(3.9. x) If $x' = A(t)\, x$ is uniformly stable [restrictively stable] and $\int^{+\infty} \|C(t)\|\, dt < +\infty$, then the system $x' = [A(t) + C(t)]\, x$ is also uniformly stable [restrictively stable] (R. CONTI [5]).

In this statement uniform, or restrictive, stability cannot be replaced by simple stability, as the following example due to O. PERRON [2] shows:

$$a_{11} = -a, \qquad a_{12} = a_{21} = 0, \qquad a_{22} = \sin \log t + \cos \log t - 2a,$$

$c_{11} = c_{12} = c_{22} = 0$, $c_{21} = \exp(-at)$, where a is a constant with $1 < 2a < 1 + 2^{-1} \exp(-\pi)$. The system $x' = A\, x$ is stable but neither uniformly, nor restrictively stable. The system $x' = (A + C)\, x$ is not stable. Indeed the general solution of the system $x' = A\, x$ is $x_1 = c_1 \exp(-at)$, $x_2 = c_2 \exp(t \sin \log t - 2at)$, and $x_1, x_2 \to 0$ as $t \to +\infty$, since $a > 1/2$. The system $x' = (A + C)\, x$ has the general solution $x_1 = c_1 \exp(-at)$, $x_2 = [\exp(t \sin \log t - 2at)] \left[c_2 + c_1 \int_0^t \exp(-\alpha \sin \log \alpha)\, d\alpha \right]$, and $\overline{\lim}\, |x_2| = +\infty$ as $t \to +\infty$ if $c_1 \neq 0$. Indeed, for $t = t_n = \exp(2n + 1/2)\pi$, $n = 1, 2, \ldots$, we have $\sin \ln t_n = 1$, and $-\sin \ln \alpha \geq 1/2$ for all $\exp(2n - 1/2)\pi \leq \alpha \leq \exp(2n - 1/6)\pi$, i.e., for all $t_n \exp(-\pi) \leq \alpha \leq t_n \exp(-2\pi/3)$. Hence, the last integral is larger than $t_n [\exp(-2\pi/3) - \exp(-\pi)] \exp[2^{-1} t_n \exp(-\pi)]$, and $|x_2| > |c_1| \exp[1 - 2a + 2^{-1} \exp(-\pi)]\, t_n + o(1)$ as $n \to +\infty$.

For other and even sharper examples see R. E. VINOGRAD [4].

Combining (3.9. vii) and (3.9. x), we obtain the result:

(3.9. xi) If all solutions of $y' = A(t)\, y$ are bounded, if

$$R \int_0^t \operatorname{tr} A(t)\, dt \geq -d > -\infty \text{ for all } t \geq 0, \quad \text{if} \quad \int^{+\infty} \|C(t)\|\, dt < +\infty,$$

then all solutions of $x' = [A(t) + C(t)]\, x$ are bounded.

In addition, under the same hypotheses, for every y, there is an x such that $x - y \to 0$ as $t \to +\infty$, and conversely, for every x there is a y for which $x - y \to 0$ as $t \to +\infty$ (A. WINTNER [3]). On this subject see also A. WINTNER [1], N. LEVINSON [6], H. WEYL [4].

The second condition above holds necessarily if $\operatorname{tr} A = 0$. In particular, as we deduce from (1.4), the boundedness of $|y| + |y'|$ for all solutions y of $y'' + a(t)\, y = 0$ implies the same fact for all solutions x

of $x''+[a(t)+c(t)]x=0$ provided $\int^{+\infty}|c(t)|\,dt<+\infty$ (A. WINTNER [3], G. ASCOLI [9]).

The Lyapunov theorem (3.1. i) can be extended without difficulty to the case where $A(t)$ is periodic (R. BELLMAN [9, 12]). Recent theorems handle the extension of Dini-Hukuhara theorem to the case where the elements of $A(t)$ are almost periodic functions. Let $M[f(t)]$ denote the usual mean value of a continuous (complex-valued) almost periodic function; that is, $M[f(t)]=\lim(1/2T)\int_{-T}^{T}f(t)\,dt$ as $T\to+\infty$, and let

$$m(h)=\overline{\lim}\int_0^t R[\operatorname{tr}A(\tau+h)]\,d\tau \quad \text{as} \quad t\to+\infty \quad \text{for any} \quad h\geq 0.$$

(3.9. xii) If $M[R\operatorname{tr}A(t)]\geq 0$, if the system $y'=A(t)y$ has solutions all bounded in $[t_0,+\infty)$, if $\overline{\lim}\,m(h)<+\infty$ as $h\to+\infty$, if $C(t)$ is of the class L; i.e., $\int^{+\infty}\|C(t)\|\,dt<+\infty$, then all solutions $x(t)$ of the system $x'=[A(t)+C(t)]x$ are bounded in $[t_0,+\infty)$ (M. MARCUS [1]).

For other statements see I. Z. SHTOKALO [2, 7, 8, 10, 12], J. FAVARD [1].

It should be mentioned here that if $f(t)$ is a real almost periodic function in $(-\infty,+\infty)$, and hence bounded, then every bounded solution $x(t)$ of the differential equation $x^{(n)}+c_1x^{(n-1)}+\cdots+c_nx=f(t)$, $c_1,\ldots,c_n$ real, is necessarily almost periodic together with its first n derivatives which, therefore, are also bounded (E. ESCLANGON [1]; cf. also E. LANDAU [1]). See a proof in G. SANSONE [16].

3.10. Matrix conditions. Given any system (3.9.1), where $A(t)$ is a matrix whose elements are real or complex-valued functions of t, let us denote by $H(t)$ the hermitian matrix $H(t)=[A(t)+\bar{A}_{-1}(t)]/2$ and by $\lambda(t)$, $\Lambda(t)$ the smallest and greatest characteristic roots of $H(t)$ necessarily real. A. WINTNER [8] has proved that every solution $x(t)$ of (3.9.1) satisfies the following double inequality:

$$\Sigma|x_i(t_0)|^2\exp\left(2\int_{t_0}^{t}\lambda(\alpha)\,d\alpha\right)\leq\Sigma|x_i(t)|^2\leq\Sigma|x_i(t_0)|^2\exp\left(2\int_{t_0}^{t}\Lambda(\alpha)\,d\alpha\right), \tag{3.10.1}$$

where Σ ranges over $i=1,\ldots,n$. (See also T. WAZEWSKI [5]; Z. BUTLEWSKI [7], T. KITAMURA [1].)

For A and x real the proof of (3.10.1) is as follows:

Let $r^2=\Sigma x_i^2(t)$, hence $2r\,dr/dt=2\Sigma x_ix_i'=2xAx$. On the other hand, by known properties of quadratic forms, we have $\lambda(t)r^2\leq xAx\leq\Lambda(t)r^2$. By comparison we also have $\lambda(t)\leq dr/r\,dt\leq\Lambda(t)$ and, by integration, (3.10.1) follows. For A and x complex, the system may be replaced by a real system of order $2n$. If $y(t)$ is a solution of the adjoint system (3.9.2), then the analogous double inequality holds where λ and Λ are replaced by $-\Lambda$ and $-\lambda$ respectively. From these remarks the following theorem follows:

(3.10. iii) If $\overline{\lim}\int_0^t\Lambda(\alpha)\,d\alpha<+\infty$, $\underline{\lim}\int_0^t\lambda(\alpha)\,d\alpha>-\infty$, then the system (3.9.1) is restrictively stable.

In particular, the contention holds if the following limits exist and

$$\int_0^{+\infty}\Lambda(\alpha)\,d\alpha<+\infty, \qquad \int_0^{+\infty}\lambda(\alpha)\,d\alpha>-\infty. \tag{3.10.2}$$

It may also be pointed out that we necessarily have

$$-\|A(t)\|\leq-\|H(t)\|\leq\lambda(t)\leq\Lambda(t)\leq\|H(t)\|\leq\|A(t)\|, \tag{3.10.3}$$

and thus conditions (3.10.2) are certainly verified if $\int^{+\infty} \|H(t)\|\, dt < +\infty$ (Z. BUTLEWSKI [7], A. ROSENBLATT [3]) or if $\int^{+\infty} \|A(t)\|\, dt < +\infty$ (see 3.8). Finally, the conditions of (3.10. iii) are certainly satisfied if $\int^{+\infty} \|A(t) + A_{-1}(t)\|\, dt < +\infty$.

3.11. Nonhomogeneous systems. Concerning nonhomogeneous differential equations and systems the following original result of O. PERRON [4, 5] (based on his previous paper [2]) should be mentioned.

(3.11. i) If $F(t)$, $f_i(t)$, $t_0 \leq t < +\infty$, $i = 1, \ldots, n$, are real continuous functions of t having finite limits $F(t) \to b$, $f_i(t) \to a_i$ as $t \to +\infty$, if the roots λ_i, $i = 1, \ldots, n$, of the equation

$$\varrho^n + a_1 \varrho^{n-1} + \cdots a_n = 0 \tag{3.11.1}$$

are real, distinct, and $\neq 0$, then the equation

$$x^{(n)} + f_1(t)\, x^{(n-1)} + \cdots + f_n(t)\, x = F(t) \tag{3.11.2}$$

has at least one solution $x(t)$ with $x(t) \to b/a_n$, $x^{(r)}(t) \to 0$, $r = 1, \ldots, n$, as $t \to +\infty$. If $\lambda_i < 0$, $i = 1, \ldots, n$, then all solutions of (3.11.2) have these properties.

This theorem has been successively generalized by A. WALTHER [1] and H. SPÄTH [1, 2]. The latter has proved the following theorems:

(3.11. ii) If the functions $f_i(t) = a_i$ are constants and $F(t)$ is continuous in $t_0 \leq t < +\infty$, and $F(t) \to b$ as $t \to +\infty$, then all solutions $x(t)$ of (3.11.2) having a finite limit as $t \to +\infty$ (if any) have the property that $x^{(r)}(t) \to 0$ as $t \to +\infty$, $r = 1, \ldots, n$. There are certainly solutions $x(t)$ having a finite limit as $t \to +\infty$, if $R(\lambda_i) \neq 0$ where λ_i, $i = 1, \ldots, n$, are the roots of the equation $\varrho^n + a_1 \varrho^{n-1} + \cdots + a_n = 0$. The same holds if $a_n \neq 0$ and, for every purely imaginary root $\varrho_i = i\tau$ (if any), τ real, of multiplicity l, $1 \leq l \leq n$, the following integrals exist

$$\int_0^{+\infty} \exp(-i\tau\xi)\, \Psi(\xi)\, d\xi, \quad \int_0^{+\infty} d\xi_1 \int_{\xi_1}^{+\infty} \exp(-i\tau\xi_2)\, \Psi(\xi_2)\, d\xi_2,$$
$$\ldots, \int_0^{+\infty} d\xi_1 \int_{\xi_1}^{+\infty} d\xi_2 \ldots \int_{\xi_{l-1}}^{+\infty} \exp(-i\tau\xi_l)\, \Psi(\xi_l)\, d\xi_l,$$

where $\Psi(t) = F(t) - b$. The same holds if $a_n = 0$ provided $b = 0$.

(3.11. iii) Suppose the functions $F(t)$ and $f_i(t)$, $i = 1, \ldots, n$, $t_0 \leq t < +\infty$, are continuous and have finite limits $F(t) \to b$, $f_i(t) \to a_i$ as $t \to +\infty$, and, if (3.11.1) has some roots $\lambda = i\tau$, τ real with multiplicities $\leq l$, $1 \leq l \leq n$, then also $\int^{+\infty} t^{b-1} s(t)\, dt < +\infty$ where $s(t) = |a_1 - f_i(t)| + \cdots + |a_n - f_n(t)|$. Under these conditions equation (3.11.2) and equation

$$y^{(n)} + a_1 y^{(n-1)} + \cdots + a_n y = F(t) \tag{3.11.3}$$

always have together solutions which are bounded together with their first $n-1$ derivatives in $[t_0, +\infty)$. If k is the number of the roots λ_i of (3.11.1) with $R(\lambda_i) < 0$, then (a) for each of the solutions y of (3.11.3) there is exactly one solution x of (3.11.2) with

$$x^{(\nu)} = y^{(\nu)} + o(1), \quad \nu = 0, 1, \ldots, n; \qquad x^{(\nu)}(t_0) = y^{(\nu)}(t_0), \quad \nu = 0, 1, \ldots, k-1;$$

(b) for each of the solutions x of (3.11.2) there is exactly one solution y of (3.11.3) with

$$y^{(\nu)} = x^{(\nu)} + o(1), \quad \nu = 0, 1, \ldots, n; \qquad x^{(\nu)}(t_0) = y^{(\nu)}(t_0), \quad \nu = 0, 1, \ldots, k-1.$$

In theorem (3.11. ii) the case $a_n = 0$, $b \neq 0$, is actually exceptional as is shown by the trivial example $x' = 1$ whose solutions $x = t + c$, c a constant, are all unbounded as $t \to +\infty$.

The following particular case of the previous theorems is of interest:

(3.11. iv) If $f_i(t) \to a_i$, $F(t) \to b$ as $t \to +\infty$, $i = 1, \ldots, n$, if $R(\lambda_i) < 0$, $i = 1, \ldots, n$, then all solutions of (3.11.2) verify the relations $x(t) \to b/a_n$, $x^{(r)}(t) \to 0$, $r = 1, 2, \ldots, n$, as $t \to +\infty$. If $a_n \neq 0$, $R(\lambda_i) \leq 0$, $i = 1, \ldots, n$, and the roots λ_i with $R(\lambda_i) = 0$ are simple, if $\int^{+\infty} |a_i - f_i(t)|\, dt < +\infty$, $i = 1, \ldots, n$, $\int^{+\infty} |b - F(t)|\, dt < +\infty$, then all solutions of (3.11.2) are bounded as $t \to +\infty$. The same holds if $a_n = 0$ provided $b = 0$.

This theorem, and other analogous ones, have been proved by a different process also for measurable functions $f_i(t)$, $F(t)$ (L. Cesari [2]) and extended to systems of differential equations (L. Cesari [2]).

The following theorem attacks the problem of behavior of the solutions of nonhomogeneous linear systems from another direction.

(3.11. v) If $f(t)$ is a vector function, $t \geq 0$, if $A(t)$ is a continuous matrix, and the system $y' = A(t)\, y$ admits of a fundamental system of solutions $Y(t)$, then a necessary and sufficient condition in order that the system $x' = A(t)\, x + f(t)$ have solutions all bounded in $[0, +\infty)$ for every vector $f(t)$ with $\int^{+\infty} \|f(t)\|\, dt < +\infty$ is that $\|(Y(t)\, Y^{-1}(t_1)\| \leq c_1$ for some c_1 and all $t \geq t_1 \geq 0$. A necessary and sufficient condition in order that the same system have solutions all bounded in $[0, +\infty)$ for every vector $f(t)$ with $\|f(t)\| \leq c_2 < +\infty$, $t \geq 0$, is that $\int_0^t \|Y(t)\, Y^{-1}(t_1)\|\, dt_1 \leq c_3 < +\infty$ for some c_3 and all $t \geq t_1 \geq 0$.

For the sufficiency of the first of these conditions see D. Caligo [2] who has improved previous conditions of O. Perron [10]. For the necessity and the remaining part of the theorem see R. Bellman [8] who has made use of Banach space methods. Analogous results hold for vector functions $f(t)$ with $\int^{+\infty} \|f(t)\|^p\, dt < +\infty$ (R. Bellman [8]).

3.12. Lyapunov's type numbers. If $f(t)$, $t_0 \leq t < +\infty$, is any continuous real or complex function of the real variable t, and a is any real number such that $f(t)\, e^{at}$ is bounded in $[t_0, +\infty)$, then $f(t)\, e^{a't}$ with $a' < a$ is also bounded in $[t_0, +\infty)$, and approaches zero as $t \to +\infty$. If $f(t)\, e^{bt}$ is unbounded in $[t_0, +\infty)$ then $f(t)\, e^{b't}$ with $b' > b$ is also unbounded in $[t_0, +\infty)$. Thus if A, B are the classes of all real numbers a, b with $f(t)\, e^{at}$ bounded, and $f(t)\, e^{bt}$ unbounded in $[t_0, +\infty)$, then (A, B) is a partition of the real field defining a real number which we shall denote by $-\lambda$ (maximum of A, or minimum of B, if both A and B are not empty). The number λ is said to be the *type number* of $f(t)$ in $[t_0, +\infty)$. It may occur that B is empty and then we say that $\lambda = -\infty$, or that A is empty and then we say that $\lambda = +\infty$. The type number λ was defined by A. Lyapunov [3] (with the opposite sign).

Examples. 1. $f(t) = 0$, $\lambda = -\infty$; 2. $f(t) = e^{\alpha t}$, α real $\lambda = \alpha$; 3. $f(t) = t^m$, m real, $\lambda = 0$; 4. $f(t) = e^{t \cos t}$, $\lambda = +1$; 5. $f(t) = e^{-t} \cos t^{-1}$, $t \geq 1$, $\lambda = -1$; 6. $f(t) = e^{t \exp(\sin t)}$, $\lambda = +e$; 7. $f(t) = e^{-t \exp(\sin t)}$, $\lambda = -1/e$; 8. $f(t) = t^t$, $t \geq 1$, $\lambda = +\infty$; 9. $f(t) = e^{-t^2}$, $t \geq 0$, $\lambda = -\infty$; 10. $f(t) = e^{t h(t)}$, $\lambda = \overline{\lim}\, R[h(t)]$ as $t \to +\infty$.

In general we may say that the type number λ of any function $f(t)$ is given by $\lambda = \overline{\lim}\, (\log |f(t)|)/t$ as $t \to +\infty$.

(3.12. i) If $\lambda_1 \geq \lambda_2 \geq \cdots \geq \lambda_n$ are the type numbers of the functions $f_i(t)$, $i = 1, 2, \ldots, n$, and λ, λ' those of $f_1 + \cdots + f_n$, $f_1, f_2, \ldots, f_n$, then $\lambda \leq \lambda_1$, $\lambda' \leq \lambda_1 + \lambda_2 + \cdots + \lambda_n$. In addition $\lambda = \lambda_1$, or $\lambda \leq \lambda_1$ according as $\lambda_1 > \lambda_2$, or $\lambda_1 = \lambda_2$.

Indeed if $\lambda_1 \geq \lambda_2 \geq \cdots$, $-a > \lambda_1$, then $f_1 e^{at} + \cdots + f e^{at} \to 0$ as $t \to +\infty$; if $\lambda_1 > \lambda_2$ and $\lambda_1 > -a > \lambda_2$, then the same expression is unbounded in $[t_0, +\infty)$ since it is the sum of one unbounded function and of $n-1$ other functions which approach zero as $t \to +\infty$. This proves the last part of (i). If $\varepsilon > 0$ is any number, then $f_1 f_2 \cdots f_n e^{(-\lambda_1 - \cdots - \lambda_n - \varepsilon)t} = \Pi f_i e^{(-\lambda_i - \varepsilon/n)t} \to 0$ as $t \to +\infty$. This proves the first part of (i).

(3.12. ii) The type number of $f(t)$ is not altered by multiplication by a function $g(t)$ with $0 < a \leq |g(t)| \leq b < +\infty$. In particular $g(t) = \text{const} \neq 0$, $g(t) = e^{ih(t)}$, $(i = \sqrt{-1}$, $h(t)$ real) have the property mentioned.

(3.12. iii) If $\lambda_1 > \lambda_2 > \cdots > \lambda_n$ are the type numbers of the functions $f_i(t)$, $i = 1, 2, \ldots, n$, none of which is identically zero, then these functions are linearly independent in $[t_0, +\infty)$.

Indeed, for every expression $f = \Sigma c_i f_i$ with at least two numbers c_i different from zero the type number is given by $\lambda_k \geq \lambda_{n-1} > -\infty$, where k is the minimum index i with $c_i \neq 0$. Thus f cannot be zero since the type number of $f = 0$ is $\lambda = -\infty$.

(3.12. iv) If $f(t) \neq 0$ in $[t_0, +\infty)$ and λ, λ' are the type numbers of $f(t)$, $[f(t)]^{-1}$, then $\lambda + \lambda' \geq 0$. Examples 6 and 7 above show that the sign $>$ may occur. If $f(t) = e^{th(t)}$, $h(t)$ real, then $\lambda + \lambda' = 0$ if and only if $\lim h(t)$ exists and is finite as $t \to +\infty$.

(3.12. v) If λ, λ', l, L are the type numbers of $f(t)$, $[f(t)]^{-1}$, $g(t)$, $f(t)\,g(t)$, where $f(t) \neq 0$ and $\lambda + \lambda' = 0$, then we have $L = \lambda + l$.

(3.12. vi) If λ, l are the type numbers of $f(t)$ and $u(t)$ where

$$u(t) = \int_{t_0}^{t} f(t)\,dt \quad \text{if} \quad \lambda \geq 0, \qquad u(t) = \int_{t}^{+\infty} f(t)\,dt \quad \text{if} \quad \lambda < 0,$$

then we have $l \leq \lambda$.

We shall denote by the type number of a vector function $f(t) = [f_1(t), \ldots, f_n(t)]$ the largest of the type numbers of its coefficients. Then theorems i, ii, iii above are extended immediately to vector functions.

3.13. First application of type numbers to differential equations.

(3.13. i) If the coefficients $a_{ij}(t)$ of the linear homogeneous system

$$x_i' = \sum_{j=1}^{n} a_{ij}(t)\,x_j, \quad i = 1, \ldots, n, \qquad \text{or} \qquad x' = A(t)\,x, \tag{3.13.1}$$

are continuous bounded real functions in $[t_0, +\infty)$, then every solution $[x_i(t), i = 1, \ldots, n]$ different from $[0, \ldots, 0]$ of (3.13.1) has a finite type number (A. Lyapunov [3]).

Proof. For any λ real let us put $z_i(t) = e^{\lambda t} x_i(t)$, $i = 1, 2, \ldots, n$. Then (3.13.1) is transformed into the system

$$dz_i/dt = \sum_{h=1}^{n} [a_{ih} + \delta_{ih}\lambda]\,z_h, \quad i = 1, 2, \ldots, n, \tag{3.13.2}$$

where $\delta_{ii} = 1$, $\delta_{ih} = 0$ for all $i \neq h$, $i, h = 1, 2, \ldots, n$. Finally we have

$$2^{-1}(d/dt)\sum_{i=1}^{n} z_i^2 = \sum_{i=1}^{n} (a_{ii} + \lambda)\,z_i^2 + \sum_{i \neq h} a_{ih} z_i z_h.$$

If λ is positive and large enough, the quadratic form in the right member is definite positive for every $t \geq t_0$, and also $\geq 2^{-1} N (z_1^2 + \cdots + z_n^2)$ for some $N > 0$. If λ is negative and large enough in absolute value, then the same quadratic form is definite negative and $\leq -2^{-1}N(z_1^2+\cdots+z_n^2)$ for some $N>0$. As a consequence, we have

$$\Sigma z_i^2 > C e^{Nt}, \quad \text{or} \quad \Sigma z_i^2 < C e^{-Nt},$$

for some constant $c>0$, according as λ is positive or negative, and large enough in absolute value. In the first case at least one z_i is unbounded; in the second case all z_i approach zero as $t\to+\infty$. It follows that the type number of $[x_1, \ldots, x_n]$ is finite.

Under the same hypotheses of (3.13. i), if $|a_{ij}(t)|\leq c$ for all $0\leq t<+\infty$, $i, j=1, \ldots, n$, then $|\lambda|\leq nc$ (O. PERRON [14]).

3.14. Normal systems of solutions. A fundamental system

$$x_h=(x_{ih},\ i=1, 2, \ldots, n),\ h=1, 2, \ldots, n, \tag{3.14.1}$$

of solutions of (3.13.1) is said to be normal if every linear combination of any number of them with non-zero coefficients has a type number equal to the largest of the characteristic numbers of the combined solutions. For instance

$$x_1=[x_{i1}=(-1)^{i-1}e^{-x},\ i=1, 2, 3],$$

$$x_2=[x_{i2}=(-2)^{i-1}e^{-2x},\ i=1, 2, 3], \quad x_3=[x_{i3}=(-3)^{i-1}e^{-3x},\ i=1, 2, 3],$$

with type numbers $\lambda_1=-1$, $\lambda_2=-2$, $\lambda_3=-3$, is a normal system of solutions of the differential system

$$x_1'=x_2, \quad x_2'=x_3, \quad x_3'=-6x_1-11x_2-6x_3.$$

The fundamental system

$$x_1=[x_{i1}=(-1)^{i-1}e^{-x}+(-2)^{i-1}e^{-2x}+(-3)^{i-1}e^{-3x}, \qquad i=1, 2, 3],$$

$$x_2=[x_{i2}=(-1)^{i-1}e^{-x}+(-2)^{i-1}e^{-2x}, \qquad i=1, 2, 3],$$

$$x_3=[x_{i3}=(-1)^{i-1}e^{-x}, \qquad i=1, 2, 3],$$

is not normal since x_{i1}, x_{i2}, x_{i3} have type numbers -1, while x_2-x_3 and x_1-x_2 have type numbers -2 and -3 respectively.

(3.14. i) Every linear differential system with continuous, bounded coefficients has some normal fundamental system of solutions (A. LYAPUNOV [3]).

By the extension to vectors of statement (3.12. iii) it follows that all solutions of (3.13.1) with distinct type numbers are linearly independent. Hence the set $[\lambda]$ of all possible distinct real numbers λ which are type numbers of at least one (non-zero) solution of system (3.13.1) [under conditions (3.13. i)] is finite and has, therefore, m, $1\leq m\leq n$, elements. We may say that these m numbers $\lambda_1>\lambda_2>\cdots>\lambda_m$ are the type numbers of system (3.13.1).

Let us denote now by N_s the maximum number of independent solutions of (3.13.1) having the type number λ_s, $s=1, 2, \ldots, n$. Then we have $N_1>N_2>\cdots>N_m$ and $N_1=n$. On the other hand, if we consider any fundamental system of solutions, say (3.14.1), and we denote by n_s the number of solutions in (3.14.1) having type number λ_s, we have $n_1+n_2+\cdots+n_m=n=N$, and

$$n_s+n_{s+1}+\cdots+n_m\leq N_s, \qquad s=2, 3, \ldots, m.$$

According to the definition given above, a system (3.14.1) is normal if and only if we have

$$n_s+n_{s+1}+\cdots+n_m=N_s \quad \text{for all } s=2, 3, \ldots, m.$$

As an immediate consequence of this fact and of the relation $n_1 + \cdots + n_m = n = N_1$, we have:

(3.14. ii) A fundamental system (3.14.1) of solutions of (3.13.1) is normal if and only if

$$n_1 = n - N_2, \quad n_2 = N_2 - N_3, \ldots, \quad n_{m-1} = N_{m-1} - N_m, \quad n_m = N_m.$$

Thus, under conditions (3.13. i), for normal fundamental systems of solutions (3.14.1) the numbers $n_1, \ldots, n_m$ do not depend on the particular system (3.14.1) but only on the system (3.13.1). They are denoted as the multiplicities of the type numbers $\lambda_1, \ldots, \lambda_m$, and the statement holds:

(3.14. iii) $$n_1 + n_2 + \cdots + n_m = n.$$

It will be often said that the differential system (3.13.1) has n type numbers $\lambda_1 \geq \lambda_2 \geq \cdots \geq \lambda_n$ where each of the $m \leq n$ distinct characteristic numbers λ_s considered above is counted n_s times.

Let us now consider the number

$$S = n_1 \lambda_1 + n_2 \lambda_2 + \cdots n_m \lambda_m.$$

(3.14. iv) A fundamental system (3.14.1) of solutions of (3.13.1) is normal if and only if S takes its minimal value (A. Lyapunov [3]).

Indeed, if we put $n_s = n_{s+1} + \cdots + n_m = N_s' \leq N_s$, $s = 1, 2, \ldots, m$, then

$$N_1' = N_1 = n, \quad n_s = N_s' - N_{s+1}', \quad s = 1, 2, \ldots, m-1, \quad n_m = N_m',$$

and

$$S = (N_1' - N_2')\lambda_1 + (N_2' - N_3')\lambda_2 + \cdots + (N_{m-1}' - N_m')\lambda_{m+1} + \\ + N_m'\lambda_m = n\lambda_1 + N_2'(\lambda_2 - \lambda_1) + \cdots + N_m'(\lambda_m - \lambda_{m-1}),$$

and thus S is minimum if and only if $N_s' = N_s$, $s = 1, 2, \ldots, m$.

Let us denote by Λ, Λ_1 $(\Lambda + \Lambda_1 \geq 0)$ the type numbers of the functions

$$F(t) = e^{\int(\Sigma p_{ss})\,dt}, \quad F^{-1}(t) = e^{-\int(\Sigma p_{ss})\,dt}.$$

(3.14. v) For every fundamental system (3.14.1) of solutions of (3.13.1) we have $S \geq \Lambda \geq -\Lambda$ (A. Lyapunov [3]).

Proof. The relation $\Lambda + \Lambda_1 \geq 0$ is a consequence of (3.12. iv). Now let us consider the determinant $\Delta = \det[x_{ih}]$ and observe that $\Delta = CF(x)$ where $C \neq 0$ is a constant [see (2.2.2)]. Thus the type number of Δ is Λ. On the other hand Δ is a sum of products and each product contains as a factor exactly one term from each of the n solutions $x_1, x_2, \ldots, x_n$. By (3.12. i) we have then that each product has a type number $\leq n_1\lambda_1 + \cdots + n_m\lambda_m = S$, and that the type number of the sum of all these products must have a type number not larger than the largest of their type numbers, in any case $\leq S$. This proves that $\Lambda \leq S$.

3.15. Regular differential systems. A differential system (3.13.1) with coefficients continuous and bounded in $[t_0, +\infty)$ is said to be *regular* if there exists at least one fundamental system (3.14.1) of solutions with $S = -\Lambda_1$. Thus, by (3.14. iv), we have $S = \Lambda = -\Lambda_1$.

Obviously a system (3.14.1) for which $S = \Lambda = -\Lambda_1$ is normal since S certainly has its minimum value. On the other hand, a system (3.14.1) may be normal and yet have $S > \Lambda \geq -\Lambda_1$.

Example. Let us consider with A. Lyapunov [3] the system

$$\begin{aligned} dx_1/dt &= x_1 \cos\log t + x_2 \sin\log t, \\ dx_2/dt &= x_1 \sin\log t + x_2 \cos\log t. \end{aligned} \tag{3.15.1}$$

A fundamental system of solutions is given by

$$x_1 = [x_{11} = e^{t \sin \log t}, \quad x_{21} = e^{t \sin \log t}],$$

$$x_2 = [x_{12} = e^{t \cos \log t}, \quad x_{22} = - e^{t \cos \log t}].$$

The solutions x_1, x_2 both have the type number $+1$, and it easily shown that this system of solutions is normal since every linear combination of them also has the type number $+1$. Thus $S = +2$. On the other hand, we have

$$F(t) = e^{\int 2 \cos \log t \, dt} = e^{t (\sin \log t + \cos \log t)},$$

and hence $\Lambda = +\sqrt{2}$ and also $\Lambda_1 = -2$. Thus $S > \Lambda > -\Lambda_1$ and the differential system (3.15.1) is not regular.

All linear systems with constant coefficients are regular in the sense described above. Also the systems whose coefficients are periodic of the same period T are regular as follows from the Floquet theory (see § 4). The class of the regular systems is remarkably wide. The system for instance

$$dx_1/dt = x_1 \cos a t + x_2 \sin b t, \quad dx_2/dt = x_1 \sin b t + x_2 \cos a t,$$

where a, b are real constant, is regular (A. LYAPUNOV [3]). The following theorem gives also an idea of the general character of the regular systems.

(3.15. i) Consider the system

$$x_i' = \sum_{h=1}^{i} a_{ih}(t)\, x_h, \quad i = 1, 2, \ldots, n,$$

where the coefficients $a_{ih}(t)$ are continuous functions of t and form a triangular matrix $A = [a_{ih}(t)]$. A necessary and sufficient condition in order that this system be regular is that the functions $e^{\int p_{ii} dt}$, $e^{-\int p_{ii} dt}$ have characteristic numbers whose sum is zero, $i = 1, 2, \ldots, n$ (A. LYAPUNOV [3]).

3.16. A relation between type numbers and generalized characteristic roots. Any two complex or real matrices $A(t)$, $B(t)$ continuous and bounded in $[0, +\infty)$ are said to be *kinematically similar* if there exists a matrix $L(t)$ with $\det L(t) \neq 0$, whose elements are absolutely continuous functions, which is bounded together with $L^{-1}(t)$ in $[0, +\infty)$, and such that $L^{-1} A L - L^{-1} L' = B$ [cf. (3.9)]. Obviously, $L^{-1}(t)$ has the same property, $A = L B L^{-1} - L(L^{-1})'$, and the kinematic similarity is an equivalence relation. If L is constant, then this relation reduces to the usual similarity (§ 2) (static similarity). The concept of kinematic similarity can be traced in A. LYAPUNOV [3]. O. PERRON [14] proved that for every matrix $A(t)$ continuous and bounded, there are infinitely many matrices $L(t)$ bounded together with $L^{-1}(t)$ and $L'(t)$ such that $B = L^{-1} A L - L^{-1} L'$ is bounded and triangular (i.e., $b_{ij} = 0$ for $i < j$). If A is real, then both L and B can be assumed to be real. S. P. DILIBERTO [1], proved that if A is real, then L can be assumed to be orthogonal. For other results see S. P. DILIBERTO [1], L. MARKUS [6], R. E. VINOGRAD [8].

Given the system $x' = A(t)\, x$, where $A(t)$ is continuous and bounded in $[0, +\infty)$, and a matrix $L(t)$ satisfying the conditions mentioned for kinematic similarity, then the transformation $x = L(t)\, y$ transforms the given system into $y' = B(t)\, y$, where $B = L^{-1} A L - L^{-1} L'$. By the consideration of normal fundamental systems of solutions (3.14) it can be proved that both systems have the same type numbers λ_s with the same multiplicities n_s (3.14), and the same numbers Λ, Λ_1 (A. LYAPUNOV [3]). Thus the numbers Λ, Λ_1, and $\lambda_1 \geq \lambda_2 \geq \cdots \geq \lambda_n$ are invariant with respect to kinematic similarity. Also, both systems above are together regular, or not. R. CONTI [6] has studied the same type of transformations in a more general setting, namely under the assumption that the elements of the matrix $A(t)$ are L-integrable in every finite interval. R. CONTI has proved that Lyapunov stability, uniform stability, and restrictive stability (3.9) are invariant with respect to the same type of transformations.

L. MARKUS [6] has studied the relations between the type numbers λ_i, $i=1, \ldots, n$, (3.14) and the generalized characteristic roots $\varrho_i(t)$, $i=1,\ldots,n$, (3.4) for systems $x'=A(t)\,x$, where $A(t)$ is a continuous bounded matrix which is supposed to be commutative with its conjugate transpose $A^*=A_{-1}^{-1}$, i.e., $AA^*=A^*A$. L. MARKUS [6] has proved the following theorem:

(3.16. i) If $m(t)$, $M(t)$ are the minimum and the maximum of the generalized characteristic roots then

$$\overline{\lim_{t\to+\infty}}\, t^{-1}\int_0^t m(s)\,ds \leq \lambda_i \leq \overline{\lim_{t\to+\infty}}\, t^{-1}\int_0^t M(s)\,ds, \quad i=1,\ldots,n.$$

R. E. VINOGRAD [4] has given examples sharpening the one of O. PERRON discussed in (3.9) under (3.9. x). More precisely he has given examples of systems

$$x'=A(t)\,x,\ x'=[A(t)+B(t)]\,x, \quad \text{with} \quad \int^{+\infty}\|B(t)\|\,dt<+\infty,$$

having different type numbers. Conditions for the equality of the type numbers of systems $x'=A(t)\,x$, $x'=B(t)\,x$, have been given by O. PERRON [2] for one matrix constant, by N.C. CETAJEV [3], and B.F. BYLOV [1, 2].

3.17. Bibliographical notes. The transformation of the system $x'=(A+C)\,x$ into the Volterra type integral (matrix) equation (3.2.2) is typical and has been used throughout this section. Conditions for boundedness of the solution of a Volterra type integral matrix equation have been given by D. CALIGO [3].

For studies on hereditary phenomena, linear and non-linear, see D. GRAFFI [9, 12], and R. BELLMAN [13]. For studies on linear systems of infinitely many equations see M. R. RESETOV [1], R. BELLMAN [4], K. PERSIDSKII [1—10], V. HARASAHAL [1, 2, 3], N. ARLEY and V. BORHSENIUS [1], W. L. HART [1], A. WINTNER [40].

§ 4. Linear systems with periodic coefficients

4.1. Floquet theory. Let

$$x'=A(t)\,x, \qquad x=(x_1,\ldots,x_n), \qquad -\infty<t<+\infty, \tag{4.1.1}$$

be a system of first order linear equations, where $A(t)$ denotes an $n\times n$ matrix whose elements are continuous periodic functions $a_{ij}(t)$ of period T; i.e., $A(t+T)=A(t)$.

A system (4.1.1) does not necessarily have periodic non-zero solutions of period T, as the trivial example $x'=a(t)\,x$, with $a(t)$ real and $a(t+T)=a(t)$, $n=1$, shows. The solutions of this equation are of the form $x=c\exp\left(\int_0^t a(t)\,dt\right)$, c a constant, and obviously x is nonzero and periodic of period T if and only if $\int_0^T a(t)\,dt=0$, $c\neq 0$.

A basic theorem is the following one.

(4.1. i) A system (4.1.1) with $A(t+T)=A(t)$ has at least one solution $x(t)$ not identically zero with

$$x(t+T) = \lambda x(t) \tag{4.1.2}$$

for all t, where $\lambda \neq 0$ is a convenient constant (real, or complex) (G. FLOQUET [1]).

Proof. Let us consider first a fundamental system

$$x_k(t), \quad k = 1, 2, \ldots, n,$$

of solutions

$$x_k(t) = \operatorname{col}[x_{1k}(t), x_{2k}(t), \ldots, x_{nk}(t)].$$

Thus $X(t) = \|x_{ik}\|$ is a $n \times n$ matrix with $\det X(0) \neq 0$. By (2.2) we have $\det X(t) = \det X(0) \cdot \exp \int_0^t \operatorname{tr} A(\tau)\, d\tau$. Hence $\det X(t+T) \neq 0$, $\det X(t) \neq 0$ for all t. As a consequence $X(t+T)$ is a fundamental system of solutions of (4.1.1) just as well as $X(t)$. In other words we have, for convenient constants c_{jk},

$$x_k(t+T) = \sum_{j=1}^{n} c_{jk}\, x_j(t), \quad k = 1, \ldots, n,$$

or

$$x_{ik}(t+T) = \sum_{j=1}^{n} c_{jk}\, x_{ij}(t), \quad i, k = 1, \ldots, n,$$

i.e., $X(t+T) = X(t) \cdot C$ where C is a $n \times n$ constant matrix. Since $\det X(t+T) = \det X(t) \cdot \det C$, we conclude that $\det C = \exp \int_0^T \operatorname{tr} A(t)\, dt \neq 0$.

Now let us discuss whether there are nonzero solutions $x(t)$ of (4.1.1) such that (4.1.2) holds for all t and some complex or real number λ.

If such a solution $x(t) = (x_1, \ldots, x_n)$ exists, then we must have

$$x(t) = \sum_{k=1}^{n} m_k\, x_k(t), \quad \text{or} \quad x_s(t) = \sum_{k=1}^{n} m_k\, x_{sk}(t), \; s = 1, \ldots, n, \tag{4.1.3}$$

for some nonzero vector $m = (m_1, \ldots, m_n)$. Then

$$\begin{aligned} x_s(t+T) &= \sum_{k=1}^{n} m_k\, x_{sk}(t+T) = \sum_{k=1}^{n} m_k \sum_{j=1}^{n} c_{jk}\, x_{sj}(t) \\ &= \sum_{j=1}^{n} \left(\sum_{k=1}^{n} c_{jk}\, m_k \right) x_{sj}(t) \end{aligned}$$

and finally by (4.1.2) and (4.1.3) the algebraic relations follow

$$\sum_{k=1}^{n} c_{jk}\, m_k = \lambda m_j, \quad j = 1, \ldots, n. \tag{4.1.4}$$

Thus λ must be a characteristic root of the matrix C. Conversely, if λ is any characteristic root of the matrix C, then system (4.1.4) certainly has a nonzero solution m, and the corresponding vector function $x(t)$ verifies equation (4.1.2).

If the characteristic roots λ_i of the matrix C are all distinct, then there are n solutions of (4.1.1) verifying (4.1.2) for $\lambda=\lambda_i$, $i=1,\ldots,n$, respectively, and they form obviously a fundamental system. Otherwise, if $\lambda_1,\ldots,\lambda_m$, $1\leq m\leq n$, are the distinct characteristic roots of matrix C then there are at least $m\geq 1$ solutions $x(t)$ of (4.1.1) verifying (4.1.2) for $\lambda=\lambda_1,\ldots,\lambda_m$. Thereby (4.1. i) is proved.

Using the same notations as above, let Y be another fundamental system of solutions of (4.1.1). Then $Y=XM$ for some constant matrix M with $\det M\neq 0$, and $X=YM^{-1}$. Then from $X(t+T)=X(t)C$ it follows $Y(t+T)=Y(t)C'$, where $C'=M^{-1}CM$, and the characteristic roots of C and C' are the same (2.1). This shows that the numbers $\lambda_1,\ldots,\lambda_m$, $1\leq m\leq n$, do not depend upon the particular fundamental system X used in the reasoning above. The numbers λ_i, $i=1,\ldots,m$, are called the *characteristic factors*, or *multipliers*, of the system (4.1.1) and, if we put $\lambda_i=e^{r_iT}$, the numbers r_i (real, or complex), $i=1,\ldots,m$, are called the *characteristic exponents* of the system (4.1.1) and are determined up to multiples of ωi, $i=\sqrt{-1}$, where $\omega=2\pi/T$. Let us observe that if $x_i(t)$ is a vector solution of (4.1.1) satisfying $x_i(t+T)=\lambda_i x_i(t)$ and we put $x_i(t)=p_i(t)\,e^{r_it}$ we deduce $p_i(t+T)=p_i(t)$, i.e., $p_i(t)$ is a periodic function of t of period T. We conclude that system (4.1.1) has at least m, $1\leq m\leq n$, solutions of the form

$$x_i(t)=p_i(t)\,e^{r_it},\qquad i=1,\ldots,m,$$

$p_i(t)$ periodic of period T, and these solutions are certainly linearly independent.

We may denote by $\lambda_1,\ldots,\lambda_n$ the multipliers, and by $r_1,\ldots,r_n$ the characteristic exponents, where each λ_i and corresponding r_i is repeated as many times as the multiplicity of λ_i. By considering the fundamental system $X(t)$ defined by $X(0)=I$, we have $C=X(T)$, and by (2.2), also

$$\exp(r_1+\cdots+r_n)\,T=\lambda_1\lambda_2\ldots\lambda_n=\det X(T)=\exp\int_0^T \operatorname{tr}A(t)\,dt,$$

and hence

$$r_1+r_2+\cdots+r_n\equiv T^{-1}\int_0^T \operatorname{tr}A(t)\,dt\quad(\operatorname{mod}\omega i).$$

The considerations above show also that, by a convenient choice of the fundamental system X we may find that the matrix C of the relation $X(t+T)=X(t)\,C$ has a canonical form (2.1), i.e., C is the direct sum of matrices $C_s=[c_{ij}]$ with

$$c_{ii}=\lambda_s',\quad i=1,\ldots,n_s;\qquad c_{i,i+1}=1,\quad i=1,2,\ldots,n_s-1,$$

$c_{ij}=0$ otherwise, where λ_s' is any one of the roots λ_i determined above and, for each root $\lambda_i, i=1, \ldots, m,$ we have $\mu_i=\Sigma n_s$ where μ_i is the multiplicity of the root λ_i. Correspondingly Σn_s denotes the sum of the orders n_s of all the matrices C_s having $\lambda_s'=\lambda_i$ in the main diagonal (companion matrices of the characteristic roots $\lambda_s, s=1, \ldots, m$).

For each of the matrices C_s of orders n_s the fundamental system X presents n_s solutions, say $x_1(t), \ldots, x_{n_s}(t)$, which verify the relations

$$\left.\begin{aligned} &x_1(t+T)=\lambda_s' x_1(t),\\ &x_2(t+T)=\lambda_s' x_2(t)+x_1(t), \ldots, x_{n_s}(t+T)=\lambda_s' x_{n_s}(t)+x_{n_s-1}(t). \end{aligned}\right\} \qquad (4.1.5)$$

If we write $C=e^{TD}$, where D is an $n\times n$ matrix, we have $X(t+T)=X(t)\,e^{TD}$, and now we may suppose that D, instead of C, has the canonical form (2.1), i.e., D is the direct sum of matrices $D_s=[d_{ij}]$ with $d_{ij}=d$, $d_{i,i+1}=1$, $d_{ij}=0$ otherwise. Then, from (2.1), $C=e^{TD}$ is the direct sum of the matrices $e^{TD_s}=[g_{ij}]$ with $g_{ii}=e^{Td}$, $g_{i,i+1}=e^{Td}\,T$, $g_{i,i+2}=e^{Td}\,T^2/2!$, ... Since we must have $e^{Td}=\lambda_s$, we have $g_{ii}=\lambda_s$, $g_{i,i+1}=\lambda_s T$, $g_{i,i+2}=\lambda_s T^2/2!, \ldots,$ and $g_{ij}=0$ for $i>j$. Relations (4.1.5) are now replaced by

$$\begin{aligned} x_k(t+T) = \lambda_s[x_k(t) + T x_{k-1}(t) + \\ + T^2 x_{k-2}(t)/2! + \cdots + T^{k-1} x_1(t)/(k-1)!], \end{aligned}$$

$k=1, 2, \ldots, n_s$. If we write $x_1(t)=p_1(t)\,e^{rt}$ where $\lambda_s=e^{rT}$ we verify immediately that $p_1(t)$ is periodic of period T. If we write

$$x_k(t) = [t^{k-1} p_1(t)/(k-1)! + t^{k-2} p_2(t)/(k-2)! + \cdots + t\,p_{k-1}(t) + p_k(t)]\,e^{rt},$$

$k=1, 2, \ldots, n_s$, where $\lambda_s=e^{rT}$, we may prove by induction that all the functions $p_1(t), \ldots, p_{n_s}(t)$ so defined are periodic of period T. More directly, if we put $X(t)=P(t)\,e^{tD}$, then $P(t)=X(t)\,e^{-tD}$, and by $X(t+T)=X(t)\,e^{TD}$, we deduce immediately $P(t+T)=P(t)$. If $M(t)=e^{tD}$, we conclude that there is a fundamental solution $X(t)$ such that

$$X(t) = P(t)\,M(t)$$

where $P(t)$ is periodic and $M(t)$ is the fundamental solution of a differential system with constant coefficients.

For a more detailed exposition of FLOQUET's theory we may refer to G. SANSONE [16], E. L. INCE [5], E. A. CODDINGTON and N. LEVINSON [3].

From the form of the fundamental solution determined above we conclude, as for systems with constant coefficients, that

(4.1. ii) All solutions $x(t)$ of system (4.1.1) approach zero as $t\to+\infty$ if and only if $|\lambda_i|<1, i=1, \ldots, m$; i.e. $R(r_i)<0$. All solutions $x(t)$ of system (4.1.1) are bounded as $t\to+\infty$ if and only if $|\lambda_i|\leq 1$, $i=1, 2, \ldots, m$ [i.e. $R(r_i)\leq 0$] and, for those λ_i (if any) with $|\lambda_i|=1$, we have $\mu_i=\nu_i$, i.e. the multiplicity μ_i of the root λ_i for the equation $\det(\lambda I-C)=0$ must be equal to the nullity of the matrix $\lambda_i I-C$.

System (4.1.1) has a periodic solution of period T if and only if there is at least one root $\lambda_i = 1$.

Let us observe that all considerations above hold even in the case where $A(t)$ is a constant matrix and then $T = 2\pi/\omega$ is arbitrary. Then the characteristic exponents r_j can be assumed to be equal to the characteristic roots ϱ_j of A (or congruous to these mod ωi).

If $A = A(t, \varepsilon)$ is any matrix periodic in t of period $T = 2\pi/\omega$, continuous in t and a parameter ε (real or complex), say for ε in some domain U containing $\varepsilon = 0$, then the multipliers $\lambda_j(\varepsilon)$ and the characteristic exponents $r_j(\varepsilon)$ are functions of ε for $\varepsilon \in U$, $j = 1, 2, \ldots, n$. The multipliers $\lambda_j(\varepsilon)$ are the characteristic roots of the matrix C. By taking the fundamental system $X = X(t, \varepsilon)$ as defined by $X(0, \varepsilon) = I$, we have $C = X(T, \varepsilon)$, and thus $\lambda_j(\varepsilon)$ are the characteristic roots of the matrix $X(T, \varepsilon)$. By (1.1) the elements of $X(T, \varepsilon)$ are continuous functions of ε, and so are the coefficients of the characteristic equation $\det[\lambda I - X(T, \varepsilon)] = 0$. In the theory of complex functions it is usually proved that the roots of an algebraic equation whose coefficients are continuous functions of a real or complex parameter can be thought of as continuous functions of the same parameter (cf. 3.5). As a consequence we may think of the n multipliers $\lambda_j(\varepsilon)$ as complex-valued, never zero (4.1. i), continuous functions of ε. Finally the characteristic exponents $r_j(\varepsilon)$ are defined (mod ωi) by $r_j(\varepsilon) = T^{-1} \ln \lambda_j(\varepsilon) = T^{-1} Ln|\lambda_j(\varepsilon)| + i \arg \lambda_j(\varepsilon)$, where $i = \sqrt{-1}$, $j = 1, \ldots, n$. Since $|\lambda_j(\varepsilon)| > 0$ we may fix arbitrarily one of the values (mod 2π) of $\arg \lambda_j(0)$ and then consider $\arg \lambda_j(\varepsilon)$ as a single-valued continuous function of ε. This shows that each determination (mod ωi) of $r_j(\varepsilon)$ can be thought of as a single-valued continuous function of ε. If, for some value of ε, say $\varepsilon = 0$, $A(t, 0)$ is a constant matrix with characteristic roots ϱ_j, we may consider that particular determination of $r_j(\varepsilon)$ for which $r_j(0) = \varrho_j$, $j = 1, \ldots, n$. If the elements of A are analytic regular functions of the complex parameter $\varepsilon \in U$, then the coefficients of the equation for the λ_j are also analytic regular functions of ε in U, and hence the functions $\lambda_j(\varepsilon)$ and finally the functions $r_j(\varepsilon)$ are analytic with at most branch points of finite order. That branch points may actually occur is shown by the trivial example $x_1' = x_1 + \varepsilon x_2$, $x_2' = x_1 + x_2$, with constant coefficients ($T > 0$ arbitrary), where $r_1(0) = r_2(0) = 1$, $r_1(\varepsilon), r_2(\varepsilon) = 1 \pm \sqrt{\varepsilon}$.

4.2. Some important applications. It is clear from the previous considerations that the determination of the numbers λ_i depend upon the knowledge of the matrix C, and thus on the knowledge of a fundamental system X of solutions of (1).

A case in which the application of the previous considerations has been particularly fruitful is the equation (Hill equation)

$$x'' + p(t)\, x = 0, \tag{4.2.1}$$

or, if $x=x_1$, $x'=x_2$, the system $x_1'=x_2$, $x_2'=-p(t)\,x_1$, where $p(t)$ is a periodic function of period T. We may choose the fundamental system X so that

$$x_{11}(0)=1, \quad x_{21}(0)=0; \quad x_{12}(0)=0, \quad x_{22}(0)=1.$$

We have, tr $A=0$, and, from $X(t+T)=X(t)\,C$, also

$$x_{i1}(t+T)=x_{i1}(t)\,c_{11}+x_{i2}(t)\,c_{21}, \qquad i=1,2,$$

$$x_{i2}(t+T)=x_{i1}(t)\,c_{12}+x_{i2}(t)\,c_{22}, \qquad i=1,2,$$

with det $C=c_{11}c_{22}-c_{12}c_{21}=e^0=1$. Thus the equation $\det|\lambda I-C|=0$ becomes $\lambda^2-(c_{11}+c_{22})\,\lambda+1=0$ and finally

$$\lambda^2-2A\,\lambda+1=0, \tag{4.2.2}$$

where $2A=c_{11}+c_{22}=x_{11}(T)+x_{22}(T)$. For $A^2<1$ this equation has roots $|\lambda|=1$ and all solutions of (4.2.1) are bounded; for $A^2>1$ equation (4.2.1) has unbounded solutions. A more detailed analysis has led to the following statement:

(4.2. i) If $p(t)\leq 0$, then equation (4.2.2) has real roots $0<\lambda_1<1<\lambda_2$ and thus (4.2.1) has infinitely many unbounded solutions in $[0,+\infty)$. If $p(t)>0$ and $\int_0^T p(t)\,dt\leq 4/T$ equation (4.2.2) has roots $\lambda_1=e^{i\theta}$, $\lambda_2=e^{-i\theta}$, $\theta>0$, and (4.2.1) has all solutions x bounded with x' in $[0,+\infty)$ (A. LYAPUNOV [3]).

Proof. Suppose first $p(t)\leq 0$. Then (4.2.1) becomes $x''=-p(t)\,x$, where $-p(t)\geq 0$, and the solution $x(t)$ with $x(0)=1$, $x'(0)=1$, has $x''(t)\geq 0$, $x'(t)\geq 1$, for all $t\geq 0$. Hence $x(t)\to+\infty$ as $t\to+\infty$, and the first part of (4.2. i) is proved. The second part of (4.2. i) is contained in a more general statement due to G. BORG [5] which will be stated and proved in (4.3).

Another proof of (4.2. i) has been given by N. E. ZUKOVSKII [2], who also proved that the solutions of (4.2.1) are all bounded provided $n^2\pi^2T^{-2}\leq p(t)\leq(n+1)^2\pi^2T^{-2}$ for all t and some $n=0,1,\ldots$ See, for the last result, also S. WALLACH [1]. N. E. ZUKOVSKII [2] proved also that $4/T$ is the best bound since if we replace it by any $\varepsilon+4/T$, $\varepsilon>0$, then the conclusion of (4.2. i) is no longer true. See also A. LYAPUNOV [9].

It is convenient sometimes to consider the equations

$$x''+\varepsilon\,p(t)\,x=0, \tag{4.2.3}$$

$$x''+[\delta+\varepsilon\,r(t)]\,x=0, \tag{4.2.4}$$

where $p(t)$, $r(t)$ are real periodic nonconstant functions, $r(t)$ of mean value zero, and ε, δ are real parameters, $\delta\geq 0$.

If $p(t)\geq 0$, $\varepsilon<0$ [or $p(t)\leq 0$, $\varepsilon>0$], then equation (4.2.3) has infinitely many unbounded solutions and thus all solutions are unstable. If $p(t)\geq 0$ and $\varepsilon>0$, then the semi-infinite interval $0<\varepsilon<+\infty$ can be divided into consecutive intervals by means of a sequence $0=\varepsilon_0<\varepsilon_1\leq\varepsilon_2<\varepsilon_3\leq\cdots$, with $\varepsilon_m\to+\infty$, where alternately, either $A^2<1$, $|\lambda_1|$,

$|\lambda_2| = 1$, all solutions of (4.2.3) are bounded and stable (intervals of stability), or $A^2 > 1$, $0 < |\lambda_1| < 1 < |\lambda_2|$, infinitely many solutions of (4.2.3) are unbounded, and all are unstable (intervals of instability). The same occurs if $p(t)$ is not of constant sign and $\varepsilon < 0$ since we may change ε into $-\varepsilon$ and $p(t)$ into $-p(t)$. If $p(t)$ is odd then the intervals above are symmetric with respect to $\varepsilon = 0$ since we may change ε into $-\varepsilon$ and t into $-t$. An analogous result holds for equation (4.2.4) where $r(t)$ is certainly not of constant sign since $r(t)$ is not constant and is of mean value zero. These results, which had been stated by A. LYAPUNOV [6, 7], have been rediscovered and deeply investigated by O. HAUPT [2]. O. HAUPT has shown that, by taking $\varepsilon \neq 0$ constant in (4.2.4), then the infinite interval $-\infty < \delta < +\infty$ is always divided into alternate intervals of stability and instability by means of a sequence $\delta_0 < \delta_1 \leqq \delta_2 < \delta_3 \leqq \cdots$, with $\delta_m \to +\infty$, the semi-infinite interval $-\infty < \delta < \delta_0$ being an interval of instability. In general for equation (4.2.4) the whole $\varepsilon\delta$-plane is divided into alternate zones of stability and instability [cf. (4.4)], the half straight line ($\varepsilon = 0$, $-\infty < \delta < 0$) being contained in a zone of instability (O. HAUPT [2]). On the lines of subdivision we have $A^2 = 1$ and equation (4.2.4) has at least one solution $x(t)$ with $x(t+T) = \pm x(t)$, hence periodic of period T, or $2T$ ("harmonics", or "subharmonics" of order 1/2).

For these and other questions see also G. HAMEL [2], H. A. KRAMERS [1], H. SCHWERDTFEGER [1], L. A. PIPES [2], G. BORG [2].

It should be noted that, in the conditions of the theorem of LYAPUNOV, $\varepsilon = 1$ belongs to the first interval of stability $(0, \varepsilon_1)$ for equation (4.2.3). In the conditions of the theorem of ZUKOVSKII, $\varepsilon = 1$ belongs to the $(n-1)$-th interval of stability $(\varepsilon_{2n}, \varepsilon_{2n+1})$ for (4.2.3). We may simply say that "$p(t)$ belongs to the first, or the $(n-1)$-th interval of stability".

4.3. Further results concerning equation (4.2.1) and extensions. We shall denote the mean value of the periodic function $p(t)$ by $p_m = m\{p(t)\} = T^{-1}\int_0^T p(t)\,dt$.

(4.3. i) Each of the following conditions may replace Lyapunov conditions of the second part of (4.2. i) ($p(t)$ not identically zero):

$$p_m \geq 0, \qquad T\int_0^T |p(t)|\,dt \leq 4, \qquad \text{(G. BORG [5])} \tag{4.3.1}$$

$$p_m \geq 0, \quad T\int_0^T p^+(t)\,dt \leq 4 \text{ with } p^+ = 2^{-1}(|p| + p) \quad \text{(M. G. KREIN [2])} \tag{4.3.2}$$

$$p_m \geq 0, \; p \leq b, \quad T\int_0^T p^+(t)\,dt \leq 4 + (9.51)\, b^{-1} T^{-2} \quad \text{(A. M. GOLDIN [1])}, \tag{4.3.3}$$

$$p(t) \text{ odd}, \qquad T\int_0^T p^2(t)\,dt < 4T^{-2}, \qquad \text{(A. LYAPUNOV [4])}, \tag{4.3.4}$$

$$p(t) \geq 0, \qquad T\int_0^T p^2(t)\,dt \leq (63.03)\, T^{-2} \qquad \text{(G. BORG [1])}. \tag{4.3.5}$$

Each of these conditions assures that the solutions of (4.2.1) are all bounded and stable and that $p(t)$ belongs to the first zone of stability.

Proof of (4.3.1). Equation (4.2.2) has either real distinct roots $\lambda_1 < 1 < \lambda_2$, or a double root $\lambda = \pm 1$, or complex conjugate roots $|\lambda_1| = |\lambda_2| = 1$. According to Floquet's theory, (4.2.1) has infinitely many unbounded solutions in the first two cases, and solutions all bounded with their first derivatives in the third case. Thus we have to prove that under (4.3.1) both first and second alternative are contradictory. Under both, there is a real solution $x(t)$ with $x(t+T) = \lambda x(t)$, and thus either $x(t) \neq 0$ in $[0, T]$ and then $x(t) \neq 0$ for all t, or $x(t)$ has two consecutive zeros a, b (cf. 5.1) with $b - a \leq T$ (and hence infinitely many zeros in $(-\infty, +\infty)$. In the first case we have $x(T) = \lambda x(0)$, $x'(T) = \lambda x'(0)$, and hence $x'(T)/x(T) = x'(0)/x(0)$. On the other hand, by (4.2.1), $(x''/x) + p(t) = 0$, and, by integration over $[0, T]$ and integration by parts, also

$$\int_0^T (x'^2/x^2)\, dt + p_m = 0,$$

a contradiction. In the second case we may suppose $x(t) > 0$ for $a < t < b$ and we have

$$4/T \geq \int_0^T |p(t)|\, dt = \int_0^T |x''/x|\, dt \geq \int_a^b |x''/x|\, dt.$$

On the other hand, if $x_0 = x(c) = \max x(t)$ for $a < t < b$, and $c - a = \alpha$, $b - c = \beta$, $\alpha + \beta = b - a$, then, by the mean value theorem, there are two points $a < u < c < v < b$ with $x'(u) = x_0/\alpha$, $-x'(v) = x_0/\beta$, and finally

$$4/T \geq \int_a^b |x''/x|\, dt > x_0^{-1} \int_a^b |x''|\, dt \geq x_0^{-1} |x'(v) - x'(u)| \geq (1/\alpha + 1/\beta) = (\alpha + \beta)/\alpha\beta.$$

Since $4\alpha\beta \leq (\alpha + \beta)^2$ we have $4/T > 4/(\alpha + \beta) \geq 4/T$, a contradiction. Thus equation (4.2.2) has complex conjugate roots and the validity of condition (4.3.1) is proved.

If M denotes the maximum of $T^{-1} \left| \int_0^T [p(u+t) - p_m]\, t\, dt \right|$ for all $0 \leq u \leq T$, then the following statement holds

(4.3. ii) If $p_m < 0$ and $\varepsilon < -M^{-2} p_m$, then $A^2 > 1$ and all solutions of (4.2.3) are unstable. If $p_m \geq 0$ and $\varepsilon < \xi/M$ where ξ is the positive root of the equation $(T p_m M^{-1} + \xi) \cdot (e^{\xi} - 1)^2 = \pi^2 \xi$, then $A^2 < 1$ and all solutions of (4.2.3) are bounded and stable. In particular this occurs if $p_m = 0$ and $\left| \int_0^T p(u+t)\, t\, dt \right| < \ln(1 + \pi)$ (A. Lyapunov [10]).

The exact determination of the intervals of stability and instability is very difficult in general. Hence criteria for stability and instability are of particular interest. The following criterion is among those proved by G. Borg [1] by a variational analysis:

(4.3. iii) If for some $n = 0, 1, \ldots$, $n^2\pi^2 T^{-2} \leq p_m \leq (n+1)^2 \pi^2 T^{-2}$, and

$$\int_0^T |p(t) - p_m|\, dt < \min\left[2p_m^{\frac{1}{2}} (T p_m^{\frac{1}{2}} - n\pi),\quad 4(n+1)\, p_m^{\frac{1}{2}} \operatorname{ctg} 2^{-1}(n+1)^{-1} T p_m^{\frac{1}{2}}\right],$$

then all solutions of (4.2.1) are bounded and stable and $p(t)$ belongs to the $(n+1)$-th zone of stability.

Other results have been obtained by R. S. Gusarova [2], in part improved by the following criteria of V. A. Yakubovic and M. G. Krein.

(4.3. iv) If c_1, c_2 are constants, and $n\pi T^{-1} \leq c_1, c_2 \leq (n+1)\pi T^{-1}$ for some $n = 0, 1, \ldots$, if

$$\int_0^T |p(t) - c_1^2|\, dt < c_1 (c_1 T - n\pi), \quad \int_0^T |p(t) - c_2^2|\, dt < 2c_2 (n+1) \operatorname{ctg} 2^{-1} (n+1)^{-1} T c_2,$$

then all solutions of (4.2.1) are bounded and stable, and $p(t)$ belongs to the n-th zone of stability (V. A. YAKUBOVIC [2, 7]).

Conditions above can be replaced by either of the following ones:

$$\left.\begin{aligned} &p(t) \geq n^2 \pi^2 T^{-2}, \quad n\pi T^{-1} \leq c < (n+1)\pi T^{-1}, \\ &\int_0^T |p(t) - c^2|\, dt \leq 2c(n+1) \operatorname{ctg} (2^{-1}(n+1)^{-1} T c), \end{aligned}\right\} \tag{4.3.6}$$

$$\left.\begin{aligned} &p(t) \leq (n+1)^2 \pi^2 T^{-2}, \quad n\pi T^{-1} < c \leq (n+1)\pi T^{-1}, \\ &\int_0^T |p(t) - c^2|\, dt < c(cT - n\pi). \end{aligned}\right\} \tag{4.3.7}$$

For $c = n\pi T^{-1}$ condition (4.3.6) reduces to $p(t) \geq n^2\pi^2 T^{-2}$, $T \int_0^T p(t) \leq n^2\pi^2 + 2\pi n(n+1) \operatorname{tg} 2^{-1}\pi(n+1)^{-1}$, which, for $n = 0$, reduces to LYAPUNOV's criterion (4.2.1). Criterion (4.3.6) includes also a criterion proved independently by F. W. SCHÄFKE [3].

(4.3. v) If $a^2 \leq p(t) \leq b^2$, and $n\pi T^{-1} \leq a \leq (n+1)\pi T^{-1}$ for some $n = 0, 1, \ldots$, if

$$\int_0^T [p(t) - a^2]\, dt < 2(n+1)(b^2 - a^2)/v_0,$$

where v_0 is the least positive root of the equation

$$a \operatorname{ctg} [2^{-1}(n+1)^{-1} a T - a v] = b \tan b v,$$

then the conclusion of (4.3. iv) holds (V. B. LIDSKII and M. G. NEYGAUS [1]).

(4.3. vi) If $0 \leq h \leq p(t) \leq H$, $h < H$, and $d = T(H-h)^{-1}(p_m - h)$, $D = T(H-h)^{-1}(H - p_m)$, if

$$4n^2 h^{-1} D^{-2} \psi^2 (H/h, d/D) < 1 < 4(n+1)^2 H^{-1} d^{-2} \psi^2 (h/H, D/d)$$

for some $n = 0, 1, \ldots$, where $\psi(s, r)$ is the least positive root of the equation $\operatorname{tg} \psi = \sqrt{s} \operatorname{ctg} (\psi r \sqrt{s})$, then the conclusion of (4.3. iv) holds (M. G. KREIN [2]).

The equation

$$x'' + q(t) x' + p(t) x = 0 \tag{4.3.8}$$

where $p(t)$, $q(t)$ are periodic of period T can be reduced to the equation $y'' + p^*(t) y = 0$ where $p^* = p - 4^{-1} q^2 - 2^{-1} q'$ by the transformation of (5.1). The following criterion can be obtained by the same method of A. LYAPUNOV:

(4.3. vii) If $p^*(t) \leq 0$ and $q_m < 2(-p_m)^{\frac{1}{2}}$, then the solutions of (4.3.8) are stable. If $p^*(t) \leq a^2$, $n\pi T^{-1} < a \leq (n+1)\pi T^{-1}$ for some $n = 0, 1, \ldots$, and

$$\int_0^T [a^2 - p^*(t)]\, dt < \min a [aT - k\pi + 4^{-1}(aT - k\pi)^{-1} (T q_m)^2],$$

where min is taken for $k = 0, 1, \ldots, n$, then the solutions of (4.3.8) are asymptotically stable (V. A. YAKUBOVIC [6, 7]).

For other criteria we refer to V. A. YAKUBOVIC, loc. cit. The case $q(t) = a = \text{const}$, is of interest. R. EINAUDI [3] proved that the condition $T p_m \tan 2^{-1} a T < 2a$

implies the asymptotic stability of the solutions. The reduction of the equation to $y''+p^*(t)\,y=0$ and the application of (4.3. ii) may yield an improvement of this condition.

A previous result of G. CALAMAI [2] has been improved by D. CALIGO [4] as follows:

(4.3. viii) If $p(t), q(t)\geq 0$, and $p(t), q(t)$ are zero at most in a set of measure zero, if

$$\int_0^T (T-t)\,p(t)\,dt\leq 1, \quad \int_0^t q(\tau)\,d\tau\leq 1-\int_0^t \tau\,p(\tau)\,d\tau,$$

$$q(t)\leq p(t)\left[1-\int_0^t (t-\tau)\,p(\tau)\,d\tau\right]\left[\int_0^t p(\tau)\,d\tau\right]^{-1}$$

$$\int_0^T q(t)\,dt\leq\left[\int_0^T p(t)\left\{T\left[1-\int_0^t (t-\tau)\,p(\tau)\,d\tau\right]+\int_0^t (t-\tau)^2 p(\tau)\,d\tau\right\}dt\right]\left[1+2T\int_0^T p(t)\,dt\right]^{-1},$$

then all solutions of (4.3.8) are stable.

Other conditions for asymptotic stability of equation (4.3.8) have been given by R. NARDINI [1].

The method of LYAPUNOV can be extended to systems of the form

$$x_1'=p_{11}(t)\,x_1+p_{12}(t)\,x_2, \qquad x_2'=p_{21}(t)\,x_1+p_{22}(t)\,x_2, \tag{4.3.9}$$

where $p_{ij}(t)$, $i, j=1, 2$, are periodic functions of period T. We prefer to refer here to the following criterion proved by V. I. BURDINA [1] by means of the method of V. A. YAKUBOVIC [3, 7].

(4.3. ix) If $\mu_0=(2T)^{-1}\int_0^T (p_{11}+p_{22})\,dt<0$, $\alpha=\cosh\mu_0 T$, and $h(t)$, $H(t)$ are the min and max characteristic roots (functions of t) of the matrix $[h_{ij}]$, where

$$h_{11}=-p_{21}, \quad h_{22}=p_{12}, \quad h_{12}=h_{21}=2^{-1}(p_{11}-p_{22}),$$

if for some $k=0, \pm 1, \pm 2, \ldots$, and $-1\leq\theta\leq 1$, we have

$$k\pi-\int_0^T h(t)\,dt\geq \arccos(1-\alpha\,\theta)\,(\alpha-\theta)^{-1},$$

$$\int_0^T H(t)\,dt-k\pi\leq \arccos(1+\alpha\,\theta)\,(\alpha+\theta)^{-1},$$

then the solutions of (4.3.9) are stable.

For systems of n-th order

$$x'=P(t)\,x,\ x=(x_1,\ldots,x_n), \qquad P(t)=[p_{ij}(t)], \tag{4.3.10}$$

where $P(t+T)=P(t)$, criteria have been proved by N. G. CETAEV [5] and CH. NOUGMANOVA [1], both based on the second method of LYAPUNOV (§ 7). We mention here the second criterion. Put $c_{ij}=m\{p_{ij}(t)\}$, $\tilde{p}_{ij}(t)=p_{ij}(t)-c_{ij}$, $i, j=1,\ldots,n$, and denote by ϱ_k, $k=1,\ldots,n$, the characteristic roots of the constant matrix $[c_{ij}]$, which are supposed to be all distinct and $\neq 0$. Then n independent linear forms $U_k=A_{k1}x_1+\cdots+A_{kn}x_n$ with constant coefficients A_{kj} can be determined in such a way that $\Delta=\det[A_{kj}]=1$, and

$$\sum_{s=1}^n (c_{s1}x_1+\cdots+c_{sn}x_n)\,(\partial U_k/\partial x_s)=\varrho_k U_k, \quad k=1,\ldots,n,$$

(cf. § 7). Denote by Δ_{kj} the minor of A_{kj} in Δ and determine the periodic functions

$$\beta_{ji}(t) = \sum_{s,r=1}^{n} \Delta_{js} \Delta_{ir} \tilde{p}_{sr}(t), \quad j, i = 1, \ldots, n,$$

$$g_{kj}(t) = (1-\lambda)(\varrho_k + \bar{\varrho}_k)^2 \delta_{kj} + (\varrho_k + \bar{\varrho}_k)\beta_{kj}(t) + (\varrho_j + \bar{\varrho}_j)\overline{\beta}_{jk}(t),$$

$k, j = 1, \ldots, n$. Then the following statements hold: (a) If the roots ϱ_k are distinct and $R[\varrho_k] < 0$, if, for some $\lambda > 0$, $F(t) = \Sigma g_{kj}(t)\, \xi_k \xi_j > 0$ for all $0 < t < T$ and all real $\xi_1, \ldots, \xi_n$ not all zero, then all solutions of system (4.3.10) are bounded and stable; (b) if for at least one k, $R[\varrho_k] > 0$ and, for some $\lambda > 0$, $F(t) > 0$ for all $0 \le t \le T$ and all real $\xi_1, \ldots, \xi_n$ not all zero, then system (4.3.10) has infinitely many unbounded solutions, and all solutions are unstable (Ch. Nougmanova [1]).

For systems of $(2n)$-th order

$$x'' + \varepsilon P(t)\, x = 0, \quad x = (x_1, \ldots, x_n), \quad P(t) = [p_{ij}(t)], \tag{4.3.11}$$

where $P(t) = P_{-1}(t)$, $P(t+T) = P(t)$, the following theorems due to M. G. Krein [1, 6] may be considered as generalizations of Lyapunov's criterion (4.2. i):

(4.3. x) If for every $0 \le t \le T$ and all real $\xi_1, \ldots, \xi_n$ not all zero we have $F(t) = \Sigma p_{ij}(t)\, \xi_i \xi_j \ge 0$, $\int_0^T F(t)\, dt > 0$, if $H(t)$ denotes the maximal characteristic root (function of t) of the matrix $P(t)$ and $0 < \varepsilon < 4T^{-1}\left(\int_0^T H(t)\, dt\right)^{-1}$, then all solutions of (4.3.11) are bounded and stable.

(4.3. xi) If for every $0 \le t \le T$ and all real $\xi_1, \ldots, \xi_n$ not all zero we have $\int_0^T F(t)\, dt > 0$, if M (a scalar) denotes the maximum characteristic root of the matrix $Q = [q_{ij}]$, $q_{ij} = \int_0^T |p_{ij}(t)|\, dt$, $i, j = 1, \ldots, n$, and $0 < \varepsilon < 4T^{-1}M^{-1}$, then all solutions of (4.3.11) are bounded and stable.

Further bibliographical notes are given in (4.6).

4.4. Mathieu equation. The case of the Mathieu equation has been discussed in great detail

$$x'' + (\delta + \varepsilon \cos 2t)\, x = 0, \tag{4.4.1}$$

where $T = \pi$ [see bibliographical notes in (4.6)]. We give below an illustration showing the zones of stability and instability (the former are thatched). For large values of ε the stable zones become very narrow and tend to curves having slope -1 for $\varepsilon > 0$. The whole picture is symmetric with respect to the δ-axis. It is interesting to note that the points $\delta = n^2$, $n = 0, 1, \ldots$, $\varepsilon = 0$, are on the boundary of zones of stability as well as of zones of instability. In other words, in every neighborhood of the points $(n^2, 0)$ there are pairs (δ, ε) for which (4.4.1) has only bounded solutions and pairs (δ, ε) for which (4.4.1) has unbounded solutions. Of course for $\varepsilon = 0$, $\delta > 0$ equation (4.4.1) has only bounded solution $c \sin\left(\sqrt{\delta}\, t + \nu\right)$, c, ν constants. The phenomenon now mentioned has been denoted as a case of "parametric instability", and also of "resonance" between the "free" oscillations of the solutions $c \sin(nt + \nu)$ of the equation $x'' + n^2 x = 0$ of frequency $n/2\pi$ and the periodic disturbance $\varepsilon \cos 2t$ of frequency $1/\pi$. It is interesting to note that this resonance occurs not only at the integer multiples of the frequency $1/\pi$, but also at the half of these multiples, say $(1/\pi)\,(n/2)$. As already observed for (ε, δ) on the lines of transition there are periodic solutions of periods π, or 2π. Besides these the equation (4.4.1) has other periodic solutions corresponding to convenient lines in the stable zones, namely those lines on which the characteristic factors $\lambda_1 = \exp(ir T)$, $\lambda_2 = \exp(-ir T)$ have $r = p/q$, p, q integers. On these lines equation (4.4.1) has periodic solutions of periods πq and such lines are everywhere dense in the stable zones.

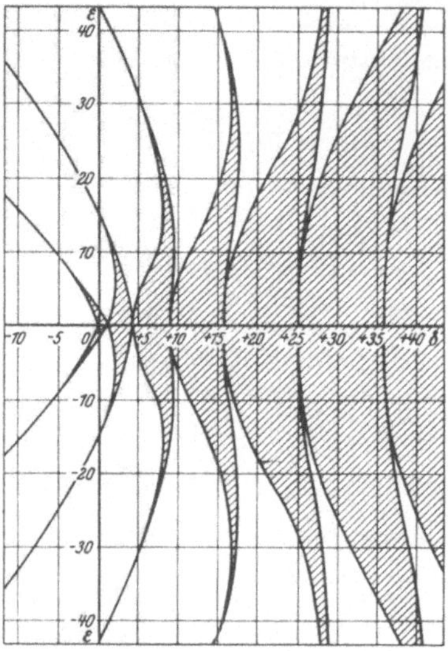

For comparison with the statements of the next section it is useful to rephrase the last results concerning ε small as follows. Consider the differential equation $x'' + (\sigma^2 + \varepsilon \cos \omega t)\, x = 0$ where σ is a fixed positive number. Thus the natural frequency is $\sigma/2\pi$ and the frequency of the periodic disturbance is $\omega/2\pi$. The transformation $\omega t = 2t_1$ brings us back to equation (4.4.1) with a new $\varepsilon_1 = 4\varepsilon/\omega^2$ and $\delta_1 = 4\sigma^2/\omega^2$. The resonance equation for ε small, say $\delta_1 = n^2$, yields $\omega = 2\sigma/n$. Thus, for ε small, resonance may occur only at frequencies close to $(2\sigma/n)\,(1/2\pi)$, $n = 1, 2, \ldots$.

Analogous facts hold for the Meissner equation $x'' + \big(\delta + \varepsilon f(t)\big)\, x = 0$, where $f(t)$ is periodic of period 2π and is defined by $f(t) = 1$ if $0 \leq t < \pi$, $f(t) = -1$ if $\pi \leq t < 2\pi$ (E. Meissner [1]). Since $f(t)$ is piecewise constant all integrations can be performed in finite form and the discussion is elementary. For other cases see L. Ricci [1], S. N. Simanov [1]. For further bibliographical references on the subject of this paragraph see (4.6).

4.5. Small periodic perturbations. Systems of the form

$$x' = [A + \varepsilon\, \Phi(t)]\, x, \tag{4.5.1}$$

with A a constant $n \times n$ matrix and $B(t)$ periodic of a given period $T = 2\pi/\omega$, have been discussed recently for ε small. Systems of this kind occur in applied mathematics (G. Krall [3]) when a given mechanical or physical system regulated by the differential system $x' = A\,x$, undergoes a periodic disturbance small in intensity modifying, according to the matrix $\varepsilon B(t)$, the terms of the matrix A. Also, systems of this kind are taken into consideration in discussing the stability of a periodic solution of a weakly nonlinear differential system (see 1.7 and 6.4).

A theorem often mentioned is the following one:

(4.5. i) If A is a constant matrix with characteristic roots ϱ_j and $R(\varrho_j)<0$, $j=1, \ldots, n$, if $\Phi(t)$ is a continuous periodic matrix of given period T, then for some $\varepsilon_0>0$ and all real or complex ε with $|\varepsilon|<\varepsilon_0$, all solutions of (4.5.1) approach zero as $t\to+\infty$.

This statement is a trivial consequence of the Lyapunov theorem (3.1. i). A direct proof in terms of Floquet's theory can be given as follows. By (4.5.1) we may think of the characteristic exponents $r_j(\varepsilon)$ of (4.5.1) as single-valued continuous functions of ε with $r_j(0)=\varrho_j$ and hence $R[r_j(0)]=R(\varrho_j)<0$. By the continuity argument we have $R[r_j(\varepsilon)]<0$, $j=1, 2, \ldots, n$, for all $|\varepsilon|\leq\varepsilon_0$ and some $\varepsilon_0>0$. Then (4.5. i) follows from (4.1. ii).

By CARATHÉODORY's theory (1.1) statement (4.5. i) can be extended to systems $x'=[A+B(t,\varepsilon)]x$, where A is a constant matrix with characteristic roots ϱ_j with $R(\varrho_j)<0, j=1, \ldots, n$, and $B(t,\varepsilon)=[b_{ij}(t,\varepsilon)]$, $i, j=1, \ldots, n$, is a matrix whose elements $b_{ij}(t,\varepsilon)$ are functions of the real variable t, periodic in t of period T, L-integrable in $[0, T]$. It should also be assumed that each $b_{ij}(t,\varepsilon)$ is a continuous function of ε at $\varepsilon=0$ for almost all t in $[0, T]$, $b_{ij}(t, 0)=0$, and $|b_{ij}(t,\varepsilon)|<\eta(t)$ almost everywhere in $[0, T]$ for some $\eta(t)$ L-integrable in $[0, T]$.

By virtue of (3.1. i) statement (4.5. i) holds as well if the elements of the matrix $\Phi(t)$ are arbitrary almost periodic functions in the sense of BOHR, since then they are continuous and bounded in $(-\infty, +\infty)$. For references on linear systems with almost periodic coefficients see (4.6).

The same argument used for (4.5. i) proves also that, if $R(\varrho_j)>0$ for some j, then for some $\varepsilon_0>0$ and all $|\varepsilon|<\varepsilon_0$, infinitely many solutions of (4.5.1) are unbounded in $[0, +\infty)$.

We will now consider systems (4.5.1) with $\Phi(t)$ periodic of a given period $T=2\pi/\omega$ and allow some of the characteristic roots ϱ_j of A to have real part zero. Neither the argument used for (4.5. i) is valid, nor is the conclusion as it is clear from the equation studied in (4.4). For systems (4.5.1) as mentioned, L. CESARI [4] has developed a convergent method of successive approximations which, independently of Floquet theory, has yielded unexpected information on the qualitative behavior of the solutions of the systems above for ε small. L. CESARI, J. K. HALE, and R. A. GAMBILL have successively developed and modified the same method for both linear and nonlinear problems [see (8.6)]. By using this method L. CESARI has also given an explicit relation for the characteristic exponents of the Floquet theory.

a) A convergent method of approximation. Solutions to a system of differential equations are usually obtained by means of successive approximations. If the solutions are functions of the independent variable t, then an important problem in applications is to determine the behavior of the solutions as t approaches ∞.

In defining a method of successive approximations, it may happen that certain terms are introduced which behave badly for large values of t, and at the same time these terms do not portray the true character of the solutions to the system of differential equations. For instance, when the solutions are expected to be periodic, it is desirable that all the successive approximations are also periodic, but it happens that terms are obtained which are not of this type. These are the terms which are generally called "secular" terms and various methods have been devised in order to eliminate or avoid these terms. These methods are called "casting out" methods of approximation (A. LINDSTEDT [1], H. POINCARÉ [4], G. DUFFING [1]) (see § 8).

In the following, we shall denote by C_ω the family of all functions which are finite sums of functions of the form $f(t) = e^{\sigma t}\varphi(t)$, $-\infty < t < +\infty$, where σ is any complex number and $\varphi(t)$ is any complex-valued function of the real variable t, periodic of period $T = 2\pi/\omega$, L-integrable in $[0, T]$. If $\varphi(t)$ has a Fourier series

$$\varphi(t) \sim \sum_{n=-\infty}^{+\infty} c_n e^{in\omega t},$$

then we shall denote the series

$$f(t) = e^{\sigma t}\varphi(t) \approx \sum_{n=-\infty}^{+\infty} c_n e^{(in\omega+\sigma)t}$$

as the series associated with $f(t)$. Moreover, we shall denote by the *mean value* $m\{f\}$ of $f(t)$ the number $m\{f\} = 0$ if $in\omega + \sigma \neq 0$ for all n, $m\{f\} = c_n$ if $in\omega + \sigma = 0$ for some n (L. CESARI [4]). We shall also make use of the following statement: If $f(t) \in C_\omega$ and $m\{f\} = 0$, then there is one and only one primitive of $f(t)$, say $F(t)$, which belongs to C_ω and such that $m\{F\} = 0$. Moreover, this primitive $F(t)$ is obtained by formal integration of the series associated with $f(t)$ (cf. J. K. HALE [1]). The mean value $m\{A\}$ of a matrix A whose elements are functions of C_ω will designate the matrix of the mean values of the elements of A.

Furthermore we will need the following remark. If σ_j, $j = 1, \ldots, n$, are n real or complex numbers with $\sigma_j \neq \sigma_h \pmod{\omega i}$, $j \neq h$, $j, h = 1, \ldots, n$, if $\varphi_j(t)$ are periodic functions of period $T = 2\pi/\omega$, L-integrable in $[0, T]$, and none of them is zero a.e. in $[0, T]$, then the n-functions $f_j(t) = e^{\sigma_j t}\varphi_j(t)$, $j = 1, \ldots, n$, are linearly independent, i.e., any linear combination of them with coefficients not all zero is not zero in a set of positive measure. This is obviously true for $n = 1$. Let us assume it is true for $n - 1$, and suppose, if possible, that $a_1 f_1(t) + \cdots + a_n f_n(t) = 0$ a.e. in $(-\infty, +\infty)$ for some constants a_j necessarily all $\neq 0$. Then we have also $a_1 f_1(t+T) + \cdots + a_n f_n(t+T) = 0$ a.e. in $(-\infty, +\infty)$ and, by obvious manipulations, also

$$\sum_{j=1}^{n-1} a_j [1 - e^{(\sigma_j - \sigma_n)T}] e^{\sigma_j t}\varphi_j(t) = 0$$

a.e. in $(-\infty, +\infty)$, with $a_j[1 - \exp(\sigma_j - \sigma_n)T] \neq 0$, $j = 1, \ldots, n-1$, a contradiction. Thus the statement above is proved. As a consequence, if for any two expressions as above we have $\Sigma a_j \exp(\sigma_j t)\varphi_j(t) = \Sigma b_k \exp(\tau_k t)\psi_k(t)$ a.e. in $(-\infty, +\infty)$, with $a_1, \ldots, a_k \neq 0$, then for any j there is a $k = k(j)$ with $\sigma_j \equiv \tau_k \pmod{\omega i}$ and $\exp(\sigma_j t)\varphi_j(t) = \exp(\tau_k t)\psi_k(t)$ a.e. in $(-\infty, +\infty)$, $a_j = b_k$.

Consider the differential system

$$y' = Ay + \varepsilon \Phi y, \qquad (' = d/dt), \tag{4.5.1}$$

where A is a constant $n \times n$ matrix, ε is a real parameter, $y = \operatorname{col}(y_1, \ldots, y_n)$, and Φ is an $n \times n$ matrix whose elements $\varphi_{uv}(t)$ are complex-valued functions, periodic of period $T = 2\pi/\omega$, $u, v = 1, \ldots, n$.

By considering an auxiliary system

$$y' = B y + \varepsilon \Phi y, \tag{4.5.2}$$

and applying a convenient method of successive approximations, we will obtain a fundamental system $Y(t) = [y_{uv}(t)]$ of solutions for a new system

$$y' = (B - \varepsilon D)\, y + \varepsilon \Phi y, \tag{4.5.3}$$

where D is a constant matrix which depends on B, Φ, ε. Then by determining B such that

$$B - \varepsilon D = A, \tag{4.5.4}$$

the fundamental system $Y(t)$ of solutions of (4.5.3) becomes a fundamental system of solutions of (4.5.1).

Throughout section (4.5) we will suppose, for the sake of simplicity, that the functions $\varphi_{uv}(t)$ have mean value $m\{\varphi_{uv}(t)\} = 0$ and Fourier series which are absolutely convergent, $\varphi_{uv}(t) = \sum_{n=-\infty}^{+\infty} c_{uvn}\, e^{in\omega t}$, $u, v = 1, \ldots, n$. J. K. HALE [2] has shown, by the same method, that these conditions can be removed. Also, let $A = \operatorname{diag}(\varrho_1, \ldots, \varrho_n)$, where

$$\varrho_u \not\equiv \varrho_v \pmod{\omega i}, \quad u \neq v, \quad u, v = 1, \ldots, n. \tag{4.5.5}$$

Put

$$\delta_0 = \min_{\substack{u, v = 1, \ldots, n \\ u \neq v}} |\varrho_u - \varrho_v|, \quad \delta = \min_{\substack{u, v = 1, \ldots, n \\ k = 0, 1, \ldots \\ |u-v| + k > 0}} |i k \omega - (\varrho_u - \varrho_v)|, \tag{4.5.6}$$

$0 < \delta \leq \delta_0$, and, if $0 < \lambda < \frac{1}{2}$ is any number, let us consider in the complex ϱ-plane, n circles $C_1, C_2, \ldots, C_n$ with radius $\lambda\delta$ and centers $\varrho_1, \ldots, \varrho_n$, respectively. Let $\tau_1, \tau_2, \ldots, \tau_n$ be n points lying in the interior or on the boundary of these circles, i.e., $\tau_u \in C_u$, $u = 1, 2, \ldots, n$. All of these points τ_u are thus distinct, and since

$$\min_{\substack{u, v = 1, \ldots, n \\ k = 0, 1, \ldots \\ |u-v| + k > 0}} |i k \omega - (\tau_u - \tau_v)| \geq \delta - 2\lambda\delta = \delta(1 - 2\lambda) > 0,$$

we have

$$\tau_u \not\equiv \tau_v \pmod{\omega i}, \quad u \neq v, \quad u, v = 1, \ldots, n. \tag{4.5.7}$$

Put $B = \operatorname{diag}(\tau_1, \ldots, \tau_n)$ and consider the differential system

$$Y' = B Y + \varepsilon \Phi Y, \tag{4.5.8}$$

where $Y = [y_{uv}(t)]$ denotes an $n \times n$ matrix. If we let $Z(t) = \operatorname{diag}(e^{\tau_1 t}, \ldots, e^{\tau_n t})$, we see that a matrix $Y(t)$ satisfying the equation

$$Y = Z + \varepsilon Z \int Z^{-1} \Phi Y\, dt \tag{4.5.9}$$

for any n^2 arbitrary constants (one for each element of the matrix) also satisfies (4.5.8).

Let us put $Y_0 = Z$, and consider the matrix

$$[p_{rs}^{(1)}] = Z^{-1} \Phi Y_0 = Z^{-1} \Phi Z = \begin{bmatrix} \varphi_{11} & e^{-\tau_1 t} \varphi_{12} e^{\tau_2 t} & \cdots & e^{-\tau_1 t} \varphi_{1n} e^{\tau_n t} \\ e^{-\tau_2 t} \varphi_{21} e^{\tau_1 t} & \varphi_{22} & \cdots & e^{-\tau_2 t} \varphi_{2n} e^{\tau_n t} \\ \cdots & \cdots & \cdots & \cdots \\ e^{-\tau_n t} \varphi_{n1} e^{\tau_1 t} & e^{-\tau_n t} \varphi_{n2} e^{\tau_2 t} & \cdots & \varphi_{nn} \end{bmatrix}$$

Each element of this matrix is contained in the class C_ω and since $\tau_u \not\equiv \tau_v \pmod{\omega i}$, $u \neq v$, $u, v = 1, \ldots, n$, each element of this matrix has mean value zero.

Therefore, from the theorem stated at the beginning of this section, there is one and only one matrix

$$[q_{rs}^{(1)}] = \int Z^{-1}(t)\, \Phi(t)\, Y_0(t)\, dt = \int Z^{-1}(t)\, \Phi(t)\, Z(t)\, dt,$$

whose elements are in C_ω and have mean value zero. Moreover, by this same theorem,

$$q_{rs}^{(1)} = \int e^{-\tau_r t} \varphi_{rs}(t)\, e^{\tau_s t}\, dt = \sum_{\substack{l=-\infty \\ l \neq 0}}^{+\infty} \frac{c_{rsl}}{-\tau_r + i\, l\, \omega + \tau_s} e^{(-\tau_r + i l \omega + \tau_s) t}.$$

Put $Y_1 = Z + \varepsilon Z \int Z^{-1} \Phi\, Y_0\, d\alpha$, where we will take the n^2 particular primitives of mean value zero. Moreover, let us consider the matrix

$$[p_{rs}^{(2)}] = Z^{-1} \Phi\, Y_1 = Z^{-1} \Phi Z + \varepsilon Z^{-1} \Phi Z \int Z^{-1} \Phi Z\, d\alpha,$$

where $p_{rs}^{(2)}$ has the associated series

$$p_{rs}^{(2)} = e^{-\tau_r t} \varphi_{rs} e^{\tau_s t} + \varepsilon \sum_{h=1}^{n} e^{-\tau_r t} \varphi_{rh} e^{\tau_h t} \int e^{-\tau_h \alpha} \varphi_{hs} e^{\tau_s \alpha}\, d\alpha$$

$$= p_{rs}^{(1)} + \varepsilon \sum_{h=1}^{n} \sum_{\substack{l_1, l_2 = -\infty \\ l_1 \neq 0,\, l_2 \neq 0}}^{+\infty} \frac{c_{rhl_1} c_{hsl_2}}{-\tau_h + i\, l_2\, \omega + \tau_s} e^{[-\tau_r + i(l_1 + l_2)\omega + \tau_s] t}.$$

Then, by definition, we have $m\{p_{rs}^{(2)}\} = 0$ if $r \neq s$,

$$m\{p_{rr}^{(2)}\} = \varepsilon \sum_{h=1}^{n} \sum_{l_1 + l_2 = 0} \frac{c_{rhl_1} c_{hrl_2}}{-\tau_h + i\, l_2\, \omega + \tau_r},$$

and, thus, $m\{p_{rr}^{(2)}\}$ may be different from zero. These terms where the mean value is different from zero are those which lead to "secular terms" (see § 8). Put $D_2 = m\{Z^{-1} \Phi\, Y_1\}$, that is, $D_2 = [d_{rs}^{(2)}]$, $d_{rs}^{(2)} = m\{p_{rs}^{(2)}\}$. The matrix D_2 is diagonal and, therefore, Z^{-1}, Z, D_2 commute. As a consequence we may write

$$[\bar{p}_{rs}^{(2)}] = Z^{-1}(\Phi - D_2)\, Y_1 = Z^{-1} \Phi\, Y_1 - Z^{-1} D_2 (Z + \varepsilon Z \int Z^{-1} \Phi Z\, d\alpha)$$
$$= Z^{-1} \Phi\, Y_1 - D_2 - \varepsilon D_2 \int Z^{-1} \Phi Z\, d\alpha,$$

and, therefore, since $m\{\Sigma\} = \Sigma m\{\ \}$, we have

$$m\{[\bar{p}_{rs}^{(2)}]\} = m\{Z^{-1} \Phi\, Y_1 - D_2\} - \varepsilon D_2\, m\{\int Z^{-1} \Phi Z\, d\alpha\}$$
$$= m\{Z^{-1} \Phi\, Y_1\} - D_2 - \varepsilon D_2\, m\{q_{rs}^{(1)}\} = 0.$$

Thus, there exists one and only one matrix

$$[q_{rs}^{(2)}] = \int Z^{-1}(t) \left(\Phi(t) - D_2\right) Y_1(t)\, dt$$

whose elements are functions of mean value zero. With this convention, put $Y_2 = Z + \varepsilon Z \int Z^{-1} (\Phi - D_2)\, Y_1\, d\alpha$. We have thus defined Y_2 in such a way that the secular terms are omitted, but it should be observed that in doing so, we have added some new terms. The reason for doing so is to obtain a solution of an equation of the form (4.5.3).

Suppose that we have repeated the above process $m - 1$ times, i.e., we have calculated successively the matrices

$$D_3 = m\{Z^{-1} \Phi\, Y_2\}, \qquad Y_3 = Z + \varepsilon Z \int Z^{-1} (\Phi - D_3)\, Y_2\, d\alpha,$$

. . . .

$$D_{m-1} = m\{Z^{-1} \Phi\, Y_{m-2}\}, \quad Y_{m-1} = Z + \varepsilon Z \int Z^{-1} (\Phi - D_{m-1})\, Y_{m-2}\, d\alpha,$$

where $D_2, \ldots, D_{m-1}$ are diagonal matrices and we shall show that the process may be continued once more.

The integrals in Y_k, $k = 1, \ldots, m-1$, are

$$[q_{rs}^{(k)}] = \int Z^{-1}(\Phi - D_k)\, Y_{k-1}\, d\alpha, \qquad q_{rs}^{(k)} = \int \sum_{t_1=1}^{n} e^{-\tau_r \alpha} (\Phi_{r t_1} - d_{r t_1})\, y_{t_1 s}^{(k-1)}\, d\alpha.$$

Let us assume that $q_{rs}^{(k)}$, $k = 1, \ldots, m-1$, has the form

$$q_{rs}^{(k)} = e^{(-\tau_r + \tau_s)t} \sum_{\substack{l=-\infty \\ l \neq 0 \text{ if } r=s}}^{+\infty} \gamma_{rsl}^{(k)}\, e^{il\omega t}, \tag{4.5.10}$$

and observe that this is true for $q_{rs}^{(1)}$. We then have

$$[p_{rs}^{(m)}] = Z^{-1} \Phi\, Y_{m-1} = Z^{-1} \Phi Z + \varepsilon Z^{-1} \Phi Z \int Z^{-1}(\Phi - D_{m-1})\, Y_{m-2}\, d\alpha,$$

and, by definition, since $\tau_u \not\equiv \tau_v \pmod{\omega i}$, $u \neq v$, we have $m\{p_{rs}^{(m)}\} = 0$ if $r \neq s$. The matrix $D_m = [d_{rs}^{(m)}] = m\{[p_{rs}^{(m)}]\} = m\{Z^{-1} \Phi\, Y_{m-1}\}$ is, therefore, a diagonal matrix. Put

$$\begin{aligned}[\bar{p}_{rs}^{(m)}] &= Z^{-1}(\Phi - D_m)\, Y_{m-1} \\ &= Z^{-1} \Phi\, Y_{m-1} - Z^{-1} D_m Z - Z^{-1} D_m Z \int Z^{-1}(\Phi - D_{m-1})\, Y_{m-2}\, d\alpha \\ &= Z^{-1} \Phi\, Y_{m-1} - D_m - \varepsilon D_m \int Z^{-1}(\Phi - D_{m-1})\, Y_{m-2}\, d\alpha.\end{aligned}$$

Moreover, we have

$$y_{rs}^{(m-1)} = \delta_{rs}\, e^{\tau_s t} + \varepsilon\, e^{\tau_r t} q_{rs}^{(m-1)} = \delta_{rs}\, e^{\tau_s t} + \varepsilon\, e^{\tau_s t} \sum_{\substack{l=-\infty \\ l \neq 0 \text{ if } r=s}}^{+\infty} \gamma_{rs}^{(m-1)}\, e^{il\omega t},$$

or,

$$y_{rs}^{(m-1)} = e^{\tau_s t}\, p_{rs}^{(m-1)}, \tag{4.5.11}$$

where $p_{rs}^{(m-1)}$ is periodic of period $T = 2\pi/\omega$, $m\{y_{rs}^{(m-1)}\} = 0$. Consequently, using (4.5.10) and (4.5.11), we see that

$$\bar{p}_{rs}^{(m)} = e^{(-\tau_r + \tau_s)t}\, Q_{rs}^{(m)} \tag{4.5.12}$$

where $Q_{rs}^{(m)}$ is periodic of period T. Therefore, $\bar{p}_{rs}^{(m)} \in C_\omega$, and since, by assumption, $m\{\int Z^{-1}(\Phi - D_m)\, Y_{m-1}\, d\alpha\} = 0$, we have

$$m\{\bar{p}_{rs}^{(m)}\} = 0, \qquad r, s = 1, \ldots, n. \tag{4.5.13}$$

Thus, there exists one and only one matrix

$$[q_{rs}^{(m)}] = \int Z^{-1}(\Phi - D_m)\, Y_{m-1}\, dt, \qquad q_{rs}^{(m)} = \int \bar{p}_{rs}^{(m)}\, dt$$

whose elements are functions contained in C_ω and of mean value zero, and the elements are of the type

$$q_{rs}^{(m)} = \sum_{\substack{l=-\infty \\ l \neq 0 \text{ if } r=s}}^{+\infty} \gamma_{rsl}^{(m)}\, e^{(-\tau_r + il\omega + \tau_s)t}.$$

We can, therefore, put $Y_m = Z + \varepsilon Z \int Z^{-1}(\Phi - D_m)\, Y_{m-1}\, d\alpha$, and, in general, define the infinite algorithm as follows:

$$\left.\begin{aligned} &D_0 = 0, \quad Y_0 = Z, \\ &D_m = m\{Z^{-1} \Phi\, Y_{m-1}\}, \\ &Y_m = Z + \varepsilon Z \int Z^{-1}(\Phi - D_m)\, Y_{m-1}\, d\alpha, \quad m = 1, 2, \ldots. \end{aligned}\right\} \tag{4.5.14}$$

The method of successive approximations just now defined is convergent (L. CESARI [4], J. K. HALE [2]), i.e., $D_m \to D$, and $Y_m \rightrightarrows Y$ uniformly with respect to t as $m \to \infty$, provided $|\varepsilon|$ is sufficiently small. Here $D = \operatorname{diag}(d_1, \ldots, d_n)$ is a diagonal matrix. By (4.5.14), as $m \to \infty$, we have

$$Y = Z + \varepsilon Z \int Z^{-1}(\Phi - D)\, Y\, d\alpha,$$

and hence, by differentiation,

$$Y' = (B - \varepsilon D + \varepsilon \Phi)\, Y.$$

It remains only to assure that we can choose the numbers $\tau_1, \ldots, \tau_n$ such that $B - \varepsilon D = A$. Here specifically we must show that the system of equations

$$f_k(\tau_1, \ldots, \tau_n; \varepsilon) \equiv \tau_k - \varepsilon\, d_k(\tau_1, \ldots, \tau_n; \varepsilon) - \varrho_k = 0, \qquad k = 1, \ldots, n, \tag{4.5.15}$$

has a unique solution $\tau_1, \ldots, \tau_n$ as functions of $\varrho_1, \ldots, \varrho_n, \varepsilon$, and also, for $|\varepsilon|$ sufficiently small, that $\tau_u \in C_u$, where $C_1, \ldots, C_n$ are the n circles mentioned at the beginning. Now it is immediately seen that $f_k(\varrho_1, \ldots, \varrho_n; 0) = 0$, $k = 1, \ldots, n$, and that the Jacobian of $f_1, \ldots, f_n$ with respect to $\tau_1, \ldots, \tau_n$, is equal to 1 for $\varepsilon = 0$, $\tau_u = \varrho_u$, $u = 1, \ldots, n$. Then, by the theorem for implicit functions in the complex field, for $|\varepsilon|$ sufficiently small there exists a solution of the system (4.5.15),

$$\tau_u = \varrho_u + \sum_{h=1}^{\infty} a_{hu}\, \varepsilon^h, \qquad u = 1, \ldots, n,$$

where a_{hu} are functions of $\varrho_1, \ldots, \varrho_n$ and these series are convergent for $|\varepsilon|$ sufficiently small and $\tau_u \in C_u$, $u = 1, \ldots, n$, for $|\varepsilon|$ sufficiently small.

A more detailed analysis shows that the elements of the s-th column of the matrix $Y(t)$ are of the form $P_{rs}(t)\, e^{\tau_s t}$, $r = 1, \ldots, n$, where the functions $P_{rs}(t)$ are periodic of period T. This assures that the n columns of $Y(t)$ are independent solutions of the given system (4.5.1), and the numbers τ_s are the characteristic exponents.

Explicit relations for the characteristic exponents are the following ones due to L. CESARI [4]:

$$\tau_r - \varepsilon\, d_r = \varrho_r, \qquad r = 1, \ldots, n, \qquad d_r = \lim_{m \to \infty} d_{rr}^{(m)}$$

where

$$\begin{aligned} d_{rr}^{(m)} = \varepsilon \sum_{p=0}^{m-2} \varepsilon^p \sum_{t_1=1}^{n} \cdots \sum_{t_{p+1}=1}^{n} \sum_{l_1 + \cdots + l_{p+2} = 0} \cdot\, c_{r t_1 l_1}^{(m)}\, c_{t_1 t_2 l_2}^{(m-1)} \cdots c_{t_{p+1}, r, l_{p+2}}^{(m-p-1)} \times \\ \times \{(-\tau_{t_1} + i(l_2 + \cdots + l_{p+2})\,\omega + \tau_r) \times \\ \times (-\tau_{t_2} + i(l_3 + \cdots + l_{p+2})\,\omega + \tau_r) \cdots (-\tau_{t_{p+1}} + i\, l_{p+2}\,\omega + \tau_r)\}^{-1}, \end{aligned}$$

$r = 1, \ldots, n$, where $c_{rsl}^{(m)} = c_{rsl}$ if $l \neq 0$, $c_{rs}^{(m)} = -d_{rs}^{(m)}$ if $l = 0$, with the convention to omit from the sums all terms for which one of the expressions in braces is zero. For other relations see the quoted paper.

For more details on the present method and extensions see L. CESARI [4], J. K. HALE [1, 2, 4, 5], R. A. GAMBILL [1, 2, 3], L. CESARI and H. R. BAILEY [1]. For its nonlinear formulation and relative bibliography see (8.6).

(b) Case $n = 2$. For $n = 2$ the systems above with ε small represent a natural generalization of the Hill, or Mathieu equation, and the overall picture described in (4.4) of possible resonance for ε small only at lattice points is a general phenomenon as the following theorems show:

(4.5. ii) Consider the differential system of order two

$$x' = A\,x + B(t, \varepsilon)\,x, \tag{4.5.16}$$

where ε is a real parameter, $A=[a_{ij}]$, $(i, j=1, 2)$ is a real constant matrix and $B(t, \varepsilon)=[b_{ij}(t, \varepsilon)]$ $(i, j=1, 2)$ is a matrix whose elements $b_{ij}(t, \varepsilon)$ are real valued functions of the real variable t, periodic in t of period $T=2\pi/\omega$, L-integrable in $[0, T]$ and each function $b_{ij}(t, \varepsilon)$ is a continuous function of ε at $\varepsilon=0$ for almost all t in $[0, T]$. Moreover, we assume $b_{ij}(t, 0)=0$ and $|b_{ij}(t, \varepsilon)|<\eta(t)$ almost everywhere in $[0, T]$, $|\varepsilon|\leq\varepsilon_0$, for some $\varepsilon_0>0$, and $\eta(t)$ L-integrable in $[0, T]$. If the characteristic roots of the matrix A are ϱ_1, ϱ_2 where either (a) $R(\varrho_j)<0$, $j=1, 2$; or (b) $\varrho_1=i\sigma$, $\varrho_2=-i\sigma$, $\sigma>0$, $m\omega\neq 2\sigma$, $m=1, 2, \ldots$, and $\int_0^T \operatorname{tr} B(t, \varepsilon)\, dt\leq 0$ for all $|\varepsilon|\leq\varepsilon_0$, then the AC solutions $x(t)$ of (4.5.16) are bounded in $[0, +\infty)$ for $|\varepsilon|$ sufficiently small (L. CESARI and J.K. HALE [1]).

(4.5. iii) Consider the second order differential equation

$$x''+b(t; \varepsilon)\, x'+a(t, \varepsilon)\, x+\sigma^2 x=0, \quad -\infty<t<+\infty \tag{4.5.17}$$

where $\sigma>0$, ε are real parameters, the functions $a(t, \varepsilon)$, $b(t, \varepsilon)$ are real functions periodic in t of period $T=2\pi/\omega$, L-integrable with respect to t in $[0, T]$ for all $|\varepsilon|\leq\varepsilon_0$, continuous function of ε at $\varepsilon=0$ for almost all t in $[0, T]$ and $a(t, 0)=b(t, 0)=0$. Moreover, assume that $\int_0^T b(t; \varepsilon)\cdot dt\geq 0$ for all $|\varepsilon|\leq\varepsilon_0$, and some $\varepsilon_0>0$, and that there exists a function $\eta(t)$, L-integrable in $[0, T]$ such that $|b(t; \varepsilon)|\leq\eta(t)$, $|a(t, \varepsilon)|\leq\eta(t)$ for almost all t in $[0, T]$ and $|\varepsilon|\leq\varepsilon_0$. If $|\varepsilon|$ is sufficiently small and $m\omega\neq 2\sigma$, $m=1, 2, \ldots$, then the (AC) solutions x of (4.5.17) are bounded with x' in $[0, +\infty)$ (L. CESARI and J.K. HALE [1]).

Proof of (4.5. ii). For the sake of simplicity we prove (4.5. ii) in the case of all $b_{ij}(t, \varepsilon)$ continuous in t and ε. Case (a) is contained in (4.5. i). Let us prove case (b). The same considerations as in (4.2) prove that the multipliers λ_1, λ_2 are the roots of the equation $\lambda^2-2A\lambda+B=0$, where $2A=x_{11}(T)+x_{22}(T)$, and $B=\det[x_{jh}(T)]$, $(j, h=1, 2)$, are continuous functions of ε. By (2.2) we have

$$B=\exp\int_0^T (a_{11}+b_{11}+a_{22}+b_{22})\, dt=\exp\int_0^T (b_{11}+b_{22})\, dt>0,$$

since $a_{11}+a_{22}=i\sigma-i\sigma=0$. For $\varepsilon=0$ we have $\lambda_1(0)=e^{i\sigma T}$, $\lambda_2(0)=e^{-i\sigma T}$, $B(0)=1$, and, if we put $A_0=A(0)$, $B_0=B(0)$, we have $e^{2i\sigma T}-2A_0 e^{i\sigma T}+1=0$, or $A_0=\cos\sigma T$. Since $m\omega\neq 2\sigma$, $m=1, 2, \ldots$, we have $\sigma T\neq m\pi$, $A_0^2-B_0=A_0^2-1<-\delta<0$ for some $\delta>0$ fixed. Since $A(\varepsilon)$, $B(\varepsilon)$ are continuous functions of ε we have $A^2(\varepsilon)-B(\varepsilon)<0$ for $|\varepsilon|$ sufficiently small. Finally, since $\lambda_1, \lambda_2=A\pm(A^2-B)^{\frac{1}{2}}$, we obtain $|\lambda_1|=|\lambda_2|=B(\varepsilon)=1$, and hence all solutions of (4.5.16) are bounded in $[0, +\infty)$. Thereby (4.5. ii) is proved. Statement (4.5. iii) is a corollary of (4.5. ii).

Theorems (4.5. ii) and (4.5. iii) show that for second order systems or equations "resonance", or "parametric instability", may occur for $|\varepsilon|$ small only at the same frequencies as for the Mathieu equation, i.e., for periodic disturbances of frequencies "close" to one of the numbers $(2\sigma/n)(1/2\pi)$. Both theorems above have been obtained in terms of the Floquet theory.

(c) *Case* $n>2$. The situation for $n>2$ may be definitely different. First of all a parametric instability may occur at far more frequencies than for $n=2$.

For instance each of the two systems of order $n=4$

$$x_1''=-\sigma_1^2 x_1+\varepsilon x_2\cos(\sigma_1\pm\sigma_2)t,\qquad x_2''=-\sigma_2^2 x_2,$$

$\sigma_1>0$, $\sigma_2>0$, $\sigma_1\neq\sigma_2$, has for every $\varepsilon\neq 0$ the corresponding particular solution $x_1=(4\sigma_1)^{-1}\varepsilon t\sin\sigma_1 t+(8\sigma_2)^{-1}(\sigma_1\mp\sigma_2)^{-1}\varepsilon\cos(\sigma_1\mp 2\sigma_2)t$, $x_2=\cos\sigma_2 t$, which is unbounded in $[0,+\infty)$. Both systems can be reduced to systems (4.5.1) by the usual transformations. These examples show that "resonance" may occur at frequencies "close" to $(\sigma_1\pm\sigma_2)(1/2\pi)$.

The following system of order $n=4$ (L. Cesari [4]) shows a more striking behavior:

$$\begin{aligned}
x_1'&=-x_2+\varepsilon x_3\cos\omega t+\varepsilon x_4\sigma_2^{-1}\sin\omega t,\\
x_2'&=\sigma_1^2 x_1+\varepsilon x_3\sigma_1\sin\omega t-\varepsilon x_4\sigma_1\sigma_2^{-1}\cos\omega t,\\
x_3'&=-x_4-\varepsilon x_1\sin\omega t+\varepsilon x_2\sigma_1^{-1}\cos\omega t,\\
x_4'&=\sigma_2^2 x_3+\varepsilon x_1\sigma_2\cos\omega t+\varepsilon x_2\sigma_2\sigma_1^{-1}\sin\omega t,
\end{aligned}$$

with ε real, $\sigma_1,\sigma_2>0$, $\omega>0$, $\omega-\sigma_1-\sigma_2\neq 0$. Its solutions can be given explicitely. Let $\delta=\omega-\sigma_1-\sigma_2$, and consider the equation $z^2-i\delta z+i\varepsilon^2=0$, or $z=\varepsilon^2(\delta+iz)^{-1}$, which has two distinct roots, one say $\alpha-i\beta$, $\alpha-i\beta\to 0$ as $\varepsilon\to 0$, with $\alpha\neq 0$ for $\varepsilon\neq 0$, and the other $i\delta-\alpha+i\beta$ which approaches $i\delta$ as $\varepsilon\to 0$. Let $\gamma_1=\beta-\sigma_1$, $\gamma_2=\beta-\sigma_2$, $\Delta_1=(-\sigma_2+\omega+\gamma_1)^2+\alpha^2$, $\Delta_2=(-\sigma_1+\omega+\gamma_2)^2+\alpha^2$, and observe that α,γ_1,γ_2 verify the relations $\alpha=\varepsilon^2\Delta_1^{-1}(-\sigma_1+\omega+\gamma_1)$, $\gamma_1+\sigma_1=\varepsilon^2\Delta_1^{-1}\alpha$, and two others which are obtained by exchanging σ_1 with σ_2, Δ_1 with Δ_2, γ_1 with γ_2. With these notations it is easy to verify that the system above has the particular solution $x_1=e^{\alpha t}\cos\gamma_1 t$, $x_2=-\sigma_1 e^{\alpha t}\sin\gamma_1 t$,

$$\begin{aligned}
x_3&=\varepsilon\Delta_1^{-1}e^{\alpha t}[(-\sigma_2+\omega+\gamma_1)\cos(\gamma_1+\omega)t-\alpha\sin(\gamma_1+\omega)t],\\
x_4&=\varepsilon\sigma_2\Delta_1^{-1}e^{\alpha t}[(-\sigma_2+\omega+\gamma_1)\sin(\gamma_1+\omega)t+\alpha\cos(\gamma_1+\omega)t].
\end{aligned}$$

Another solution can be obtained by replacing cos by sin, and sin by $-$cos. Two other solutions can be obtained by replacing α by $-\alpha$, σ_1 by σ_2, γ_1 by γ_2, Δ_1 by Δ_2, x_1, x_2 by x_3, x_4. Obviously two of these solutions are unbounded in $[0,+\infty)$ for $\varepsilon\neq 0$, and this occurs for every value of ω (for $\varepsilon=0$ all solutions are bounded). Also, every pair $(\omega,0)$, $\omega>0$, is an element of accumulation of pairs (ω',ε), $\omega'>0$, $\varepsilon\neq 0$, for which the system above has solutions which are unbounded in $[0,+\infty)$. In other words, "resonance" or "parametric instability" occurs at every frequency $\omega/2\pi$ of the small periodic disturbance (total instability). Another example of this phenomenon is given below.

Conditions have been given under which instability for ε small occurs only at discrete frequencies as for $n=2$.

(4.5. iv) Consider the system

$$x''+A(\varepsilon)\,x+\varepsilon B(t,\varepsilon)\,x=0, \quad -\infty<t<+\infty, \tag{4.5.18}$$

where ε is a real parameter, $x=(x_1,\dots,x_n)$, $A=\operatorname{diag}(\sigma_1^2,\dots,\sigma_n^2)$, $\sigma_i=\sigma_i(\varepsilon)$ continuous real functions of ε at $\varepsilon=0$; $B(t,\varepsilon)$ an $n\times n$ matrix whose elements $b_{ij}(t,\varepsilon)$ are real, periodic in t of period $T=2\pi/\omega$, L-integrable in $[0,T]$, continuous in ε at $\varepsilon=0$ and $|b_{ij}(t,\varepsilon)|<\eta(t)$ for all $|\varepsilon|\leqq\varepsilon_0$ and some $\varepsilon_0>0$, and $\eta(t)$ is L-integrable in $[0,T]$. If either (a) $B(t,\varepsilon)=B(-t,\varepsilon)$, or (b) $B(t,\varepsilon)=B_{-1}(t,\varepsilon)$; if $m\omega\neq\sigma_i(0)\pm\sigma_j(0)$, $m\omega\neq 2\sigma_i$, $i\neq j$, $i,j=1,\dots,n$, $m=0,1,\dots$, then for $|\varepsilon|$ sufficiently small the AC solutions of (4.5.18) are all bounded in $(-\infty,+\infty)$ (L. Cesari [4], J.K. Hale [2]).

For instance the system

$$x_1''+\sigma_1^2x_1=\varepsilon\,x_2\cos\omega t, \qquad x_2''+\sigma_2^2x_2=\varepsilon\,x_1\cos\omega t,$$

satisfies the hypotheses of theorem (4.5. iv) [both conditions (a) and (b)] and thus parametric instability may occur at most at frequencies "close" to $(\omega/2\pi)$ for which $m\omega=\sigma_1\pm\sigma_2$, $2\sigma_1$, $2\sigma_2$, $m=1,2,\dots$.

R.A. Gambill [1] has shown that conditions (a) and (b) can be included in the following one:

(c) $B(t,\varepsilon)=[b_{ij}]$ is the sum $B=B_0+C$ of two $n\times n$ analogous matrices B_0 and C of which B_0 is the direct sum of matrices $B_{01},\dots,B_{0j}$ of orders n_j with $n_1+\cdots+n_j=n$, each B_{0j} satisfying either condition (a), or condition (b) and C is a matrix whose terms not identically zero are all above (or all below) the ones of B_0.

Theorem (4.5. iv) holds finally also for analogous systems of the type

$$x''+A(\varepsilon)\,x+\varepsilon\,B(t,\varepsilon)\,x+\varepsilon\,C(t,\varepsilon)\,x'=0,$$

where $x=(x_1,\dots,x_n)$, $A=\operatorname{diag}(\sigma_1^2,\dots,\sigma_n^2)$, under the following condition:

(d) $B=(B_{11},B_{12};B_{21},B_{22})$, $C=(C_{11},C_{12};C_{21},C_{22})$ are $n\times n$ matrices, B_{11}, C_{11} are $\mu\times\mu$ matrices, B_{22}, C_{22} are $\nu\times\nu$ matrices, $\mu+\nu=n$, and B_{11}, B_{22}, C_{21}, C_{12} are even in t, and B_{12}, B_{21}, C_{11}, C_{22} are odd in t (J.K. Hale [4]).

The last statement includes the case (a) above, as well as other cases considered by R.A. Gambill [1], but does not include the cases (b), (c) above. Other results have been obtained by J.K. Hale [5] for linear systems of differential equations of first and second order some of which include the case of a characteristic root $\varrho=0$ of A of any multiplicity.

All these results follow from the convergent method described in (a) developed and successively modified by L. CESARI [4], J. K. HALE [2, 4, 5] and R. A. GAMBILL [1, 2]. Nevertheless, results have been proved also by direct argument and Floquet's theory (L. CESARI and J. K. HALE [1], J. K. HALE [7]) [(4.5 ii), (4.5. iii), and (4.5. v) which contains statement (d)]. Statement (c) could be obtained as a consequence of (a) and (b). See also J. MOSER [7], I. M. GELFAND and V. B. LIDSKII [1].

The theorems above show that cases of "resonance", i.e., of "parametric instability" may occur for a periodic disturbance $\varepsilon B(t, \varepsilon) x$ of frequency $\omega/2\pi$ if some multiple $m\omega$ of ω is "close" to any one of the numbers $\sigma_i \pm \sigma_j$, $2\sigma_i$, $i \neq j$, $i, j = 1, 2, \ldots, n$.

R. A. GAMBILL [2] has analyzed in detail the coefficients of the development in series of ε of the characteristic exponents [see (a)] and has obtained criteria for parametric instability. From these it is clear that if conditions similar to (a), or (b), or (c) are not satisfied a parametric instability at all frequencies is likely to occur (total instability).

R. A. GAMBILL has shown that a sequence of expressions, say M_1, $M_2, \ldots$, can be written in terms of $\varepsilon, \sigma_1, \ldots, \sigma_n$, so that any one of these being $\neq 0$ implies the phenomenon of total instability described above.

For instance the system (L. CESARI [4], R. A. GAMBILL [2])

$$x_1'' + \sigma_1^2 x_1 = \varepsilon x_2 \sin \omega t, \qquad x_2'' + \sigma_2^2 x_2 = \varepsilon x_1 \cos \omega t, \tag{4.5.19}$$

presents total instability, i.e., for every three numbers $\omega > 0, \sigma_1 > 0$, $\sigma_2 > 0$, any neighborhood of $(\sigma_1, \sigma_2, \omega)$ contains a triad $(\sigma_1', \sigma_2', \omega')$ for which (4.5.19) has unbounded solutions no matter how small $\varepsilon \neq 0$ is (even with $\sigma_1' = \sigma_1, \sigma_2' = \sigma_2$ provided a certain relation is not satisfied).

We shall now state and prove a theorem which extends (4.5. iv, d) and whose proof is quite elementary. The following few definitions are needed. A function $f(t)$ is said to be essentially even (with respect to a constant γ) if $f(t+\gamma) = f(-t)$, essentially odd if $f(t+\gamma) = -f(-t)$. As it is immediately seen the two concepts above coincide with the usual evenness and oddness when the translation $t_1 = t + \gamma/2$ is made. For real functions $f(t)$ periodic of period $T = 2\pi/\omega$, L-integrable in $[0, T]$ with Fourier coefficients a_0, a_k, b_k, $k = 1, 2, \ldots$, the following cases, among others, are of interest: $f(t)$ is essentially even with (a) $\gamma = 0$ if $f(t) = f(-t)$, (b) $\gamma = T/2$ if $a_{2k-1} = b_{2k} = 0, k = 1, 2, \ldots$, (c) $\gamma = T/4k$ if $f(t) = a(\cos k\omega t + \sin k\omega t)$; $f(t)$ is essentially odd with (a') $\gamma = 0$ if $f(t) = -f(-t)$, (b') $\gamma = T/2$ if $a_{2k} = b_{2k-1} = 0, k = 1, 2, \ldots$, (c') $\gamma = T/4k$ if $f(t) = a\cos 2k\omega t + b \sin 4k\omega t$.

Now let us consider the differential system of order $2n+1$ made up of n second order and one first order differential equations

$$\left.\begin{aligned} &Dx + B(t, \varepsilon)x + C(t, \varepsilon)x' = 0,\\ &Dx = \operatorname{col}[x_j'' + \sigma_j^2 x_j,\ j = 1, \ldots, n,\ x_{n+1}'], \end{aligned}\right\} \tag{4.5.20}$$

where $x=(x_1, x_2, \ldots, x_{n+1})$, $\sigma_j=\sigma_j(\varepsilon)>0$ are functions continuous in ε at $\varepsilon=0$, $B(t,\varepsilon)$, $C(t,\varepsilon)$ are $(n+1)\times(n+1)$ matrices whose elements are real periodic in t of period $T=2\pi/\omega$, L-integrable in $[0, T]$, continuous in ε at $\varepsilon=0$, and $|b_{jh}(t,\varepsilon)|, |c_{jh}(t,\varepsilon)|\leq\eta(t)$ for all t and $|\varepsilon|\leq\varepsilon_0$, and $\eta(t)$ is L-integrable in $[0, T]$, $b_{jh}(t,0)=0$, $c_{jh}(t,0)=0$ for all t, $c_{j,n+1}(t,\varepsilon)=0$ for all t and ε, $j, h=1,\ldots,n+1$. We shall suppose $B=[B_{uv}]$, $C=[C_{uv}]$, $u, v=1, 2, 3$, where the matrices B_{11}, C_{11} are $m\times m$, $0\leq m\leq n$, B_{22}, C_{22} are $(n-m)\times(n-m)$, and B_{33}, C_{33} are 1×1 (C_{13}, C_{23}, C_{33} are identically zero). For $\varepsilon=0$ the system above has constant coefficients with characteristic roots $\varrho_{2j-1}=i\sigma_j$, $\varrho_{2j}=-i\sigma_j$, $j=1,\ldots,n$, $\varrho_{2n+1}=0$, and we shall assume that $\varrho_j\not\equiv\varrho_h \pmod{\omega i}$, $j\neq h$, $j, h=1, 2, \ldots, 2n+1$.

(4.5. v) If all elements of the matrices B_{11}, B_{22}, B_{23}, B_{31}, C_{12}, C_{21}, C_{32} are essentially even, and all elements of the matrices B_{12}, B_{21}, B_{13}, B_{32}, B_{33}, C_{11}, C_{22}, C_{31} are essentially odd, all with respect to the same constant γ, then, for all $|\varepsilon|\leq\varepsilon_1$ and some $\varepsilon_1>0$, all solutions of (4.5.20) are bounded in $(-\infty, +\infty)$ (J. K. HALE [7]).

Proof. By the previous remark we may assume $\gamma=0$. The transformation $x_j=z_j$, $x_j'=z_{n+j}$, $j=1,\ldots,n$, $x_{n+1}=z_{2n+1}$, leads to the system

$$z'=A\,z+D(t,\varepsilon)\,z, \tag{4.5.21}$$

where $A=[A_{jh}]$, $D=[D_{jh}]$, $j, h=1, 2, \ldots, 5$, where the matrices A_{11}, D_{11}, A_{33}, D_{33} are $m\times m$, A_{22}, D_{22}, A_{44}, D_{44} are $(n-m)\times(n-m)$, and A_{55}, D_{55} are 1×1. If E_m denotes the unit matrix of type $m\times m$, we have $A_{13}=E_m$, $A_{24}=E_{n-m}$, $A_{31}=\operatorname{diag}(-\sigma_1^2,\ldots,-\sigma_m^2)$, $A_{42}=\operatorname{diag}(-\sigma_{m+1}^2,\ldots,-\sigma_n^2)$, and all other $A_{jh}=0$. We have also $D_{1h}=D_{2h}=0$, $h=1, 2, \ldots, 5$, $D_{j1}=-B_{j-2,1}$, $D_{j2}=-B_{j-2,2}$, $D_{j3}=-C_{j-2,1}$, $D_{j4}=-C_{j-2,2}$, $D_{j5}=-B_{j-2,3}$, $j=3, 4, 5$. If we replace t by $-t$ in (4.5.21) and $w_j(t)=z_j(-t)$, $w_{n+j}(t)=-z_{n+j}(-t)$, $j=1, 2, \ldots, m$, $w_j(t)=-z_j(-t)$, $w_{n+j}(t)=z_{n+j}(-t)$, $j=m+1,\ldots,n$, $w_{2n+1}(t)=-z_{2n+1}(-t)$, then (4.5.21) is transformed into an analogous system

$$w'=A\,w+D^*(t,\varepsilon)\,w, \tag{4.5.22}$$

where $w=(w_1,\ldots,w_{2n+1})$, $D^*=[D_{jh}^*]$, $j, h=1, 2, \ldots, 5$, and $D_{1h}^*=D_{2h}^*=0$, $h=1, 2, \ldots, 5$, $D_{jh}^*(t,\varepsilon)=D_{jh}(-t,\varepsilon)$ for $j+h$ even, $j=3, 4, 5$, $h=1, 2$, and $j+h$ odd, $j, h=3, 4, 5$, $D_{jh}^*(t,\varepsilon)=-D_{jh}(-t,\varepsilon)$ for $j+h$ odd, $j=3, 4, 5$, $h=1, 2$, and $j+h$ even, $j, h=3, 4, 5$. By virtue of the hypotheses we can verify one by one the relations $D_{jh}^*(t,\varepsilon)=D_{jh}(t,\varepsilon)$ for all $j, h=1, 2, \ldots, 5$. Thus system (4.5.22) coincides with system

$$z'(t)=A\,z(t)+D(t,\varepsilon)\,z(t),$$

that is, with system (4.5.21).

By Floquet's theory we know that the characteristic exponents of system (4.5.21) can be thought of as functions of ε, say $r_j(\varepsilon)$, continuous at $\varepsilon=0$, with $r_j(0)=\varrho_j$, $j=1, 2, \ldots, 2n+1$. Here we have $\varrho_{2j-1}=i\sigma_j$, $\varrho_{2j}=-i\sigma_j$, $j=1,\ldots,n$, $\varrho_{2n+1}=0$. Let $d>0$ denote the minimum of $|im\omega-(\varrho_j-\varrho_h)|$ for all $j, h=1, 2, \ldots, 2n+1$, $m=0, 1, \ldots$, $|j-h|+m>0$, and C_j the circle of center ϱ_j and radius $d/4$, $j=1, 2, \ldots, 2n+1$. We may take ε_1, $0<\varepsilon_1<\varepsilon_0$, such that $r_j(\varepsilon)\in C_j$ for all $|\varepsilon|<\varepsilon_1$, $j=1, 2, \ldots, 2n+1$. Then $|im\omega-(r_j-r_h)|\geq d-2(d/4)=d/2>0$, and hence $r_j(\varepsilon) \neq r_h(\varepsilon) \pmod{\omega i}$ for all $j \neq h$, $j, h = 1, 2, \ldots, 2n+1$, and also $J(r_j) \neq 0$, $j = 1, 2, \ldots, 2n$, and hence $r_{2j-1} = \bar{r}_{2j}$, $j = 1, 2, \ldots, n$, r_{2n+1} real. Thus (4.5.21) has its characteristic exponents all uncongruous $(\mathrm{mod}\,\omega i)$ and hence its multipliers are all distinct. Consequently (4.5.21) has a fundamental system of solutions X of the form $X(t) = [\exp(r_h t)\, p_{jh}(t)]$, where $p_{jh}(t)$ are periodic functions of period $T = 2\pi/\omega$. Thus every solution $z(t) = [z_j(t)]$ of (4.5.21) is of the type $z_j(t) = \Sigma\, m_h \exp(r_h t)\, p_{jh}(t)$, where Σ ranges over $h = 1, 2, \ldots, 2n+1$, the m_h are constants, and we may take $z(t)$ in such a way that all $m_h \neq 0$. The corresponding vector $w(t)$ is then given by $w_j(t) = \varepsilon_j \Sigma\, m_h \exp(-r_h t)\, p_{jh}(-t)$, where $\varepsilon_j = \pm 1$, and hence

$$\varepsilon_j \Sigma\, m_h \exp(-r_h t)\, p_{jh}(-t) = \Sigma\, \mu_k \exp(r_k t)\, p_{jk}(t)$$

for all t, all j, and convenient constants μ_k. By the remark at the beginning of (a) this implies that for every h there is an integer $k = k(h)$ with $-r_h = r_k$, $h = 1, 2, \ldots, 2n+1$. Since, for $|\varepsilon| \leq \varepsilon_1$, we have $r_h \in C_h$ for the same h, we must have $r_{2j-1} = -r_{2j}$ together with $r_{2j-1} = \bar{r}_{2j}$, $j = 1, 2, \ldots, n$, and $r_{2n+1} = 0$. Thus $r_{2j-1} = i\tau_j$, $r_{2j} = -i\tau_j$, $j = 1, 2, \ldots, n$, $r_{2n+1} = 0$, for all $|\varepsilon| \leq \varepsilon_1$, for some $\tau_1, \ldots, \tau_n$ real, distinct, and not zero. This assures that all solutions of system (4.5.21) are bounded in $(-\infty, +\infty)$ and the same occurs for the solutions of system (4.5.20) together with their first derivatives.

Theorem (4.5. v) holds also if $b_{n+1,j} = 0$ for all t and j, and then (4.5.20) is a system of n second order differential equations. Then, if cases (a), (a′) are verified, theorem (4.5. v) reduces to (4.5. iv, d). If the cases (b), (b′) are verified, and $c_{jh} = 0$ for all t and j, h, theorem (4.5. v) reduces to a particular theorem proved by M. Golomb [3]. Note that in veryfying that the conditions of (4.5. v) are satisfied it may be convenient to reorder first both equations and unknowns.

We shall consider now a real system (4.5.1) where Φ is periodic of period $T = 2\pi/\omega$, and the constant matrix A has μ characteristic roots with real parts negative (real or complex, distinct or not) and ν pairs of purely imaginary conjugate roots (or $\nu - 1$ such pairs and one zero root), the latter all two by two uncongruous $(\mathrm{mod}\,\omega i)$. L. Cesari and H. R. Bailey [1], by means of the method described in (a) have given criteria assuring that all solutions approach zero as $t \to +\infty$. The symmetry conditions of (4.5. iv), (4.5. v) are now replaced by inequalities. Namely explicit expressions are given $M_1, M_2, \ldots, M_\nu$, in terms of the Fourier coefficients of the elements of the matrix Φ such that the inequalities $M_1 < 0, \ldots, M_\nu < 0$ imply that all solutions approach zero as $t \to +\infty$, while any one inequality $M_j > 0$ implies that infinitely many solutions are unbounded in $[0, +\infty)$. For the sake of brevity only one of the statements and a few examples are given below.

Consider the system of one first order and n second order differential equations

$$\left.\begin{aligned} &D x + \varepsilon \Phi(t)\, x = 0, \\ &D x = \mathrm{col}\,(x_1', x_j'' + c_j x_j' + d_j x_j, \quad j = 2, \ldots, n), \end{aligned}\right\} \qquad (4.5.23)$$

where $x = (x_1, \ldots, x_n)$, Φ is an $n \times n$ matrix, $c_j > 0$, $d_j > 0$, $j = 2, \ldots, n$, ε real, $n \geq 1$, φ_{ih} real periodic functions of period $T = 2\pi/\omega$, L-integrable in $[0, T]$ with Fourier coefficients a_{jhl}, b_{jhl}, $l = 0, 1, 2, \ldots, j, h = 1, 2, \ldots, n$, $b_{jh0} = 0$.

(4.5. vi) If either $a_{110} \neq 0$, $\varepsilon a_{110} > 0$, or $a_{110} = 0$, $M < 0$, $\varepsilon \neq 0$, then for $|\varepsilon|$ sufficiently small all solutions of (4.5.23) approach zero as $t \to +\infty$. If either $a_{110} \neq 0$, $\varepsilon a_{110} < 0$, or $a_{110} = 0$, $M > 0$, $\varepsilon \neq 0$, then for $|\varepsilon|$ sufficiently small, infinitely many solutions of (4.5.23) are unbounded in $[0, +\infty)$. In both cases M is given by

$$M = \sum_{j=2}^{n} \sum_{l=0}^{\infty} (p_{jl} C_{1jl} + q_{jl} D_{1jl}),$$

$$p_{j0} = d_j^{-1}, \quad q_{j0} = 0, \quad m_{jl} = [d_j^2 + l^2 \omega^2 (c_j^2 - 2d_j) + l^4 \omega^4]^{-1},$$

$$p_{jl} = 2^{-1} m_{jl} (d_j - l^2 \omega^2), \quad q_{jl} = 2^{-1} m_{jl} l \omega c_j,$$

$$C_{1jl} = a_{1jl} a_{j1l} + b_{1jl} b_{j1l}, \quad D_{1jl} = a_{j1l} b_{1jl} - b_{j1l} a_{1jl},$$

$$j, h = 1, \ldots, n, \quad l = 0, 1, 2, \ldots.$$

For instance the system

$$x_1' + \varepsilon x_1 \cos t + \varepsilon x_2 \sin t = 0,$$

$$x_2'' + 3x_2' + 2x_2 - \varepsilon x_1 \sin t + \varepsilon x_2 \cos t = 0,$$

has $\varphi_{11} = \cos t$, $\varphi_{12} = \sin t$, $\varphi_{21} = -\sin t$, $\varphi_{22} = \cos t$, $d_2 = 2$, $c_2 = 3$, and hence $C_{121} = -1$, $D_{121} = 0$, $M = -(20)^{-1} < 0$. All solutions of this system approach zero as $t \to +\infty$ for all $|\varepsilon| \neq 0$ sufficiently smyll. For the analogous system where φ_{21} is replaced by $\varphi_{21} = \cos t$, we have $C_{121} = 0$, $D_{121} = 1$, $M = 3(20)^{-1} > 0$ and infinitely many solutions are unbounded in $[0, +\infty)$ no matter how small is $|\varepsilon| \neq 0$. As a second example the system

$$x_1'' + 2x_1 + \varepsilon x_1 \sin t + \varepsilon x_2 \cos t = 0,$$

$$x_2'' + 2x_2' + x_2 + \varepsilon x_1 \cos t + \varepsilon x_2 \cos t = 0,$$

by a criterion analogous to (4.5. vi) can be proved to have all solutions approaching zero as $t \to +\infty$ for all $|\varepsilon|$ sufficiently small. By replacing $\varphi_{21} = \cos t$ by $\varphi_{21} = \sin t$, a system is obtained which has infinitely many unbounded solutions in $[0, +\infty)$ for $|\varepsilon| \neq 0$ small (see L. Cesari and H. R. Bailey [1]). For problems with multiple roots see C. Imaz [1] who used the variant of the present method given in (8.5).

4.6. Bibliographical notes. For further research on linear systems with periodic coefficients and the concept of uniform stability see K. Persidskii [6], L. Ermolaev [1], G. Ascoli [1].

Concerning Mathieu equation and related topics excellent books exist referring in detail to the very wide literature on the subject: M. J. O. Strutt [4], N. W. McLachlan [3], and the books already mentioned by J. J. Stoker [1], and N. Minorski [6]. For further studies on the Hill and Mathieu equations see in particular: M. J. O. Strutt [1, 3, 5, 6, 7, 8], G. Horvay [1], G. Horvay and S. W. Yuan [1], A. P. Proskuryakov [1], E. L. Ince [1, 2, 3, 4], B. van der Pol and M. J. O. Strutt [1] and, also for applications, S. Goldstein [1], N. W. McLachlan [1, 2, 5, 8], S. Goldstein and N. W. McLachlan [1], T. Kato [3].

For further extensions of the Lyapunov criteria, besides those already referred to in (4.3), see A. V. Yurovskii [1], M. G. Krein [3, 4, 5], M. I. Elsin [1, 2, 3], V. I. Burdina [1, 2], A. M. Goldin [1], M. Ya. Leonov [1, 2, 3, 4, 5, 6], S. Wallach [1], M. Zlamal [1]. See also the detailed report of V. M. Starzinskii [5].

A. Wintner [14] has extended as follows the Lyapunov criterion for the equation $x'' + f(t)x = 0$, where f is supposed to be periodic of period 1. Let $p_m = m[f]$, and $|f(t)| \leq M$. Then, if $\Sigma M^n/(2n-1)! < p_m \leq 1$, or $\Sigma M^n/(2n+1)! < 4 - \mu \leq 2$, where Σ ranges on $n \geq 2$, then the solutions are all bounded and stable as $t \to +\infty$.

Results analogous to (4.5. iii) for even periodic coefficients only have been given by E. Mettler [1]. Theorems concerning the asymptotic expansion of the solutions of the same systems have been given by W. Haacke [1, 2]. For theorem of boundedness and for questions of resonance for systems with periodic coefficients as mentioned in (4.4), (4.5) see B. P. DEMIDOVIC [4], G. S. GORELIK [1, 2, 3], G. S. GORELIK and G. HINTZ [1], V. V. STEPANOV [1]. For topological properties of the family of the solutions of a linear system with periodic coefficients see N. V. ADAMOV [1—6].

Systems of the type $x' = [A + \varepsilon B(t)]x$, where A is constant matrix with characteristic roots with negative real parts, ε a small parameter, and $B(t)$ a matrix whose elements are quasi periodic functions of t, have been discussed in a long series of papers by I. Z. SHTOKALO [1—14]. The case where the elements of B are finite sums of trigonometric polynomials with incommensurable periods is particularly investigated and series expansions for the solutions are given, which are asymptotic to zero if ε is sufficiently small. On this subject, besides the classical papers of H. POINCARÉ (particularly [6]) see also P. BOHL [2], R. H. CAMERON [1, 2], J. FAVARD [1], S. BOCHNER [2]. On linear equations with almost periodic coefficients see also C. R. PUTNAM and A. WINTNER [1], A. WINTNER [1], A. HALANAY [1].

§ 5. The second order linear differential equation and generalizations

5.1. Oscillatory and nonoscillatory solutions. A great deal of research has been dedicated to the differential equation

$$x'' + f(t)\,x = 0, \tag{5.1.1}$$

or to its generalization

$$x'' + g(t)\,x' + f(t)\,x = 0. \tag{5.1.2}$$

We have already referred to many of their properties in §§ 3, 4. Equation (5.1.2) is often reduced to equation (5.1.1) by means of the usual transformation $x = y \exp\left(-2^{-1}\int_0^t g(s)\,ds\right)$ which takes (5.1.2) into $y'' + F(t)\,y = 0$ where $F = f - 4^{-1}g^2 - 2^{-1}g'$.

Let us suppose first $f(t)$ continuous for $t \geqq t_0$, and let us consider only nonzero solutions of (5.1.1). Then any two independent solutions x_1, x_2 are never zero together, and each has only zeros of multiplicity one (consequence of the theorems of uniqueness). As a consequence, the same zeros form a discrete set in every finite interval.

More precisely let us suppose that $f(t)$ is above some constant $a > 0$. Then it is known that any solution $x(t)$ of (5.1.1) (non identically zero) has infinitely many zeros for $t \geqq t_0$, say $t_1 < t_2 < t_3 < \cdots < t_n < \cdots$ and $t_n \to \infty$ as $n \to \infty$ (consequence of Sturm comparison theorem; see, e.g., G. SANSONE [16], Vol. 2, or E. L. INCE [5], Chap. 10). Also, between any two consecutive zeros t_{n-1}, t_n of $x(t)$ there is one and only one zero, say t'_n, of $x'(t)$, where $|x(t)|$ takes its maximum, say m_n in $[t_{n-1}, t_n]$, and $x(t)$ has alternate signs in the intervals (t_{n-1}, t_n). On the other hand each point t_n is a maximum for $|x'(t)|$ in $[t'_n, t'_{n+1}]$, say μ_n, and $x'(t)$ has alternate signs in the intervals (t'_{n-1}, t'_n). Finally if we suppose that $f(t)$ is also monotone nondecreasing in $[t_0, +\infty)$, then $m_1 \leqq m_2 \leqq \cdots$, $\mu_1 \geqq \mu_2 \geqq \cdots$; if we suppose that $f(t)$ is monotone nonincreasing in

$[t_0, +\infty)$, then $m_1 \geq m_2 \geq \cdots, \mu_1 \leq \mu_2 \leq \cdots$. Thus for $f(t)$ monotone, also the sequences $[m_n]$, $[\mu_n]$ are monotone and it is proved that their limits $m_n \to m$, $\mu_n \to \mu$ are finite (G. SANSONE [16], Vol. 2, p. 25) and, if $f(t) \to \alpha$, then $\mu = \alpha m$. This last result holds even if $f(t)$ is of bounded variation in $[t_0, +\infty)$, or if $\log A(t)$ is of bounded variation in some $[t_0, +\infty)$. In particular, they hold if $f(t)$ is AC in $[t_0, +\infty)$ and $\int^{+\infty} |f'(t)|\, dt < +\infty$. For all results above and others we may refer to A. KNESER [1, 3], M. PICONE [1], G. ASCOLI [1, 2], B. GAMBIER [1], T. SATO [1], R. CACCIOPPOLI [1], W. M. CHEPELEFF [1]. A general exposition is given in G. SANSONE ([16], Vol. 2, p. 25—33).

We have already mentioned (3.9) that if $\varphi(t)$ is a measurable function with $\int^{+\infty} |\varphi(t)|\, dt < +\infty$, and the solutions of the equation $x'' + f(t)\, x = 0$ are bounded with their first derivatives, then the same holds for the solutions of $x'' + [f(t) + \varphi(t)]\, x = 0$. From all the statements above it follows, for instance, that the solutions of the equation $x'' + (A^2 + b/t + m(t)/t^2)\, x = 0$ are all bounded with their first derivatives as $t \to +\infty$ where A, b are real constant and $m(t)$ is bounded. More generally, all solutions of the equation $x'' + [a^2 + \varphi(t) + \psi(t)]\, x = 0$ are bounded with their first derivatives provided $\int^{+\infty} |\psi|\, dt < +\infty$, $\varphi(t)$ is of bounded variation in $[t_0, +\infty)$, and $\varphi(t) \to 0$ as $t \to +\infty$.

The following simple criterion is due to L. A. GUSAROV [1]: If the AC function $f(t)$ has derivative $f'(t)$ of bounded variation in $[t_0, +\infty)$, and $0 < a^2 \leq f(t) \leq b^2 < +\infty$ for some constants a, b, then all solutions of (5.1.1) are bounded in $[t_0, +\infty)$ with their first derivatives. U. BARBUTI [3] has extended these criteria as follows:

(5.1. i) If $f(t) = g(t) + \varphi(t) + \psi(t)$, where $0 < a^2 \leq g(t) \leq b^2 < +\infty$, $\varphi(t) \to 0$ as $t \to +\infty$, $g(t)$ is AC in every finite interval, $g'(t)$ and $\varphi(t)$ are of bounded variation in $[t_0, +\infty)$, and $\int^{+\infty} |\psi(t)|\, dt < +\infty$, then all solutions of (5.1.1) are bounded in $[t_0, +\infty)$ with their first derivatives.

We refer to U. BARBUTI [3] for the proof and for a comparison of these criteria. Other results will be mentioned in (5.4).

The following statements are also of interest. A function $f(t)$, $t_0 \leq t < +\infty$, is said to be of class L^p if $\int^{+\infty} |f|^p\, dt < +\infty$. If $|f(t)| < M$ and all solutions of equation (5.1.1) belong to the class L^2, then all solutions of the equation $x'' + [f(t) + \psi(t)]\, x = 0$ belong to L^2 provided $|\psi(t)| < c$ for all $t \geq t_0$ and some constant $c > 0$ (T. CARLEMAN [2], H. WEYL [2], R. BELLMAN [2]). Extensions are known for classes L^p (R. BELLMAN [2]).

5.2. FUBINI's theorems. Concerning the asymptotic behavior of the solutions of (5.1.1) the following considerations due to G. FUBINI [1] shall be mentioned. If we suppose that $f(t) \to a^2 > 0$, and we replace the independent variable t by $t = t_1/a$, equation (5.1.1) is transformed into

$$x'' + \big(1 - q(t)\big)\, x = 0, \tag{5.2.1}$$

where $q(t) \to 0$ as $t \to +\infty$. It is easy to verify that the expression $x = Z_1(t)\, e^{it} + Z_2(t)\, e^{-it}$ satisfies (5.2.1) if $Z_1(t), Z_2(t)$ satisfy the differential system

$$Z_1' = (2i)^{-1}[Z_1 + Z_2 e^{-2it}]\, q(t), \quad Z_2' = -(2i)^{-1}[Z_2 + Z_1 e^{2it}]\, q(t),$$

and that solutions of this system verifying the conditions $Z_1(t) \to \alpha_1$, $Z_2(t) \to \alpha_2$ as $t \to +\infty$ (if they exist) should verify the system of linear integral equations

$$\left.\begin{aligned} Z_1(t) &= \alpha_1 + (2i)^{-1} \int_{+\infty}^{t} [Z_1(\tau) + Z_2(\tau)\, e^{-2i\tau}]\, q(\tau)\, d\tau \\ Z_2(t) &= \alpha_2 - (2i)^{-1} \int_{+\infty}^{t} [Z_2(\tau) + Z_1(\tau)\, e^{2i\tau}]\, q(\tau)\, d\tau. \end{aligned}\right\} \tag{5.2.2}$$

Under the additional hypothesis that $\int^{+\infty} |q(t)|\, dt < +\infty$, the usual theory for integral equations can be applied to (5.2.2) and thus the Liouville method of successive approximations leads to a solution of (5.2.1) of the form

$$x = e^{it}\Big[\alpha_1 + \sum_{n=1}^{\infty} u_n(t)\Big] + e^{-it}\Big[\alpha_2 + \sum_{n=1}^{\infty} v_n(t)\Big], \tag{5.2.3}$$

where u_n, v_n are defined by

$$\left.\begin{aligned} u_0(t) &= \alpha_1, \quad u_n(t) = (2i)^{-1} \int_{+\infty}^{t} [u_{n-1}(\tau) + v_{n-1}(\tau)\, e^{-2i\tau}]\, q(\tau)\, d\tau \\ v_0(t) &= \alpha_2, \quad v_n(t) = -(2i)^{-1} \int_{+\infty}^{t} [v_{n-1}(\tau) + u_{n-1}(\tau)\, e^{2i\tau}]\, q(\tau)\, d\tau. \end{aligned}\right\} \tag{5.2.4}$$

If $\alpha = \max(|\alpha_1|, |\alpha_2|)$ and ε, $0 < \varepsilon < 1$, any given number, we may determine a $c > 0$ such that $\int_t^{+\infty} |q(t)|\, dt < \varepsilon$ for all $t \geq c$. Then we may prove by induction that $|u_n(t)| \leq \alpha\varepsilon^n$, $v_n(t) \leq \alpha\varepsilon^n$, for all $t \geq c$, $n = 1, 2, \ldots$. Thus series (5.2.3) converges uniformly in $[c, +\infty)$ and its sum satisfies (5.2.1). A more rapid convergence is assured if $|q(t)| \leq m t^{-1-a}$, m, a positive constants. Then we have $|u_n(t)|, |v_n(t)| \leq \alpha (m a t^{-a})_n (n!)^{-1}$. The considerations above prove

(5.2. i) If $\int^{+\infty} |q(t)|\, dt < +\infty$, then given any two arbitrary constants α_1, α_2 there exists a solution $x(t)$ of (5.2.1) behaving as $\alpha_1 e^{it} + \alpha_2 e^{-it}$ as $t \to +\infty$ and $x(t)$ is given by (5.2.3) in some interval $[c, +\infty)$ (G. Fubini [1]).

The series (5.2.3) is of interest. For instance, if we consider the Bessel equation

$$(d^2 w/dz^2) + z^{-1}(dw/dz) + (h^2 - n^2 z^{-2})\, w = 0, \tag{5.2.5}$$

h, n constants, $h \neq 0$, and we put $x = w z^{\frac{1}{2}}$, $t = hz$, we obtain equation (5.2.1) with $q(t) = kt^{-2}$, $k = (4n^2 - 1)/4$, and the considerations above apply with $a = 1$.

The equation (5.1.1) when $f(t) \to -a^2 < 0$ can be reduced to the equation

$$x'' - (1 + q(t))\, x = 0, \tag{5.2.6}$$

where $q(t) \to 0$ as $t \to +\infty$. Under the additional hypothesis that $\int^{+\infty} |q(t)|\, e^{2t}\, dt < +\infty$, the solutions of equations (5.2.6) can be developed in an asymptotic expansion

$$x = e^t [\alpha_1 + \Sigma u_n(t)] + e^{-t} [\alpha_2 + \Sigma v_n(t)],$$

where α_1, α_2 are constant, and u_n, v_n are defined by an inductive process analogous to the one above (G. Sansone [16], Vol. 2, p. 51).

The equation (5.2.1) when $f(t) \to 0$ is asymptotic to the equation $y'' = 0$. The solutions $x(t)$ do not behave always as the solutions $\alpha_1 t + \alpha_2$ of $y'' = 0$, as can be shown by examples. For instance, the equation

$$x'' + [(1/4t) + (3/16t^2)]\, x = 0$$

has the general solution $C_1 t^{\frac{1}{4}} \cos \sqrt{t} + C_2 t^{\frac{1}{4}} \sin \sqrt{t}$. The condition $|f(t)| < c\, t^{-\alpha}$ for all $t \geq t_0$ where $c > 0$, $\alpha > 3$ are constant, assures that the solutions $x(t)$ of $x'' + f(t)\, x = 0$ have the asymptotic developments

$$x(t) = t [\alpha_1 + \Sigma u_n(t)] + [\alpha_2 + \Sigma v_n(t)],$$

analogous to the previous ones (G. Sansone [16], Vol. 2, p. 46—48). Here are other theorems assuring that the solutions of $x'' + f(t)\, x = 0$ behave as the ones of $y'' = 0$. If $\int^{+\infty} t\, |f(t)|\, dt < +\infty$, then $x'(t)$ has a finite limit as $t \to +\infty$, and the general solution of (5.1.1) is asymptotic to $c_0 t + c_1$, for some c_0, c_1 constants (O. Haupt [1], E. Hille [1]).

The same occurs if $\int^{+\infty} f(t)\, dt$ is convergent and $\int^{+\infty} \varphi(t)\, dt < +\infty$, where $\varphi(t) = \max \left| \int_v^{+\infty} f(u)\, du \right|$ for $t \leq v < +\infty$ (A. Wintner). Thus under these hypotheses the solutions of (5.1.1) are nonoscillatory. For instance the solutions of $x'' + t^{-3} x = 0$ are nonoscillatory.

It may be pointed out here that the equation $x'' + m t^{-2} x = 0$ has nonoscillatory solutions if $m < \frac{1}{4}$, has oscillatory solutions if $m > \frac{1}{4}$; namely $x = c_1 t^{\alpha_1} + c_2 t^{\alpha_2}$, α_1, α_2 real and positive if $m < \frac{1}{4}$; or $x = c_1 t^{\frac{1}{2}} \cdot \cos \beta \log t + c_2 t^{\frac{1}{2}} \sin \beta \log t$, β real and positive, if $m > \frac{1}{4}$ (c_1, c_2 arbitrary constants).

In the case $f(t) < 0$ we may write equation (5.1.1) in the form $x'' = A(t)\, x$ with $A(t) > 0$. Then it is easy to show that if at a point t_0 we have $x(t) \geq 0$, $x'(t) > 0$, then $x(t) \to +\infty$, $x'(t) \to +\infty$ as $t \to +\infty$

where $x(t)$ is increasing and positive for $t \geq t_0$. If at a point t_0 we have $x(t) \leqq 0$, $x'(t) < 0$, then $x(t) \to -\infty$, $x'(t) \to -\infty$ as $t \to +\infty$, where $x(t)$ is decreasing and negative for $t \geq t_0$. A more stringent analysis shows that if $\int^{+\infty} A(t)\,dt < +\infty$, then for every pair (α, β) there is one and only one solution $x(t)$ with $x(\alpha) = \beta$ and $x(t) \to 0$ as $t \to +\infty$, while for all other solutions with $x(\alpha) = \beta$ either $x(t) \to +\infty$, $x'(t) \to +\infty$, or $x(t) \to -\infty$, $x'(t) \to -\infty$ as $t \to +\infty$. A proof on the line of the previous ones is given by G. SANSONE ([16], Vol. 2, p. 49). For a more general statement see A. MAMBRIANI [1].

5.3. Some transformations. In addition to the transformation discussed at the beginning of (5.1), the following ones are of interest:

(a) Given the equation $x'' \pm f^2(t)\,x = 0$, $f(t) > 0$, $(') = d/dt$, let $s = \int_0^t f(t)\,dt$. Then the equation is transformed into the equation $x'' + [f'(t)/f^2(t)]\,x' \pm x = 0$, where now $(') = d/ds$ [Liouville transformation]. By applying now the transformation of (5.1), i.e., in the present case, $y = x f^{\frac{1}{2}}(t)$, we have the equation $d^2y/ds^2 + \{\pm 1 - 2^{-1}(d/ds)\,[f'(t)/f^2(t)] - 4^{-1}[f'(t)/f^2(t)]^2\}\,y = 0$.

(b) By applying $x = \exp\left(\int_0^t y\,dt\right)$, equation (5.1.1) is transformed into the Riccati equation $y' + y^2 + f(t) = 0$.

(c) By putting $y_1 = x$, $y_2 = x'$, equation (5.1.1) is transformed into the system $y_1' = y_2$, $y_2' = -f(t)\,y_1$, and by putting $y_1 = \varrho\cos\theta$, $y_2 = \varrho\sin\theta$, we have

$$\varrho'/\varrho = 2^{-1}[1 - f(t)]\sin 2\theta, \quad \theta' = -2^{-1}[1 + f(t)] + 2^{-1}[1 - f(t)]\cos 2\theta.$$

For some applications of the Liouville transformation see S. A. SCHELKUNOFF [1], and L. BRILLOUIN [2].

5.4. BELLMAN's and PRODI's theorems. By using transformation (b) and by applying a method of successive approximations to the ensuing Riccati equation the following two theorems have been proved:

(5.4. i) (R. BELLMAN [11]) If $f(t) \to 0$ as $t \to +\infty$; if $\int f^2(t)\,dt < +\infty$, there are two solutions x_1, x_2 of the equation $x'' - [1 + f(t)]\,x = 0$ having the following asymptotic form

$$x_i = \exp(-1)^i \left[t + 2^{-1}\int_0^t f(t)\,dt + o(1)\right], \quad i = 1, 2.$$

(5.4. ii) (R. BELLMAN [11]) If $f(t) \to 0$ as $t \to +\infty$, and if $\int^{+\infty} |f^3(t)|\,dt < +\infty$, then there are two solutions x_1, x_2 of the equation $x'' - [1 + f(t)]\,x = 0$ having the asymptotic form

$$x_i = \exp(-1)^i \left[t + 2^{-1}\int_0^t f(\alpha)\,d\alpha - 4^{-1}\int_0^t f(\alpha)\,d\alpha \int_0^\alpha f(\beta)\,d\beta\, e^{2\alpha - 2\beta} + o(1)\right],$$
$$i = 1, 2.$$

Both statements (5.4. i) and (5.4. ii) had been obtained by the method mentioned above and under the additional hypothesis $f(t) \to 0$ by R. BELLMAN [11]. The present slightly more general form has been obtained by the same method and a more elaborate analysis by G. ASCOLI [8]. Statement (5.4. i) had been previously obtained also by P. HARTMANN [5] (under similar hypotheses and analogous method) and by A. WINTNER [16] under more restrictive conditions.

Concerning the equation $x'' + [1 + f(t)]\,x = 0$ the following theorems hold

(5.4. iii) If $f(t) \to 0$, if the integrals $\int_{t_0}^{t} f(\alpha)\,d\alpha$, $\int_{t_0}^{t} f(\alpha)\sin 2\alpha\,d\alpha$, $\int_{t_0}^{t} f(\alpha)\cos 2\alpha\,d\alpha$, are bounded, if $\int_{t_0}^{t} \left| f(\alpha) \int_{\alpha}^{t} \sin(t-\alpha)\sin(\alpha-\beta)\,f(\beta)\,d\beta \right| d\alpha < K$ for all $t \geq t_0$ and some t_0, then all solutions of the equation $x'' + [1 + f(t)]\,x = 0$ are bounded with their first derviatives (G. PRODI [2]).

(5.4. iv) If $f(t) = a(t)\,b(t)$, $a(t)$ monotone with $\int^{+\infty} a^2(t)\,dt < +\infty$, if $\int_{t}^{s} b(\alpha)\sin 2\alpha\,d\alpha$, $\int_{t}^{s} b(\alpha)\cos 2\alpha\,d\alpha$ are bounded for all t and s, $t_0 \leq t < s$, then all solutions of the equation $x'' + [1 + f(t)]\,x = 0$ are bounded with their first derivatives (U. BARBUTI [1]).

For other research on the same subject see G. ASCOLI [1, 2, 3, 7, 8], U. BARBUTI [2, 3]. In particular, U. BARBUTI [2] has given conditions assuring that some solutions of the same equation considered in (5.5. iv) are unbounded in $[0, +\infty)$.

5.5. The case $f(t) \to +\infty$. The case where $f(t) \to +\infty$ has been also widely studied G. PRODI [1] has recently proved that if $f(t)$ is nondecreasing and $f(t) \to +\infty$ as $t \to +\infty$, then equation (5.1.1) admits always of at least one solution $x(t)$ (not identically zero) with $x(t) \to 0$ as $t \to +\infty$. Concerning the question as to whether all solutions of (5.1.1) have this property positive results are known under weak conditions of regular growth of $f(t)$. A first result of A. WIMAN [2] reads as follows:

(5.5. i) If $f(t)$, $t_0 \leq t < +\infty$, is continuous and increasing in $[t_0, +\infty)$ with continuous derivative and $f(t) > 0$, $f'(t) > 0$, $f(t) \to +\infty$ as $t \to +\infty$; if given any two constants k, $m > 0$ there is a t_1, $t_0 \leq t_1 < +\infty$, such that for all $|h| \leq k/A^{\frac{1}{2}}(t)$ we have

$$|f(t \pm h) - f(t)| \leq m\,|f(t)|, \qquad |f'(t \pm h) - f(t)| \leq m\,|f'(t)|, \qquad (5.5.1)$$

then for every solution $x(t)$ of (5.1.1) we have $\lim x(t) = 0$, $\overline{\lim}\,|x'(t)| = +\infty$ as $t \to +\infty$. For the zeros t_n of $x(t)$ we have $t_n \to 0$, $t_n - t_{n-1} \to 0$ as $n \to \infty$.

Under these hypotheses we have $\lim x(t) = 0$, $\overline{\lim}\,|x(t)| > 0$, as $t \to +\infty$ for every nonzero solution $x(t)$ of (5.1.1). Thus the solution $x = 0$ of (5.1.1) is certainly not stable in the sense of LYAPUNOV, though it is stable in the sense considered in (1.5), and in the sense of ROUTH (1.5).

The functions $c\,t^n$, $n > -2$, satisfy the condition of growth given in (5.5. i), but more general conditions have been devised. According to G. ARMELLINI [1], L. TONELLI [4], G. SANSONE [18], a concept of regular growth can be introduced as follows. Let θ_n denote a sequence $\theta_n = [a_1, a_2, \ldots]$ of points $t_0 \leq a_n < a_{n+1}$, with $a_n \to +\infty$ as $n \to \infty$. Let $f(t)$ be a positive, nondecreasing function of t in $[t_0, +\infty)$ with $f(t) \to +\infty$. We shall say that $f(t)$ *grows intermittently* in $[t_0, +\infty)$ if for every $\varepsilon > 0$ there is a sequence θ_n with

$$\sum_{n=1}^{\infty} [f(a_{2n+1}) - f(a_{2n})] < +\infty, \qquad \overline{\lim_{n \to +\infty}} \sum_{k=1}^{n} (a_{2k} - a_{2k-1}) \big/ a_n \leq \varepsilon. \qquad (5.5.2)$$

If this does not occur then we say that $f(t)$ *grows regularly* in $[t_0, +\infty)$. The last condition (5.5.2) can be expressed by saying that the sequence of the odd intervals (a_{2k-1}, a_{2k}), $k = 1, 2, \ldots$, has (upper) density $\leq \varepsilon$ in $[t_0, +\infty)$.

Then the following theorem holds:

(5.5. ii) If $f(t)$, $t_0 \leq t < +\infty$, is positive, nondecreasing, with a continuous derivative, and $f(t) \to +\infty$, if $\log f(t)$ grows regularly in $[t_0, +\infty)$, then for every solution $x(t)$ of (5.1.1) we have $x(t) \to 0$ as $t \to +\infty$ (G. ARMELLINI [1], L. TONELLI [4], G. SANSONE [18]). Theorem (5.3. ii) was stated by G. ARMELLINI [1] and then proved independently by L. TONELLI [4] and G. SANSONE [18].

Other independent conditions have been given by G. SANSONE [18], a corollary of one of which reads as follows:

(5.5. iii) If $f(t)$, $t_0 \leqq t < +\infty$, is positive, nondecreasing, with a continuous derivative, and $\liminf f'(t) \geqq A > 0$; if $\int^{+\infty} dt/f(t) = +\infty$, then for every solution $x(t)$ of (5.3.1) we have $x(t) \to 0$ as $t \to +\infty$.

In all cases much more detailed information has been obtained concerning both the behavior of $x(t)$, and of the zeros of $x(t)$ and $x'(t)$. Besides the articles already quoted, see M. BIERNACKI [1], H. MILLOUX [1], A. WIMAN [1, 2], Z. BUTLEWSKI [3]. See, in particular, A. MEIR, D. WILLETT, and J. S. W. WONG, Michigan Math. J. **14**, 47–52 (1967).

For some research on the solutions of the equation $x^{(n)} + f(t)\,x = 0$ for $n > 2$, see M. BIERNACKI [2, 4, 5].

M. BIERNACKI [1] proved that if $f(t)$ is positive continuous nondecreasing and $f(t) \to +\infty$, then $(t_n' - t_n)/(t_{n+1} - t_n)$ approaches $\frac{1}{2}$ as n, avoiding certain "exceptional" values, tends to ∞. A. BIELECKI [1] gave an example which shows that this statement does not hold for $n \to \infty$ without restrictions.

5.6. Solutions of class L^2. In the following lines we shall consider a differential equation of the form

$$L\,x + \lambda\,x = 0, \tag{5.6.1}$$

where λ is a complex parameter, L a (selfadjoint) differential operator of the form $Lx = -(px')' + qx$, and we will suppose that $p(t)$, $q(t)$ are real-valued continuous functions in $[0, +\infty)$, and that $p(t)$ is positive and has a continuous derivative there.

A solution $x(t)$ of (5.6.1) is said to be of class L^2 if $\int_0^{+\infty} |x(t)|^2\,dt < +\infty$. We may say also that $x(t)$ satisfies condition L^2 at infinity. If, for a given L, (5.6.1) has two independent solutions both of class L^2, then every solution of (5.6.1) for the same λ, is of class L^2, as it is immediately seen.

The following theorem is particularly important

(5.6. i) If, for some λ_0, all solutions of (5.6.1) are of class L^2, then for every λ, all solutions of (5.6.1) are of class L^2.

Proof. If $u(t)$, $v(t)$, $0 \leqq t < +\infty$, are two independent solutions of (5.6.1) for $\lambda = \lambda_0$, then, by multiplying one of them by a constant, if necessary, we may suppose that $[u(0)\,v'(0) - u'(0)\,v(0)]\,p(0) = 1$. Then by writing (5.6.1) in the form $Lx + \lambda_0 x = (\lambda_0 - \lambda)\,x$, and as an application of the formulas (2.2.2), (2.2.4), we have, for every $\lambda \neq \lambda_0$ and $t \geqq 0$,

$$x(t) = c_1 u(t) + c_2 v(t) + (\lambda_0 - \lambda) \int_a^t [u(t)\,v(\tau) - u(\tau)\,v(t)]\,x(\tau)\,d\tau, \tag{5.6.2.}$$

where a is any constant value of t, and c_1, c_2 arbitrary constants. By SCHWARZ inequality we have, for $t \geqq \alpha$,

$$\left| \int_a^t [u(t)\,v(\tau) - u(\tau)\,v(t)]\,x(\tau)\,d\tau \right| \leqq M\big(|u(t)| + |v(t)|\big) \left(\int_a^t |x(\tau)|^2\,d\tau \right)^{\frac{1}{2}},$$

provided a is chosen so large that $\int_a^{+\infty} |u|^2\,d\tau \leqq M^2$, $\int_a^{+\infty} |v|^2\,d\tau \leqq M^2$. By (5.6.2) and

MINKOWSKI inequality we have now

$$\left(\int_a^t |x|^2\,d\tau\right)^{\frac{1}{2}} \leq (|c_1|+|c_2|)\,M + 2\,|\lambda-\lambda_0|\,M^2\left(\int_a^t |x|^2\,d\tau\right)^{\frac{1}{2}}.$$

If a is chosen so large that $2\,|\lambda-\lambda_0|\,M^2 < \frac{1}{2}$, we conclude that

$$\left(\int_a^t |x|^2\,d\tau\right)^{\frac{1}{2}} \leq 2\,(|c_1|+|c_2|)\,M,$$

for every $t \geq a$. Thus $x(t)$ is of class L^2.

Statement (5. 6. i) implies that either (a) for every λ all solutions of (5.6.1) are of the class L^2, or (b) for every λ at most one nontrivial solution of (5.6.1) is of class L^2.

(5.6. ii) In the alternative (b) for every λ with $J(\lambda) \neq 0$, there is exactly one nontrivial solution of (5.6.1) which is of class L^2.

We omit the proof of (5.6. ii).

For every fixed α, $0 \leq \alpha < \pi$, we shall now denote by $u(t;\lambda,\alpha)$, $v(t;\lambda,\alpha)$ the solutions of (5.6.1) defined by the following initial conditions

$$u(0)=\sin\alpha,\quad p(0)\,u'(0) = -\cos\alpha;\quad v(0)=\cos\alpha,\quad p(0)\,v'(0)=\sin\alpha. \tag{5.6.3}$$

Then every solution $x(t;\lambda)$ of (5.6.1) [up to a multiplicative constant and with exception of $v(t;\lambda)$] can be written in the form $x(t;\lambda)=u+mv$ for some $m=m(\lambda)$. For every number $b>0$ we shall now consider also the boundary condition at $t=b$,

$$\cos\beta\cdot x(b) + \sin\beta\cdot p(b)\,x'(b) = 0, \tag{5.6.4}$$

where β is a given number, $0 \leq \beta < \pi$.

Once α and β are fixed we may ask for what value of m the solution $x(t) = u+mv$ satisfies condition (5.6.4). It is immediately found

$$m = -\,[\cot\beta\,u(b) + p(b)\,u'(b)]\,[\cot\beta\,v(b) + p(b)\,v'(b)]^{-1}.$$

If $z=\cot\beta$, and α,β,λ,b are fixed, then m is given in the form

$$m = -\,(A\,z+B)/(C\,z+D).$$

Thus when z describes the real axis, m describes a circumference C_b in the complex m-plane. It can be proved that, if $b<b'$, then C_b contains $C_{b'}$ in its interior. Thus, if $b\to+\infty$, C_b has a limit which is either a proper circle, say C_∞, or a single point m_∞.

(5.6. iii) In the alternative (a) discussed before we have $C_b\to C_\infty$; in the alternative (b) we have $C_b\to m_\infty$ as $b\to+\infty$.

We omit the proof of (5.6. iii). Because of this theorem, in the alternative (a) it is said that the operator L [or equation (5.6.1)] is of the *limit-circle type at infinity*, and in the alternative (b) it is said that L is of the *limit-point type at infinity* (H. WEYL [2]). There are very simple criteria which assure that a given opertor L is of the limit-point type at infinity. A few of them are given below.

(5.6. iv) If $L=-(p\,x')'+q\,x$, and, for some positive function $M(t)$ and positive constants k_1, k_2, we have, for all large t,

$$q(t) \geq -\,k_1 M(t),\qquad \int_t^{+\infty}(p\,M)^{-\frac{1}{2}}\,dt = +\infty,\qquad |p^{\frac{1}{2}}(t)\,M'(t)\,M^{-\frac{3}{2}}(t)| \leq k_2,$$

then L is of the limit-point type at infinity.

The same occurs if either $q(t) \geq -k$ and $\int_0^{+\infty} p^{-\frac{1}{2}}(t)\,dt = +\infty$, or if $p(t) = 1$, $q(t) \geq -kt^2$.

5.7. Parseval relation for functions of class L^2. If we consider now the differential equation (5.6.1) with $p(t)$, $q(t)$ real, together with the boundary conditions

$$\sin\alpha\, x(0) - \cos\alpha\, p(0)\, x'(0) = 0, \quad \cos\beta\, x(b) + \sin\beta\, p(b)\, x'(b) = 0, \tag{5.7.1}$$

where $0 \leq \alpha, \beta < \pi$, and $0 < b$, then, since L is self-adjoint, it is known that there is a sequence λ_{bn}, $n = 1, 2, \ldots$, of real eigenvalues and a corresponding complete system $\theta_{bn}(t)$, $n = 1, 2, \ldots$, of orthogonal normalized eigenfunctions. Then for $\lambda = \lambda_{bn}$, the function $v(t; \lambda_{bn})$ does satisfy the first of the conditions (5.7.1) and hence $\theta_{bn}(t) = c_{bn} v(t; \lambda_{bn})$ for some nonzero constant c_{bn}. Then the usual Parseval relation assures that

$$\int_0^b |f(t)|^2\,dt = \sum_{n=1}^{\infty} |c_{bn}|^2 \left| \int_0^b f(t)\, v(t; \lambda_{bn})\,dt \right|^2, \tag{5.7.2}$$

for every function $f(t)$, $a \leq t \leq b$, under the usual condition of L^2-integrability of $f(t)$ in $[a, b]$.

We may suppose $f(t)$ defined in $[0, +\infty)$ and zero in $[b, +\infty)$. We shall denote by $g(\lambda)$ the function $g(\lambda) = \int_0^{+\infty} f(t)\, v(t; \lambda)\,dt$, and by $\varrho_b(\lambda)$ the monotone nondecreasing step function of λ, zero at the origin, having at the points $\lambda = \lambda_{bn}$ a right jump given by $|c_{bn}|^2$, and constant otherwise. Then (5.7.2) can be written in the form

$$\int_0^{+\infty} |f(t)|^2\,dt = \int_{-\infty}^{+\infty} |g(\lambda)|^2\,d\varrho_b(\lambda), \tag{5.7.3}$$

and $\varrho_b(\lambda)$ is the so called spectral function for the operator L with the boundary conditions (5.7.1). It has been proved that as $b \to +\infty$ the function $\varrho_b(\lambda)$ converges (for all but countably many λ) toward a monotone nondecreasing function of λ, $\varrho(\lambda)$, which is not necessarily a step function. The function $\varrho(\lambda)$ can be thought of as the spectral function for a boundary value problem in the infinite interval $[0, +\infty)$ with condition L^2 at infinity replacing the usual condition at $t = b$. We mention here only the following statement

(5.7. i) If L is of the limit-point type at infinity, if $f(t)$ is any function of the class L^2, then there exists a function $g(\lambda)$, $-\infty < \lambda < +\infty$, such that

$$\lim_{b\to\infty} \int_{-\infty}^{+\infty} \left| g(\lambda) - \int_0^b f(t)\, v(t; \lambda) \right|^2 dt = 0 \tag{5.7.4}$$

and the Parseval type relation holds

$$\int_0^{+\infty} |f(t)|^2\,dt = \int_{-\infty}^{+\infty} |g(\lambda)|^2\,d\varrho(\lambda).$$

In addition the identity holds

$$\lim_{a\to+\infty} \int_0^{+\infty} \left| f(t) - \int_{-a}^{a} g(\lambda)\, u(t; \lambda)\,d\varrho(\lambda) \right|^2 dt = 0. \tag{5.7.5}$$

If we write relations (5.7.4) and (5.7.5) as

$$g(\lambda) = \int_0^{+\infty} f(t)\, u(t; \lambda)\,dt, \quad f(t) = \int_{-\infty}^{+\infty} g(\lambda)\, u(t; \lambda)\,d\varrho(\lambda),$$

where the convergence and the equalities have to be understood in the sense explained above, then the similarity of the present situation with Fourier expansion and corresponding completeness theorem becomes evident.

We shall now denote by *spectrum* S the set S of all values λ at which $\varrho(\lambda)$ is not constant. Then the spectrum S of an operator L is a closed set of real numbers λ, and S may well contain isolated points and complete intervals. The spectrum S can be thought of as the set of the cluster points of the spectrum of the boundary value problems determined by (5.6.1) and conditions (5.7.1) as $b\to\infty$. Also the spectrum S can be thought of as the set of eigenvalues for the problem determined by (5.6.1), condition L^2 at infinity, and a linear homogeneous condition at $t=0$,

$$\sin\alpha\, x(0) - \cos\alpha\, p(0)\, x'(0) = 0. \tag{5.7.6}$$

(5.7. ii) If L is of the limit-point type at infinity, then the spectrum S does not depend upon the particular initial condition (5.7.6) (H. Weyl [2]).

A theorem analogous to (5.7. i) holds also in the limit-circle case, where only a further condition at infinity is needed to assure a proper selection of the eigenfunctions. The detailed statements are omitted.

5.8. Some properties of the spectrum S. Let λ denote a real parameter and let $q=q(t)$ be, for instance, a bounded real-valued continuous function on $0\leq t<+\infty$. Then the differential equation

$$x''+\left(\lambda+q(t)\right)x=0 \tag{5.8.1}$$

is of the limit-point type (5.6) and the spectrum S of (5.8.1) [see (5.7)], that is, the set of cluster points of the spectrum of the boundary value problem determined by (5.8.1) and a linear homogeneous boundary condition $\alpha x(0)+\beta x'(0)=0$, $\alpha^2+\beta^2=1$, at $t=0$, is independent of the particular boundary condition chosen.

In case $q(t)$ is periodic of some period T in $-\infty<t<+\infty$ [see (4.2)], the set S is identical with the closure of the set of the λ-values for which all solutions of (5.8.1) are bounded in $-\infty<t<+\infty$ (see A. Wintner [39] and S. Wallach [4]).

If $(-\infty,\lambda_0)+\Sigma(\lambda_k,\lambda^k)$ denotes the canonical decomposition of the open complement of the set S, and $q(t)$ is supposed to be bounded and uniformly continuous in $[0,+\infty)$, then the lengths $\lambda^k-\lambda_k$ tend to 0 as $k\to+\infty$. This occurs in particular if $q(t)$ is almost periodic. If $q(t)$ is periodic then the intervals (λ_k,λ^k) are the intervals in which a zone of instability mentioned in (4.2) and in (4.4) is intersected by the straight line $\varepsilon=1$ parallel to δ-axis (λ-axis in the present notations). Various estimates of the rapidity of convergence of the limit relation $\lambda^k-\lambda_k\to 0$, depending upon the smoothness of $q(t)$, have been obtained by P. Hartman and C. R. Putman [2] and by P. Hartman [14].

In general, when $q(t)$ is supposed only bounded, the solutions of (5.8.1) when λ is not in S behave like the solutions of a Hill equation in the interior of the regions of instability; thus, in both cases, there exists exponentially "large" and "small" solutions on the half-line $0\leq t<+\infty$ (see C. R. Putman [5], P. Hartman [8]). Furthermore, it is possible to define for every λ not in S a "characteristic exponent" $L(\lambda)$, somewhat analogous to the one defined when $q(t)$ is periodic [see (4.1)] and to obtain estimates of the size of $L(\lambda)$ as $\lambda\to+\infty$ (P. Hartman and C. R. Putnam [2]).

5.9. Bibliographical notes. For the classical Sturm comparison theory [1] of second order ordinary differential equations, we refer to the book of G. Sansone [16], where the theories of M. Picone [1] and G. Mammana [1, 2] are developed in detail. See also the books of M. Bocher [1] and E. L. Ince [5]. It is interesting to note that Sturm research and successive developments in the last century are the forerunners of the modern approach on stability problems and asymptotic behavior (qualitative theory of differential equations). STURM indeed pointed out in his memoir [1] that so many problems lead to second order linear differential equations for which formal solutions by series do not allow us to read off the asymptotic behavior, the zeros, the oscillatory character as $t \to +\infty$, of the same solutions. Instead, the direct study of the differential equation may give a great deal of information which is often all that is needed. For recent extended research on the comparison theory, also including nonlinear cases, see G. SANSONE [1—9, 18], M. PICONE [1], Z. BUTLEWSKI [1—9], F. V. ATKINSON [3, 5], J. H. BARRETT [1], L. AMERIO [1], E. MARCHENTE [1]. See also L. P. BURTON [1], S. CORONATO [1], M. M. CRUM [1], M. I. ELSIN [1—9], E. GAGLIARDO [1, 2, 3], L. GIULIANO [2], E. KAMKE [2, 3], G. CALAMAI [2], D. CALIGO [4, 6], R. CACCIOPPOLI and A. GHIZZETTI [1]. Nonoscillation theorems have been given recently by P. HARTMAN and A. WINTNER [10, 20, 21, 23, 24, 25], E. HILLE [1], and A. WINTNER [28, 29]. See also A. P. SVARCMAN [1], M. G. KREIN [2], A. M. GOLDIN [1], R. L. STERNBERG [1], Y. SUYAMA [1].

For research on oscillation theorems, properties of the zeros, and Sturm-Liouville theory, see R. L. POTTER [1], M. I. SEROV [1], I. M. SOBOL [3], A. WINTNER [15], E. C. TITCHMARSH [2], L. D. NIKOLENKO [1], F. V. ATKINSON [1], M. BIERNACKI [1—6], J. MIKUSINSKI [1, 2, 3], W. B. FITE [1].

For asymptotic properties of the solutions as $t \to \infty$ see M. YA. LEONOV [1], F. V. ATKINSON [3], R. GOSSE [1], U. RICHARD [3], M. CIMINO [1, 2], I. P. GINSBURG [1], R. NARDINI [2].

G. BORG [2, 3] has investigated the inverse Sturm problem, namely the question to determine the function $f(t)$ when the eigenvalues are known. The solution in general is not unique, but it is unique under convenient restrictions. See also L. A. CUDOV [1], N. LEVINSON [11, 12, 14, 15].

The theory of the second order linear equations in the complex field has been widely investigated by W. LEIGHTON [1] and subsequently by Z. NEHARI [1], and C. TAAM [1—7]. See also N. S. KOSLYAKOV [1].

The adiabatic theory for differential equations has been investigated by D. GRAFFI [1] and A. WINTNER [6, 9, 10].

The existence of nonzero solutions bounded, or L, or L^2, or L^p integrable in $(0, +\infty)$ for the equation $x'' + \varepsilon f(t)\, x = 0$ and the connected eigenvalue problem has been discussed by A. WINTNER [18, 27], P. HARTMAN [1—13], P. HARTMAN and A. WINTNER [2, 3, 4, 6, 11, 15], C. R. PUTNAM [1], P. HARTMAN and C. R. PUTNAM [1—4], V. B. LIDSKII [1], T. MANACORDA [4], D. B. SEARS [1].

The case of $f(t) \to \infty$ has been further investigated by G. TREVISAN [1], H. MILLOUX [1], G. PRODI [1, 3].

The equation $x'' + (a + b t^{\alpha} \sin \beta t)\, x = 0$, $\alpha < 0$, a, b, β constants, has been discussed by G. ASCOLI [4]. For general theorems on this equation and other analogous ones, see U. BARBUTI [1], G. PRODI [2].

For recent research on the zeros of the oscillatory solutions of equations $x'' + f(t)\, x = 0$ associating the consecutive zeros with the elements of a group see O. BORUVKA [1], M. GREGUS [1].

Concerning the asymptotic development of the solutions for t large we should mention here the extensive research of R. E. LANGER [1—8] which will be referred to in § 10. See also M. HUKUHARA [2], M. HUKUHARA and H. NAGUMO [1], P. HARTMAN [1].

Chapter III

Nonlinear systems

§ 6. Some basic theorems on nonlinear systems and the first method of Lyapunov

6.1. General considerations. We shall consider mainly nonlinear systems of the form

$$x' = A(t)\, x + f(t, x), \tag{6.1.1}$$

where $A(t)$ is a square matrix whose elements are constants, or functions of t continuous for $t \geqq 0$, and where $f(t, x)$ is a vector function of t and x, continuous for all $t \geqq 0$ and $x \in U$, where U is a neighborhood of $x = 0$ in E_n. We shall also suppose $f(t, 0) = 0$ for all t so that $x = 0$ is a trivial solution of system (6.1.1).

The hypotheses above assure the local existence for every $u \in U$ of a solution $x(t; 0, u)$ satisfying the initial condition $x(0; 0, u) = u$, and $x(t; 0, u)$ then exists in a maximal interval $[0, \bar{t})$, where $0 < \bar{t} \leqq +\infty$, and $\bar{t}$ generally depends upon $u \in U$ (see example 1 in (1.3)). The conditions above do not assure uniqueness. In some of the following statements we shall suppose that $f(t, x)$ satisfies the condition

$$\|f(t, x)\| = o(\|x\|) \quad \text{as} \quad \|x\| \to 0$$

uniformly with respect to t, i.e., there exists some function $\eta(\sigma) > 0$, $\eta(\sigma) \to 0$ as $\sigma \to 0$, such that $\|f(t, x)\| \leqq \|x\|\, \eta(\|x\|)$. Even this hypotheses does not assure uniqueness. Also, we shall often consider systems of the form

$$x' = A(t)\, x + \varepsilon f(t, x),$$

where ε is a parameter whose absolute value is sufficiently small. These systems are said to be weakly nonlinear.

Let $x(t)$, $t_0 \leqq t < \bar{t}$, denote any solution of system (6.1.1), let $y(t)$, $t_0 \leqq t < +\infty$, denote the solution of system $x' = A\, x$, with $y(t_0) = x(t_0)$, let $Y(t)$ denote the fundamental system of solutions of $x' = A\, x$ with $Y(t_0) = I$. Then, by (2.2) we have

$$x(t) = y(t) + Y(t) \int_{t_0}^{t} Y^{-1}(\alpha)\, f[\alpha, x(\alpha)]\, d\alpha, \tag{6.1.2}$$

for all $t_0 \leqq t < \bar{t}$. In other words $x(t)$ satisfies a nonlinear Volterra integral equation (in matrix form), and conversely every solution $x(t)$ of (6.1.2) verifies (6.1.1).

6.2. A theorem of existence and uniqueness. We shall now suppose A to be a constant matrix. Then $Y(t)\, Y^{-1}(\tau) = Y(t - \tau)$ (2.2) and (6.1.2) becomes

$$x(t) = y(t) + \int_{t_0}^{t} Y(t - \alpha)\, f[\alpha, x(\alpha)]\, d\alpha. \tag{6.2.1}$$

The following theorem is both an existence theorem in $[t_0, +\infty)$ of all solutions $x(t, 0, u)$ with $\|u\|$ sufficiently small, and a theorem of asymptotic stability of the same solutions, in particular of the trivial solution $x=0$, $t\geq t_0$ of (6.1.1).

(6.2. i) If A is a constant matrix whose characteristic roots all have negative real parts, if $\|f(t, x)\| = o(\|x\|)$ uniformly in $0\leq t<+\infty$, then every solution $x(t)$ of the system

$$x' = A\,x + f(t, x), \tag{6.2.2}$$

with $\|x(0)\|$ sufficiently small, exists in $[0, +\infty)$ and the solution $x=0$ of (6.2.2) is asymptotically stable (at the right) (A. Lyapunov [3]).

Proof. The proof is similar to the one given for linear systems in (3.2). Indeed by (6.2.1) with $t_0=0$ we have

$$\|x(t)\| \leq \|y(t)\| + \int_0^t \|Y(t-\alpha)\|\,\|f(\alpha, x(\alpha))\|\,d\alpha$$

for all $0\leq t<\bar{t}$. By (2.3. iii) there exist constants $c_1, c_2, a>0$ such that $\|y(t)\|\leq c_1 \|y(0)\|e^{-at}$, $\|Y(t)\|\leq c_2\,e^{-at}$, with $y(0)=x(0)$. On the other hand, given any $m>0$ there is a $d>0$ such that for all $t\geq 0$ and $\|x\|\leq d$ we have $\|f(t, x)\|\leq m\|x\|$. If we suppose $\|x(0)\|<d$, we can certainly take $\bar{t}>0$ such that $\|x(t)\|<d$ for all $0\leq t<\bar{t}$, and hence

$$\|x(t)\|\leq c_1\|x(0)\|\,e^{-at} + \int_0^t m\,c_2\,e^{-a(t-\alpha)}\|x(\alpha)\|\,d\alpha$$

for all $0\leq t<\bar{t}$, and then

$$\|x(t)\|\,e^{at}\leq c_1\|x(0)\| + \int_0^t m\,c_2\,e^{a\alpha}\|x(\alpha)\|\,d\alpha.$$

By (3.2. i) we have

$$\|x(t)\|\,e^{at}\leq c_1\|x(0)\|\,e^{mc_2 t},$$

and hence $\|x(t)\|\leq c_1 e^{(mc_2-a)t}\|x(0)\|\leq c_1\|x(0)\|$ provided $mc_2<a$.

Thus, if we suppose $\|x(0)\|<dc_1^{-1}/2$, we have $\|x(t)\|<d/2$ for all $0\leq t<\bar{t}$. This implies that $x(t)$ always remains in the sphere of center $x=0$ and radius $d/2$ of E_n and thus always remains in the inside of the set U. By the remark at the end of (1.1) we conclude that $x(t)$ can be continued for all $t\geq 0$. Thus the solution $x=0$ satisfies condition (α) of (1.2), and, since $d>0$ can be taken arbitrarily small, also condition (β) of (1.2). In other words, $x=0$ is stable (at the right) in the sense of Lyapunov. On the other hand $mc_2<a$ implies that $\|x(t)\|\to 0$ as $t\to+\infty$, and thus $x=0$ satisfies condition (β') of (1.2), and hence, it is asymptotically stable. This completes the proof. For the present proof cf. E. A. Coddington and N. Levinson [3], N. Levinson [10], R. Bellman [6].

Under the same conditions as (6.2. i) let us denote by λ_i, $i=1,\ldots,n$, the characterictic roots of A and let $\lambda=\max R(\lambda_i)$. Hence $\lambda<0$.

Under the conditions as (6.2. i) the solutions $x(t)$ of (6.2.2) for which $\|x(t)\|\to 0$ as $t\to+\infty$ have type number $\leq\lambda$, i.e., by (3.12) $\overline{\lim}\,\log\|x(t)\|/t\leq\lambda$, as $t\to+\infty$ (A. Lyapunov [3]).

Proof. In the proof of (6.2.i) the number a is any number with $0<a<-\lambda$ (2.3. iii). We may have $a=-\lambda-\varepsilon$ for any $0<\varepsilon<-\lambda$. For ε small this may imply that the constants c_1, c_2 are large, and m small, since $m\,c_2 < a$. We will take m so small that we have also $m\,c_2 < \varepsilon$. Finally the number d will be very small. If $x(t) \to 0$ as $t \to +\infty$ there will be a $t_0 > 0$ such that $\|x(t)\| \leq d/c_1^{-1}/2$ for all $t \geq t_0$. Now if we repeat the reasoning above where $t = 0$ is replaced by $t = t_0$ we conclude that $\|x(t)\| < C\,e^{(m c_2 - a)t} \leq C^{(\lambda + 2\varepsilon)t}$, thus $\overline{\lim}\, \log \|x(t)\|/t \leq \lambda + 2\varepsilon$, and since ε is arbitrary, by (3.12) it is proved that the type number of $x(t)$ is $\leq \lambda$. For the present proof see E. A. Coddington and N. Levinson [3].

The condition $\|f(t, x)\| = o(\|x\|)$ can be remarkably reduced. It is sufficient to assume that $\|f(t, x)\| \leq k\|x\|$ for some constant $k > 0$, all $\|x\|$ sufficiently small and all t, and that, given $\varepsilon > 0$, there exist $\delta > 0$, $T > 0$, such that $\|f(t, x)\| \leq \varepsilon\|x\|$ for all $\|x\| \leq \delta$, $t \geq T$. Thus A may be replaced by a matrix $A + B(t)$ where $B(t) \to 0$ as $t \to +\infty$.

Also, it is sufficient to assume that $\|f(t, x)\| \leq k\|x\| + \|x\|^{1+a} t^b$, for some constants $k, a, b > 0$, all $\|x\|$ sufficiently small and all t, and that, given $\varepsilon > 0$ there exists $\delta > 0$, $T > 0$ such that $\|f(t, x)\| \leq \varepsilon\|x\| + \|x\|^{1+a} t^b$, for all $t \geq T$, $\|x\| < \delta$.

(6.2. ii) Under the same conditions of (6.2. i) for $f(t, x)$, the solution $x = 0$ of (6.2.2) is certainly unstable if at least one of the characteristic roots of A has real part positive (A. Lyapunov [3]). A proof can be given in the same terms as the previous proofs.

Finally all theorems above hold even if we replace the constant matrix A by a periodic matrix $A(t)$, and the characteristic roots of A by the characteristic exponents of the periodic linear system $y' = A(t)\,y$ (4.1). The reason is the same as in (3.9), i.e., that a convenient substitution $x = Py$, with P periodic $|\det P|, |\det P^{-1}| \leq M$, transforms also the present nonlinear system in an analogous one where $A(t)$ is replaced by a constant matrix (A. Lyapunov [3]). A further extension where $A(t)$ is any continuous bounded matrix and the characteristic roots are replaced by the type numbers (3.12) is not possible (R. E. Vinograd [11]).

Finally it must be added that if exactly k of the n characteristic roots of A [or characteristic exponents of the periodic system $x' = A(t)\,x$] have real parts negative, and the remaining $n - k$ have real parts positive, $1 \leq k < n$, then the solution $x = 0$ is certainly unstable (see above) but is conditionally stable, namely is conditionally asymptotically stable with respect to a family of solutions depending upon exactly k parameters (a family of "dimension" k) (A. Lyapunov [3]). For a recent proof see E. A. Coddington and N. Levinson [3].

Remark. Theorem (6.2. i) is one of the Lyapunov theorems we have mentioned in (1.7). We can say now that if $f(x, t)$ satisfies one of the conditions above, and A has no characteristic root with zero real part, then (a) either $x = 0$ is stable for $x' = A\,x$, thus $R(\lambda_i) < 0$ for all $i = 1, \ldots, n$, and $x = 0$ is asymptotically stable for $x' = A\,x$ and $x' = A\,x + f$ as well; or (b) $x = 0$ is unstable for $x' = A\,x$, thus $R(\lambda_i) > 0$ for some i, and $x = 0$ is unstable for $x' = A\,x$ and $x' = A\,x + f$ as well. The same holds if $A(t)$ is periodic of period T where characteristic exponents replace characteristic roots.

An important variant of (6.2. i) is the following

(6.2. iii). Consider the system

$$x' = A x + f(x, \omega t) + \varepsilon b(\omega t), \tag{6.2.3}$$

where $x = (x_1, \dots, x_n)$, A is a constant $n \times n$ matrix whose characteristic roots have all negative real parts. Suppose that $b(t)$ is a continuous vector function periodic of period 1 and of mean value zero, ε, ω constants, and that $f(x, t)$ is a continuous vector function, periodic in t of period 1, such that for any $\eta > 0$ there is a $\delta > 0$ with $\|f(x, t) - f(y, t)\| \leq \eta \|x - y\|$ for all $\|x\|, \|y\| \leq \delta$, and all t. Suppose $f(0, t) = 0$. Then for $|\varepsilon|$ sufficiently small, or for ω sufficiently large, system (6.2.3) has a periodic solution $x = p(t)$ of period $1/\omega$. Moreover, there exists a constant M such that $\|p(t)\| \leq M \varepsilon(1 + \omega)$, and the solution $p(t)$ is stable. More precisely there exists an $\varepsilon > 0$ such that $\|x(t_0) - p(t_0)\| < \varepsilon$ implies $x(t) - p(t) \to 0$ as $t \to +\infty$ provided $\varepsilon(1 + \omega)$ is sufficiently small (A.B. Farnell, C.E. Langenhop, N. Levinson [1]).

We omit the proof. Theorem (6.2. iii) can be applied, e.g., to the second order equation

$$x'' + f(x, x')\, x' + g(x) = \varepsilon \sin \omega t,$$

where $f(0, 0) > 0$ and $g'(0) > 0$. For statements analogous to the one above see R. Faure [1] and H.A. Antosiewicz [2].

We conclude this section with two theorems which may be considered as extensions of Dini-Hukuhara theorem [see (3.3. ii) and (3.9. x)] to nonlinear systems.

(6.2. iv) If the linear system $y' = A(t)\, y$ is restrictively stable [see (3.9)], if $c(t)$, $t \geq 0$, is a nonnegative function with $\|f(t, x)\| \leq c(t)\, \|x\|$, $\int_0^{+\infty} c(t)\, dt < +\infty$, then there is a constant $K > 0$ such that, for every solution $x(t)$ of system (6.1.1) we have $\|x(t)\| \leq K \|Y(t)\|\, \|x(0)\|$.

Proof. By (6.2.1) and (3.9) with $t \geq t_0$ we have $\|Y(t)\| \leq M$, $\|Y^{-1}(\alpha)\| \leq M$ for all t and α, and

$$x(t) = y(t) + \int_0^t Y(t)\, Y^{-1}(\alpha)\, f(\alpha, x(\alpha))\, d\alpha,$$

where $y(t) = Y(t)\, y(0)$, $y(0) = x(0)$. Hence

$$x(t) = Y(t) \left[x(0) + \int_0^t Y^{-1}(\alpha)\, f(\alpha, x(\alpha))\, d\alpha \right],$$

$$\|x(t)\| \leq \|Y(t)\| \left[\|x(0)\| + \int_0^t \|Y^{-1}(\alpha)\|\, \|Y(\alpha)\| \cdot c(\alpha)\, \|x(\alpha)\|\, \|Y(\alpha)\|^{-1}\, d\alpha \right],$$

and, if $u(t) = \|x(t)\|\, \|Y(t)\|^{-1}$, $t \geq 0$, we have

$$u(t) \leq \|x(0)\| + \int_0^t M^2 c(\alpha) \cdot u(\alpha)\, d\alpha.$$

By (3.2.1) we have now successively

$$u(t) \leqq \|x(0)\| \exp \int_0^t M^2 c(\alpha)\, d\alpha,$$

$$\|x(t)\| \leqq K \|Y(t)\| \, \|x(0)\|,$$

for some constant $K > 0$. Thereby (6.2. iv) is proved.

(6.2. v) If the linear system $y' = A(t)\, y$ is uniformly stable [see (3.9)] in $[0, +\infty)$, if $c(t)$, $t \geqq 0$, is a nonnegative function with $\|f(t, x)\| \leqq c(t)\, \|x\|$, $\int_0^{+\infty} c(t)\, dt < +\infty$, then all solutions $x(t) = x(t; 0, x_0)$ of (6.1.1) exist in $[0, +\infty)$ and there exists some constant $K > 0$ such that for every solution $x(t)$ of (6.1.1) we have $\|x(t)\| \leqq K \|x(0)\|$. In addition, the solution $x(t) = 0$ of (6.1.1) is stable and uniformly stable.

Proof. By (6.2.1) and (3.9) we have $\|Y(t)\| \leqq M$, $\|Y(t)\, Y^{-1}(\alpha)\| \leqq M$ for all $0 \leqq \alpha \leqq t$, and

$$x(t) = Y(t)\, x(0) + \int_0^t Y(t)\, Y^{-1}(\alpha)\, f(\alpha, x(\alpha))\, d\alpha,$$

$$\|x(t)\| \leqq \|Y(t)\|\, \|x(0)\| + \int_0^t \|Y(t)\, Y^{-1}(\alpha)\|\, \|f(\alpha, x(\alpha))\|\, d\alpha$$

$$\leqq M \|x(0)\| + \int_0^t M\, c(\alpha)\, \|x(\alpha)\|\, d\alpha.$$

By (3.2.1) we have finally

$$\|x(t)\| \leqq M \|x(0)\| \exp \int_0^t M\, c(\alpha)\, d\alpha \leqq K \|x(0)\|,$$

for some constant $K > 0$. Let us prove that the solution $x(t) = 0$ of (6.1.1) is stable. Indeed, $\|x(t)\| \leqq K \|x(0)\|$; hence, given $\varepsilon > 0$, if we take $\|x(0)\| \leqq \delta = \varepsilon/K$, then $\|x(t)\| \leqq \varepsilon$ for all $t \geqq 0$. Let us prove that the solution $x(t) = 0$ of (6.1.1) is uniformly stable. Given $\varepsilon > 0$ there is some $\delta > 0$ such that for every $t_0 \geqq 0$ and solution $y(t)$, $t \geqq t_0$, of $y' = A(t)\, y$ with $\|y(t_0)\| \leqq \delta$ we have $\|y(t)\| \leqq \varepsilon/K$ for all $t \geqq t_0$. Now let $x(t)$, $t \geqq t_0$, be any solution of (6.1.1) with $\|x(t_0)\| \leqq \delta$, and let $y(t)$, $t \geqq t_0$, denote the solution of $y' = A(t)\, y$ with $y(t_0) = x(t_0)$. Then for $t \geqq t_0$ and again by (3.2.1) we have successively

$$x(t) = y(t) + \int_{t_0}^t Y(t)\, Y^{-1}(\alpha)\, f((\alpha, x(\alpha))\, d\alpha,$$

$$\|x(t)\| \leqq (\varepsilon/K) + \int_{t_0}^t M\, c(\alpha)\, \|x(\alpha)\|\, d\alpha,$$

$$\|x(t)\| \leqq (\varepsilon/K) \exp \int_0^t M\, c(\alpha)\, d\alpha \leqq \varepsilon.$$

Thereby (6.2. v) is proved.

It should be remarked here that, under the same hypotheses of (6.2. v), if the system $y' = A(t)y$ is asymptotically stable, then all solutions $x(t)$ of (6.1.1) approach zero as $t \to +\infty$. We omit the proof. For theorems (6.2. iv), (6.2. v), and other evaluations of the solutions of system (6.1.1) see M. GOLOMB [1].

Remark. In (1.2) we have defined stability in the sense of LYAPUNOV (at the right) by means of requirements (α) and (β), and asymptotic stability (at the right) by means of requirements (α), (β), and (β'). Thus asymptotic stability implies Lyapunov stability. On the other hand, example (1.3, no. 3) assures that Lyapunov stability does not imply asymptotic stability. Now the question arises concerning the relationship between requirements (α), (β) on one side, and requirements (α), (β') on the other side. These requirements could be thought of as defining two concepts of stability, namely the usual stability in the sense of LYAPUNOV, and a "weak" asymptotic stability. The same example above shows that (α), (β) do not imply (α), (β'). For homogeneous linear systems (1.4.1) requirements (α), (β') imply (β). Indeed every solution of a fundamental system of solutions of (1.4.1) approaches zero as $t \to +\infty$, hence it is bounded in $[0, +\infty)$, and this assures stability in the sense of LYAPUNOV (3.9. i). In general, (α), (β') do not imply (β), and thus (α), (β), and (α), (β') constitute two independent concepts of stability. This can be shown by the following elementary example.

Consider the system $x_1' = f(t, x_1, x_2)$, $x_2' = g(t, x_1, x_2)$, $t \geq 0$, of order $n = 2$, given, in polar coordinates r, φ, by

$$r'/r = h'/h, \qquad \varphi' = 0, \qquad (') = d/dt,$$

where $x_1 = r\cos\varphi$, $x_2 = r\sin\varphi$, $h = h(t, \varphi) = (1 + t^3\sin^2\varphi)(1 + t + t^4\sin^4\varphi)^{-1}$, $t \geq 0$. The solution satisfying the initial condition $r = r_0 \geq 0$, $\varphi = \varphi_0$, is unique and given by $r(t) = r_0 h(t, \varphi_0)$, $\varphi = \varphi_0$, $t \geq 0$. In particular $r = 0$, i.e. $x_1 = x_2 = 0$, is a solution. Also, for every r_0, φ_0, we have $r(t) \to 0$ as $t \to +\infty$, and hence, the solution $x_1 = x_2 = 0$ satisfies requirements (α), (β'). On the other hand, for $0 < \varphi < \pi/2$, $r_0 > 0$, $t = t_\varphi = (\sin\varphi)^{-\frac{3}{2}}$, we have

$$r(t_\varphi) = r_0[1 + (\sin\varphi)^{-\frac{5}{2}}][1 + (\sin\varphi)^{-\frac{3}{2}} + (\sin\varphi)^{-2}]^{-1},$$

and $r(t_\varphi) \to +\infty$ as $\varphi \to 0$. Thus the solution $x_1 = x_2 = 0$ does not satisfy requirement (β), i.e., it is not stable in the sense of LYAPUNOV. By actual computations it is easy to verify that $f = x_1 h'/h$, $g = x_2 h'/h$, are continuous functions of t, x_1, x_2 for all $t \geq 0$, x_1, x_2, and that $f(t, 0, 0) = g(t, 0, 0) = 0$.

For an example of an autonomous system presenting the same behavior see R.E. VINOGRAD [11]. For other considerations concerning the concept of stability see, e.g., (3.9), (7.3), and J.L. MASSERA [2].

6.3. Periodic solutions of periodic systems. Let us suppose that a given vector function $F(t, x)$, $x = (x_1, \ldots, x_n)$, $F = (F_1, \ldots, F_n)$, whose components F_i, together with their first partial derivatives $\partial F_i/\partial x_j$, are continuous in (t, x) for all $-\infty < t < +\infty$, and x of an open set S of the x-space E_n. Let us suppose that $F(t, x)$ is periodic of period T, say $F(t+T, x) = F(t, x)$ for all $-\infty < t < +\infty$, $x \in S$, and that the differential system

$$x' = F(t, x) \tag{6.3.1}$$

has some periodic solution $x = p(t)$, $-\infty < t < +\infty$, say $p(t+T) = p(t)$, with $p(t) \in S$ for all t. Then $x = p(t)$ defines a closed path C in E_n (orbit) and to discuss the stability of the solution $x = p(t)$ we shall consider the variational equation of $x = p(t)$ as in (1.7). Thus, if $x = p(t) + u$, and we suppose $F(t, x)$ continuous with its first partial derivatives for $-\infty < t < +\infty$, $x \in S$, we have (1.7)

$$u' = A(t)\,u + f(t, u),$$

where the elements a_{ij} of the matrix $A(t)$ are continuous functions of t, $-\infty < t < +\infty$, $f(t, u)$ is a continuous vector function of (t, u), $-\infty < t < +\infty$, $u \in S_0$, where S_0 is a neighborhood of the origin in the u-space $\bar{E}_n$, $f(t, 0) = 0$ for all t, and

$$A(t + T) = A(t), \qquad f(t + T, u) = f(t, u).$$

The following theorem is a consequence of the Lyapunov theorem (6.2. i) and its corollaries discussed in (6.2):

(6.3. i) If the characteristic exponents of the linear system $x' = A(t)\,x$ have all negative real parts, then (a) the solution $x = p(t)$ is asymptotically stable for (6.3.1); (b) the orbit C: $x = p(t)$ is asymptotically stable.

Let us observe that T need not be the minimum period either of F or of $p(t)$.

Let us denote by T now the minimum period of $F(t, x)$. Then if T is also the minimum period of $p(t)$, $p(t)$ may be called a *harmonic*. Let $p > 1$, $q > 1$ denote integers with no common divisor. If pT, or T/p, or pT/q is now the smallest period of $p(t)$, then $p(t)$ may be called a subharmonic of order p, or a superharmonic of order $1/p$, or a ultra-subharmonic of order p/q respectively (J. J. Stoker [1]). More simply, according to J.L. Massera [3] and other authors, we may denote by harmonics all constant solutions, and all solutions $p(t)$ which have period T even if their minimum period is T/p, $p = 1, 2, \ldots$, and by subharmonics of order q those solutions $p(t)$ which have period qT, $q > 1$, and no smaller period which is a multiple of T, even if their minimum period is qT/p, $p = 1, 2, \ldots$, p, q with no common divisor.

It has been pointed out by J. L. Massera [3] recently that a periodic solution $p(t)$ of a periodic system (6.3.1) need not have a period T in rational ratio with T. Indeed it may occur that the orbit C defined by $p(t)$ is completely contained in the locus K of E_n where $F(t, x)$ actually does not depend on t. Then $F(t, x) = \varphi(x)$ for all t and $x \in K$, and hence $x' = \varphi(x)$ on C. Obviously $p(t)$ may be of any period T'.

Let us consider again system (6.3.1) where $F(t, x)$ is continuous for all t, $-\infty < t < +\infty$, and $x \in E_n$, and where F has T as its minimum period, and satisfies conditions assuring a uniqueness theorem. For systems of order $n = 1$, J.L. Massera [3] has proved that, if a solution $x(t, x_0, 0)$ exists for all $t \geqq 0$, and is bounded in $[0, +\infty)$, then the system has at least one harmonic solution (the same is true for any $n \geqq 1$ if the system is linear). For $n = 1$ every (Bohr) almost periodic solution is actually a harmonic solution (J.L. Massera [3]).

For $n \geqq 2$ an example of Massera shows that the existence of subharmonics does not imply the existence of harmonics. Nevertheless, for $n = 2$, if all solutions

$x(t, x_0, 0)$ exist for all $t \geqq 0$, and one is bounded in $[0, +\infty)$, then there exists at least one harmonic solution (J. L. MASSERA [3]).

For $n > 1$ the numbers $N(q)$, $q = 1, 2, \ldots$, of harmonic ($q = 1$) and subharmonic ($q = 2, 3, \ldots$) solutions of (6.3.1) have been investigated by N. LEVINSON [5] and J.L. MASSERA [1, 3]. Mild conditions (N. LEVINSON [5]) assure that $N(q)$ is an even multiple of q, $q \geqq 2$. On the other hand, for $n = 2$, J.L. MASSERA [3] has proved that, given arbitrary numbers $N(q)$, $N(1) \geqq 1$, $N(q) \geqq 0$ a multiple of q, $q = 2, 3, \ldots$, there exists a system (6.3.1) with $n = 2$ such that (a) all solutions exist for all $t \geqq 0$ and (b) the system has exactly $N(1)$ harmonics and $N(q)$ subharmonics of order q, $q = 2, 3, \ldots$. If $N(1) = 0$ and not all $N(q)$ are zero, there exists a system (6.3.1), with $n = 2$, such that (b) holds. Finally, for $n > 2$, given any set of integers $N(q) \geqq 0$, $q = 1, 2, \ldots$, $N(q)$ a multiple of q for $q \geqq 2$, there exists a system (6.3.1) with $n > 2$, which possesses exactly $N(1)$ harmonics and $N(q)$ subharmonics of order q. If all $N(q)$ are zero, a system exists which has a bounded almost periodic solution but no periodic solution (J.L. MASSERA [3]).

6.4. Periodic solutions of autonomous systems. Here $F(x)$ denotes a vector function, $F(x) = [F_1(x), \ldots, F_n(x)]$, whose components are continuous with their first partial derivatives $F_{ij} = \partial F_i / \partial x_j$ in an open set S of E_n. Let us suppose that the autonomous differential system

$$x' = F(x) \tag{6.4.1}$$

is known to have a periodic solution $x = p(t)$ of some period T with $p(t) \in S$ for all t, $p(t) = [p_1(t), \ldots, p_n(t)]$. The variational system (1.7) is then

$$u' = A(t)\, u + f(t, u), \qquad A = [a_{ij}(t)],$$

with $a_{ij} = F_{ij}[p(t)]$. From $p_i'(t) = F_i[p(t)]$, $i = 1, \ldots, n$, by differentiation, we obtain

$$p_i''(t) = \sum_{j=1}^{n} F_{ij}[p(t)]\, p_j'(t), \quad i = 1, \ldots, n,$$

i.e., the linear system $y' = A(t)\, y$ has a periodic solution, say $y = p'(t)$ of period T. This implies that $\lambda = 1$ is one of the characteristic factors and $r = 0$ one of the characteristic exponents of the periodic linear system $y' = A(t)\, y$ (4.1). Thus the hypotheses of (6.3. i) cannot be verified in the autonomous case. The contention of (6.3. i), too, does not hold in the autonomous case. Indeed every vector function $p(t + \gamma)$, γ a constant (phase) is a solution of (6.4.1) and $p(t+\gamma) - p(t)$ does not approach zero as $t \to +\infty$ no matter how small $\gamma \neq 0$. Thus $p(t)$ is not asymptotically stable. As a matter of fact in most cases $p(t)$ is not even stable [see examples 2 and 3 of (1.9)]. Let us denote by C the orbit defined by $x = p(t)$. The following important theorem replaces (6.3. i):

(6.4. i) If $n - 1$ of the characteristic exponents of $y' = A(t)\, y$ have real parts negative, then $x = p(t)$ presents asymptotic orbital stability and even asymptotic phase for system (6.4.1).

In terms of (1.8) this means that there is an $\varepsilon_0 > 0$ such that every solution $x = x(t; t_0, x_0)$ with $\{x_0, C\} < \varepsilon_0$ verifies the relations

$$\{x(t; t_0, x_0), C\} \to 0 \quad \text{as} \quad t \to +\infty,$$

$$x(t; t_0, x_0) - p(t+\gamma) \to 0 \quad \text{as} \quad t \to +\infty,$$

where $\gamma = \gamma(t_0, x_0)$ is a convenient number (asymptotic phase) (A. LYAPUNOV [3]).

The proof of (6.4. i) is too involved to be given here and we refer to A. LYAPUNOV [3], or to some recent exposition (e.g., E. A. CODDINGTON and N. LEVINSON [3], or S. LEFSCHETZ [2]).

Remark. The main results above (6.1–6.4) are all essentially due to A. LYAPUNOV who proved them under conditions of analyticity of the functions F_i, or f_i, which have been removed.

The proofs given by LYAPUNOV were based on a process of successive approximations which is similar to the LIOUVILLE process of successive approximations for integral equations and constitute what is known as "the first method of LYAPUNOV". This method has had and still has a number of applications and important ramifications.

For conditions closer to the ones considered above see D. M. GROBMAN [1, 2, 3].

In the following lines we will give some technical information on the first method of LYAPUNOV.

6.5. A method of successive approximations and the first method of LYAPUNOV. *(a) The differential system.* Let us consider the nonlinear system

$$x' = A(t)\, x + P(t, x), \quad t_0 \leq t < +\infty, \tag{6.5.1}$$

where the elements $a_{ih}(t)$ of the matrix $A = [a_{ih}]$ are continuous bounded functions of t in $[t_0, +\infty)$ and the components $P_i(t, x_1, \ldots, x_n)$ of P are analytic functions of $x_1, \ldots, x_n$ for every t, $t_0 \leq t < +\infty$. We shall suppose that the development of P_i in power series of $x_1, \ldots, x_n$ begins with terms of degree at least two

$$P_i(t, x_1, \ldots, x_n) = \sum_{h=2}^{\infty} \sum_{\substack{h_1 + \cdots + h_n = h \\ h_1, \ldots, h_n \geq 0}} p_{i h_1 \ldots h_n}(t)\, x_1^{h_1} \ldots x_n^{h_n}, \quad i = 1, \ldots, n, \tag{6.5.2}$$

and that the coefficients $p_{i h_1 \ldots h_n}$ of these developments satisfy a relation of the form

$$|p_{i h_1 \ldots h_n}(t)| < M / C^{h_1 + \cdots + h_n} \tag{6.5.3}$$

for all t, $t_0 \leq t < +\infty$, where M, C are constants independent of $t, i, h_1, \ldots, h_n$. Thus, if $0 < B < C$, the series (6.5.2) converges absolutely and uniformly in the set $S = [t_0 \leq t < +\infty, \ \|x\| \leq B]$.

(b) The case of a finite interval. Let $[t_0, t_1]$ be a given finite interval, let $a = (a_1, \ldots, a_n)$ denote any given vector, $\|a\| < B$. Let us inquire as to whether the solution $x(t) = x(t; t_0, a)$ of (6.5.1) exists in $[t_0, t_1]$, verifies the relation $\|x(t)\| < B$ in $[t_0, t_1]$, and can be determined as the sum of a uniformly convergent series of the form

$$x = x^{(1)} + x^{(2)} + \cdots + x^{(m)} + \cdots, \tag{6.5.4}$$

i.e.,

$$x_i = x_i^{(1)} + x_i^{(2)} + \cdots + x_i^{(m)} + \cdots, \quad i = 1, 2, \ldots, n.$$

Let us first substitute such a series formally in (6.5.1) by admitting differentiation term by term and the possibility of developing the power series P_i into new power series of monomial terms of the form

$$c(t) \left(x_{i_1}^{(m_1)}\right)^{h_1} \left(x_{i_2}^{(m_2)}\right)^{h_2} \ldots \left(x_{i_k}^{(m_k)}\right)^{h_k}, \tag{6.5.5}$$

where $c(t)$ denotes a function of t, and where $k \geq 1$, $h_1, \ldots, h_k \geq 0$, $m_1, \ldots, m_k \geq 1$, $1 \leq i_1, \ldots, i_k \leq n$, are all integers. Let us denote by w the number $w = m_1 h_1 + \cdots + m_k h_k$, or the weight of the term (6.5.5). Let us denote by $R_i^{(m)}$ the finite sum of all terms of weight $w = m$ in the development of P_i. If $R^{(m)}$ is the vector $(R_1^{(m)}, \ldots, R_n^{(m)})$ then P_i is formally given by the series

$$P = R^{(2)} + R^{(3)} + \cdots + R^{(m)} + \cdots,$$

since there are no terms (6.5.5) of weight < 2. Finally, system (6.5.1) is formally solved if

$$dx^{(1)}/dt = A(t)\, x^{(1)}, \tag{6.5.6}$$

$$dx^{(m)}/dt = A(t)\, x^{(m)} + R^{(m)}, \quad m = 2, 3, \ldots, \tag{6.5.7}$$

as it is seen by formal addition of all these relations and formal differentiation of the series (6.5.4).

Obviously, $R^{(m)}$ depends only on $x^{(1)}, x^{(2)}, \ldots, x^{(m-1)}$, and hence (6.5.7) may be used to define successively $x^{(1)}, x^{(2)}, \ldots$.

For instance we may require that the vectors $x^{(m)}(t)$ satisfy the initial conditions

$$x^{(1)}(t_0) = a, \quad a = (a_1, a_2, \ldots, a_n), \quad x^{(m)}(t_0) = 0, \quad m = 2, 3, \ldots, \tag{6.5.8}$$

and then, formally, the sum $x(t)$ of the series (6.5.4) can be expected to satisfy the differential system (6.5.1) and the initial conditions $x(t_0) = a$.

Let us observe that the expressions $R_i^{(m)}$, $i = 1, 2, \ldots, n$, can be really determined in particular cases, one after the other in the order $m = 2, 3, 4, \ldots$, though their actual determination may be a tedious and lengthy process and no general expression has been found for them. As a matter of fact there will be no need of the explicit expressions $R_i^{(m)}$. The formulas (6.5.4), (6.5.6), (6.5.7) define a method of successive approximations. Let us denote by

$$x_h = (x_{ih},\ i = 1, 2, \ldots, n), \quad h = 1, 2, \ldots, n, \tag{6.5.9}$$

a fundamental system of solutions of the linear homogeneous system

$$x' = A(t)\, x, \tag{6.5.10}$$

.e., of system (6.5.6). Then by (6.5.7) and (6.5.8) we have that

$$x_i^{(m)} = \sum_{j,h=1}^{n} x_{ih}(t) \int_{t_0}^{t} \frac{\Delta_{jh}}{\Delta} R_j^{(m)}\, d\alpha, \quad i = 1, 2, \ldots, n, \tag{6.5.11}$$

for every $m = 2, 3, \ldots$, where Δ is the determinant $\Delta = \det[x_{ih}]$, and Δ_{jh} is the cofactor of x_{ih} in Δ (cf. 2.2).

A. Lyapunov [3] proved that given any finite interval $[t_1, t_2]$ there is a constant A, $0 < A < B$, such that for all initial values $a = (a_1, \ldots, a_n)$ with $\|a\| \leq A$, the series (6.5.4) converges absolutely and uniformly for all $t_1 \leq t \leq t_2$ and $|a| \leq A$, and $x(t) = [x_i(t), i = 1, \ldots, n]$ is the unique solution of (6.5.1) satisfying $x(t_0) = a$. In addition, A. Lyapunov proved that the same $x_i(t)$ can be thought of as sums of power series in $a_1, \ldots, a_n$, absolutely and uniformly convergent for all $t_1 \leq t \leq t_2$, $|a| \leq A$,

$$x_i(t) = \sum_{h=1} \sum_{m_1 + \cdots + m_n = h} A^{(i)}_{m_1 \ldots m_n}(t)\, a_1^{m_1} \ldots a_n^{m_n}, \quad i = 1, \ldots, n.$$

All $R_i^{(m)}$ and $x_i^{(m)}$ are forms of degree m in $a_1, \ldots, a_n$.

(c) The case of the infinite interval. Let us consider now system (6.5.1) and let us inquire whether a constant b, $0 < b \leq B$, can be determined such that, for every $a = (a_1, a_2, \ldots, a_n)$ with $\|a\| < b$, the solution $x(t) = x(t; t_0, a)$ of (6.5.1) exists

in $[t_0, +\infty)$ verifies the relation $\|x(t)\| < B$ in $[t_0, +\infty)$, and can be determined as the sum of a series of the form (6.5.4) uniformly convergent in $[t_0, +\infty)$.

The process in (b) cannot be repeated without convenient restrictions in the system (6.5.1) and modifications in the way in which the successive approximations are defined. We shall suppose (1) that the coefficients $a_{ih}(t)$ are bounded in $[t_0, +\infty)$; (2) that the linear system (6.5.1) is regular (3.12) and thus the fundamental system (6.5.9) can be supposed to be regular ($S - \Lambda = -\Lambda_1$); (3) that the type numbers λ_i of system (6.5.10) are all negative.

If we suppose for a moment that the integrands in the formulas (6.5.10) have type numbers all negative we shall define the method of successive approximations by means of the formulas (6.5.6), (6.5.7), (6.5.8) and

$$x_i^{(m)} = \sum_{j,h=1}^{n} x_{ih}(t) \int_{+\infty}^{t} \frac{\Delta_{jh}}{\Delta} R_j^{(m)}\, d\alpha, \qquad i = 1, 2, \ldots, m. \tag{6.5.12}$$

The interval of integration $(+\infty, t]$ is taken in harmony with (3.12. vi). Let λ_0 be the largest of the type numbers λ_i of (6.5.10). Then Δ^{-1} has the type number $-\Lambda$ and each Δ_{ij} has type number $\leq \Lambda - \lambda_0$, and each $R_j^{(2)}$ has type number $\leq 2\lambda_0$. Thus each term $(\Delta_{ij}/\Delta) R_{ij}^{(2)}$ has a type number $\leq (\Lambda - \lambda_0) - \Lambda + 2\lambda_0 = \lambda_0$, and $x_i^{(2)}$ has a type number $\leq 2\lambda_0$. By induction, it is easy to prove that each $x_i^{(m)}$ has a type number $\leq m\lambda_0$. Thus the last requirement above is automatically satisfied.

A. Lyapunov proved (under the conditions above) that there is a number A_1, $0 < A_1 < B$, such that for all $\|\alpha\| \leq A_1$, these series are absolutely and uniformly convergent in $t_0 \leq t < +\infty$ and $\|\alpha\| \leq A_1$, and give a solution $x(t)$ of (6.5.1) with $\|x(t)\| < B$ for all $t_0 \leq t < +\infty$, satisfying the condition $x_i(t_0) = a_i$, $i = 1, \ldots, n$.

(d) The first method of Lyapunov. It consists in the systematic use of the process above. An analysis of the asymptotic behavior of these solutions as $t \to +\infty$ is then possible by a discussion of the corresponding series. For systems with constant coefficients a_{ih}, S. Lefschetz has greatly simplified the discussion of the expressions $R_i^{(m)}$ and thus the proof of the convergence of the method.

6.6. Some results of Bylov and Vinograd. R. E. Vinograd [5] has given examples of systems of the form $x' = A(t)x + f(t, x)$, with $f(t, 0) = 0$, $\|f(t, x) - f(t, x_1)\| \leq g(t)\|x - x_1\|$ and $\int^{+\infty} g(t)\, dt < +\infty$, presenting type numbers different from the ones of the linear system $x' = A(t)x$. R. E. Vinograd [5] has given also conditions for their equality. Analogous questions for weakly nonlinear system have been answered by D. M. Grobman [1, 2, 3]. R. F. Bylov [3] has discussed the following question. Consider nonlinear systems of the form

$$x_i' = P_{ii}(t)\, x_i + F_i(t, x), \qquad i = 1, \ldots, n, \tag{6.6.1}$$

where $x = (x_1, \ldots, x_n)$, and

$$F_i(t, 0) = 0, \qquad |F_i(t, x') - F_i(t, x')| < \delta\, |x - x'|, \tag{6.6.2}$$

$i = 1, \ldots, n$. The linear system

$$x_i' = P_{ii}(t)\, x:, \qquad i = 1, \ldots, n, \tag{6.6.3}$$

is supposed to be regular (3.15) with coefficients bounded and such that the limit $\lim t^{-1} \int_0^t P_{ii}(\tau)\, d\tau = \lambda_i$ exists as $t \to +\infty$. Let $\Lambda = \max \lambda_i$. System (6.6.1) is said to be *stable beyond the greatest type number* if for every $\varepsilon > 0$ there exists $\delta > 0$ such that for every $F_1, \ldots, F_n$ satisfying (6.6.2) the type numbers of system

(6.6.1) are $\leq \Lambda + \varepsilon$. Then it is proved that this type of stability of system (6.6.3) occurs if and only if, for every $\alpha > 0$ and continuous bounded function $g(t)$ with $t^{-1}\int_0^t g(\xi)\,d\xi \to \Lambda$ as $t \to +\infty$, there is a constant $C = C(\alpha)$ and a continuous bounded function $f_\alpha(t)$ with $\overline{\lim}\, t^{-1}\int_0^t f(\xi)\,d\xi \leq \alpha$ as $t \to +\infty$ such that $A \leq BC$, where

$$A = \exp\int_0^t [P_{ii} - g]\,d\tau, \qquad B = \exp\left[\alpha(t-\tau) + \int_\tau^t f_\alpha\,d\xi\right]$$

for all i and $t \geq \tau \geq 0$.

6.7. The theorems of Bellman. Consider first a linear system of the form

$$y' = A(t)\,y + \varphi(t), \tag{6.7.1}$$

where $y = (y_1, \ldots, y_n)$, $A(t)$ a given $n \times n$ matrix continuous in $0 \leqq t < +\infty$, $\varphi(t)$ an arbitrary vector function of t. Now consider the nonlinear system

$$x' = A(t)\,x + \Phi(t, x), \tag{6.7.2}$$

where $\Phi(t, x)$ is any vector function, continuous in (t, x) for all $t \geqq 0$ and $x \in E_n$. The question arises concerning the connection between the boundedness in $[0, +\infty)$ of the solutions of system (6.7.1) for arbitrary functions $\varphi(t)$, and the existence in $[0, +\infty)$ and boundedness of the solutions of system (6.7.2). An analogous question can be asked concerning solutions which are L-integrable, or L^2-integrable in $[0, +\infty)$. First results for boundedness have been obtained by O. Perron [10]. In connection with the results already mentioned for linear systems (3.11) and by the use of Banach space methods, R. Bellman [8] has proved the following theorem.

(6.7. i) A necessary condition in order that (6.7.2) possess only bounded solutions for every vector function $\Phi(t, x)$ with either

(a) $\|\Phi(t, x)\| \leqq M$ for $t \geqq 0$ and x arbitrary, or

(b) $\|\Phi(t, x)\| \leqq f_1(t), \quad \int^{+\infty} f_1(t)\,dt < +\infty$, or

(c) $\|\Phi(t, x)\| \leqq f_2(t), \quad \int^{+\infty} f_2^2(t)\,dt < +\infty$,

is that the linear system (6.7.1) has only bounded solutions for all vector functions $\varphi(t)$ satisfying respectively the condition

(a′) $\|\varphi(t)\| \leqq N$ for $t \geqq 0$, or

(b′) $\int^{+\infty} \|\varphi(t)\|\,dt < +\infty$, or

(c′) $\int^{+\infty} \|\varphi(t)\|^2\,dt < +\infty$.

Conversely, if condition (a'), or (b'), or (c') is satisfied, and the vector function $\Phi(t, x)$ respectively satisfies the further condition

(a'') $\|\Phi(t, x)\| \leqq c_1 \|x\|$, or

(b'') $\|\Phi(t, x)\| \leqq c_1 f_3(t) \|x\|$, $\int^{+\infty} f_3(t)\, dt < +\infty$, or

(c'') $\|\Phi(t, x)\| \leqq c_1 f_4(t) \|x\|$, $\int^{+\infty} f_4^2(t)\, dt < +\infty$,

for all $t \geqq 0$ and c_1 sufficiently small, whenever $\|x\| \leqq c_2$, c_2 sufficiently small, then every solution of (6.7.2) for which $\|x(0)\|$ is sufficiently small, is bounded in $[0, +\infty)$. In particular under conditions (a'), (a'') we have $\|x\| \to 0$ as $t \to +\infty$.

For a more abstract formulation of this statement and applications to difference equations, see the same paper of R. BELLMAN [8].

6.8. Invariant measure. Let us consider the autonomous system

$$x' = F(x), \tag{6.8.1}$$

where $x = (x_1, \ldots, x_n)$, $F = (F_1, \ldots, F_n)$, and the components F_j of F are continuous functions in E_n with their first partial derivatives $\partial F_j/\partial x_h$, $j, h = 1, \ldots, n$. Let us consider any point $x_0 \in E_n$ and the corresponding solution $x(t) = x(t, x_0, 0)$ issued from x_0, which certainly exists in some open interval (t_1, t_2) with $t_1 < 0 < t_2$. It is clear that also $x(t_0 + t, x_0, t_0)$ is a solution, where t_0 is an arbitrary constant. For every $t \in (t_1, t_2)$ there exists a neighborhood U of x_0 such that, for every $u \in U$, also the solution $x(t, u, 0)$ exists in some interval containing $[0, t]$. Thus $x(t, u, 0)$ may be thought of as defining a mapping T, or T_t, from U into E_n. By the theorems of existence, uniqueness, and continuous dependence upon the initial values (1.1), it follows that for every t, the mapping T_t: $x = x(t, u, 0)$, $u \in U$, is continuous and one-one, i.e., $u_1 \neq u_2$, implies $T(u_1) \neq T(u_2)$ and vice versa. Also, the general relation holds $T_t T_s = T_{t+s}$. If we consider any open set $V_0 \subset E_n$ and the corresponding solutions $x(t, u, 0)$, $u \in V_0$, which we suppose to exist in some interval (t_1, t_2) with $t_1 < 0 < t_2$, then, for every $t \in (t_1, t_2)$, the mapping $T = T_t$ maps V_0 into an open set V_t. We shall denote by $|V|$ the Lebesgue measure (volume) of an open set $V \subset E_n$. If $u = (u_1, \ldots, u_n) \in V_0$, then $x = T(u)$, $x = (x_1, \ldots, x_n)$, is given by functions $x_j = x_j(u_1, \ldots, u_n)$, $j = 1, \ldots, n$, which, by (1.1), are continuous with their first partial derivatives in V_0. Let us denote as usual by Jacobian $J(u)$, or $J(t, u)$, of the transformation T the determinant of the square matrix $[\partial x_j/\partial u_h]$, $j, h = 1, \ldots, n$. Finally, we shall denote as $\operatorname{div} F$ the scalar $\operatorname{div} F = (\partial F_1/\partial x_1) + \cdots + (\partial F_n/\partial x_n)$.

(6.8. i) If $\operatorname{div} F = 0$, then $|V_t| = V_0$ for every $t \in (t_1, t_2)$; that is, T_t is a measure preserving transformation.

Proof. Let $dx = dx_1 \ldots dx_n$, $du = du_1 \ldots du_n$. Then we have

$$|V_t| = \int_{V_t} dx = \int_{V_0} J(t, u)\, du.$$

By differentiating with respect to t we have

$$(d/dt)\,|V_t| = \int_{V_0} (d/dt)\, J(t, u)\, du,$$

where dJ/dt can be thought of as the sum of n determinants D_j, $j = 1, \ldots, n$, each D_j being obtained by differentiation of the row $\partial x_j/\partial u_h$, $h = 1, \ldots, n$, of J. Since

$$(d/dt)\,(\partial x_j/\partial u_h) = (\partial/\partial u_h)\,(dx_j/dt),$$

each D_j is the Jacobian of $x_1, \ldots, x_{j-1}, \varphi_j, x_{j+1}, \ldots, x_n$, with respect to $u_1, \ldots, u_n$, where $\varphi_j = dx_j/dt$. On the other hand $\varphi_j = F_j(x)$, and

$$\partial\varphi_j/\partial u_h = \sum_{s=1}^{n} (\partial F_j/\partial x_s)\,(\partial x_s/\partial u_h).$$

Thus each D_j is the sum of n determinants, all of which are zero but the one with $s = j$, and we have $D_j = J(\partial F_j/\partial x_j)$, and

$$dJ/dt = D_1 + \cdots + D_n = J \operatorname{div} F.$$

Finally, we have

$$(d/dt)\,|V_t| = \int_{V_0} (dJ/dt)\, du = \int_{V_0} J \operatorname{div} F\, du = 0,$$

and $|V_t|$ is a constant.

Obviously, $\operatorname{div} F = 0$ is also a necessary condition in order that the volume $|v_t|$ of the image v_t of any part v_0 of V_0 be constant with respect to t. Also, by set theory, the invariant measure theorem (6.8. i) can be extended to Borel measurable sets $v \subset E_n$.

Theorem (6.8. i) is particularly important since the system of $n = 2m$ Hamiltonian equations (1.6) regulating a mechanical conservative system with constraints independent of t does verify the condition $\operatorname{div} F = 0$, where now E_{2m} is the space of the $2m$ Hamiltonian coordinates p_i, q_i. This is evident from (1.6.2) since $\operatorname{div} F = S - S = 0$, where S is the sum of the second order derivatives $\partial^2 H/\partial p_i\,\partial q_i$, $i = 1, \ldots, n$. Note that, for an oscillator regulated by the equation $d^2x/dt^2 + f(x) = 0$, x a scalar, by putting $x_1 = x$, $x_2 = dx/dt$, we have the system $x_1' = x_2$, $x_2' = -f(x_1)$, and hence, $\operatorname{div} F = 0$.

A closed bounded set $M \subset E_n$ is said to be an invariant set for a system (6.8.1) provided for every $u \in M$ the solution $x(t, u, 0)$ exists for all $-\infty < t < +\infty$, and $x(t, u, 0) \in M$ for all t.

(6.8. ii) If $\operatorname{div} F = 0$, if M is an invariant set for (6.8.1), and v_0 is any part of M which is Borel measurable and has positive measure $|v_0| > 0$, then there are sequences t_m, $m = 1, 2, \ldots$, with $t_m \to +\infty$ as $m \to +\infty$ ($t_m \to -\infty$) and $v_0 v_{t_m} \neq 0$, $m = 1, 2, \ldots$.

Proof. Let us consider for instance the sequence $t_m = m$, $m = 1, 2, \ldots$, and put $v_m = T_m(v_0)$. If the sets $v_0, v_1, v_2, \ldots$, were all disjoint, then we would have $+\infty > |M| > |v_0| + |v_1| + \cdots + |v_{m-1}| = m v_0$ for every $m \geqq 1$, a contradiction. Thus, there are two m', m'' with $m' < m''$, $v_{m'} v_{m''} \neq 0$, and hence, if $m_1 = m'' - m'$, also $v_0 v_{m_1} \neq 0$, by applying the mapping T_t to both $v_{m'}, v_{m''}$ with $t = -m'$. The same reasoning holds for the sequence v_{hK}, $h = 1, 2, \ldots$, with $K > m_1$, and we deduce that pairs $v_{h'K}, v_{h''K}$, $h' < h''$, have points in common, and the same holds for v_0 and v_{m_2} with $m_2 = (h' - h'')K > m_1$. By indefinite repetition of this reasoning where K takes successively values $K > m_1, K > m_2, \ldots$, we prove (6.8. ii).

If a mechanical system has an invariant set M in the space of the Hamiltonian coordinates, then (6.8. ii) holds. If we think of v_0 as any small neighborhood of an initial state x_0 then, according to POINCARÉ, (6.8. ii) can be interpreted by saying that the probability is unity that an arbitrary motion returns infinitely often to the neighborhood of its initial state. He called this property of mechanical systems "stability in the sense of POISSON". A motion having this property is usually said to be recurrent. For developments and applications of the concept of recurrency, and the wide bibliography on the subject see, e.g., G.D. BIRKHOFF [25].

Statement (6.8. i), by a hydrodynamic analogy, is often referred to as characterizing an "incompressible" motion, or flow. An important class of properties of measure invariant continuous transformations is studied in ergodic theory. See on the subject, e.g., E. HOPF [2]. The family T_t of transformations T_t, $-\infty<t<+\infty$, verifying $T_s T_t = T_{s+t}$, can be thought of as a one parameter transformation group acting on a topological space. For recent extensions of incompressibility and recurrent properties to general transformation groups in topological spaces, see W.H. GOTTSCHALK and G.A. HEDLUND [1].

The recurrency property of solutions of dynamical systems raises the question as to whether these solutions are necessarily periodic or almost periodic. This is not the case in general. Nevertheless the following positive statement seems pertinent here:

(6.8. iii) If $\operatorname{div} F = 0$, if a solution $x(t)$ of (6.8.1) exists for all $-\infty<t<+\infty$, is bounded and (Lyapunov) stable at both sides (1.2), then $x(t)$ is (Bohr) almost periodic (C.R. PUTNAM and L.L. HELMS [1]).

For other results connecting recurrency, periodicity, and almost periodicity, see G.D. BIRKHOFF [7, 20, 25], P. FRANKLIN [1], I.G. MALKIN [1], C.R. PUTNAM [12, 20], P. HARTMAN and A. WINTNER [26], N. BOGOLYUBOV [1], N. BOGOLYUBOV and N. KRYLOV [20, 21, 26, 27, 29, 30], A. MARKOV [1], W. STEPANOV and A. TIHONOV [1], V.V. NEMICKII and V.V. STEPANOV [1].

It may be interesting to mention that the stability at both sides of a point of equilibrium $x_0 \in E_n$, i.e., a point where $F(x_0) = 0$ (1.5), is related to the concept of invariant set by the following statement:

(6.8. iv) An equilibrium point $x_0 \in E_n$ is (Lyapunov) stable at both sides if and only if there exists a sequence of invariant sets M_n, $n = 1, 2, \ldots$, closing down to x_0, i.e., $M_n \supset M_{n+1}$, and x_0 is the only point of the intersection set $M_1 M_2 M_3 \ldots$.

For the existential analysis of periodic solutions (closed orbits) of dynamical systems a variational method has been devised, based on the fact that these orbits satisfy principles of minimum (e.g., the principle of the least action). On this subject see E.T. WHITTAKER [1], the subsequent rigorous analysis of L. TONELLI [1, 2] and A. SIGNORINI [1,2], the further research of G.D. BIRKHOFF [14] and W. DAMKÖHLER [1] and the exposition of G.D. BIRKHOFF [25], and the more general analysis of L. TONELLI [3, 6, 7] and M. MORSE [1].

6.9. Differential equations on a torus. We refer briefly here to some basic results, mainly due to P. BOHL [4] and A. DENJOY [1], which have had recently important applications (see 8.11). Consider a single differential equation

$$x' = f(t, x), \qquad (') = d/dt, \tag{6.9.1}$$

where $f(t, x)$ is a real continuous function of the real variables t and x, periodic of period 1 with respect to t and x, i.e., $f(t, x+1) = f(t+1, x) = f(t, x)$ for all t and x. We suppose also that conditions are satisfied assuring the uniqueness of the solution $x(t, x_0, t_0)$ through each point (x_0, t_0). Because of the double periodicity of f, we may think of $f(t, x)$ as defined on the surface of a torus. Indeed, the transformation $u = (a + b\cos 2\pi x)\cos 2\pi t$, $v = (a + b\cos 2\pi x)\sin 2\pi t$, $w = b \sin 2\pi x$, maps the (x, t)-plane into a torus S of E_3, and all points $(t+n, x+m)$, $m, n = 0, \pm 1, \pm 2, \ldots$, into the same point (u, v, w) of the torus S. This same transformation will enable us to map every solution $x(t, x_0, t_0)$ of (6.6.1) into a curve on the torus, which we will denote as an orbit on S.

If $x(t, u, 0)$ is any solution of (6.9.1), it is clear that $x(t, u+1, 0) = x(t, u, 0) + 1$ is also a solution, and that, for every fixed t, $x(t, u, 0)$ is a continuous strictly increasing function of u (a consequence of the uniqueness). Thus for any fixed t, say $t = 1$, we may consider the function of u, $x = \varphi(u) = x(1, u, 0)$, as defining a mapping of the straight line $t = 0$ onto the straight line $t = 1$, with $\varphi(u+1) = \varphi(u)$. The mapping φ is one-one, sense preserving, and continuous together with its inverse φ^{-1}. The same mapping induces on the torus S a mapping, say T, of the main section $C = [v = 0,\ (u-a)^2 + w^2 = b^2]$ onto itself. If P is any point of C, then TP denotes the image of P on C, and we may consider also all the iterate images $T^n P$, $n = 0, \pm 1, \pm 2, \ldots$, of P, corresponding to the values $x = \varphi_n(u) = x(n, u, 0)$ of x. The group relation $T^m T^n P = T^{m+n} P$ obviously holds.

A first basic theorem on the subject states: For every real number u the limit $\varrho = \lim \varphi_n(u)/n$ as $|n| \to +\infty$ exists, is finite, and independent of u. Thus the number ϱ is a property in the large of the given system (6.9.1). A first property of ϱ is given by the following statement: The number ϱ is rational if and only if (6.9.1) has a solution $x(t, u, 0)$ with $x(n, u, 0) = u + m$ for some integers $m, n = 0, \pm 1, \pm 2, \ldots$. In other words, ϱ is rational if and only if some power T^n of T has a fixed point on C. Finally we may say, ϱ is rational if and only if there is on S some closed orbit. For ϱ rational it is said that (6.9 1) presents the *periodic case*.

For the sake of simplicity we may assume now that f is sufficiently smooth, say f has continuous first partial derivative f_x with respect to x, though a much weaker hypothesis would suffice (A. DENJOY [1]). We shall assume also that ϱ is irrational. Then ϱ can be interpreted as a "rotation number", in the sense that T can be interpreted as effecting a rotation R of an angle $2\pi\varrho$ of C. More precisely, there is a homeomorphism h of C onto itself such that $T=h^{-1}Rh$. A more detailed description of the behavior of the orbits on the torus is given by the following result of P. BOHL: There is a continuous function $F(t, v)$, periodic of period 1 with respect to t and v, such that the relation holds

$$x(t, u, 0) = t\varrho + c + F(t, t\varrho + c), \tag{6.9.2}$$

where c is a constant. More precisely, for every constant c, (6.9.2) is the solution of (6.9.1) with $u=c+F(0, c)$, and, conversely, for every solution $x(t, u, 0)$ of (6.9.1) there is a constant c for which (6.9.2) holds. It should be added finally that for every $P\in C$ the sequence $T^n P$, $n=1, 2, \ldots$, has the whole of C as its set of cluster points. For ϱ irrational, it is said that (6.9.1) presents the *ergodic case*.

On the same subject, and particularly, for extensions of these results to systems of differential equations, see P. BOHL [4], A. DENJOY [1], A.N. KOLMOGOROV [1], H. KNESER [1], C.L. SIEGEL [1], E.A. CODDINGTON and N. LEVINSON [3].

6.10. Bibliographical notes. The extension of the previous theory to systems of infinitely many equations and unknowns is complete now. See V. HARASAHAL [1, 2], K. PERSIDSKII [7, 9], M. R. RESETOV [1, 2], H. I. IBRASEV [1], S. GORSIN [1, 2, 3]. YU. A. RYABOV [1] has recently proved the convergence of the formal series of LYAPUNOV for periodic series of autonomous systems in one case left undiscussed by LYAPUNOV. For other research in the line of the first method, see V. A. TROICKII [1], B. TULEGENOV [1].

The theory of automatic regulators monitored by a system of equations of the type

$$x_i' = \sum_{j=1}^{n} a_{ij} x_j + \alpha_i \xi, \quad i = 1, \ldots, n,$$

$$\xi' = \sum_{j=1}^{n} \beta_j x_j + f(\xi),$$

corresponds to one single regulator. The stability and behavior of the solutions of systems of this type has been discussed by M. A. AIZERMAN [1, 2, 3, 4], A. M. LETOV [1, 3, 4, 5, 6], P. V. BROMBERG [1], B. V. BULGAKOV [1], B. V. BULGAKOV and M. Z. LITVIN-SEDOL [1], in the line of the first, second method of LYAPUNOV, and other ideas, on which we shall return later (§ 8). A. I. LURE [6, 7, 9, 10] has extended the theory particularly in the line of the first method of LYAPUNOV for the case of many organs of regulation. On this subject see also the books A. I. LURE [8] and A. M. LETOV [7].

§ 7. The second method of LYAPUNOV

7.1. The function V of LYAPUNOV. We consider a differential system

$$x_i' = f_i(t, x_1, \ldots, x_n), \quad i = 1, 2, \ldots, n, \tag{7.1.1}$$

or $x' = f(t, x)$, where the functions f_i are real and continuous in the set $S_0 = [t_0 \leq t < +\infty, \|x\| \leq b]$ of the real variables $t, x_1, \ldots, x_n$, and some condition is satisfied assuring the uniqueness of the solutions $x(t; t_0, x_0)$ at every point (t_0, x_0) of S_0. We shall suppose also $f_i(t; 0, \ldots, 0) = 0$, $i = 1, \ldots, n$, and, therefore, $x = 0$ is a solution of (7.1.1).

We will now consider a function $V(t, x_1, \ldots, x_n) = V(t, x)$ real and continuous in some set $S = [T \leq t < +\infty, \|x\| \leq B]$ with $T \geq t_0$, $0 < B \leq b$, and we will suppose that V has the following further properties: (a) V has first partial derivatives $\partial V/\partial t$, $\partial V/\partial x_s$, $s = 1, 2, \ldots, n$, continuous in S; (b) $V(t, 0, \ldots, 0) = 0$ for all $t \geq T$.

A function V as above is said to be *positive (negative) semidefinite* if we have $V \geq 0$ $[V \leq 0]$ in some set S. A function $W(x_1, \ldots, x_n)$ as above, independent of t, is said to be *positive (negative) definite* if $W > 0$ $[< 0]$ for all $x \neq 0$, $\|x\| \leq B$. A function $V(t, x_1, \ldots, x_n) = V(t, x)$, depending on t, is said to be *positive (negative) definite* if there exists a positive definite function $W(x_1, \ldots, x_n)$ such that $V \geq W$ $[-V \geq W]$ in some set S.

For instance $V = x_1^2 + x_2^2 + 2x_1 x_2 \cos t$ is semidefinite positive. Both the functions $V = t(x_1^2 + x_2^2) - 2x_1 x_2 \cos t$, and $V = x_1^2 + x_2^2 + 2x_1 x_2 \cos t \sqrt{x_1^2 + x_2^2}$ are definite positive. In the former we shall suppose for instance $t \geq 2$, and in the latter we shall suppose for instance $\|x\| < \frac{1}{2}$, where $x = (x_1, x_2)$.

A function V as above is said to be *bounded* if there is a constant $M > 0$ such that $|V| < M$ in some set S. A bounded function V is said to have an *infinitesimal upper bound* if, given $\varepsilon > 0$, there exists an $h > 0$ such that $|V| < \varepsilon$ for all $t \geq T$ and $\|x\| < h$.

For instance $V = \sin(x_1 + \cdots + x_n)\, t$ is bounded but has no infinitesimal upper bound. The function $V = (x_1 + \cdots + x_n) \sin t$ is bounded and has an infinitesimal upper bound.

Given any function V in a set S let us denote by V' the following function of $t, x_1, \ldots, x_n$ in S:

$$V' = V'(t, x_1, \ldots, x_n) = V'(t, x) = \sum_{i=1}^{n} \frac{\partial V}{\partial x_i} f_i(t, x_1, \ldots, x_n) + \frac{\partial V}{\partial t}.$$

Obviously if $x(t)$, $t_1 \leq t \leq t_2$, is any solution of (7.1.1) in an interval $[t_1, t_2)$ with $T \leq t_1 < t_2$ and $\|x(t)\| \leq B$, then we may consider the function $V(t) \equiv V[t, x(t)]$ of t only, $t_1 \leq t \leq t_2$, obtained by replacing $x(t)$ in V, and we have

$$\frac{dV}{dt} = \sum_{i=1}^{n} \frac{\partial V}{\partial x_i} x_i' + \frac{\partial V}{\partial t} = V'[t, x(t)].$$

Thus V' is the derivative of V along the solution $x(t)$ of (7.1.1). For the sake of simplicity we shall say that V' is the derivative of V.

7.2. The theorems of LYAPUNOV. (7.2. i) If a function V exists which is definite and whose derivative V' is a semidefinite function whose sign is contrary to that of V, then the solution $x=0$ of (7.1.1) is stable (A. LYAPUNOV [3]).

Proof. We may suppose V definite positive. By hypothesis a function $W(x_1, \ldots, x_n)$ and constants $T \geq t_0$, $0 < B \leq b$, exist such that $V \geq W > 0$ for all $t \geq T$, $\|x\| \leq B$, $x \neq 0$. In addition, $V' \leq 0$ for all $t \geq T$, $\|x\| \leq B$. If we consider any solution $x(t) = x(t, T, a)$, not identically zero, with $\|a\| < B$, then the solution $x(t)$ certainly exists in some interval $[T, t_1)$, $T < t_1$, namely it exists as long as we have $\|x(t)\| < B$. Thus we may suppose either $t_1 = +\infty$, or $T < t_1 < +\infty$, and then t_1 is the first point $t_1 > T$ where $\|x(t_1)\| = B$. For all t with $T \leq t \leq t_1$ $[T \leq t < +\infty$ if $t_1 = +\infty]$ we have then

$$V(t, x(t)) - V(T, a) = \int_T^t V'(u, x(u))\, du \leq 0,$$

and thus also $0 < V(t, x(t)) \leq V(T, a)$ since we cannot have $x(t) = 0$ for any t.

Given any $\varepsilon > 0$, $0 < \varepsilon \leq B$, let us consider the set I of all $x = (x_1, \ldots, x_n)$ such that $\varepsilon \leq \|x\| \leq B$. Let us denote by μ the minimum of W in I. Since 0 is not in I and I is compact, we have $\mu > 0$. Now let us determine a constant λ, $0 < \lambda < \varepsilon$, such that $V(T, a) < \mu$ for all $\|a\| \leq \lambda$. Until now we have used only the continuity of V and the condition $V(T, 0) = 0$. If $x(t) = x(t; T, a)$ with $\|a\| < \lambda$, we have for all $T \leq t < t_1$,

$$\mu > V(T, a) \geq V[t, x(t)] \geq W[x(t)],$$

and from here we deduce that $\|x(t)\| < \varepsilon$ for the same t. Indeed, this relation is true for $t = T$ and thus it is valid in an interval $[T, t_2]$. It cannot occur that we have $\|x(t_2)\| = \varepsilon$ at a first point $t_2 \geq T$ since then we would have

$$\mu > V(T, a) \geq V[t_2, x(t_2)] \geq W[x(t_2)] \geq \mu,$$

a contradiction. This proves that the relation $\|x(t)\| < \varepsilon$ holds for all $T \leq t \leq t_2$, and thus $\|x(t)\| < B$. As a consequence we have $t_2 = t_1 = +\infty$, the solution $x(t) = x(t; T, a)$ exists in $[T, +\infty)$ and $\|x(t)\| < \varepsilon$ for all $t \geq T$. This proves that the solution $x=0$ of (7.1.1) is stable in the sense of LYAPUNOV.

(7.2. ii) If a function V exists which is definite, and has an infinitesimal upper bound, if the derivative V' is also a definite function whose sign is contrary to that of V, then the solution $x=0$ of (7.1.1) is asymptotically stable (A. LYAPUNOV [3]).

Proof. We may suppose V definite positive and V' definite negative. By hypotheses, there are certain positive definite functions $W(x)$, $W'(x)$, independent of t, such that $V \geq W$, $-V' \geq W'$ for all $t \geq T$, $\|x\| \leq B \leq b$, where T, B are constants. By (7.2. i) we know that there exists some λ, $0 < \lambda < B$, such that all solutions $x(t) = x(t, T, a)$ with $\|a\| < \lambda$, exist in $[T, +\infty)$ and $\|x(t)\| < B$ for all $t \geq T$. Suppose, if possible, that for such a solution we could have $V[t, x(t)] \geq l > 0$ for all $t \geq T$. Then, since V has an infinitesimal upper bound, there exists an h, $0 < h < B$, such that $\|x\| < h$, implies $0 \leq V < l$ for all $t \geq T$. We conclude that the solution $x(t)$ satisfies the relation $\|x(t)\| > h$ for all $t \geq T$. If $\mu > 0$ is the minimum of W' for all $l \leq \|x\| \leq B$, we have $W'[x(t)] \geq \mu$ for all $t \geq T$ and finally $-V' \geq W' \geq \mu$, $V(t) \leq V(T) - \mu(t-T)$. This implies $V(t) < 0$ for all t large enough, a contradiction. This assures that $V(t)$ takes (positive) values as small as we want for large t, and since $V(t)$ is nonincreasing we have $V(t) \to 0$ as $t \to +\infty$. Thus we have also $W[x(t)] \to 0$ as $t \to +\infty$ and finally $\|x(t)\| \to 0$ as $t \to +\infty$.

(7.2. iii) If a function V exists such that (1) V has an infinitesimal upper bound; (2) V' is definite; (3) there exists $t_1 \geq T$ such that for each $t \geq t_1$ and $h > 0$ there is at least one a with $\|a\| < h$ and $V(t, a)$ of the same sign as V', then the solution $x = 0$ of (7.1.1) is unstable (A. LYAPUNOV [3]).

Proof. We may suppose V' definite positive. By hypothesis, there exists a positive definite function $W(x)$ independent of t such that $V' \geq W$ for all $t \geq T$ and $\|x\| < B$, T, B constants. Since V is bounded we may suppose T, B such that $|V| < L$ for some $L > 0$ and all $t \geq T$, $\|x\| < B$. In addition, since $V' > 0$ for the same t and $x \neq 0$, and $V(t) - V(t_1) = \int_{t_1}^{t} V'(t)\, dt$, we have $V(t) > V(t_1)$ for all $t \geq t_1 \geq T$. Given $\varepsilon > 0$, by condition (3) above, there are $t_1 \geq T$ and a with $\|a\| < \min[B, \varepsilon]$, such that $V_0 = V(t_1, a) > 0$. Let $x(t) = x(t; t_1, a)$. Since V has an infinitesimal upper bound there exists a number $\lambda > 0$ such that $\|x\| < \lambda$, $t \geq t_1$, imply $|V(t, x)| < V_0$. The solution $x(t)$ defined above exists in an interval $[t_1, t_2)$, $t_1 < t_2 < +\infty$, and we will suppose that t_2 is the first point $t_2 > t_1$ at which $\|x(t_2)\| = B$, if such a point exists; otherwise take $t_2 = +\infty$. In $[t_1, t_2)$ the function $V(t)$ is nondecreasing, hence $V(t) \geq V(t_1) = V_0 > 0$ and hence $\|x(t)\| \geq \lambda$ for all $t_1 \leq t < t_2$. If now $\mu > 0$ is the minimum of W for all $\|x\| \leq B$, $\max \|x\| \geq \lambda$, we have $V'(t) \geq W > \mu > 0$ and hence $V(t) \geq V(t_1) + \mu(t - t_1)$. Since the last expression approaches $+\infty$ as $t \to +\infty$ while $V(t) \leq L$ for all $t \geq t_1$, $\|x\| \leq B$, we conclude that a first point t_2, $t_1 < t_2 < +\infty$, exists where $\|x(t_2)\| = B$. Thus no matter how small ε we take, the solution $x(t; t_1, a)$ with $\|a\| < \varepsilon$ will satisfy the relation $\|x(t)\| = B$ at some $t = t_2 > t_1$. We conclude that the solution $x = 0$ of (7.1.1) is unstable.

Example. Consider the differential system

$$\frac{dx_i}{dt} = \frac{\partial V}{\partial x_i}, \quad i = 1, 2, \ldots, n, \tag{7.2.1}$$

where V is any function V as in (7.1). Then

$$V' = \sum_{i=1}^{n} \left(\frac{\partial V}{\partial x_i}\right)^2,$$

and thus V' is semidefinite positive. If V is definite negative, then by (7.2. i) the solution $x = 0$ of (7.2.1) is stable; if V is definite positive and also V' is supposed to be definite, then, by (7.2. iii), $x = 0$ is unstable.

Example. Consider the differential system

$$\frac{d^2 x_i}{dt^2} = -\frac{\partial W}{\partial x_i}, \quad i = 1, 2, 3, \tag{7.2.2}$$

which may be interpreted as to describe the motion of a unit mass in the potential field $W = W(x_1, x_2, x_3)$. We assume $W(0, 0, 0) = 0$, and we write (7.2.2) as

$$\frac{dx_i}{dt} = x_{3+i}, \qquad \frac{dx_{3+i}}{dt} = -\frac{\partial W}{\partial x_i}, \quad i = 1, 2, 3.$$

If we put $V = W(x_1, x_2, x_3) + 2^{-1}(x_4^2 + x_5^2 + x_6^2)$, then

$$V' = \sum_{i=1}^{3} \left(\frac{\partial W}{\partial x_i} x_{3+i} - x_{3+i} \frac{\partial W}{\partial x_i}\right) = 0,$$

hence V' is semidefinite. If V is positive definite (i.e., W is positive definite, and hence the potential energy W has a minimum at $x = 0$), then the solution $x_i = 0$, $i = 1, 2, 3$, of (7.2.2) is stable. In this example the Lyapunov function V is the total energy of the system.

7.3. More recent results. The theorems proved in (7.2) have been widely improved by V. MARACKOV, I. G. MALKIN, and J. L. MASSERA in recent years and put into forms which have been proved as both necessary and sufficient for stability, or instability. The problem of determining a V-function for systems (7.2.1) for which the solution $x=0$ is known to be stable, is called the inverse problem. A few concepts should be discussed first.

The solution $x=0$ of (7.1.1) is said to be *equiasymptotically stable* if there exists a $\delta>0$ fixed, and, for any $\varepsilon>0$, a $T=T(\varepsilon)\geq 0$ such that $|x_0|\leq\delta$, $t\geq T(\varepsilon)$ imply $\|x(t; x_0, 0)\|<\varepsilon$ (J. L. MASSERA [2]). In other words, $x\to 0$ as $t\to+\infty$ uniformly with respect to the initial values x_0 at $t=0$. Examples show that equiasymptotic stability is actually a stronger requirement than asymptotic stability (J. L. MASSERA [2]). Nevertheless, the same author has proved that for all systems (7.1.1) of order $n=1$, for all systems of any order which are periodic, or autonomous, equiasymptotic stability and asymptotic stability are equivalent.

Concerning the vector function V the following requirements, due to I. G. MALKIN [1], can be taken into consideration.

Assumption M: given $\varepsilon''>0$ and $T'\geq 0$ there are certain numbers $0<\varepsilon'=\varepsilon'(\varepsilon)\leq\varepsilon''$, and $T''=T''(\varepsilon'', T')\geq T'$ such that for all x', t', x'', t'' with $\|x'\|\leq\varepsilon'$, $t'\leq T'$, $\|x''\|\geq\varepsilon''$, $t''\geq T''$, we have $V(t', x')\leq V(t'', x'')$.

Assumption ML: There is a positive definite function V^* such that $V'+V^*\to 0$ as $t\to+\infty$ uniformly with respect to all x verifying a relation of the form $0<\lambda\leq\|x\|\leq\delta$.

A positive definite function V having an infinitesimal upper bound satisfies necessarily assumption M. A function V having a derivative V' which is negative definite satisfies necessarily condition ML.

I. G. MALKIN [1] proved first that in theorem (6.2. i) the hypotheses can be replaced by the weaker ones that V is positive definite, has an infinitesimal upper bound, and satisfies ML. Later J. L. MASSERA [2] showed that the same assumptions of (7.2. i), and even the weaker ones of I. G. MALKIN assure actually a stronger conclusion, namely that $x=0$ is asymptotically stable and even equiasymptotically stable:

(7.3. i) If a positive definite function V exists which has an infinitesimal upper bound, and V' satisfies ML, then the solution $x=0$ is equiasymptotically stable (J. L. MASSERA [2]).

A further stronger conclusion has been proved by I. G. MALKIN [21] for linear systems. In another direction theorem (7.2. ii) has been extended by V. MARACKOV and by J. L. MASSERA as follows:

(7.3. ii) If $f(t, x)$ is bounded, and a positive definite function V exists for which V' is negative definite, then the solution $x=0$ is asymptotically stable (V. MARACKOV [1]).

(7.3. iii) If system (7.1.1) is periodic, or autonomous, and a positive definite function V exists for which V' is negative definite, then the solution $x=0$ is asymptotically (equiasymptotically) stable (V. MARACKOV [1], J.L. MASSERA [2]).

(7.3. iv) If a function V exists which is positive definite, and satisfies condition M, and V' is negative definite, then $x=0$ is equiasymptotically stable (J.L. MASSERA [2]).

J.L. MASSERA has shown by examples that the boundedness condition for f cannot be removed in general in (7.3. ii) and that, in general, the hypotheses (7.3. ii) do not assure equiasymptotic stability, while, as (7.3. iii) states, this is always the case for periodic, or autonomous systems, for which, after all, asymptotic and equiasymptotic stability coincide. The following statement, due to J.L. MASSERA, shows that (7.3. iii) is actually a necessary and sufficient condition for asymptotic stability:

(7.3. v) If the system (7.1.1) is periodic, or autonomous, and the solution $x=0$ is equiasymptotically (asymptotically) stable, then a positive definite function V exists such that V' is negative definite (J.L. MASSERA [2]).

For linear systems statements (7.3. v) can be completed in the sense that V is proved to satisfy condition M, and thus providing the converse of (7.3. iv) for linear systems.

In another direction theorem (7.2. i) has been improved. I.G. MALKIN [1] has shown that the existence of a function V is a necessary and sufficient condition for uniform stability.

For linear systems with periodic coefficients N.G. CETAEV [5] has shown that a function V always exists, namely a quadratic form. CH. NOUGMANOVA [1] has then obtained by the same method, a criterion for stability and instability. For linear systems with general continuous coefficients, A.D. GORBUNOV [2] has proved, by another process, that a suitable function V always exists (a quadratic form).

Let us consider now systems of the type already considered in 6:

$$x' = A\,x + P(t, x), \qquad x = (x_1, \ldots, x_n),$$

where A is a $n \times n$ matrix whose coefficients are continuous bounded functions of t, and where P is a vector function of t and x whose components are power series of $x_1, \ldots, x_n$, convergent for all $|x| < k$ for some $k > 0$ (uniformly with respect to $t \geq 0$), whose coefficients are continuous bounded functions of t. I. G. MALKIN [1] has proved that, if a positive definite function V exists with V' negative definite, then there is also such a function V which is a quadratic form in $x_1, \ldots, x_n$, and

$$V \geq a^2 \Sigma\, x_i^2, \qquad V' \leq -b^2 \Sigma\, x_i^2,$$

for some $a, b > 0$.

If A is a constant or periodic matrix having k characteristic roots equal to zero, the theorems of § 6 do not allow us to infer that the zero solution is stable, and various methods have been proposed (see A. LYAPUNOV [3]). I. G. MALKIN [2] has given a method with which the question whether $x=0$ is stable or unstable can be answered by discussing a convenient derived system of order k.

7.4. A particular partial differential equation. Let us consider the partial differential equation

$$\sum_{i=1}^{n} (a_{i1} x_1 + a_{i2} x_2 + \cdots + a_{in} x_n) \frac{\partial V}{\partial x_i} = \lambda V, \tag{7.4.1}$$

where the coefficients a_{ih} are given real constants, where λ denotes an indeterminate constant, and V a homogeneous polynomial of some degree m with indeterminate coefficients $\xi_1, \xi_2, \ldots, \xi_N$. Here $N=N(m)$ denotes the number of the coefficients ξ_i of V. Then (7.4.1) can be reduced to a system of N algebraic linear homogeneous equations in the N coefficients ξ_i of the form

$$\sum_{h=1}^{N} \alpha_{ih} \xi_h = \lambda \xi_i, \quad i = 1, 2, \ldots, N, \tag{7.4.2}$$

where by α_{ih} we denote N^2 convenient forms in the coefficients a_{ih} with constant integral positive coefficients. Finally (7.4.1) has a nonzero solution if and only if λ verifies the algebraic equation of degree N

$$D_m(\lambda) = \det[\alpha_{ih} - \lambda \delta_{ih}] = 0, \tag{7.4.3}$$

where $\delta_{ii}=1$, $\delta_{ih}=0$, $i \neq h$, $i, h = 1, 2, \ldots, N$. For $m=1$, equation (7.4.3) is the characteristic equation of the matrix $[a_{ih}]$, i.e.

$$D_1(\lambda) = \det[a_{ih} - \lambda \delta_{ih}] = 0. \tag{7.4.4}$$

Let us denote by $\lambda_1, \lambda_2, \ldots, \lambda_n$ the n roots of the equation (7.4.4), that is, the characteristic roots of the matrix $[a_{ih}]$. A. LYAPUNOV has discussed equation (7.4.3) for every $m \geq 1$ and has proved the following theorems:

(7.4. i) For every $m>1$ the roots of $D_m(\lambda)=0$ are contained in the formula $\lambda = m_1\lambda_1 + m_2\lambda_2 + \cdots + m_n\lambda_n$, with nonnegative integers m_i such that $m_1 + m_2 + \cdots + m_n = m$.

(7.4. ii) For every $m>1$, if $D_m(0) \neq 0$, then there exists one and only one form V of degree m such that

$$\sum_{i=1}^{n} (a_{i1} x_1 + \cdots + a_{in} x_n) \frac{\partial V}{\partial x_i} = U,$$

where U is any given form of degree m. The condition $D_m(0) \neq 0$ is certainly satisfied if $\Sigma m_i \lambda_i \neq 0$ for all positive integers m_i with $\Sigma m_i = m$.

(7.4. iii) If $R(\lambda_i) < 0$, $i = 1, 2, \ldots, n$, and U is a definite form of even degree m, then the form V of the same degree m defined in (7.4. ii) is also definite and of sign contrary to U.

Indeed if $x = x(t)$ is any nonzero solution of the linear system $x' = Ax$ then $\|x\| \to 0$ as $t \to +\infty$. On the other hand, we have $dV/dt = U$ and thus $dV/dt > 0$ if U is positive; $dV/dt < 0$ if U is negative. All this is possible only if V is definite and of sign contrary to that of U.

(7.4. iv) If $R(\lambda_i) > 0$ for some $i = 1, 2, \ldots, n$, and U is a definite form of even degree m, then the V of the same degree m defined in (7.4. ii) is not definite.

7.5. Autonomous systems. We shall consider again differential systems of the form

$$x' = A\,x + P(x)\,, \tag{7.5.1}$$

where $A = [a_{ih}]$ and each component P_i of P has a development in power series of $x_1, \ldots, x_n$ with coefficients $P_{ih_1\ldots h_n}$, $h_1 + \cdots + h_n \geqq 2$, verifying a relation $|P_{ih_1\ldots h_n}| < M\,C^{-(h_1+\cdots+h_n)}$ with M and C positive constants independent of $i, h_1, \ldots, h_n$. Hence the same series converge absolutely and uniformly for $\|x\| < B$ where $0 < B < C$ is any constant.

New proof of theorem (6.2. i). Let us consider the following partial differential equation

$$\sum_{i=1}^{n} (a_{i1}\,x_1 + \cdots + a_{in}\,x_n)\,\frac{\partial V}{\partial x_i} = x_1^2 + \cdots + x_n^2. \tag{7.5.2}$$

By $R(\lambda_i) < 0$, $i = 1, 2, \ldots, n$, and (7.4. iii) it follows that there exists a quadratic form $V = V(x_1, \ldots, x_n)$ definite negative satisfying (7.5.2) and thus

$$V' = \sum_{i=1}^{n} \frac{\partial V}{\partial x_i}\,[a_{i1}\,x_1 + \cdots + a_{in}\,x_n + P_i(x)] = x_1^2 + \cdots + x_n^2 + \sum_{i=1}^{n} P_i(x)\,\frac{\partial V}{\partial x_i}\,.$$

Certainly V' is positive for all x with $x \neq 0$, $\|x\| < b$, where $b \leq B$ is a positive constant sufficiently small. Then, by (7.2. i) the solution $x = 0$ of system (7.5.1) is stable, and, by (7.2. ii) it is also asymptotically stable. Thus theorem (6.2. i) is proved again for system (7.5.1) by the second method of LYAPUNOV.

New proof of theorem (6.2. ii). If $R(\lambda_i) > 0$ for some $i = 1, 2, \ldots, n$, and $D_2(0) \neq 0$, then, by (7. 4. ii), a quadratic form V satisfying (7.5.2) exists and, by (7.4. iv), V is not definite. Since V is a quadratic form, V takes then positive as well as negative values for vectors $x \neq 0$ in norm $\|x\|$ as small as we want. By (7.2. iii) the solution $x = 0$ of system (7.5.1) is not stable. Thereby theorem (6.2. ii) is proved again for system (7.5.1) under the additional hypothesis $D_2(0) \neq 0$. This condition is certainly satisfied if $\Sigma m_i \lambda_i \neq 0$ for all noनnegative integers m_i with $\Sigma m_i = 2$.

7.6. Bibliographical notes. For further criteria of stability and instability involving a V-function, besides the research of I. G. MALKIN, V. MARACKOV, and J. L. MASSERA already mentioned, see N. G. CETAEV [1—9], G. N. DUBOSIN [3], A. P. DUVAKIN and A. M. LETOV [1], A. D. GORBUNOV [1—4], S. V. KALININ [1, 2], N. N. KRASOVSKII [1—10], A. I. LURE [8], A. A. LEBEDEV [1, 2], A. D. MAIZEL [1].

For the inverse problem, besides the papers of I. G. MALKIN and J. L. MASSERA, A. D. GORBUNOV and V. MARACKOV, see N. N. KRASOVSKII [7], E. A. BARBASIN [4], and N. G. CETAEV [5, 9], T. YOSHIZAWA [5], J. KURZWEIL [1, 2].

For a method similar to the one of this § see N. P. ERUGIN [9].

For reduction principles similar to the one of I. G. MALKIN mentioned at the end of (7.3), see E. I. DYHMAN [1, 2], N. G. CETAEV [5], S. GORSIN [1, 2].

For applications of the first and second method of LYAPUNOV to the question of the stability of regulators, see A. I. LURE [1—10], A. M. LETOV [1—7], R. A. SPASSKII [1], A. F. LURE and M. G. FIALKO [1], A. ANDRONOV and N. N. BAUTIN [1], M. A. AIZERMAN [1—4], E. A. BARBASIN and N. N. KRASOVSKII [1], B. V. BULGAKOV [1], B. V. BULGAKOV and M. Z. LITVIN-SEDOL [1], P. V. BROMBERG [1].

For a general discussion of the asymptotic stability region, i.e., the region of all points (x_0, t_0) such that $x(t; x_0, t_0) \to 0$ as $t \to +\infty$, see L. A. KUNIN [1], K. PERSIDSKII [1], E. A. BARBASIN and N. N. KRASOVSKII [1, 2], N. N. KRASOVSKII [2, 5, 13], S. GORSIN [2]. In particular V. A. PLISS [1] has given a necessary and sufficient condition in order that the region above is the whole space (stability in the large). On the same subject see also N. P. ERUGIN [6, 8, 13], B. A. ERSOV [1, 2],

V. V. NEMICKII [9], R. E. VINOGRAD [11], T. YOSHIZAWA [6, 7, 8]. For extensions of LYAPUNOV's method to system of countably many equations see K. PERSIDSKII [1—10], H. I. IBRASEV [1], E. GORSIN [3], L. ERMOLAEV [1]. For recent expositions see H. A. ANTOSIEWICZ [6], W. HAHN [1].

§ 8. Analytical methods

We shall attempt to give some information on various results and analytical methods concerning existence and approximation of periodic solutions for nonlinear systems. Of particular interest are the so-called weakly nonlinear systems which contain a parameter ε which is supposed to be "small", and whose limit system for $\varepsilon = 0$ is linear.

Methods more topological in character will be referred to in § 9.

8.1. Introductory considerations. We may consider systems of the form

$$x' = f(x, t), \quad \text{or} \quad x_i' = f_i(x, t), \quad x = (x_1, \ldots, x_n), \quad i = 1, \ldots, n, \tag{8.1.1}$$

possibly containing a parameter, and if they are weakly nonlinear, of the form

$$x' = A\,x + \varepsilon f(x, t), \quad \text{or} \quad x_i' = \sum_{h=1}^{n} a_{ih}\, x_h + \varepsilon f_i(x, t), \quad i = 1, \ldots, n, \tag{8.1.2}$$

where A may be constant, or periodic. It may be convenient to consider equations of the form

$$x'' + g(x, x')\, x' + f(x, x') = F(t), \tag{8.1.3}$$

possibly containing a parameter, or of the form

$$x'' + \sigma^2 x = \varepsilon f(x, x', t, \varepsilon), \tag{8.1.4}$$

or systems of second order differential equations

$$x_j'' + \sigma_j^2 x_j = \varepsilon f_j(x_1, \ldots, x_n, x_1', \ldots, x_n', t, \varepsilon), \quad j = 1, \ldots, n, \tag{8.1.5}$$

where the functions f_j are all periodic of a period T, or independent of t (autonomous case).

Every solution of the equation

$$x'' + \sigma^2 x = 0, \tag{8.1.6}$$

is of the form $x = A \cos \sigma t + B \sin \sigma t$, where A, B are constants determined by the initial conditions; hence every solution of (8.1.6) is periodic. Although for ε "small" the solutions of (8.1.4) may differ very little from the solutions of (8.1.6) we cannot expect that (8.1.4) has *periodic* solutions in the neighborhood of an arbitrary periodic solution of (8.1.6), but only in certain special neighborhoods of the solutions of that reduced equation; e.g., for a given f, there may be no *periodic* solution of (8.1.4) in the neighborhood of $A = 1$, $B = 5$, say, no matter how small ε, $(\varepsilon \neq 0)$, while in the neighborhood of $A = 2$, $B = 7$, say, there may exist periodic solutions of (8.1.4) for every $\varepsilon < \frac{1}{100}$. This important point makes the problem of existence of periodic solutions perhaps more difficult than might at first be anticipated.

Some other aspects of the intricate question of the characterization and determination of the periodic solutions are better illustrated by the following examples, for which we will refer to other paragraphs of the book for more details. Numerous examples are considered in (8.5).

(a) The equation $x''+ax'+bx=0$, a, b positive constants, has no nonzero periodic solutions since all nonzero solutions are asymptotic to the zero solution as $t\to+\infty$. The addition of a periodic term, say $C\sin\omega t$, leads to the equation $x''+ax'+bx=C\sin\omega t$, C, ω constants, which has exactly one periodic solution $x=A\cos(\omega t+\gamma)$, A, γ constants depending on a, b, C, ω, and all others solutions are asymptotic to this one as $t\to+\infty$ (cf. 2.4).

(b) The equation $x''+\sin x=0$ (pendulum equation) has three types of solutions: (α) the trivial solution $x=0$ and the constant solutions $x=k\pi$, $k=\pm1, \pm2, \ldots$; (β) the periodic solutions $x=P(a, t+\gamma)$, a, γ constants, of period $T=T(a)$ depending upon the "amplitude" a (true oscillations of the pendulum); (γ) the unbounded solutions whose "amplitude" approaches $+\infty$ as $t\to+\infty$ (complete revolutions of the pendulum) (cf. 8.10).

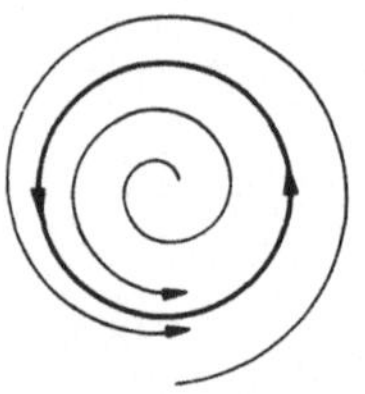

(c) The system $x_1'=-x_2$, $x_2'=x_1$ has all its solutions periodic of period 2π, namely $x_1=A\cos(t+\gamma)$, $x_2=A\sin(t+\gamma)$, A, γ arbitrary constants. If $\varepsilon>0$ is any constant, the system

$$x_1'=-x_2+\varepsilon x_1(1-x_1^2-x_2^2),\quad x_2'=x_1+\varepsilon x_2(1-x_1^2-x_2^2),$$

presents the following periodic solutions: (α) the zero solution $x_1=x_2=0$; (β) the solutions $x_1=\cos(t+\gamma)$, $x_2=\sin(t+\gamma)$, γ an arbitrary constant. All other solutions are not periodic, and $x_1^2+x_2^2\to1$ as $t\to+\infty$. To see this let us consider polar coordinates r, θ with $x_1=r\cos\theta$, $x_2=r\sin\theta$. Then we have the system $r'=\varepsilon r(1-r^2)$, $\theta'=1$, and the solution with $r(0)=a>0$, $\theta(0)=b$, is given by $r=e^{\varepsilon t}(-1+a^{-2}+e^{2\varepsilon t})^{-\frac{1}{2}}$, $\theta=b+t$, and $r\to1$, $\theta\to+\infty$ as $t\to+\infty$.

The illustration gives an idea of how the solutions spiral toward $x_1^2+x_2^2=1$ as $t\to+\infty$. In other words, if the "system" is taken out of its position of equilibrium $x_1=x_2=0$, it does not return to it, but oscillations start and approach the periodic solution $x_1=\cos t$, $x_2=\sin t$.

Systems of this sort are called selfstarting systems, or selfoscillatory systems, and the oscillations produced are called autooscillations.

The van der Pol equation $x''+\varepsilon(x^2-1)\,x'+x=0$, $\varepsilon>0$, i.e., if $x_1=x$, $x_2=-x'$, the system

$$x_1'=-x_2,\qquad x_2'=x_1+\varepsilon x_2(1-x_1^2),\tag{8.1.7}$$

has an analogous behavior [cf. remarks after (8.7. i), (9.1. ii), and (9.2. ix)]. The only periodic nonzero solution cannot be given in closed form as in the example above but, for ε small has a period close to 2π and amplitude close to 2, say $x\approx2\cos(t+\gamma)$. This equation regulates the free oscillations of a feedback electrical circuit with triode (8.14) and was discussed first by VAN DER POL [1] in 1924. We will return to it often in the following sections.

8.2. Method of LINDSTEDT. If we suppose ε "small", it seems natural to try a solution of (8.1.4) of the form

$$x=x_0(t)+\varepsilon x_1(t)+\varepsilon^2x_2(t)+\cdots,\tag{8.2.1}$$

a power series in ε, with coefficients functions of t. In general, if x in (8.2.1) is thought of as a vector, then we solve (8.1.5).

One way to assure that x in (8.2.1) is periodic is to require that each x_n, $n = 0$, 1, ..., is periodic. This requirement may not be so easy to fulfill however, for on substituting the right member of (8.2.1) into (8.1.4) and equating the coefficients of the successive powers of ε, some members of the resulting set of recursive linear differential equations may not have a periodic solution. In some one of the equations, there may be a term in the solution of the form, say, $t\cos t$. This is said to be a *secular* term. The term in the differential equation which leads to such an expression should be taken into consideration, and the method of LINDSTEDT consists of casting out these terms as they appear in the successive linear differential equations. The problem which then arises is the convergence of (8.2.1) to a solution of (8.1.4), and POINCARÉ has shown by an example that, in general, the series (8.2.1) obtained by the method of LINDSTEDT may not converge.

This may occur in situations where the solutions of the recursive differential equations lead to integrands of the form $e^{(m_1\alpha_1+\cdots+m_k\alpha_k)t}$ where the linear combination $\Sigma m_i\alpha_i$ of k fixed real or complex numbers α_i with integral coefficients $m_i \gtreqless 0$, may assume values arbitrarily small (e.g., $k=2$, α_1/α_2 real irrational) and thus the corresponding integrals, $(\Sigma m_i\alpha_i)^{-1} e^{(\Sigma m_i\alpha_i)t}$ present denominators which may not be bounded away from zero ("small divisors").

On the other hand, it may happen that x in (8.2.1) is periodic without each coefficient x_n being periodic. That is, one might be able to develop some conditions for periodicity and force the series solution (8.2.1) to conform to these conditions. Then, if the series (8.2.1) converges to a solution of (8.1.4), and if the periodicity conditions are satisfied, then one is assured of a periodic solution. This is basically the method of POINCARÉ. Before giving an example let us say something of the expected period of the oscillation. We speak now only of equations of the form (8.1.4), but the discussion is easily extended to systems of the form (8.1.5). Consider first the autonomous case, i.e., the case in which f in (8.1.4) is independent of t. Then (8.1.4) has the form

$$x'' + \sigma^2 x = \varepsilon f(x, x', \varepsilon). \tag{8.2.2}$$

For $\varepsilon = 0$, the periodic oscillation has period $2\pi/\sigma$. We may expect that for ε small ($\varepsilon \neq 0$), the periodic oscillation (if one exists) has period $(2\pi/\sigma) + O(\varepsilon)$, that is, the period may depend on ε. Hence, we may expect that the frequency of the perturbed periodic solution has the form $\tau/2\pi$, with $\tau = \sigma + \tau_1\varepsilon + \tau_2\varepsilon^2 + \cdots$, a power series in ε. Next, consider the case where f in (8.1.4) does depend explicitly on t and $f(x, x', t, \varepsilon) = f\left(x, x', t + \frac{2\pi}{\omega}, \varepsilon\right)$. Then, we may expect that a periodic solution of (8.1.4) (if one exists) for ε small ($\varepsilon \neq 0$), has period $2\pi/\omega$ or an integral multiple of $2\pi/\omega$. However, for $\varepsilon = 0$, the period of the solutions is $2\pi/\sigma$, hence we expect either $\omega = \sigma$, or $\omega = \sigma + O(\varepsilon)$. It may even be possible to obtain a periodic solution of (8.1.4) in case ω is "far away" from σ, e.g., $\sigma = \sqrt{2}$, $\omega = 1$, but then this solution could be expected to have amplitude $a = o(\varepsilon)$, e.g., we would take as our first approximation $x_0(t) = 0$. We consider this case somewhat later.

Let us now apply LINDSTEDT's method to the van der Pol equation

$$x'' + x = \varepsilon(1 - x^2)\,x', \quad (' = d/dt). \tag{8.2.3}$$

We do not know the frequency of the periodic solution of (8.2.3) for ε small ($\varepsilon \neq 0$), however from our former discussions, we suspect that the frequency of the linear oscillation is near $1/2\pi$. Hence, in (8.2.3) let us replace the independent variable t, by $\theta = \omega t$. Equation (8.2.3) then becomes

$$\omega^2 x'' + x = \varepsilon\,\omega(1 - x^2)\,x', \quad (' = d/d\theta). \tag{8.2.4}$$

We now try to determine ω so that (8.2.4) has a periodic solution $x(\theta)$ of period 2π, i.e., $x(\theta) = x(\theta + 2\pi)$, then the corresponding solution x^* of (8.2.3) will have period $2\pi/\omega$, i.e., $x^*(t) = x^*\left(t + \frac{2\pi}{\omega}\right)$, and if our suspicions are correct, we will find $\omega = 1 + O(\varepsilon)$. We assume for x, and ω, in (8.2.4) the power series

$$x = x_0(\theta) + \varepsilon x_1(\theta) + \varepsilon^2 x_2(\theta) + \cdots, \qquad \omega = \omega_0 + \varepsilon\omega_1 + \varepsilon^2\omega_2 + \cdots. \tag{8.2.5}$$

We also assume that the solution $x(\theta)$ of (8.2.4) has period 2π, and that $x'(0) = 0$. This last requirement is not restrictive, effecting only a shift of the origin along the θ axis.

Substituting (8.2.5) in (8.2.4), we obtain

$$\left.\begin{aligned} &(\omega_0^2 + 2\varepsilon\omega_0\omega_1 + \cdots)(x_0'' + \varepsilon x_1'' + \cdots) + x_0 + \varepsilon x_1 + \cdots \\ &= \varepsilon(\omega_0 + \varepsilon\omega_1 + \cdots)(1 - x_0^2 - 2\varepsilon x_0 x_1 - \cdots)(x_0' + \varepsilon x_1' + \cdots) \end{aligned}\right\} \tag{8.2.6}$$

and, equating powers of ε, also

$$\left.\begin{aligned} &\text{(a)}\quad \omega_0^2 x_0'' + x_0 = 0, \\ &\text{(b)}\quad \omega_0^2 x_1'' + x_1 = -2\omega_0\omega_1 x_0'' + \omega_0(1 - x_0^2)x_0', \\ &\quad \cdots\cdots\cdots\cdots\cdots\cdots \end{aligned}\right\} \tag{8.2.7}$$

The general solution to (8.2.7a) is

$$x_0 = A_0 \cos(\theta/\omega_0) + B_0 \sin(\theta/\omega_0), \tag{8.2.8}$$

and, satisfying the conditions $x_0(\theta) = x_0(\theta + 2\pi)$, $x_0'(0) = 0$, we have $\omega_0 = 1$, $B_0 = 0$, and hence $x_0 = A_0 \cos\theta$.

We will determine A_0, ω_1 in the next step. Equation (8.2.7b) now becomes

$$\left.\begin{aligned} x_1'' + x_1 &= 2\omega_1 A_0 \cos\theta - A_0(1 - A_0^2\cos^2\theta)\sin\theta \\ &= 2\omega_1 A_0\cos\theta - A_0(1 - A_0^2/4)\sin\theta + (A_0^3/4)\sin 3\theta. \end{aligned}\right\} \tag{8.2.9}$$

The terms $2\omega_1 A_0 \cos\theta$, and $A_0(1 - A_0^2/4)\sin\theta$ in the right member of (8.2.9) lead to terms in the solution of (8.2.9) of the form $K_1\theta\cos\theta + K_2\theta\sin\theta$, which are not periodic. These are the so-called secular terms. Hence, following LINDSTEDT, we choose $A_0 = 2$, $\omega_1 = 0$. The general solution of (8.2.9) then is

$$x_1 = A_1\cos\theta + B_1\sin\theta - (\tfrac{1}{4})\sin 3\theta.$$

The constant B_1 is determined by the condition $x_1'(0) = 0$. The constants A_1 and ω_2 are determined in the next step, and so on. Hence we obtain

$$x = 2\cos\omega t + O(\varepsilon), \qquad \omega = 1 + O(\varepsilon),$$

where we may well replace t by $t + \gamma$, since (8.2.3) is autonomous. LINDSTEDT's method is also discussed by N. MINORSKY [8] and J. J. STOKER [1].

8.3. Method of POINCARÉ. Let us now look at the method of POINCARÉ already mentioned in (8.2). Consider the equation

$$x'' + x = \varepsilon f(x, x'), \tag{8.3.1}$$

where $f(x, x')$ is a polynomial in x, x', or a power series in x, x' convergent for all x, x'. For $\varepsilon = 0$, the solution of (8.3.1) with $x'(0) = 0$, is $x(t, A) = A\cos t$. Let us assume that periodic solutions of (8.3.1) exist for ε small ($\varepsilon \neq 0$), and let $x = x(t, A, \varepsilon)$ be the solution.

For $t = 0$, let us write this solution in the form $x(0, A, \varepsilon, \beta) = x(0, A) + \beta(\varepsilon) = A + \beta$, which defines the function β, and gives the initial amplitude of the oscillation for all ε sufficiently small. Certainly $\beta(0) = 0$. POINCARÉ's method consist in developing the solution $x(t, A, \varepsilon, \beta)$ as a power series of ε and β, that is

$$x(t, A, \varepsilon, \beta) = A\cos t + \alpha_1(t)\varepsilon + \alpha_2(t)\beta + \alpha_3(t)\varepsilon\beta + \cdots. \tag{8.3.2}$$

The fotmal solution with $x(0, A, \varepsilon, \beta) = A + \beta$, $x'(0, A, \varepsilon, \beta) = 0$, may be obtained as follows: Substitute the right member of (8.3.2) into (8.3.1) and equate coefficients of like powers of ε, β, obtaining a set of recursive linear differential equations of the form

$$\alpha_j'' + \alpha_j = g_j(t), \quad j = 1, 2, \ldots. \tag{8.3.3}$$

From the initial prescribed conditions, and the form of the series (8.3.2), we have the following initial conditions for each equation (8.3.3)

$$\left.\begin{aligned} \alpha_2(0) = 1, \quad \alpha_1(0) = \alpha_3(0) &= \alpha_4(0) = \cdots = 0, \\ \alpha_1'(0) = \alpha_2'(0) &= \alpha_3'(0) = \cdots = 0. \end{aligned}\right\} \tag{8.3.4}$$

Poincaré [6, vol. 1, ch. 2] shows that the expansion (8.3.2) converges uniformly in any finite time interval $0 < t < t_1$ provided $|\varepsilon|$, $|\beta|$ are sufficiently small. We observe here that the dominant term in (8.3.2) has period 2π, and yet, we expect the solution for ε small to have period 2π plus terms of order ε.

Clearly, the condition for periodicity of (8.3.2) is

$$\begin{aligned} \Phi(\tau, A, \varepsilon, \beta) &\equiv x(0, A, \varepsilon, \beta) - x(2\pi + \tau, A, \varepsilon, \beta) = 0, \\ \chi(\tau, A, \varepsilon, \beta) &\equiv x'(0, A, \varepsilon, \beta) - x'(2\pi + \tau, A, \varepsilon, \beta) = 0, \end{aligned} \tag{8.3.5}$$

where $\tau = \tau(\varepsilon)$ is the correction for the period and has to be determined. Certainly $\tau(0) = 0$.

Since $\Phi_{\varepsilon=0} = 0$, $\chi_{\varepsilon=0} = 0$ are identically satisfied for all A, we may write

$$\Phi = \varepsilon\, \Phi_1(\tau, A, \varepsilon, \beta), \qquad \chi = \varepsilon\, \chi_1(\tau, A, \varepsilon, \beta).$$

If $\Phi_1 = 0$, $\chi_1 = 0$ can be solved for $\tau(\varepsilon)$, $\beta(\varepsilon)$ such that $\tau(0) = 0$, $\beta(0) = 0$, then the problem is solved, i.e. we have a sufficient condition for periodicity.

In particular, if there exist τ_0, A_0 such that $\Phi_1 = 0$, $\chi_1 = 0$ for $\tau = \tau_0$, $A = A_0$, $\varepsilon = 0$, and if the Jacobian J of Φ_1, χ_1 with respect to τ, A at $\tau = \tau_0$, $A = A_0$, $\varepsilon = 0$ is $\neq 0$, then $\tau(\varepsilon)$, $A(\varepsilon)$ can be determined. Relations (8.3.5), or other equivalent, are often called Poincaré's periodicity conditions [also, in another situation, bifurcation equations; cf. remarks after (9.2. ix)].

The solution finally obtained is of the form

$$x(t) = \sum_{k=0}^{\infty} \{A_k(\varepsilon) \cos k\,[\omega(\varepsilon)\, t] + B_k(\varepsilon) \sin k\,[\omega(\varepsilon)\, t]\}, \tag{8.3.6}$$

where both the amplitude and frequency are analytic functions of ε. The periodic function (8.3.6) is analogous to the series obtained by Lindstedt's method. Let us expand the right member of (8.3.6) in Taylor series about the point $\varepsilon = 0$ obtaining

$$\left.\begin{aligned} x(t) = {} & \sum_{k=0}^{\infty} [A_k(0) \cos k\,t + B_k(0) \sin k\,t] + \\ & + \varepsilon \sum_{k=0}^{\infty} [A_k'(0) \cos k\,t + B_k'(0) \sin k\,t - A_k(0)\, \omega'(0)\, k\,t \sin k\,t \\ & + B_k(0)\, \omega'(0)\, k\,t \cos k\,t] + \varepsilon^2 \sum_{k=0}^{\infty} \cdots. \end{aligned}\right\} \tag{8.3.7}$$

The method of Poincaré yields the solution in the form (8.3.7). In general, the first secular terms appear as coefficients of ε. The presence of secular terms does not destroy the periodicity of $x(t)$ in (8.3.7), if the full power series is considered. However, the successive approximations are not periodic, but may be of the form

$$x(t) = A_0 \cos t + \varepsilon\, A_1\, t \cos t + \cdots.$$

The secular terms in the expansions of POINCARÉ will not appear if one is able to obtain the corrected frequency ω. For instance, the following function, whose expansion shows secular terms, is actually periodic:

$$\left.\begin{aligned}&\cos(1+\gamma\,\varepsilon)\,t = \cos t\cos\gamma\,\varepsilon\,t - \sin t\sin\gamma\,\varepsilon\,t\\&= \left(1-(\gamma\,\varepsilon\,t)^2/2+(\gamma\,\varepsilon\,t)^4/24+\cdots\right)\cos t - \left(\gamma\,\varepsilon\,t-(\gamma\,\varepsilon\,t)^3/6+\cdots\right)\sin t.\end{aligned}\right\} \quad (8.3.8)$$

If $1+\gamma\varepsilon=\omega$ is the corrected frequency in equation (8.3.1), then a term in the solution of (8.3.1) by the method of LINDSTEDT may contain the left member of (8.3.8) $\cos(1+\gamma\varepsilon)t$, whereas the corresponding term in the solution of (8.3.1) by POINCARÉ's method may contain say $\gamma^2 t^2 \cos t$.

In case the second member of (8.3.1) is $f(x, x', t)$ with f periodic of period $T=2\pi/\omega$, the considerations above can be repeated by taking τ equal to ω, or to a fixed rational multiple of ω.

POINCARÉ was interested in astronomical phenomena and his solutions were suited for such applications since the effect of the secular terms is not particularly felt because of the relatively slow motions of planets. However, a thermionic generator with a frequency of several megacycles per second goes through stages in a few seconds, corresponding to those which an astronomical system passes in millions of years. Thus a casting out method of the secular terms as the one of LINDSTEDT seems better suited for the study of nonlinear oscillations. For research on these methods see, besides the papers of A. LINDSTEDT and H. POINCARÉ, also W. M. H. GRAVES [1], G. E. H. REUTER [1], W. D. MACMILLAN [1], and the book of I. G. MALKIN [9]. In particular for the method of POINCARÉ see N. MINORSKY ([6], pp. 138—165) and, in a more general form, E. A. CODDINGTON and N. LEVINSON ([2], pp. 20—35; [3], pp. 356—370). J. A. NOHEL [1] has given criteria for stability and instability of the periodic solutions obtained by the present method. This method, of the perturbation type, is distinct from the one in (8.6).

8.4. Method of KRYLOV and BOGOLYUBOV, and VAN DER POL. The solution of (8.1.4) for $\varepsilon=0$ may be written in either of the two forms

$$\text{(a)}\;\; x = A\cos\sigma t + B\sin\sigma t, \qquad \text{(b)}\;\; x = C\cos(\sigma t+\Phi), \quad (8.4.1)$$

where A, B, C, Φ are constants determined by the initial conditions. For $\varepsilon\neq 0$ but small, a periodic solution of (8.1.4) should retain the same form as (8.4.1) (a) or (b), provided we consider the quantities A, B, or C, not as constants but as certain "slowly varying" functions of time to be determined. The method of KRYLOV and BOGOLYUBOV starts with equation (8.4.1 b) and the method of VAN DER POL starts with equations (8.4.1 a). We shall mention only a few points of the former method which has been developed more extensively. The VAN DER POL method as far as first approximation is concerned follows the same lines.

We shall suppose that f does not depend on t or on ε, and by changing the sign of f, we will write (8.1.4) in the form

$$x'' + \sigma^2 x + \varepsilon f(x, x') = 0. \quad (8.4.2)$$

We first observe that for $\varepsilon=0$, (8.4.2) has the solution

$$\text{(a)}\;\; x = a\sin(\sigma t+\delta), \qquad \text{(b)}\;\; x' = a\,\sigma\cos(\sigma t+\delta), \quad (8.4.3)$$

where the amplitude a and the phase δ are constants. Now let us consider both a and δ as functions of t and let us determine them in such a way that both (8.4.3) (a) and (b) hold and represent a solution of (8.4.2). From (8.4.3a) we deduce

$$x' = a' \sin(\sigma t + \delta) + a\,\delta' \cos(\sigma t + \delta) + a\,\sigma \cos(\sigma t + \delta),$$

and, by comparison with (8.4.3b), we conclude that

$$a' \sin(\sigma t + \delta) + a\,\delta' \cos(\sigma t + \delta) = 0. \qquad (8.4.4)$$

From (8.4.3b) we deduce now

$$x'' = a'\,\sigma \cos(\sigma t + \delta) - \sigma\, a\,\delta' \sin(\sigma t + \delta) - \sigma^2 a \sin(\sigma t + \delta),$$

and, by replacing in (8.4.2), also

$$\left.\begin{aligned} \sigma a' \cos(\sigma t + \delta) - \sigma a\,\delta' \sin(\sigma t + \delta) = -\\ - \varepsilon f[a \sin(\sigma t + \delta),\ a\,\sigma \cos(\sigma t + \delta)]. \end{aligned}\right\} \qquad (8.4.5)$$

By (8.4.4) and (8.4.5) we obtain

$$\left.\begin{aligned} a' &= -\varepsilon\,\sigma^{-1} f[a \sin(\sigma t + \delta),\ a\,\sigma \cos(\sigma t + \delta)] \cos(\sigma t + \delta),\\ \delta' &= \varepsilon\, a^{-1} \sigma^{-1} f[a \sin(\sigma t + \delta),\ a\,\sigma \cos(\sigma t + \delta)] \sin(\sigma t + \delta). \end{aligned}\right\} \qquad (8.4.6)$$

Thus, instead of the equation (8.4.2) we have now the pair of equations (8.4.6). For ε "small" relations (8.4.6) show that a, and δ are "slowly" varying functions as expected.

Let us consider now for a moment the two following functions of the auxiliary variable u, periodic in u of period 2π:

$$f(a \sin u,\ a\,\sigma \cos u) \cos u, \qquad f(a \sin u,\ a\,\sigma \cos u) \sin u. \qquad (8.4.7)$$

Under usual conditions of measurability and summability we may consider the Fourier series of the functions above, say

$$f(a \sin u,\ a\,\sigma \cos u) \cos u = A_0(a) + \Sigma\,[A_n(a) \cos n u + B_n(a) \sin n u],$$

$$f(a \sin u,\ a\,\sigma \cos u) \sin u = C_0(a) + \Sigma\,[C_n(a) \cos n u + D_n(a) \sin n u],$$

where the sums are extended over all $n = 1, 2, \ldots$, and where

$$A_0(a) = (2\pi)^{-1} \int_0^{2\pi} f(a \sin u,\ a\,\sigma \cos u) \cos u\, du,$$

$$C_0(a) = (2\pi)^{-1} \int_0^{2\pi} f(a \sin u,\ a\,\sigma \cos u) \sin u\, du,$$

are the "mean values" of the functions (8.4.1) in $(0, 2\pi)$. The first approximation of the method of KRYLOV and BOGOLYUBOV consists in replacing the second members of equations (8.4.6) by their "mean values" in the sense indicated above,

$$da/dt = -\varepsilon\,\sigma^{-1} A_0(a), \qquad d\delta/dt = \varepsilon\, a^{-1} \sigma^{-1} C_0(a). \qquad (8.4.8)$$

If we put $\varphi = \sigma t + \delta$, we have $\varphi' = \sigma + \delta'$ and the second equation (8.4.8) is replaced by $d\varphi/dt = \sigma + \varepsilon a^{-1}\sigma^{-1}C_0(a)$. Thus equation (8.4.2) has the solution (first approximation)

$$x = a \sin \varphi,$$

where

$$da/dt = -\varepsilon(2\pi\sigma)^{-1}\int_0^{2\pi} f(a\sin u, a\sigma\cos u)\cos u\, du,$$

$$d\varphi/dt = \sigma + \varepsilon(2\pi a\sigma)^{-1}\int_0^{2\pi} f(a\sin u, a\sigma\cos u)\sin u\, du.$$

If f does not contain x', that is $f(x, x') = f(x)$, then we have

$$da/dt = -\varepsilon(2\pi\sigma)^{-1}\int_0^{2\pi} f(a\sin u)\cos u\, du = F(a\sin u)\Big|_0^{2\pi} = 0,$$

where F is a primitive of f, and

$$d\varphi/dt = \omega(a) = \sigma + \varepsilon(2\pi a\sigma)^{-1}\int_0^{2\pi} f(a\sin u)\sin u\, du.$$

Thus a is a constant, $\varphi = \omega(a)\,t + d$, and the oscillatory solutions (first approximation) $x = a\sin(\omega(a)\,t + d)$ have a frequency $\omega(a)/2\pi$ depending on the constant amplitude a and an arbitrary constant phase d. In the phase plane (x, x') the solutions appear as a family of cycles (ellipses with centers at the origin).

Example 1. Consider the reduced equation of the oscillations of the pendulum

$$x'' + g l^{-1}(x - x^3/6) = 0, \qquad (8.4.9)$$

where the term in parenthesis differs from the term $\sin x$ of the exact equation $x'' + gl^{-1}\sin x = 0$ by less than 10^{-3} if $|x| < \pi/6$ ($30°$ oscillations). Thus $\varepsilon = 1$, $\sigma^2 = gl^{-1}$, $f(x) = -gl^{-1}x^3/6$, and

$$\omega(a) = \sigma + \varepsilon(2\pi a\sigma)^{-1}\int_0^{2\pi}(-gl^{-1}/6)\,a^3\sin^4 u\, du = \sqrt{g/l}\,(1 - a^2/16).$$

Equation (8.4.9) has the solutions $x = a\sin(\omega(a)\,t + d)$, a, d constants (first approximation), whose period T, by retaining only terms in a^2 is given by

$$T = 2\pi\sqrt{l/g}\,(1 + a^2/16).$$

These two terms are the same as those which are obtained by solution of the equation $x'' + gl^{-1}\sin x = 0$ by an exact process.

Example 2. Consider the van der Pol equation (8.2.3) with $\varepsilon > 0$, where the parameter ε is supposed small. Thus $\sigma = 1$, $f(x, x') = (x^2 - 1)\,x'$,

$$da/dt = -\varepsilon(2\pi)^{-1}\int_0^{2\pi}(a^2\sin^2 u - 1)\,a\cos^2 u\, du = \varepsilon(a/2)(1 - a^2/4),$$

$$d\varphi/dt = 1 + \varepsilon(2\pi a)^{-1}\int_0^{2\pi}(a^2\sin^2 u - 1)\,a\cos u\sin u\, du = 1,$$

and by integration

$$a = a_0 e^{2^{-1}\varepsilon t}[1 + 4^{-1} a_0^2 (e^{\varepsilon t} - 1)]^{-\frac{1}{2}}, \qquad \varphi = t + d,$$

where a_0, d are arbitrary constants. Thus for $a_0 = 2$ we have $a = 2 =$ constant, and (8.2.3) has the periodic solutions (first approximation):

$$x = 2 \sin (t + d).$$

For $a_0 \neq 2$, we have the further solutions (first approximation):

$$x = a_0 e^{2^{-1}\varepsilon t}[1 + 4^{-1} a_0^2 (e^{\varepsilon t} - 1)]^{-\frac{1}{2}} \sin (t + d),$$

whose amplitude approaches 2 as $t \to +\infty$. Thus as far as the first approximation is concerned, the periodic solution above is asymptotically stable. We cannot go here into the refinement of the first approximation, into the successive second, third, and n-th approximations of the method of KRYLOV and BOGOLYUBOV, and into the consequent process of linearization which is so widely used in questions of applications (see N. BOGOLYUBOV and N. KRYLOV [13, 33, 34], N. MINORSKY [6]).

The method of KRYLOV and BOGOLYUBOV has been widely discussed. In particular B. V. BULGAKOV [6], N. V. BUTENIN [1, 2], A. A. KRASOVSKII [1], M. A. KRASNOSELSKII and M. G. KREIN [1] have contributed to this theory. For a recent and exhaustive exposition of the method and related proofs see N. BOGOLYUBOV and YU. A. MITROPOLSKY [1]. See also for applications of the method, L. S. GOLDFARB [1], A. M. KAC [2], H. KAUDERER [1], R. V. HOHLOV [1], V. O. KONONENKO [1], F. S. LOS [1], N. W. MCLACHLAN [1], N. MINORSKY [10], YU. A. MITROPOLSKII [3—7], J. I. YORISH [1], F. G. FRIEDLANDER [1], S. FURUYA [1].

8.5. A convergent method for periodic solutions and existence theorems. This method has been successively developed and modified by L. CESARI, J. K. HALE and others. The method is based on a process of successive approximations whose convergence has been proved. In particular see L. CESARI [4—9], J. K. HALE [1—12], R. A. GAMBILL and J. K. HALE [1], L. CESARI and J. K. HALE [1, 2], H. R. BAILEY and R. A. GAMBILL [1], H. R. BAILEY [1], W. R. FULLER [1], from which papers most of the considerations below have been drawn. A number of general existence theorems for cycles of autonomous systems and for harmonic and subharmonic solutions of periodic systems has been obtained by using this method, and we will list some of them below.

A first form of the method, as it was devised for linear systems (L. CESARI [1]) was given in (4.5). A variant for nonlinear analytic systems was studied by J. K. HALE [12] and R. A. GAMBILL and J. K. HALE [1]. A variant valid for general Lipschitzian systems was studied by L. CESARI [5]. For the case of systems containing time lags see W. R. FULLER [1]. The method was then given a new aspect by L. CESARI [6] for strongly nonlinear problems, that is, not of the perturbation type. The form given below for perturbation problems was studied by J. K. HALE [10, 11] in harmony with the mentioned extension studied by L. CESARI [6] for strongly nonlinear problems.

We shall consider systems of differential equations of the form

$$y' = Ay + \varepsilon f(t, y, \varepsilon), \tag{8.5.1}$$

where ε is a small parameter, A is an $n\times n$ real or complex constant matrix whose elements may depend on ε, $y=\mathrm{col}(y_1,\ldots,y_n)$, $f=\mathrm{col}(f_1,\ldots,f_n)$, and each f_j is a continuous function of y and ε, periodic in t of period $2\pi/\omega$, integrable with respect to t in $[0, 2\pi/\omega]$ (or independent of t and then (8.5.1) is autonomous, and T is arbitrary).

For the present exposition we shal assume that there are positive constants R, L, ε_0 such that

$$\|f(t,y,\varepsilon)\|\leq L,\qquad \|f(t,y,\varepsilon)-f(t,z,\varepsilon)\|\leq L\,\|y-z\| \tag{8.5.2}$$

for all $-\infty<t<+\infty$, all $|\varepsilon|\leq\varepsilon_0$, and all n-vectors y, z with $\|y\|, \|z\|\leq R$. Here $\|y\|$ denotes, as in (2.1), the sum of the absolute values of the components of y, though what is stated below holds as well for the Euclidean norm. We could have supposed (8.5.1) of the form $y'=Ay+f(t,y,\varepsilon)$ with $f(t,y,0)=0$ for all t and y, but we restricted ourselves to the case (8.5.1) for simplicity. Also, analogous considerations hold for systems $y'=Ay+f(t,y)$ if $f(t,0)=0$ and $\|f(t,y)\|\leq L\|y\|\,o(\|y\|)$.

(a) *The determining equation.* For the sake of simplicity, let us first suppose that the $n\times n$ constant matrix A has a diagonal form, namely $A=\mathrm{diag}(\varrho_1,\ldots,\varrho_n)$, $\varrho_1,\ldots,\varrho_n$ complex constants.

Also we consider mainly the periodic case, pointing out during the discussion the changes which have to be made for the autonomous case. In case f is periodic of period $2\pi/\omega$ in t, we may expect that periodic solutions of (8.5.1) will have period $T_1=2\pi\, m/\omega$ where m is a positive integer, and the numbers ϱ_j will be equal or "close" to numbers of the form $i\tau_j=i k_j\,\omega/m_j$, with k_j, m_j integers, different from zero and relatively primes, $m_j>0$, $j=1,\ldots,n$, (or $k_j=0$, $m_j=1$), and hence $\varrho_j=i\tau_j+o(\varepsilon)$ (as $\varepsilon\to 0$), and $m=m_1\ldots m_n$. (The case where only p of the numbers ϱ_j are "close" to numbers $i\tau_j$, $j=1,\ldots,p$, (and $m=m_1\ldots m_p$), will be considered below at the end of (b), and does not present difficulties. Thus, we rewrite system (8.5.1) in the form

$$y'=By+\varepsilon\, q(t,y,\varepsilon), \tag{8.5.3}$$

where $B=\mathrm{diag}(i\tau_1,\ldots,i\tau_n)$, $q=(q_1,\ldots,q_n)$, and $\varepsilon q_j=\varepsilon f_j+[\varrho_j(\varepsilon)-i\tau_j]\,y_j$, $j=1,\ldots,n$.

We will use a method of successive approximations to obtain a periodic solution $y=y(t,a,\tau,\omega,\varrho)$ of period $T=2\pi m/\omega$, $m=m_1\ldots m_n$, of the system

$$y'=By+\varepsilon\, q(t,y,\varepsilon)-\varepsilon\, e^{Bt}D(a,\tau,\omega,\varrho), \tag{8.5.4}$$

where $a=(a_1,\ldots,a_n)$, $\tau=(\tau_1,\ldots,\tau_n)$, $\varrho=(\varrho_1,\ldots,\varrho_n)$, where $a_1,\ldots,a_n$ are complex constants (determining essentially the initial amplitude and phase of the solution), and $D=\mathrm{col}(D_1,\ldots,D_n)$. The solution $y(t,a,\tau,\omega,\varrho)$ so obtained is then a periodic solution of (8.5.1) if we can choose either $\varrho_1,\ldots,\varrho_n$, or ω, or $(a_1,\ldots,a_n)$, so that the equations

$$D_j(a,\tau,\omega,\varrho)=0,\qquad j=1,\ldots,n, \tag{8.5.5}$$

are satisfied (determining equations). The reason that we obtain a solution of (8.5.4) though we start with (8.5.3) is that in the successive approximations we cast out certain terms among which are all the secular terms, and the functions $D_j(a,\tau,\omega,\varrho)$ actually correspond to all these terms cast out.

The numbers ϱ_j to be equal or close to numbers $i\tau_j=ik_j\,\omega/m_j$, $j=1,\ldots,n$, correspond to the fact that the j-th component of the first approximation is periodic of least period $2\pi\, m_j/k_j\,\omega$, that is, the dominant term in the j-th component of the solution of (8.5.4) for ε small, has frequency k_j/m_j times the natural frequency ω of the forcing terms in $f(t, y, \varepsilon)$. The solution will certainly have the period $T = 2\pi\, m/\omega$, $m = m_1 \ldots m_n$, but this may not be the least period. If $k_j = m_j = 1$, then the dominant term in the solution has the same frequency as the forcing terms, and the solution is called a *harmonic*; if $k_j = 1$, $m_j > 1$, then the dominant term in the solution has frequency $1/m_j$ times the frequency of the forcing terms, and the solution is called a *subharmonic of order* m_j; if $k_j > 1$, $m_j = 1$ then the dominant term in the solution has frequency k_j times the frequency of the forcing terms, and the solution is called a *ultraharmonic of order* k_j; if $k_j > 1$, $m_j > 1$, k_j, m_j relatively prime, then the dominant term in the solution has frequency k_j/m_j times the frequency of the forcing terms, and the solution is called a *ultrasubharmonic*. Analogous definitions hold if we allow k_j to assume negative integer values.

In case (8.5.1) is autonomous, we still write it in the form (8.5.3). All τ_j will now be rational multiples of the same frequency ω, which is to be considered as one of the unknowns in the solution of the determining equations (8.5.5).

(b) *The method of successive approximations.* Let us first consider the simplest case, where $A = 0$ is the zero $n \times n$ matrix, that is, systems (8.5.1) and (8.5.3) have the form

$$y' = \varepsilon\, q(t, y, \varepsilon). \tag{8.5.6}$$

Then the limit system as $\varepsilon \to 0$ is $y' = 0$ whose solutions are all of the form $y = a = (a_1, \ldots, a_n)$, $a_1, \ldots, a_n$ arbitrary constants.

If $g(t)$ denotes any periodic function of period $T = 2\pi/\omega$, L-integrable in $[0, T]$, say $g(t) \sim \Sigma\, c_l \exp(il\omega t)$, where Σ is extended over all $l = 0, \pm 1, \pm 2, \ldots$, then we shall denote by $Pg(t)$ the mean value of $g(t)$, that is, $Pg = c_0$. If $Pg = c_0 = 0$, then there is a unique primitive $G(t)$ of $g(t)$ which is absolutely continuous and periodic in t of period T and has mean value zero, namely $G(t) = \Sigma'(il\omega)^{-1} c_l \exp(il\omega t)$, where Σ' is extended over all $l \neq 0$. In harmony with (4.5) we shall denote $G(t)$ by Hg, or $\int g(t)\, dt$. If we denote by I the identity operator, we have $g - c_0 = (I - P)\, g = \Sigma'\, c_l \exp(il\omega t)$. It is proved in (J. K. Hale [1]; L. Cesari [5]) that for $Pg = 0$, we have also $\|Hg\| \leq K\, T^{-1} \int_0^T \|g\|\, du$, for some constant K independant of g. Also, we have obviously $\|Pg\| \leq \max \|g(t)\|$ and $\|(I - P)\, g(t)\| \leq 2 \max \|g(t)\|$. If now $g(t) = (g_1, \ldots, g_n)$ denotes a periodic vector function of period T whose components are L-integrable in $[0, T]$, we shall denote by Pg the constant vector $Pg = (Pg_1, \ldots, Pg_n)$, and, if $Pg = 0$, we shall denote by Hg, or $\int g\, dt$, the uniquely determined vector function $(Hg_1, \ldots, Hg_n)$.

Let us now define the method of successive approximations for systems (8.5.6). If we denote by $a = Py$ the unknown mean value of the periodic solution $y(t)$ of (8.5.6), then it is natural to take as a first approximation the constant vector $y^{(0)} = a = (a_1, \ldots, a_n)$. If we replace $y^{(0)} = a$ for y in the second member ot (8.5.6), then we obtain the system $y' = \varepsilon q(t, y^{(0)}, \varepsilon)$. The second member is a periodic vector function whose mean value $\varepsilon\, Pf(t, y^{(0)}, \varepsilon) = \varepsilon\alpha = (\varepsilon\alpha_1, \ldots, \varepsilon\alpha_n)$ may not be zero. Obviously, the formal integration of such a system would produce secular terms $\varepsilon\alpha_j t$ which are not periodic. If we cast out these terms we have then to integrate the system $y' = \varepsilon(I - P)\, q(t, y^{(0)}, \varepsilon)$, and it is natural to take as a second approximation the unique primitive of $\varepsilon(I - P)\, q$, which is periodic and has mean value a, namely, $y^{(1)} = a + \varepsilon \int (I - P)\, q(u, y^{(0)}, \varepsilon)\, du$. This process actually can be iterated, that is, we define a process of successive approximations by taking $y^{(0)} = a = (a_1, \ldots, a_n)$,

$$y^{(m)} = a + \varepsilon \int (I - P)\, q(u, y^{(m-1)}(u), \varepsilon)\, du, \qquad m = 1, 2, \ldots. \tag{8.5.7}$$

As we prove below, if we take any $0 < r < R$, and assume $\|a\| \leq r$, then there is an ε_1, $0 < \varepsilon_1 \leq \varepsilon_0$, such that (i) for all $|\varepsilon| \leq \varepsilon_1$ the successive approximations are all vector functions periodic of period T, mean value a, and $\|y^{(m)}\| \leq R$; (ii) $y^{(m)}(t)$ converges uniformly in $(-\infty, +\infty)$ as $m \to \infty$ toward an absolutely continuous vector function $y(t)$ which is periodic in T, has mean value a with $\|y\| \leq R$, satisfies the integral system $y = a + \varepsilon \int (I - P)\, q(u, y(u), \varepsilon)\, du$, and consequently the differential system $y' = \varepsilon q(t, y, \varepsilon) - \varepsilon D$, $D = \mathrm{col}(D_1, \ldots, D_n)$, $D = Pq(t, y(t), \varepsilon)$. Thus $y(t)$ satisfies (8.5.6) if and only if $D = 0$, that is, if the finite system of equations $D_j = 0$, $j = 1, \ldots, n$, are satisfied (determining equations). Analogous considerations hold for systems $y' = Ay + f(t, y)$ if $\|f(t, y)\| \leq L \|y\|\ \mathrm{o}(\|y\|)$, $\|a\| \leq r$, and r is sufficiently small.

Let us now go back to the general case which we have written in the form (8.5.3) with $B = \mathrm{diag}(i\tau_1, \ldots, i\tau_n)$, $\tau_j = k_j \omega / m_j$, k_j, m_j integers, $m_j > 0$, $k_j \geq 0$, or $k_j = 0$. The change of variables, $y = e^{Bt} z$, $z = (z_1, \ldots, z_n)$, or $y_j = z_j \exp(i\tau_j t)$, transforms system (8.5.3) into the system

$$Z' = \varepsilon Q(t, z, \varepsilon), \tag{8.5.8}$$

with $Q(t, z, \varepsilon) = e^{Bt} q(t, e^{Bt} z, \varepsilon)$. In (8.5.8) Q is periodic of period $T = 2\pi m_1 \ldots m_n / \omega$, and system (8.5.8) is of the type (8.5.6). We may apply the process (8.5.7) to system (8.5.8), or, equivalently, we may return to the variables y, and then obtain the method of successive approximations in terms of the given system (8.5.3):

$$y^{(0)} = e^{Bt} a = (a_1 e^{i\tau_1 t}, \ldots, a_n e^{i\tau_n t}),$$

$$y^{(m)} = e^{Bt} a + \varepsilon e^{Bt} \int (I - P)\, e^{-Bu} q(u, y^{(m-1)}(u), \varepsilon)\, du,$$

or, in component form,

$$\left.\begin{aligned} y_j^{(m)} &= a_j e^{i\tau_j t} + \varepsilon e^{i\tau_j t} \int (I - P)\, e^{-i\tau_j u} q_j(u, y^{(m-1)}(u), \varepsilon)\, du, \\ j &= 1, \ldots, n, \qquad m = 1, 2, \ldots. \end{aligned}\right\} \tag{8.5.9}$$

In terms of the original system (8.5.1) and the functions f_j, we have

$$\left.\begin{aligned} y_j^{(m)} &= a_j e^{i\tau_j t} + e^{i\tau_j t} \int (I - P)\, e^{-i\tau_j u} \times \\ &\quad \times [\varepsilon f_j(u, y^{(m-1)}(u), \varepsilon) + (\varrho_j(\varepsilon) - i\tau_j)\, y_j^{(m-1)}]\, du, \\ j &= 1, \ldots, n, \qquad m = 1, 2, \ldots. \end{aligned}\right\} \tag{8.5.10}$$

The convergence of the process is proved below in (c). The limit vector function $y = (y_1, \ldots, y_n)$ satisfies the integral equation that we obtain from (8.5.9) by replacing $y^{(m)}$ and $y^{(m-1)}$ by y. By differentiation we verify that y satisfies the differential system (8.5.4) with

$$D = \mathrm{col}(D_1, \ldots, D_n) = P[e^{-Bu} q(u, y(u), \varepsilon)],$$

or

$$D_j = P[e^{-i\tau_j u} q_j(u, y(u), \varepsilon)], \qquad j = 1, \ldots, n,$$

or

$$\varepsilon D_j = P\{e^{-i\tau_j u}[\varepsilon f_j(u, y(u), \varepsilon) + (\varrho_j(\varepsilon) - i\tau_j)\, y_j]\}.$$

If we write y_j in the form $y_j = a_j e^{i\tau_j t} + \varepsilon Y_j$, then equation $\varepsilon D_j = 0$ takes the form (for $a_j \neq 0$)

$$i\tau_j - \varepsilon F_j(a, \tau, \omega, \varrho, \varepsilon) = \varrho_j, \qquad j = 1, \ldots, n, \tag{8.5.11}$$

where

$$F_j = a_j^{-1} T_1^{-1} \int_0^{T_1} e^{-i\tau_j u} [f_j(u, y(u), \varepsilon) + (\varrho_j - i\tau_j)\, Y_j(u)]\, du,$$

and $T_1 = 2\pi m / \omega$, $m = m_1 \ldots m_n$. We may prefer to write equations (8.5.11) in the form

$$\varepsilon E_j = a_j (i\tau_j - \varrho_j), \qquad j = 1, \ldots, n, \tag{8.5.12}$$

where $E_j = a_j F_j$.

We shall now discuss equations (8.5.5), or (8.5.11). Suppose first the system (8.5.1) is periodic. The functions f_j are periodic of period $T = 2\pi/\omega$, and we consider ω as known. If $\varrho_j = \varrho_j(\varepsilon)$ are given by (8.5.11), that is, satisfy $\varepsilon E_j = a_j(i\tau_j - \varrho_j)$, $j = 1, \ldots, n$, then system (8.5.1) has a periodic solution of period $T = 2\pi\, m/\omega$, $m = m_1 \ldots m_n$, of the form

$$y_j = a_j e^{i k_j \omega t m_j^{-1}} + \varepsilon W_j(t, a_1, \ldots, a_n, \varepsilon), \qquad j = 1, \ldots, n, \tag{8.5.13}$$

where $|W_j| \leqq M$, for all $|\varepsilon| \leqq \varepsilon_1$ and some $0 < \varepsilon_1 \leqq \varepsilon_0$. Conversely, if system (8.5.1) has a periodic solution of the form (8.5.13), then the determining equations (8.5.12) are satisfied (see J. K. HALE [11]). Assume now that the number ϱ_j are fixed, with $\varrho_j(0) = i\tau_j = i k_j \omega m_j^{-1}$, and $\varrho_j(\varepsilon) = i\tau_j + \varepsilon\chi_j(\varepsilon)$, $j = 1, \ldots, n$. Then equations (8.5.12) become

$$\left.\begin{aligned} E_j = T_1^{-1}\int_0^{T_1} e^{-i\tau_j u}\,[f_j(u, y(u), \varepsilon) + \chi_j(\varepsilon) Y_j(u)]\,du = -\,a_j\chi_j(\varepsilon), \\ j = 1, \ldots, n, \end{aligned}\right\} \tag{8.5.14}$$

where $E_j = E_j(a, \tau, \omega, \varrho, \varepsilon)$. If the system of equations

$$E_j(a, \tau, \omega, \varrho, 0) = -\,a_j\chi_j(0), \qquad j = 1, \ldots, n, \tag{8.5.15}$$

has a solution $a_{10}, \ldots, a_{n0}$, and if the Jacobian $\partial E/\partial a$ exists and is different from zero at $a_0 = (a_{10}, \ldots, a_{n0})$, then, by the theorem on implicit functions, there is a unique solution $a = a(\varepsilon)$ of system (8.5.14) for all $|\varepsilon| \leqq \varepsilon_1$ and some $0 < \varepsilon_1 \leqq \varepsilon_0$, with $a(0) = a_0$, and $a(\varepsilon)$ is a continuous function of ε at $\varepsilon = 0$. Then (8.5.1) has a periodic solution of the form above.

We next consider the case where (8.5.1) is autonomous. Here we shall think of ω as unknown, though, however, the first alternative of the periodic case still holds. That is, if for a given period, and with dominant terms in the solution of a given period, say the dominant term of x_j is to have period $2\pi\, m_j/k_j\,\omega$, then the ϱ_j can be chosen as in equation (8.5.11), and (8.5.1) will then have the periodic solution with prescribed period. In general, ω is among the unknowns, and thus we have to determine $a_1, \ldots, a_n$ as functions of ε (or some of the first set of variables as functions of the remaining ones and ε). If we assume $\varrho_j(0) = i k_j \sigma m_j^{-1}$ for some $\sigma > 0$, and hence $\varrho_j(\varepsilon) = i k_j \sigma m_j^{-1} + \varepsilon\chi_j(\varepsilon)$, and if we assume $\omega = \sigma + \varepsilon\beta(\varepsilon)$, and $\tau_j = i k_j \omega m_j^{-1}$, $j = 1, \ldots, n$, then equations (8.5.12) become

$$\begin{aligned} E_j &= T_1^{-1}\int_0^{T_1} e^{-i\tau_j u}[f_j(u, y(u,\varepsilon), \varepsilon) + \varepsilon(\chi_j(\varepsilon) - i k_j\beta(\varepsilon)\, m_j^{-1}) Y_j(u, \varepsilon)]\,du \\ &= a_j(i k_j\beta(\varepsilon)\, m_j^{-1} - \chi_j(\varepsilon)), \qquad j = 1, \ldots, n, \end{aligned}$$

and the same equations for $\varepsilon = 0$ become

$$T_1^{-1}\int_0^{T_1} e^{-i\tau_j u} f_j(u, y(u, 0), 0)\,du = a_j(i k_j\beta(0)\, m_j^{-1} - \chi_j(0)).$$

These equations replace (8.5.14) and (8.5.15) in the autonomous case.

Let us assume now that only p of the complex numbers ϱ_j are equal, or "close" to numbers $i k_j\omega/m_j$, (or $i k_j\sigma/m_j$), say for $j = 1, \ldots, p$, $0 \leqq p \leqq n$, while $\varrho_j(0) \neq k\omega i/m_1 \ldots m_p$, $j = p+1, \ldots, n$, $k = 0, \pm 1, \ldots$. These inequalities then hold for $\varrho_j(0)$ replaced by $\varrho_j(\varepsilon)$ and all $|\varepsilon|$ sufficiently small. In particular, we may have $R(\varrho_j(0)) \neq 0$, or we may have $R(\varrho_j(0)) = 0$ and $\varrho_j(0) \not\equiv 0 \pmod{i\sigma/m_1 \ldots m_p}$. We write now the system in the form

$$y' = By + \varepsilon q(t, y, \varepsilon),$$

where $B = \operatorname{diag}(i\tau_1, \ldots, i\tau_p, \varrho_{p+1}(\varepsilon), \ldots, \varrho_n(\varepsilon))$. The only periodic solution of period $T_1 = 2\pi\, m_1 \ldots m_p/\omega$ of the linear system $y' = By$ are now of the form

$y_1 = a_1 e^{i\tau_1 t}, \ldots, y_p = a_p e^{i\tau_p t}, y_{p+1} = \cdots = y_n = 0$. If $g(t) = (g_1, \ldots, g_n)$ is any vector function whose first p components are periodic functions of period T integrable in $[0, T]$, and the remaining $n-p$ components are functions of the same type multiplied by $e^{-i\varrho_j t}$ respectively, $j = p+1, \ldots, n$, then we take for P the operator $Pg = (Pg_1, \ldots, Pg_p, 0, \ldots, 0)$. Thus Pg is the same mean value defined in (4.5). We may now defined the method of successive approximations by the same relations (8.5.9) where now $\tau_{p+1}, \ldots, \tau_n$ are replaced by $\varrho_{p+1}(\varepsilon), \ldots, \varrho_n(\varepsilon)$, and the last $n-p$ integrals (8.5.9) are the unique primitives of period T_1 defined in (4.5). The limit vector function $y(t)$ is then periodic in t of period T_1, and satisfies a system of the form $y' = By + \varepsilon q - \varepsilon e^{Bt} D$, where $D = \mathrm{col}(D_1, \ldots, D_p, 0, \ldots, 0)$, and $D_1, \ldots, D_p$ are constant. Then there are only p determining equations $D_j = 0$, $j = 1, \ldots, p$. The case where $A = \mathrm{diag}(B_1, B_2)$ with $B_1 = \mathrm{diag}(\varrho_1(\varepsilon), \ldots, \varrho_p(\varepsilon))$, $\varrho_j(\varepsilon) = i\tau_j + o(\varepsilon)$, as above, and the linear system $z' = B_2 z$ has no periodic solution of period T_1 but the trivial one, can be treated analogously. For more details see J. K. HALE [10, 11] and L. CESARI [5].

The same process which leads to existence theorems for harmonic and subharmonic solutions of periodic systems, leads also to analogous theorems for autonomous systems. For the sake of brevity we list in (d) and (e) some typical statements, starting with the autonomous systems. The statements are obtained by transforming the given systems into systems of first order differential equations of the form (8.5.1) or analogous ones, by applying the method above, and by a successive detailed analysis of the determining equations and successive approximations, which cannot be given here.

(c) *Proof of the convergence of the process of successive approximations.* In the particularly simple situation chosen for system (8.5.3) it is not difficult to prove the convergence of the method. Actually it is enough to consider system (8.5.6) and the process defined in (8.5.7). We shall assume that q satisfies relations (8.5.2) for given R, L, ε_0, that r has been chosen, $0 < r < R$, and $\|a\| \leq r$. Let S be the space of all vector functions $g(t)$ periodic of period $T = 2\pi/\omega$, continuous in $(-\infty, +\infty)$, with norm $\nu g = \max \|g(t)\|$ for all t, and let us consider the operator

$$\mathfrak{T} g = a + \varepsilon \int (I-P)\, q(u, g(u), \varepsilon)\, du.$$

Let S^* be the set of all $g(t) \in S$ defined by $S^* = [g \in S \mid Pg = a, \nu g \leq R]$. Obviously $\mathfrak{T}: S^* \to S$. We have also

$$\begin{aligned}\nu(\mathfrak{T} g) &\leq \|a\| + |\varepsilon|\, K T^{-1} \int_0^T \|(I-P)\, q(u, g(u), \varepsilon)\|\, du \\ &\leq \|a\| + 2|\varepsilon|\, K T^{-1} \int_0^T \|q(u, g(u), \varepsilon)\|\, du \\ &\leq \|a\| + 2|\varepsilon|\, KL,\end{aligned}$$

and then $\nu(\mathfrak{T} g) \leq r + (R-r) = R$ for all $|\varepsilon| \leq \varepsilon_1 = \min[\varepsilon_0, (R-r)/2KL]$. Also, we have $P(\mathfrak{T} g) = a$, and thus $\mathfrak{T}: S^* \to S^*$ for $|\varepsilon| \leq \varepsilon_1$. Also, for two elements $g, h \in S^*$ we have

$$\begin{aligned}\nu(\mathfrak{T} g - \mathfrak{T} h) &= |\varepsilon|\, \nu \int (I-P)\, (q(u, g(u), \varepsilon) - q(u, h(u), \varepsilon))\, du \\ &\leq |\varepsilon|\, K T^{-1} \int_0^T \|(I-P)\, (q(u, g(u), \varepsilon) - q(u, h(u), \varepsilon))\|\, du \\ &\leq 2|\varepsilon|\, K T^{-1} \int_0^T \|q(u, g(u), \varepsilon) - q(u, h(u), \varepsilon)\| \\ &\leq 2|\varepsilon|\, KL\, \nu(g-h) \leq 2^{-1} \nu(g-h)\end{aligned}$$

for all $|\varepsilon| \leq \varepsilon_2 = \mathrm{Min}[\varepsilon_1, (4KL)^{-1}]$. Thus $\mathfrak{T}: S^* \to S^*$, is a contraction. The completeness of the space S^* is well known, and thus, by Banach fixed point theorem, there is a unique fixed element in S^*, say $y(t) \in S^*$, such that $\mathfrak{T} y = y$, and $\nu(y^m - y) \to 0$

as $m \to \infty$. Also $\mathfrak{T}y = y$ implies that

$$y(t) = a + \varepsilon \int (I - P)\, q(u, y(u), \varepsilon)\, du,$$

hence $y(t)$ is absolutely continuous, and, by differentiation, we have $y' = q(t, y, \varepsilon) + \varepsilon D$, with $D = Pq(t, y(t), \varepsilon)$, as stated. The proof of the convergence in the extended cases considered at the end of (b), is analogous, and we refer to the papers already quoted.

(d) *Some existence theorems for periodic solutions (cycles) of autonomous systems.* By the method described above the following theorem, among others, has been proved. This theorem has a number of applications.

(8.5.i) Consider the system of equations

$$\left.\begin{aligned} &x_1'' + \sigma_1^2 x_1 = \varepsilon f_1(x, x', \varepsilon), \\ &x_j'' + 2\alpha_j x_j' + \sigma_j^2 x_j = \varepsilon f_j(x, x', \varepsilon), \qquad j = 2, 3, \ldots, n, \\ &x_j' + \beta_j x_j = \varepsilon f_j(x, x', \varepsilon), \qquad j = n+1, \ldots, N, \end{aligned}\right\} \tag{8.5.16}$$

where $x = (x_1, \ldots, x_N)$, $x' = (x_1', \ldots, x_n')$, $1 \leqq n \leqq N$, all f_j are continuous in ε and Lipschitzian in $x_1, \ldots, x_N, x_1', \ldots, x_n'$ for all $|x_j| \leqq R$, $j = 1, \ldots, N$, $|x_j'| \leqq R$, $j = 1, \ldots, n$, $0 \leqq \varepsilon \leqq \varepsilon_0$, and some $R > 0$, $\varepsilon_0 > 0$. Assume $\sigma_j > 0$, $j = 1, \ldots, n$, $\beta_j > 0$, $j = n+1, \ldots, N$, and either $\alpha_j \neq 0$, $\sigma_j^2 - \alpha_j^2 > 0$, or $\alpha_j = 0$ and $m\sigma_1 \neq \sigma_j$ for all $m = 0, 1, \ldots$, $(j = 2, \ldots, n)$. If there is a number λ_0, $0 < \lambda_0 < R$, satisfying the equation $Q_{10}(\lambda) = 0$, and $Q_{10}(\lambda)$ changes sign at $\lambda = \lambda_0$, let us determine β_0 by the equation $\beta\lambda + P_{10}(\lambda) = 0$. Then, for all $0 \leqq \varepsilon \leqq \varepsilon_1$ (and some $0 < \varepsilon_1 \leqq \varepsilon_0$), system (8.5.16) has a periodic solution of period $\tau = 2\pi/\omega$ of the form

$$\left.\begin{aligned} x_1(t, \varepsilon) &= \lambda(\varepsilon)\, \omega^{-1}(\varepsilon) \sin(\omega(\varepsilon)\, t + \theta) + \varepsilon W_1(\omega(\varepsilon)\, t + \theta, \varepsilon) \\ x_j(t, \varepsilon) &= \varepsilon W_j(\omega(\varepsilon)\, t + \theta, \varepsilon), \qquad j = 2, \ldots, N, \end{aligned}\right\} \tag{8.5.17}$$

where θ is arbitrary, $\lambda = \lambda_0 + O(\varepsilon)$, $\omega = \sigma + \beta_0 \varepsilon + \varepsilon\, O(\varepsilon)$, $\sigma = \sigma_1$, $|W_j| \leqq M$, $j = 1, \ldots, N$. Above P_{10} and Q_{10} are given by the trigonometric integrals:

$$P_{10} = (\sigma/2\pi) \int_0^{2\pi/\sigma} f_1(x_0, x_0', 0) \sin \sigma t\, dt = 0,$$

$$Q_{10} = (\sigma/2\pi) \int_0^{2\pi/\sigma} f_1(x_0, x_0', 0) \cos \sigma t\, dt = 0,$$

where $x_{10} = \lambda\sigma^{-1} \sin \sigma t$, $x_{10}' = \lambda \cos \sigma t$, $x_{0j} = 0$, $j = 2, \ldots, N$, $x_{0j}' = 0$, $j = 2, \ldots, n$. Note that, if $Z_1(x_1, x_1')$ denotes the function $Z_1(x_1, x_1') = f_1(x_1, 0, \ldots, 0, x_1', 0, \ldots, 0, 0)$, then there is always a decomposition

$$Z_1(x_1, x_1') = Z_{11}(x_1, x_1') + Z_{12}(x_1, x_1') + Z_{13}(x_1, x_1') + Z_{14}(x_1, x_1'),$$

where $Z_{11}[Z_{12}]$ is even in x_1 and odd [even] in x_1', $Z_{13}[Z_{14}]$ is odd in x_1 and odd [even] in x_1'. Then we have $f_1(x, x', \varepsilon) = Z_{11} + Z_{12} + Z_{13} + Z_{14} + g_1(x, x', \varepsilon)$, and $Z_{12}, Z_{13}, Z_{14}, g_1$ have no bearing in the determination of Q_{10} and hence of λ_0, since

$$Q_{10} = (\sigma/2\pi) \int_0^{2\pi/\sigma} Z_{11}(\lambda\sigma^{-1}\sin\sigma t, \lambda\cos\sigma t)\cos\sigma t\, dt.$$

In particular, if Z_{11} is a polynomial in x_1, x_1', hence

$$Z_{11} = \sum a_{hk}\, x_1^{2h}\, x_1'^{2k-1},\ h \geqq 0,\ k \geqq 1,$$

then

$$Q_{10} = 2^{-1}\sum a_{hk}(2h)!\,(2k)!\,[h!\,k!\,(h+k)!]^{-1}\sigma^{-2h}(2^{-1}\lambda)^{2h+2k-1}, \quad (8.5.18)$$

(cfr. J. K. HALE [3], L. CESARI [5]).

If, for instance, $a_{01} \neq 0$, and a_{01} and the coefficient of maximal power of λ in Q_{10} are of opposite signs, then certainly Q_{10} has a simple positive root, and (8.5.16) has a periodic solution as above. The same statement above holds if $\sigma_1(\varepsilon)$, $\alpha_j(\varepsilon)$, $\sigma_j(\varepsilon)$, $\beta_j(\varepsilon)$ are continuous functions of ε [$\sigma(\varepsilon)$ with continuous derivative], if $\sigma_j(0) > 0$, $\beta_j(0) > 0$, and either $\alpha_j(0) \neq 0$, $\sigma_j^2(0) - \alpha_j^2(0) > 0$, or $\alpha_j(0) = 0$ and $m\sigma_1(0) \neq \sigma_j(0)$ for all $m = 0, 1, \ldots$, $(j = 2, \ldots, n)$.

As a first application of (8.5.i) let us consider the van der Pol equation

$$x'' + x = \varepsilon(1 - x^2)\, x'. \quad (8.5.19)$$

Thus $n = N = 1$, $x = x_1$, $x' = x_1'$, $f = (1 - x^2)x'$, $f = Z_{11}$, $\sigma = 1$, $a_{01} = 1$, $a_{11} = -1$, hence, by (8.5.i), there exists a cycle of the form $x = \lambda\omega^{-1}\sin(\omega t + \theta) + O(\varepsilon)$. By (8.5.18) we have $Q_{10} = 2^{-1}(1 - \lambda^2/4)$, and hence $\lambda_0 = 2$, and $\lambda = 2 + O(\varepsilon)$, $\omega = 1 + O(\varepsilon)$. Actually, we have $P_{10} = 0$, and hence $\beta_0 = 0$ and $\omega = 1 + \varepsilon O(\varepsilon)$.

A representation of the cycle with any degree of accuracy can be obtained by explicit application of the method discussed in (b, c). We shall limit ourselves to the second approximation. In harmony with (a) let us write equation (8.5.19) in the form

$$x'' + \omega^2 x = \varepsilon(1 - x^2)\, x' + (\omega^2 - 1)\, x,$$

and take $\omega = 1 + \varepsilon\beta(\varepsilon)$. By the transformation $y_1 = i\omega x + x'$, $y_2 = i\omega x - x'$, this equation is reduced to the system $y_1' = i\omega y_1 + \varepsilon f$, $y_2' = -i\omega y_2' - \varepsilon f$, where

$$f = 2^{-1}[1 + 2^{-2}\omega^{-2}(y_1 + y_2)^2](y_1 - y_2) - i\beta\omega^{-1}(1 - 2^{-1}\varepsilon\beta)(y_1 + y_2).$$

As a first approximation we take $y_1^{(0)} = a_1 e^{i\omega t}$, $y_2^{(0)} = a_2 e^{-i\omega t}$, and, if we want to reach the form (8.5.17) for the solution, we take $a_1 = \lambda$, $a_2 = -\lambda$. Then we have

$$y_1^{(1)} = \lambda e^{i\omega t} + \varepsilon e^{i\omega t}\int (I - P)\, e^{-i\omega u} f\, du,$$
$$y_2^{(1)} = -\lambda e^{i\omega t} - \varepsilon e^{-i\omega t}\int (I - P)\, e^{i\omega u} f\, du,$$

where $y_1^{(0)}$, $y_2^{(0)}$ replace y_1, y_2 in the expression of f above. The computations for $y_1^{(1)}$ yield immediately

$$\begin{aligned} y_1^{(1)} = \lambda e^{i\omega t} + \varepsilon e^{i\omega t}\int (I - P)\{&[2^{-1}\lambda - 2^{-3}\omega^{-2}\lambda^3 - i\beta\omega^{-1}(1 + 2^{-1}\varepsilon\beta)\lambda] + \\ &+ 2^{-3}\omega^{-2}\lambda^3 e^{2i\omega u} + [2^{-1}\lambda - 2^{-3}\omega^{-2}\lambda^3 + i\beta\omega^{-1}(1 + 2^{-1}\varepsilon\beta)\lambda]\, e^{-2i\omega u} + \\ &+ 2^{-3}\omega^{-2}\lambda^3 e^{-4i\omega u}\}\, du, \end{aligned}$$

where the operator $I - P$ means that we have to disregard the mean value of the integrand, that is, the first bracket. This expression, equated to zero, gives the first approximation to the determining equation

$$2^{-1}\lambda - 2^{-3}\omega^{-2}\lambda^3 - i\omega^{-1}\beta(1 + 2^{-1}\varepsilon\beta)\,\lambda = 0.$$

The same computations for $y_2^{(1)}$ lead to the same equation with i replaced by $-i$. Then, the expressions for $y_1^{(1)}$, $y_2^{(1)}$ simplify and we obtain

$$y_1^{(1)} = \lambda e^{i\omega t} + \varepsilon(\lambda^3/16 i\omega^3)(e^{3i\omega t} - 2^{-1}e^{-3i\omega t}),$$
$$y_2^{(1)} = -\lambda e^{i\omega t} - \varepsilon(\lambda^3/16 i\omega^3)(2^{-1}e^{3i\omega t} - e^{-3i\omega t}),$$

and hence, for $x = (2i\omega)^{-1}(y_1 + y_2)$, we obtain

$$x^{(1)} = \lambda\omega^{-1}\sin\omega t - \varepsilon(\lambda^3/32\omega^4)\cos 3\omega t.$$

The determining equations above reduce to the real ones already obtained, hence

$$\lambda_0 = 2\omega, \qquad \beta_0 = 0, \qquad \omega_0 = 1 + \varepsilon O(\varepsilon), \qquad \lambda = 2 + O(\varepsilon).$$

As another application of (8.5.i) let us consider the system

$$\left.\begin{aligned} x'' + x &= \varepsilon(1 - x^2 - y^2)x' + \varepsilon f(x, y, x', y'), \\ y'' + 2y &= \varepsilon(1 - x^2 - y^2)y' + \varepsilon g(x, y, x', y'), \end{aligned}\right\} \tag{8.5.20}$$

where $f(-x, 0, y, 0) = -f(x, 0, y, 0)$, $g(0, -y, 0, y') = -g(0, y, 0, y')$. We have $n = N = 2$. If we take $\sigma_1 = 1, \sigma_2 = 2^{\frac{1}{2}}$, $\alpha_2 = 0$, then $Z_{11} = (1 - x^2)x'$, and the formula above for Q_{10} gives $Q_{10} = 2^{-1}(1 - \lambda^2/4) = 0$. Hence, (8.5.20) has a periodic solution of the form

$$x = \lambda(\varepsilon)\omega^{-1}(\varepsilon)\sin(\omega(\varepsilon)t + \theta) + \varepsilon W_1(\omega(\varepsilon)t + \theta, \varepsilon), \qquad y = \varepsilon W_2(\omega(\varepsilon)t + \theta, \varepsilon)$$
$$\lambda(\varepsilon) = 2 + O(\varepsilon), \qquad \omega(\varepsilon) = 1 + O(\varepsilon), \qquad \theta \text{ arbitrary}.$$

If we take $\sigma_1 = 2^{\frac{1}{2}}$, $\sigma_2 = 1$, $\alpha_2 = 0$, (that is, we interchange the office of the two equations (8.5.20)), then $Z_{11} = (1 - y^2)y'$, $Q_{10} = 2^{-1}(\lambda - \lambda^3/8)$ and we deduce from (8.5.i) that (8.5.20) has also a periodic solution of the form

$$x = \varepsilon W_1(\omega(\varepsilon)t + \theta, \varepsilon), \quad y = \lambda(\varepsilon)\omega^{-1}(\varepsilon)\sin(\omega(\varepsilon)t + \theta) + \varepsilon W_2(\omega(\varepsilon)t + \theta, \varepsilon),$$
$$\lambda(\varepsilon) = 2^{\frac{3}{2}} + O(\varepsilon), \qquad \omega(\varepsilon) = 2^{\frac{1}{2}} + O(\varepsilon), \qquad \theta \text{ arbitrary}.$$

(8.5.ii) Consider the system

$$\left.\begin{aligned} x_j'' + \sigma_j^2 x_j &= \varepsilon f_j(x, x', \varepsilon), \qquad j = 1, 2, \ldots, n, \\ x_j' &= \varepsilon f_j(x, x', \varepsilon), \qquad j = n + 1, \ldots, N, \end{aligned}\right\} \tag{8.5.21}$$

where $x = (x_1, \ldots, x_N)$, $x' = (x_1', \ldots, x_n')$, $0 \leqq n \leqq N$, the functions f_j are continuous in ε and Lipschitzian in $x_1, \ldots, x_N, x_1', \ldots, x_n'$ for $|x_j| \leqq R$, $j = 1, \ldots, N$, $|x_j'| \leqq R$, $j = 1, \ldots, n$, $0 \leqq \varepsilon \leqq \varepsilon_0$, and all f_j are odd in $(x_1, \ldots, x_n)$. Take $0 < r_1 < r_2 < R$, and assume that all σ_j are positive constants, or continuous functions $\sigma_j(\varepsilon)$ of ε with continuous first derivative, with $\sigma_j(0) > 0$, $j = 1, \ldots, n$, and $\sigma_j(0) \neq k\sigma_1(0)$ for all $j = 2, \ldots, n$, and $k = 1, 2, \ldots$. Then there is an ε_1, $0 < \varepsilon_1 \leqq \varepsilon_0$, such that, for all $0 \leqq \varepsilon \leqq \varepsilon_1$, system (8.5.21) has periodic solutions of period $T = 2\pi/\omega$, of the form

$$\left.\begin{aligned} x_1 &= \lambda\omega^{-1}\sin\omega t + \varepsilon W_1(\omega t, \varepsilon), \quad x_j = \varepsilon W_j(\omega t, \varepsilon), \quad j = 2, \ldots, n, \\ x_j &= \eta_{j-n} + \varepsilon W_j(\omega t, \varepsilon), \qquad\qquad j = n + 1, \ldots, N, \end{aligned}\right\} \tag{8.5.22}$$

$x_1, \ldots, x_n$ odd in t, $x_{n+1}, \ldots, x_N$ even in t, $|W_j| \leqq M$, $j = 1, \ldots, N$, where $\lambda, \eta_1, \ldots, \eta_{N-n}$ are arbitrary constants, $r_1 \leqq \lambda \leqq r_2$, $|\eta_j| \leqq r_2$, $j = 1, \ldots, N - n$, and $\omega = \omega(\lambda, \eta_1, \ldots, \eta_{N-n}, \varepsilon)$ is a continuous function of its arguments with $\omega(\lambda, \eta_1, \ldots, \eta_{N-n}, 0) = \sigma_1(0)$. In (8.5.22) we can replace t by $t + \vartheta$, ϑ an arbitrary constant, since (8.5.21) is autonomous. Thus (8.5.21) has one $(N - n + 2)$-parameter family of periodic solutions (J. K. Hale [3], L. Cesari [5]).

The same statement above holds for all $|\lambda| \leqq r_2$ provided $f(0, x_2, \ldots, x_n, 0, x_2', \ldots, x_N', \varepsilon) = 0$.

The same statement above holds if, for some m, $0 \leq m \leq n$, and $u = (x_1, \ldots, x_m)$, $v = (x_{m+1}, \ldots, x_n)$, $w = (x_{m+1}, \ldots, x_N)$, we have

$$f_j(u, -v, w, -u', v', \varepsilon) = f_j(u, v, w, u', v', \varepsilon), \qquad j = 1, \ldots, m,$$
$$f_j(u, -v, w, -u', v', \varepsilon) = -f_j(u, v, w, u', v', \varepsilon), \qquad j = m+1, \ldots, N,$$

and then $x_1, \ldots, x_m, x_{n+1}, \ldots, x_N$ are even in t, $x_{m+1}, \ldots, x_n$ are odd in t, and $x_1 = \lambda_1 \omega^{-1} \cos \omega t + \varepsilon W_1(\omega t, \varepsilon)$ if $m > 1$.

In particular, if all $\sigma_j(0)$, $j = 1, \ldots, n$, are distinct, and $\sigma_j(0) \neq k \sigma_h(0)$ for all $j \neq h$, $j, h = 1, \ldots, n$, $k = 1, 2, \ldots$, then, for every $h = 1, \ldots, n$, system (8.5.21) has a periodic solution of period $T = 2\pi/\omega$, of the form

$$x_h = \lambda_h \omega^{-1} \sin \omega t + \varepsilon W_h(\omega t, \varepsilon), \quad x_j = \varepsilon W_j(\omega t, \varepsilon), \quad j \neq h, j = 1, \ldots, n,$$
$$x_j = \eta_{j-n} + \varepsilon W_j(\omega t, \varepsilon), \qquad j = n+1, \ldots, N,$$

where $\lambda_h, \eta_1, \ldots, \eta_{N-n}$ are arbitrary constants, $r_1 \leq \lambda_h \leq r_2$, $|\eta_1|, \ldots, |\eta_{N-n}| \leq r_2$, $\omega = \omega_h(\lambda_h, \eta_1, \ldots, \eta_{N-n})$, and $\omega = \sigma_h(0)$ for $\varepsilon = 0$. By replacing t by $t + \vartheta$, ϑ an arbitrary constant, we conclude that (8.5.21) has, in the present situation, n families of periodic solutions, each depending on $N - n + 2$ parameters.

As an application of (8.5.ii) consider the autonomous third order differential equation

$$y''' + \sigma^2 y' = \varepsilon f(y, y', y'', \varepsilon), \tag{8.5.23}$$

where $f(y, -y', y'', \varepsilon) = -f(y, y', y'', \varepsilon)$. By the transformation $x_1 = y'$, $x_2 = \sigma^{-2} y'' + y$, equation (8.5.23) is reduced to the system

$$x_1'' + \sigma^2 x_1 = \varepsilon f(-\sigma^{-2} x_1' + x_2, x_1, x_1', \varepsilon),$$
$$x_2' = \varepsilon f(-\sigma^{-2} x_1' + x_2, x_1, x_1', \varepsilon),$$

and theorem (8.5.ii) applies with $m = 0$, $n = 1$, $N = 2$. Equation (8.5.23) has a three-parameter family of even periodic solutions of the form $y = -\sigma^{-2} \lambda \cos(\omega t + \theta) + \eta + \varepsilon W(\omega t + \theta, \varepsilon)$, where λ, η, θ are arbitrary constants, and $\omega = \omega(\lambda, \eta, \varepsilon)$, $\omega(\lambda, \eta, 0) = \sigma$.

(e) *Some existence theorems for periodic solutions of periodic systems.* By the method described in (b, c) the following general theorems, among others, have been proved.

(8.5.iii) Consider the system of equations

$$\left.\begin{aligned} x_j'' + 2\alpha_j x_j' + \sigma_j^2 x_j &= \varepsilon f_j(x, x', t, \varepsilon), \quad j = 1, \ldots, n, \\ x_j' + \beta_j x_j &= \varepsilon f_j(x, x', t, \varepsilon), \quad j = n+1, \ldots, N, \end{aligned}\right\} \tag{8.5.24}$$

where $x = (x_1, \ldots, x_N)$, $x' = (x_1', \ldots, x_n')$, and $0 \leq n \leq N$. Assume that $\alpha_j, \beta_j, \sigma_j$ are real constants, or continuous functions of ε, and all f_j are real continuous functions of ε, uniformly Lipschitzian in x, x', and periodic in t of a fixed period $T = 2\pi/\omega$, and L-integrable in $[0, T]$, for $|x_j| \leq R, j = 1, \ldots, N$, $|x_j'| \leq R, j = 1, \ldots, n$, $|\varepsilon| \leq \varepsilon_0$, $-\infty < t < +\infty$, $(\omega, \varepsilon_0, R > 0)$. Assume that, for some ν, μ, $0 \leq \nu \leq n \leq \mu \leq N$, we have: (i) $\alpha_j(0) = 0$, $\sigma_j(0) = \tau_j = k_j \omega / m_j$, k_j, m_j positive relatively prime integers, for all $j = 1, \ldots, \nu$; (ii) either $\alpha_j(0) = 0$, $\sigma_j(0) > 0$, $\sigma_j(0) \neq k\omega/m$, $m = m_1 \ldots m_\nu$, $k = 1, 2, \ldots$, or $\alpha_j(0) \neq 0$, $\sigma_j(0) > 0$, $\gamma_j^2 = \sigma_j^2 - \alpha_j^2 > 0$, for all $j = \nu+1, \ldots, n$; (iii) $\beta_j(0) \neq 0$ for all $j = n+1, \ldots, \mu$, (iv) $\beta_j(0) = 0$ for all $j = \mu+1, \ldots, N$. We shall assume that the derivatives $\alpha_j'(0)$, $\beta_j'(0)$, $\sigma_j'(0)$ exist, and the functions P_{j0}, Q_{j0}, R_{j0} below have first order continuous partial derivatives in $\lambda_1, \ldots, \lambda_\nu$, $\theta_1, \ldots, \theta_\nu$, $\eta_1, \ldots, \eta_{N-\mu}$. If the $2\nu + (N - \mu)$ finite equations in $\lambda_1, \ldots, \lambda_\nu$, $\theta_1, \ldots, \theta_\nu$, $\eta_1, \ldots, \eta_{N-\mu}$:

$$P_{j0}(\lambda, \theta, \eta) = \lambda_j \alpha_j'(0), \qquad j = 1, \ldots, \nu,$$
$$Q_{j0}(\lambda, \theta, \eta) = \lambda_j \sigma_j'(0), \qquad j = 1, \ldots, \nu,$$
$$R_{j0}(\lambda, \theta, \eta) = \eta_j \beta_j'(0), \qquad j = 1, \ldots, N - \mu,$$

have a solution $\lambda_{j0}, \theta_{j0}, \eta_{j0}$ with Jacobian different from zero at $(\lambda_{j0}, \theta_{j0}, \eta_{j0})$, then system (8.5.24) has a periodic solution of period $T_1 = 2\pi\, m/\omega$ of the form

$$\left.\begin{aligned} x_j(t,\varepsilon) &= \lambda_j \tau_j^{-1} \sin(\tau_j t + \vartheta_j) + \varepsilon W_j(t,\varepsilon), & j &= 1, \ldots, \nu, \\ x_j(t,\varepsilon) &= \varepsilon W_j(t,\varepsilon), & j &= \nu+1, \ldots, \mu, \\ x_j(t,\varepsilon) &= \eta_{j-\mu} + \varepsilon W_j(t,\varepsilon), & j &= \mu+1, \ldots, N, \end{aligned}\right\} \tag{8.5.25}$$

where $\lambda_j = \lambda_{j0} + O(\varepsilon)$, $\theta_j = \theta_{j0} + O(\varepsilon)$, $j = 1, \ldots, \nu$, $\eta_j = \eta_{j0} + O(\varepsilon)$, $j = 1, \ldots, N-\mu$, $|W_j| \leqq M$, for all $|\varepsilon| \leqq \varepsilon_1$ and some $0 < \varepsilon_1 \leqq \varepsilon_0$. The functions P_{j0}, Q_{j0}, R_{j0} are given by

$$P_{j0} = T_1^{-1} \int_0^{T_1} f_{j0} \cos(\tau_j t + \theta_j)\, dt,$$

$$Q_{j0} = T_1^{-1} \int_0^{T} f_{j0} \sin(\tau_j t + \theta_j)\, dt, \qquad j = 1, \ldots, \nu,$$

$$R_{j0} = T^{-1} \int_0^{T_1} f_{j0}\, dt, \qquad j = \mu + 1, \ldots, N,$$

where f_{j0} denotes f_j when the arguments ε, x_j, x_j' are replaced by $\varepsilon = 0$ and by $x_j(t, 0)$, $x_j'(t, 0)$ (cfr. L. CESARI and J. K. HALE [2], L. CESARI [5]).

As an example of (8.5.iii) let us consider the harmonic solution of the van der Pol equation with a forcing term

$$x'' + \omega^2 x = \varepsilon(1 - x^2)\, x' + \varepsilon p\, \omega \cos(\omega t + \alpha), \tag{8.5.26}$$

$p, \omega > 0$, α constants, $\varepsilon > 0$ a small parameter. By (8.5.iii) with $\nu = n = \mu = N = 1$, we have only to consider the two equations $P_0 = 0$, $Q_0 = 0$, or

$$P_0 = T^{-1} \int_0^T \{[1 - \lambda^2 \omega^{-2} \sin^2(\omega t + \theta)]\, \lambda \cos(\omega t + \theta) + p\,\omega \cos(\omega t + \alpha)\} \times$$
$$\times \cos(\omega t + \theta)\, dt = 0,$$

$$Q_0 = T^{-1} \int_0^T \{[1 - \lambda^2 \omega^{-2} \sin^2(\omega t + \theta)]\, \lambda \cos(\omega t + \theta) + p\,\omega \cos(\omega t + \alpha)\} \times$$
$$\times \sin(\omega t + \theta)\, dt = 0,$$

or

$$\lambda^3 - 4\omega^2 \lambda - 4p\,\omega^3 \cos(\alpha - \theta) = 0, \; p \sin(\alpha - \theta) = 0, \tag{8.5.27}$$

hence $\theta = \alpha$, and $\lambda^3 - 4\omega^2 \lambda - 4p\,\omega^3 = 0$ (since $\theta = \alpha + \pi$ corresponds to the change of λ into $-\lambda$). The equation in λ has certainly a simple positive root $\lambda_0 > 0$, and thus (8.5.26) has a periodic solution of period $T = 2\pi/\omega$ of the form

$$x(t, \varepsilon) = \lambda(\varepsilon)\, \omega^{-1} \sin(\omega t + \alpha) + \varepsilon W(t, \varepsilon), \tag{8.5.28}$$

with $|W(t, \varepsilon)| \leqq M$, $\lambda = \lambda_0 + O(\varepsilon)$, for all $|\varepsilon| \leqq \varepsilon_1$ and some $\varepsilon_1 > 0$.

To improve the first approximation (8.5.26) we may use the process of successive approximations (b, c). By the transformation $y_1 = i\omega x + x'$, $y_2 = i\omega x - x'$, equation (8.5.26) is reduced to the system $y_1' = i\omega y_1 + \varepsilon f$, $y_2' = -i\omega y_2 - \varepsilon f$, where

$$f = 2^{-1}[1 + 2^{-2}\omega^{-2}(y_1 + y_2)^2](y_1 - y_2) + p\omega \cos(\omega t + \alpha).$$

If we take as a first approximation $y_1^{(0)} = \lambda e^{i(\omega t + \vartheta)}$, $y_2^{(0)} = -\lambda e^{-i(\omega t + \vartheta)}$, (corresponding to (8.5.28) with λ, θ undetermined), then the second approximation is given by

$$y_1^{(1)} = \lambda e^{i(\omega t + \theta)} + \varepsilon e^{i(\omega t + \theta)} \int (I - P)\, e^{-i(\omega u + \theta)} f\, du,$$
$$y_2^{(1)} = -\lambda e^{-i(\omega t + \theta)} - \varepsilon e^{-i(\omega t + \theta)} \int (I - P)\, e^{i(\omega u + \theta)} f\, du,$$

where $y_1^{(0)}$, $y_2^{(0)}$ replace y_1, y_2 in the expression of f above. The computations for $y_1^{(1)}$ yield immediately

$$y_1^{(1)} = \lambda e^{i(\omega t+\vartheta)} + \varepsilon e^{i(\omega t+\vartheta)} \int (I-P)\{[2^{-1}\lambda - 2^{-3}\omega^{-2}\lambda^3 + 2^{-1}p\,\omega\, e^{i(\alpha-\vartheta)}] + \\ + 2^{-3}\omega^{-2}\lambda^3 e^{2i(\omega u+\theta)} + 2^{-3}\omega^{-2}\lambda^3 e^{-4i(\omega u+\theta)} + \\ + [2^{-1}\lambda - 2^{-3}\omega^{-2}\lambda^3 + 2^{-1}p\,\omega\, e^{-i(\alpha-\theta)}]\, e^{-2i(\omega u+\theta)}\}\, du,$$

where the operation $I-P$ means that we have to disregard the mean value of the integrand, that is, the first bracket. This expression, equated to zero, gives the first approximation to the determining equation

$$2^{-1}\lambda - 2^{-3}\omega^{-2}\lambda^3 + 2^{-1}p\,\omega\, e^{i(\alpha-\theta)} = 0.$$

The same computations for $y_2^{(1)}$ lead to the same equation with i replaced by $-i$. Then, the expressions for $y_1^{(1)}$, $y_2^{(1)}$ simplify, and we obtain

$$y_1^{(1)} = \lambda e^{i(\omega t+\theta)} + \varepsilon(\lambda^3/16 i\,\omega^3)(e^{3i(\omega t+\theta)} - 2^{-1} e^{-3i(\omega t+\theta)})$$
$$y_2^{(1)} = -\lambda e^{-i(\omega t+\theta)} - \varepsilon(\lambda^3/16 i\,\omega^3)(2^{-1} e^{3i(\omega t+\theta)} - e^{-3i(\omega t+\theta)})$$

and hence, for $x = (2i\,\omega)^{-1}(y_1 + y_2)$, we obtain

$$x^{(1)} = \lambda\omega^{-1}\sin(\omega t+\theta) - \varepsilon(\lambda^3/32\omega^4)\cos 3(\omega t+\theta).$$

The determining equation above and its complex conjugate yield again the real determining equations (8.5.27) already obtained.

(8.5.iv) Consider the system of equations

$$\left.\begin{aligned} x_j'' + \sigma_j^2 x_j &= \varepsilon f_j(u, v, w, u', v', \varepsilon, t) \qquad j = 1, \ldots, n, \\ x_j' &= \varepsilon f_j(u, v, w, u', v', \varepsilon, t), \qquad j = n+1, \ldots, N, \end{aligned}\right\} \tag{8.5.29}$$

where $x = (x_1, \ldots, x_N) = (u, v, w)$, $x' = (x_1, \ldots, x_n) = (u, v)$, and $u = (x_1, \ldots, x_m)$, $v = (x_{m+1}, \ldots, x_n)$, $w = (x_{n+1}, \ldots, x_N)$, for some $0 \leq m \leq n$, where σ_j are constants, or continuous functions of ε, and where all f_j are real continuous functions of ε, uniformly Lipschitzian in x, x', periodic in t of a fixed period $T = 2\pi/\omega$, L-integrable in $[0, T]$, for all $|x_j| \leq R$, $j = 1, \ldots, N$, $|x_j'| \leq R$, $j = 1, \ldots, n$, $|\varepsilon| \leq \varepsilon_0$, ($R$, ω, ε_0 positive constants), and satisfy

$$f_j(u, -v, w, -u', -v', \varepsilon, -t) = f_j(u, v, w, u', v', \varepsilon, t), \qquad j = 1, \ldots, m,$$
$$f_j(u, -v, w, -u', -v', \varepsilon, -t) = -f_j(u, v, w, u', v', \varepsilon, t), \qquad j = m+1, \ldots, N.$$

Assume that $\sigma_j(0) = \tau_j = k_j\omega/m_j$, k_j, m_j positive relatively prime integers, and the derivatives $\sigma_j'(0)$ exist, $j = 1, \ldots, n$. Let $\eta_1, \ldots, \eta_{N-n}$ be arbitrary numbers, $|\eta_j| \leq r < R$, $j = 1, \ldots, N-n$, and assume that the $N-n$ equations in $\lambda_1, \ldots, \lambda_n$,

$$Q_{j0}(\lambda, \eta) = \lambda_j\,\sigma_j'(0), \qquad j = 1, \ldots, n, \tag{8.5.30}$$

have a solution $\lambda_{10}, \ldots, \lambda_{n0}$ with Jacobian different from zero at $\lambda_{10}, \ldots, \lambda_{n0}$, and $|\lambda_{j0}| \leq r < R$, $j = 1, \ldots, n$. Then system (8.5.29) has a periodic solution of period $T_1 = 2\pi\, m/\omega$, $m = m_1 \ldots m_n$, of the form

$$\left.\begin{aligned} x_j(t, \varepsilon) &= \lambda_j \tau_j^{-1}\cos\tau_j t + \varepsilon W_j(t, \varepsilon), \qquad j = 1, \ldots, m, \\ x_j(t, \varepsilon) &= \lambda_j \tau_j^{-1}\sin\tau_j t + \varepsilon W_j(t, \varepsilon), \qquad j = m+1, \ldots, n, \\ x_j(t, \varepsilon) &= \eta_{j-n} + \varepsilon W_j(t, \varepsilon), \qquad j = n+1, \ldots, N, \end{aligned}\right\} \tag{8.5.31}$$

with $x_j(-t, \varepsilon) = x_j(t, \varepsilon)$, $j = 1, \ldots, m$, and $j = n+1, \ldots, N$, and $x_j(-t, \varepsilon) = -x_j(t, \varepsilon)$ for $j = m+1, \ldots, n$, where $|W_j(t, \varepsilon)| \leq M$, $j = 1, \ldots, N$, for all $|\varepsilon| \leq \varepsilon_1$ and some $\varepsilon_1 > 0$ (an $(N-n)$-parameter family of periodic solutions). In (8.5.30) the expression of Q_{j0} is the same as given in (8.5.iii) (L. Cesari and J. K. Hale [2], J. K. Hale [6]).

As a first application of (8.5.iv) consider the third order equation

$$y''' + \sigma^2 y' = \varepsilon f(y, y', y'', t, \varepsilon), \tag{8.5.32}$$

where f is periodic in t of period $2\pi/\omega$, $\sigma = \sigma(\varepsilon)$, $\sigma(0) = k\omega/m$, k, m positive integers, and $f(y, -y', y'', -t, \varepsilon) = -f(y, y', y'', t, \varepsilon)$. By the transformation $x_1 = y'$, $x_2 = \sigma^{-2} y'' + y$, equation (8.5.32) is reduced to the system

$$x_1'' + \sigma^2 x_1 = \varepsilon f(-\sigma^{-2} x_1' + x_2, x_1, x_1', t, \varepsilon),$$
$$x_2' = \varepsilon \sigma^{-2} f(-\sigma^{-2} x_1' + x_2, x_1, x_1', t, \varepsilon),$$

and theorem (8.5.iv) applies with $m = 0$, $n = 1$, $N = 2$. Let η be an arbitrary constant, and assume that the (unique) determining equation $Q_0(\lambda, \eta) = \lambda \sigma'(0)$ has a simple solution $\lambda_0 = \lambda_0(\eta) \neq 0$. Then, equation (8.5.32) has an even periodic solution of period $T = 2\pi/\omega$ of the form $y(t, \varepsilon) = -\sigma^{-2} \lambda \cos(k m^{-1} \omega t) + \eta + \varepsilon W(t, \varepsilon)$. In the determining equation, $Q_0(\lambda, \eta)$ is given by

$$Q_0(\lambda, \eta) = T^{-1} \int_0^T f[y(t, 0), y'(t, 0), y''(t, 0), t, 0] \sin(k m^{-1} \omega t)\, dt.$$

For the sake of brevity we omit a series of other theorems of existence, all proved by the method in (b, c). A number of examples have been discussed in detail by the process above, or variants of it, by R. A GAMBILL and J. K. HALE [1], H. BAILEY [1], and J. K. HALE [10, 11]. For instance, the following examples have been discussed:

a) $x'' + c x' + \sigma^2 x = B \cos 2\omega t + \varepsilon \alpha \cos 2\omega t \cdot x + \varepsilon b x^3$,

an equation containing as particular cases the non-linear Mathieu equation with large forcing terms ($\alpha \neq 0$, $b \neq 0$, $c = 0$), the Duffing equation ($\alpha = 0$, $c = 0$, $b \neq 0$), and the Duffing equation with damping ($\alpha = 0$, $c > 0$, $b \neq 0$);

b) $x'' + c x' + \sigma^2 x = \varepsilon p_0(x) + \varepsilon p_1(x) x' + \varepsilon b \cos(\omega t + \alpha)$,

an equation of the Liénard type with $p_0(x)$, $p_1(x)$ polynomials in x;

c) $x'' + x = \varepsilon(1 - x^{2n}) x' + \varepsilon p \omega \cos(\omega t + \alpha)$,

an equation of the van der Pol type with n an integer;

d) $x'' + \sigma^2 x = \varepsilon[3\nu \cos 2t - x^2] - \varepsilon^2(\lambda x' + \mu x) - \varepsilon^3 \mu x^2$,

an equation for which some results had been already obtained by G. E. H. REUTER [1];

e) $x'' + \sigma_1^2 x = \varepsilon \alpha x + \varepsilon A \cos t \cdot x + \varepsilon \beta x^3 + \varepsilon \gamma x y^2$,
$y'' + \sigma_2^2 y = \varepsilon \delta y + \varepsilon B \cos \omega t \cdot y + \varepsilon \mu y^3 + \varepsilon \nu x^2 y$,

a system of nonlinear Mathieu equations with ω rational;

f) $x'' + 4x = \cos t + \varepsilon[\cos 3t \cdot x + k y^3 + b x]$,
$y'' + y = \varepsilon[\cos 2t \cdot y + c x^3]$

a system of two nonlinear Mathieu equations in which one has a large forcing term.

(f) *Stability*. In (d), (e) above we have listed a number of theorems of existence of periodic solutions of nonlinear systems of the perturbation type. The same method discussed in (a), (b), (c), the results mentioned in (4.5) concerning linear systems with periodic coefficients, and the theorems (6.2.i), (6.3.i), (6.4.i) can be combined to give final criteria for the asymptotic stability of the same solutions for periodic systems, and the asymptotic orbital stability of the corresponding solutions for autonomous systems (see H. R. BAILEY and R. A. GAMBILL [1], E. W. THOMPSON [1], and particularly J. K. HALE [9]). The complete results cannot be given here. For the autonomous case they extend a criterion of A. ANDRONOV and A. WITT for $n=2$ (see [6], or N. MINORSKY [5], pp. 153–158). For periodic systems they can be thought of as extending a criterion of L. MANDELSTAM and N. PAPALEXI [1] (see also N. MINORSKY [6]) for one second order equation. For autonomous systems, under the conditions of (8.5. i), a sufficient condition for asymptotic orbital stability as $t\to+\infty$ of the periodic solution (8.5.17) is that $\varepsilon>0$ and

$$s_j=\int_0^{T_1} f_{jx'_j}(x(t,0),x'(t,0),0)\,dt<0, \qquad j=1,\ldots,n,$$

where $f_{jx'_j}$ is the partial derivative of f_j with respect to x'_j.

For periodic systems, under conditions of (8.5. iii), a sufficient condition for asymptotic stability as $t\to+\infty$ of the periodic solution (8.5.25) is that $\varepsilon>0$, $s_1>0$, $\gamma>0$, $\beta_j>0$, $j=n+1,\ldots,N$, and either $\alpha_j>0$, or $s_j<0$ for every $j=1,\ldots,n$. Here s_j denotes the expression analogous to the one above, and γ is given by

$$\begin{aligned}\gamma=\Big[T^{-1}\int_0^T f_{jx'_1}\,dt\Big]^2+\Big[(\sigma_1 T)^{-1}\int_0^T f_{1x'_1}\,dt\Big]^2-\\ -\Big[(\sigma T)^{-1}\int_0^T f_{1x_1}\cos 2\sigma t\,dt-T^{-1}\int_0^T f_{1x'_1}\sin 2\sigma t\,dt\Big]^2-\\ -\Big[(\sigma T)^{-1}\int_0^T f_{1x_1}\sin 2\sigma t\,dt-T^{-1}\int_0^T f_{1x'_1}\cos 2\sigma t\,dt\Big]^2.\end{aligned}$$

By applying these criteria it is immediate, for instance, that the periodic solution of the van der Pol equation (8.5.19) is asymptotically orbitally stable. Analogously, it is possible to see that the periodic solution of the van der Pol equation with forcing term (8.5.26) is asymptotically stable (e.g., for $0<p\leqq 1$). See J. K. HALE [9, 10, 11] for further theorems and examples, and also, H. R. BAILEY and R. A. GAMBILL [1], and H. R. BAILEY [1].

Remark. The approach given in (a), (b), (c) for periodic solutions of nonlinear differential systems of the perturbation type, can be thought of as a particular case of the process devised by L. CESARI [6] for strongly nonlinear differential systems.

For other approaches not dissimilar from the one discussed above, see E. A. BARBASIN [1–4], Y. SIBUYA [1–2], J. A. NOHEL [1], K. L. STELLMACHER [1], S. N. SIMANOV [1]. Cf. also I. G. MALKIN [9, 11, 22], A. I. LURE [5], E. WEBER [1], W. WASOW [14]. For more details on the present approach, and a comparison with a number of these analogous developments, see J. K. HALE [11].

8.6. The perturbation method. This method applies to a class of problems different from the ones we have considered in the previous sections (8, 2–5). Indeed the conditions of the theorems below (8.6. i, ii) are not satisfied when applied to the problems considered above.

(a) Let

$$x'_i=f_i(t,x_1,\ldots,x_n,\varepsilon), \quad i=1,\ldots,n, \quad \text{or } x'=f(t,x,\varepsilon), \tag{8.6.1}$$

be, as usual, a differential system containing a parameter ε. Let f be a vector function of t, x, ε, whose components f_i are continuous with their partial derivatives $f_{ij}=\partial f_i/\partial x_j$ for all $-\infty<t<+\infty$, $x\in S$, $|\varepsilon|\leq\varepsilon_0$, where ε_0 is some number $\varepsilon_0>0$ and S some open set of the x-space E_n. Thus (8.5.1) satisfies local conditions of existence and uniqueness (1.1. i, ii). Suppose now that (8.6.1) is periodic, of some period T, i.e., $f(t+T, x, \varepsilon) = f(t, x, \varepsilon)$ for all t, $x \in S$, $|\varepsilon| \leq \varepsilon_0$. We shall also suppose that system

$$x' = f(t, x, 0) \tag{8.6.2}$$

has some known periodic solution of period T, say $x = p(t)$, $p(t+T) = p(t)$, $-\infty < t < +\infty$, and that $p(t) \in S$ for all t.

We shall now ask whether system (9.6.1) has also a periodic solution $x = x(t, \varepsilon)$ of the same period T and all ε sufficiently small, say $|\varepsilon| \leq \varepsilon_1$, $0 < \varepsilon_1 \leq \varepsilon_0$, with $x(t, \varepsilon) \in S$ for all t and $|\varepsilon| \leq \varepsilon_1$ and $x(t, \varepsilon) \rightrightarrows p(t)$ as $\varepsilon \to 0$ uniformly.

If the answer is affirmative we shall say that $x(t, \varepsilon)$ for ε small is a *perturbation* of the periodic solution $p(t)$ of system (8.6.2) (H. Poincaré).

Let us observe that, in the applications, f is generally analytic in x and ε, though this is not needed in the present general discussion. Let us observe also that T need not be the least period of either f or p.

We may now, in harmony with (1.7), consider the linear variational system of (8.6.2) relative to the periodic solution $p(t)$ of period T, say

$$v' = A(t)\, v, \qquad A(t) = [a_{ij}(t)], \tag{8.6.3}$$

where $a_{ij}(t) = f_{ij}(t, p(t), 0)$, $i, j = 1, \ldots, n$, and $f_{ij} = \partial f_i / \partial x_j$. System (8.6.3) is then a linear homogeneous system with periodic coefficients of period T. In the terms of § 4 we may now associate with (8.6.3) its characteristic factors λ_s, $s = 1, \ldots, n$. If $V(t) = [v_{ij}(t)]$ is the fundamental system of solution of (8.6.3) with $V(0) = I$, then the factors are the characteristic roots of $V(T)$, i.e., they are the roots of the equation $\det[\lambda I - V(T)] = 0$. Also, we know that system (8.6.3) admits of a periodic solution $q(t)$ of period T if and only if $\lambda = 1$ is one of the factors.

The relation between the perturbation of the periodic solution $p(t)$ and the linear variational system relative to $p(t)$, and system (8.6.2) is given by the following statement:

(8.6. i) If system (8.6.3) has no periodic solution of period T then for all ε sufficiently small there exists one and only one periodic solution $x(t, \varepsilon)$ of (8.6.1) of period T, belonging to a sufficiently small neighborhood S_0 of $p = p(t)$ in S, and $x(t, \varepsilon) \rightrightarrows p(t)$ as $\varepsilon \to 0$.

Proof. Let us denote by U a neighborhood of the point $p(0)$ in E_n, and we may suppose $p(0) = 0$. Since $p(t) \in S$, hence $p(0) \in S$, and since S is open we may suppose $U \subset S$. For every point $u = (u_1, \ldots, u_n) \in U$ let us denote by $x(t; 0, u, \varepsilon)$ the solution of (8.6.1) with $x(0; 0, u, \varepsilon) = u$. This solution exists in a neighborhood of $t = 0$. Since $p(t)$ exists in $(-\infty, +\infty)$, certainly $x(t; 0, u, \varepsilon)$ exists in $[0 \leq t \leq T)$ and is in S provided $u \in U$, $|\varepsilon| \leq \varepsilon_1$, U is a sufficiently small neighborhood of $p(0)$ and ε_1, $0 < \varepsilon_1 \leq \varepsilon_0$, is a number sufficiently small (1.1. iii). Thus

$$(d/dt)\, x_i(t; 0, u, \varepsilon) = f_i[t, x(t; 0, u, \varepsilon), \varepsilon], \qquad i = 1, \ldots, n, \tag{8.6.4}$$

for all $0 \leq t \leq T$, $|\varepsilon| \leq \varepsilon_1$, $u \in U$. Since the functions $f_i(t, x, \varepsilon)$ have partial derivatives f_{ij} continuous for all $x \in S$, $|\varepsilon| \leq \varepsilon_0$, $-\infty < t < +\infty$, we conclude by (1.1. iv) that the functions $x_i(t; 0, u, \varepsilon)$ have continuous partial derivatives $x_{ij} = \partial x_i / \partial u_j$

for all $0 \leq t \leq T$, $u \in U$, $|\varepsilon| \leq \varepsilon_1$. By (8.6.4) and differentiation with respect to u_h we have

$$(d/dt)\, x_{ih}(t; 0, u, \varepsilon) = \sum_{j=1}^{n} f_{ij}[t, x(t; 0, u, \varepsilon), \varepsilon]\, x_{jh}(t; 0, u, \varepsilon), \tag{8.6.5}$$

$i = 1, \ldots, n$, $h = 1, \ldots, n$. Obviously $x(t; 0, 0, 0) = p(t)$, and $f_{ij}[t, p(t), 0] = a_{ij}(t)$. Thus by (8.6.5) for $\varepsilon = 0$, $u = 0$, we have

$$(d/dt)\, x_{ih}(t; 0, 0, 0) = \sum_{j=1}^{n} a_{ij}(t)\, x_{ih}(t, 0, 0, 0), \qquad i = 1, \ldots, n, \qquad h = 1, \ldots, n.$$

Thus $[x_{ih}(t; 0, 0, 0), i = 1, \ldots, n]$, $h = 1, \ldots, n$, are n solutions of system (8.6.3). On the other hand $x_i(0, 0, u, 0) = u_i$, and hence $x_{ih}(0, 0, 0, 0) = \delta_{ih}$. Thus if $X(t; 0, u, \varepsilon) = [x_{ij}(t; 0, u, \varepsilon)]$, $X(t; 0, 0, 0)$ is the fundamental solution of system (8.6.3), with $X(0; 0, 0, 0) = I$.

Suppose now that system (8.6.3) has no periodic solution of period T. Then $\lambda_s \neq 1$, $s = 1, \ldots, n$, and hence $\det(I - X(T, 0, 0, 0)) \neq 0$. Thus we can restrict both U and $\varepsilon_1 < 0$ in such a way that

$$\det\big(I - X(T, 0, u, \varepsilon)\big) \neq 0$$

for every $u \in U$ and $|\varepsilon| \leq \varepsilon_1$.

Now the solution $x(t; 0, u, \varepsilon)$, $u \in S$, $|\varepsilon| \leq \varepsilon_1$, is periodic of period T if and only if $x(T; 0, u, \varepsilon) = x(0; 0, u, \varepsilon)$, where $x(0; 0, u, \varepsilon) = u$; i.e., if and only if the following conditions of periodicity are satisfied

$$x_i(T; 0, u, \varepsilon) = u_i, \qquad i = 1, \ldots, n. \tag{8.6.6}$$

If we consider (8.6.6) as a system of equations in the unknowns $u_1, \ldots, u_n$, containing the parameter ε, we see that (8.6.6) for $\varepsilon = 0$ has the trivial solution $u = p(t)$, and its Jacobian $J(u, \varepsilon)$ is given by

$$J(u, \varepsilon) = \det(I - \partial x_i/\partial u_j) = \det\big(I - X(T, 0, u, \varepsilon)\big) \neq 0,$$

for all $u \in U$, $|\varepsilon| \leq \varepsilon_1$. Thus system (8.6.6) has one and only one solution $u = u(\varepsilon) \in U$ for all $|\varepsilon| \leq \varepsilon_1$ provided U and ε_1 are conveniently restricted, and $u(\varepsilon) \to 0$ as $\varepsilon \to 0$. Thus $x = x(t; 0, u(\varepsilon), \varepsilon)$ is a periodic solution of (1) and $x\big(t; 0, u(\varepsilon), \varepsilon\big) \rightrightarrows p(t)$ as $\varepsilon \to 0$ in $[0, T]$ as well as in $(-\infty, +\infty)$, because of the periodicity. If S_0 is the (open set) covered by all the trajectories $x = x(t; 0, u, \varepsilon)$, $u \in U$, then S_0 is a neighborhood of $p = p(t)$ in E_n, S_0 contains the ∞-many closed curves $x = x\big(t; 0, u(\varepsilon), \varepsilon\big)$, $|\varepsilon| < \varepsilon_1$, and no other periodic solution of period T may be contained in S_0 (since such a solution should pass through U at a point $u = u(\varepsilon)$ for some $|\varepsilon| \leq \varepsilon_1$.

Remark. If all the characteristic roots of $V(T)$ (multipliers, § 4) are in modulus less than one (i.e., all characteristic exponents have negative real parts) then systems (8.6.3) has no periodic solutions of period T and the periodic solution $x(t, \varepsilon)$ of (8.6.1) is asymptotically stable [see (6.3. i)] (cf. also E. Coddington and N. Levinson [3]).

(b) Autonomous systems. We shall consider now an autonomous system

$$x' = f(x, \varepsilon), \qquad f = \big(f_i(x_1, \ldots, x_n; \varepsilon), \quad i = 1, \ldots, n\big), \tag{8.6.7}$$

having for $\varepsilon = 0$ the periodic solution $x = p(t) = \big(p_1(t), \ldots, p_n(t)\big)$ of period T_0 and we shall ask whether system (8.6.7) for $|\varepsilon| \neq 0$ small, has a periodic solution $x(t, \varepsilon)$ which approaches $p(t)$, and whose period $T(\varepsilon) \to T_0$ as $\varepsilon \to 0$.

Here too, as in (a), we shall consider the linear variational system

$$v' = A(t)\, v \tag{8.6.8}$$

relative to the system $x' = f(x, 0)$ and its periodic solution $p(t)$. Thus

$$A(t) = [a_{ij}(t)], \qquad A(t+T_0) = A(t),$$
$$a_{ij}(t) = f_{ij}(p_1(t), \ldots, p_n(t); 0), \qquad f_{ij} = \partial f_i / \partial x_j.$$

Let us observe that

$$p_i'(t) = f_i[p_1(t), \ldots, p_n(t), 0], \qquad i = 1, 2, \ldots, n,$$

and hence, by differentiation with respect to t, we have

$$(d/dt)\, p_i''(t) = \sum_{j=1}^{n} f_{ij}[p_1(t), \ldots, p_n(t); 0]\, p_j'(t).$$

Thus (8.6.8) has the periodic solution $p'(t) = [p_1'(t), \ldots, p_n'(t)]$. As before we shall denote by $V(t) = [v_{ij}(t)]$ the fundamental system of solutions of (8.6.8) with $V(0) = I$, and by λ_s the multipliers (§ 4), i.e. the characteristic roots of $V(T_0)$. The existence of the periodic solution $p'(t)$ of period T_0 assures that at least one of the multipliers λ_s is equal to one.

The following main theorems hold:

(8.6. ii) If one is a simple characteristic root of $V(T_0)$, then for small $|\varepsilon|$ the system (8.6.7) has a periodic solution $q = q(t, \varepsilon)$ of period $T(\varepsilon)$. Both $q(t, \varepsilon)$, $T(\varepsilon)$ are continuous functions and $q(t, 0) = p(t)$, $T(0) = T_0$. The solution $q(t, \varepsilon)$ is also uniquely determined for $|\varepsilon|$ small.

(8.6. iii) If the remaining $n-1$ characteristic roots of $V(T_0)$ are all in modulus less than 1, then (8.6. ii) holds, and $q(t, \varepsilon)$ is orbitally stable for small $|\varepsilon|$.

For brevity's sake we must omit the proofs (see E. A. Coddington and N. N. Levinson [2, 3]). In the first quoted paper further results are given with applications.

In connection with the perturbation method see H. Poincaré [6] and also S. Lefschetz [1, 6] who has given an algorithm for the determination of the periodic solutions. For other theoretical research on the method see E. Hopf [1], and B. V. Bulgakov [10]. For the same method applied to periodic solutions in a torus space see A. N. Kolmogorov [2]. For the same method applied to the Schroedinger equation see H. Kallman and M. Päsler [1], and applied to Sturm-Liouville type problems see S. M. Rytov and M. E. Zhabotinsky [1]. For a series of different applications see A. Basch [1], G. F. Carrier [1], S. Fifer [1], T. C. Huang [1], H. Jounin [1], A. A. Kruming [1], A. M. Letov [2], M. E. Levenson [1], C. Obi [5], A. Rosenblatt [4], F. K. Rubert [1], G. V. Savinov [1], M. Urabe [1]. M. E. Zhabotinski [1—3]. For an application of the method to partial differential equations see J. J. Stoker [2].

8.7. The Liénard equation and its periodic solutions. A great deal of research has been dedicated to the second order real differential equation (Liénard equation)

$$x'' + f(x, x')\, x' + g(x) = 0, \tag{8.7.1}$$

which comprehends as particular cases the equations

$$x'' + f(x)\, x' + g(x) = 0, \tag{8.7.2}$$

$$x'' + f(x)\, x' + x = 0, \tag{8.7.3}$$

the latter having actually been investigated by A. LIÉNARD. Equation (8.7.3) contains as a particular case the van der Pol equation

$$x'' + \varepsilon\,(x^2 - 1)\, x' + x = 0,$$

already mentioned in (8.1), and (8.7.1) contains the Rayleigh equation

$$x'' - (a - b\, x'^2)\, x' + x = 0,$$

which can be reduced to the van der Pol equation by taking x' as the new unknown, by differentiation and manipulation. All these equations are very important since they are concerned with substained self-excited oscillations in the technique, particularly radio circuits and electronics. We shall see in (8.14) that a feedback radio circuit with triode leads naturally to the van der Pol equation.

We shall first state and prove a theorem due to N. LEVINSON and O. K. SMITH concerning the existence and uniqueness of a periodic solution (cycle) of equation (8.7.2). In (8.8) we shall state and prove an oscillation theorem due to D. GRAFFI for equation (8.7.1), and, in (8.9), we shall state the corresponding theorem of the same author concerning the existence of periodic solutions of (8.7.1). In (8.10) we shall discuss the nonlinear free oscillations of the equation $x'' + g(x) = 0$. In § 9, by means of both topological and analytical tools we will resume the discussion of the Liénard equation and we will prove (9.4; 9.5) an existence theorem due to N. LEVINSON for periodic solutions of (8.7.1). Meanwhile we shall refer as much as possible, in the limited space, to the great deal of research on the subject.

We shall assume that the coefficients $f(x), g(x)$ of equation (8.7.2) are defined in $(-\infty, +\infty)$, that $F(x) = \int_0^x f(x)\, dx$, $G(x) = \int_0^x g(x)\, dx$, and that $f(x)$ is continuous and $g(x)$ differentiable. The theorem below is proved by a modification of the argument used by A. LIÉNARD for the case $g = x$. We shall suppose also that a uniqueness theorem holds.

(8.7. i) Suppose $f(x)$ even, $F(x) < 0$ for $0 < x < x_0$, $F(x) > 0$ for $x > x_0$ and some $x_0 > 0$; suppose $g(x)$ odd, and $g(x) > 0$ for $x > 0$; suppose $F(x)$ monotone increasing for $x > x_0$, and $F(x) \to +\infty$, $G(x) \to +\infty$ as $x \to +\infty$. Then the equation

$$x'' + f(x)\, x' + g(x) = 0,$$

has a unique nonzero periodic solution (N. LEVINSON and O. K. SMITH [1]).

Proof. By taking $v = x'$, we have $x'' = v\,(dv/dx)$, and (8.7.2) is transformed into

$$dt/dx = 1/v, \qquad v\,[(dv/dx) + f(x)] + g(x) = 0.$$

By taking $y = v + F(x)$, we obtain finally

$$dt/dx = 1/[y - F(x)], \qquad [y - F(x)]\,(dy/dx) + g(x) = 0. \tag{8.7.4}$$

A nonzero unique periodic solution (of period T) for (8.7.2) is equivalent to a nonzero closed integral curve C for (8.7.4) in the xy-plane where T is given by the curvilinear integrals

$$T = \int_C dx/v = \int_C dx/[y - F(x)] = \int_C dy/g(x).$$

Since x and v cannot be zero together we conclude that T is always finite.

In the second equation (8.7.4) both $F(x)$ and $g(x)$ are odd. Hence the equation does not change by changing (x, y) into $(-x, -y)$. As a consequence, any arc solution of (8.7.4) in the $x \geq 0$ half plane whose end points on the y-axis are two symmetric points $(0, y_0)$, $(0, -y_0)$, $y_0 > 0$, can be completed, by symmetry with respect to the origin, into a closed integral curve. Conversely, any closed integral curve, say $H(x, y) = 0$, made up of two arcs in the $x \geq 0$ and $x \leq 0$ half planes and ending at two points of the y-axis, $(0, y_1)$, $(0, y_2)$, $y_1 < 0 < y_2$, is actually symmetric with respect to the origin in the xy-plane. Indeed, in the contrary case, the two solutions $H(x, y) = 0$, $H(-x, -y) = 0$, would be distinct and intersecting at some point (x, y), which is impossible. It follows that a periodic solution must intersect the y-axis at two symmetric points $(0, y_0)$, $(0, -y_0)$, $y_0 > 0$.

Let $L(x, y) = (\frac{1}{2})\,y^2 + G(x)$. Then $L(x, y)$ is even both with respect to x and to y. If an integral curve C crosses the y-axis at two points $(0, y_1)$, $(0, y_2)$, $y_2 < 0 < y_1$, then C is closed if and only if $L(0, y_1) = L(0, y_2)$. Hence we need discuss only arcs of integral curves in the $x \geq 0$ half-plane.

From (8.7.4) we deduce that integral curves have negative slopes for $y > F(x)$; positive slopes for $y < F(x)$; slope ∞ for $y = F(x)$ (see illustration).

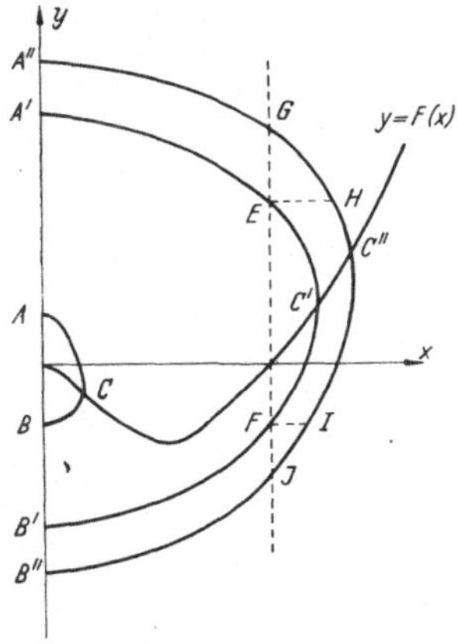

From (8.7.4) we have $y\,dy + g(x)\,dx = F(x)\,dy$, and hence

$$dL(x, y) = F(x)\,dy$$

or, by (8.7.4), also

$$dL(x, y) = -F(x)\,g(x)\,dx/[y - F(x)].$$

We consider now an integral curve lying in the half plane $x \geq 0$ and starting at a point $A = (0, y_0)$, $y_0 > 0$. Thus A is in the region $y > F(x)$, where $dx > 0$, $dy < 0$. Since $y(x)$ decreases the integral curve must intersect the locus $y = F(x)$ at some point $C = (x_1, y_1)$, $x_1 > 0$. In the region $y < F(x)$, $x \geq 0$, we have $dx < 0$, $dy < 0$, and $y(x)$ decreases. The integral curve cannot cross again the locus $y = F(x)$. Indeed, should this occur at some $(\bar{x}, \bar{y})$, $0 < \bar{x} < x_1$, $\bar{y} = F(\bar{x}) = y(\bar{x})$, with $F(x) - y > 0$ in a right neighborhood of $\bar{x}$, then we would have $|f(x)| < M$, $g(x) \geq m > 0$, $F(x) - y \to 0$, $dy/dx \to +\infty$, $f(x) - dy/dx \to -\infty$, $F(x) - y < 0$ in a right neighborhood of $\bar{x}$, a contradiction. For $\bar{x} = 0$, then $F(x)$ is decreasing at $x = 0$, y is increasing at $x = 0$, and thus $F(x) - y < 0$ in a right neighborhood of $\bar{x} = 0$, again a contradiction. Thus the integral curve remains in the region $y < F(x)$. Since for $0 \leq x \leq x_1 - \varepsilon$, $\varepsilon > 0$, we have $F(x) - y > \delta > 0$, $0 \leq g(x) \leq M$, we have $dy/dx < M/\delta$. Thus the integral curve reaches the y-axis at some point $(0, y_2)$, $y_2 < 0$. Concerning the arc $A\,C$, note that in any interval $0<\varepsilon\leqq x\leqq x_0$, and for $0<y_0\leqq 1$, we have $y\leqq 1$, $g(x)\geqq m>0$, $F(x)\geqq -M$, $0<y-F\leqq 1+M$, and finally $dy/dx<-m(1+M)^{-1}$. Thus, if y_0 is sufficiently small, we certainly have $x_1<x_0$, and the entire arc ACB is in the strip $0\leqq x\leqq x_0$.

If an integral curve ACB starts at $(0, y)$, $y>0$, with an y small enough so that $\overset{\frown}{ACB}$ is completely contained in the strip $0\leq x\leq x_0$, then $F(x)<0$, $dy<0$ along $\overset{\frown}{ACB}$, and hence $dL>0$. Thus $L_B-L_A>0$. This is true for all the solutions for which the point C of intersection with the curve $y=F(x)$ is in the strip $0<x<x_0$.

Let us consider now any two integral curves $A''GHC''IJB''$, $A'EC'FB'$, intersecting the curve $y=F(x)$ in points C'', C' beyond the strip $0<x<x_0$. Since $y-F(x)$ on $A''G$ is larger than $y-F(x)$ on $A'E$, and $dx>0$, $F(x)<0$, on both arcs, we conclude that dL has on $A''G$ values all smaller than the corresponding values on $A'E$; hence $L_G-L''_A<L_E-L'_A$. On the arc GH we have $dy<0$, $F(x)>0$, and hence $dL<0$, $L_H-L_G<0$. The values of $F(x)>0$ on HI are all larger than on EF, and $dy<0$; hence dL has on HI values all smaller than dL on EF, and $L_I-L_H<L_F-L_E$. On the arc IJ we have $dy<0$, $F(x)>0$, and hence $dL<0$, $L_J-L_I<0$. The values of $y-F(x)<0$ on JB'' are all smaller than the values of $y-F(x)<0$ on FB', and $F(x)<0$, $dx<0$; hence the values of dL on JB'' are all smaller than the values of dL on FB'; hence $L_{B''}-L_J<L_{B'}-L_F$. By adding all these five relations we obtain

$$L_{B''}-L_{A''}<L_{B'}-L_{A'},$$

and hence $\overline{0B}''-\overline{0A}''<\overline{0B}'-\overline{0A}'$.

All this implies that $\overline{0B}-\overline{0A}>0$ as long as C is in $0<x\leq x_0$. When C moves beyond this strip, then $0B-0A$ is a decreasing function. Thus, if there is a periodic solution, i.e., a point on the y-axis with $\overline{0B}-\overline{0A}=0$, it is clear that there could be only one.

To prove that a periodic solution actually exists we have only to prove that $0B-0A$ actually becomes negative if B moves outwardly along the y-axis. To do this we have to analyze the positive and negative variations of L along ACB. The analysis above implies that L actually increases along the arcs $A''G$ and JB'', and actually decreases along $GHC''IJ$. But the increases of L on $A''G$ and JB'' are decreasing as A'' moves outwardly, and hence are bounded. On the other hand, if A'' is far enough such that $A''GC''B''$ cuts the straight line $x=2x_0$ in two points say K, L, then from (8.7.4) we have

$$|dy|\leq g(x)\,dx/[y-F(2x_0)],$$

and, if $y_0=y(2x_0)$,

$$(\overline{0A}''^2/2)-(y_0^2/2)-(\overline{0A}''-y_0)\,F(2x_0)\leq G(2x_0),$$

and, successively,

$$(\overline{0A}''-y_0)\,[(\overline{0A}''+y_0)/2-F(2x_0)]\leq G(2x_0),$$

$$\overline{0A}''-y_0<2G(2x_0)/[\overline{0A}''-2F(2x_0)].$$

Thus $y_0\to+\infty$ as $\overline{0A}''\to+\infty$. Since integral curves do not intersect, the distance $\overline{KL}$ between K and L must increase and $\overline{KL}\to+\infty$ as $\overline{0A}''\to+\infty$. Since $dL=F(x)\,dy$, and $x>2x_0$, $F(x)>F(2x_0)$, $dL<F(2x_0)\,dy$, we have along KL, by integration,

$$L_L-L_K<-F(2x_0)\cdot\overline{KL}$$

and $L_L-L_K\to-\infty$ as $\overline{KL}\to+\infty$, i.e., as $\overline{0A}''\to+\infty$. This assures that for $\overline{0A}''$ large enough the negative variation of L along $A''C''B''$ becomes predominant with respect to the positive variation, hence $L_{B''}-L_{A''}$ becomes negative, i.e.,

$\overline{0B''} - \overline{0A''}$ will be <0 for $\overline{0A''}$ large enough. Since $\overline{0B} - \overline{0A}$ is >0 for $\overline{0A}$ small, and <0 for $\overline{0A}$ large we conclude that $\overline{0B} - \overline{0A} = 0$ for some $\overline{0A}$, and, as we have noticed, this will occur only for one convenient $\overline{0A}$. The integral curve through A is closed, and the theorem is thereby proved.

Remark. Under the same hypotheses of (8.7. i) the unique nonzero periodic solution of (8.7.2) is asymptotically orbitally stable (1.8) [see remarks after (9.2. ix)]. Note that the van der Pol equation $x'' + \varepsilon(x^2 - 1)x' + x = 0$ with $\varepsilon > 0$ satisfies the hypotheses of (8.7. i).

8.8. An oscillation theorem for the equation (8.7.1). We shall consider equation (8.7.1) in the interval $0 \leq t < +\infty$, under quite weak hypotheses on f and g. Nevertheless we shall always suppose that a theorem of uniqueness holds, i.e., for given x_0, x_0', the solution $x(t; t_0, x_0, x_0')$ of (8.7.1) with $x(t_0) = x_0$, $x'(t_0) = x_0'$, is unique.

(8.8. i) If f, g are continuous functions of x, x' and $g(0) = 0$, $xg(x) > 0$ for all $x \neq 0$, $g(x) \to \infty$ as $x \to \pm\infty$; if $f(x, x') \geq -M$ for some $M > 0$ and all x, x'; if $f(x, x') < 0$ for all $|x|$, $|x'| < \delta$ and some $\delta > 0$, and $f(x, x') > 0$ for all $|x| > \Delta$, $|x'| > \Delta$ and some $\Delta > \delta$, then every nonzero solution of (8.7.1) is oscillatory in $[0, +\infty)$. If $[x_{1n}]$, $[x_{2n}]$ are the sequences of the minimum and maximum values of $x(t)$, then $x_{1n} < 0$, $x_{2n} > 0$, $n = 1, 2, \ldots$, $\underline{\lim}\, x_{1n} < 0$, $\overline{\lim}\, x_{2n} > 0$, as $n \to +\infty$ (D. Graffi [3]).

Proof. Obviously $x = 0$ is a solution of (8.7.1). We shall consider only nonzero solutions. By the hypothesis concerning uniqueness, x and x' are never zero together. Thus $x' = 0$ implies $x \neq 0$ and $x'' = -g(x) \neq 0$, x'' of the sign of $-x$. Consequently, consecutive points where $x' = 0$ are alternatively points of maximum and of minimum, and there $x(t)$ has opposite signs.

Let us prove that every solution $x(t, 0, x_0)$ exists in $[0, +\infty)$. Indeed, if $[0, \bar{t})$, $\bar{t} < +\infty$, were its maximum interval of existence, then by (1.1) we should have either $x \to +\infty$, or $x \to -\infty$ as $t \to \bar{t}$. Suppose $x \to +\infty$, and hence $x > \Delta$ in some $[t_0, \bar{t})$. Then $x'(t)$ may be zero at most once and thus $x'(t) > 0$ in some $(t_1, \bar{t})$. If $0 < x' < \Delta$, then $f(x, x') \geq -M$, $f(x, x')x' \geq -M\Delta$; if $x' > \Delta$, then $f(x, x') > 0$. In either case $f(x, x')x' + g(x) \to +\infty$ as $t \to \bar{t}$, and thus $x'' < 0$, x' decreasing in some $(t_2, \bar{\ })$, and finally $x(t) \leq x(t_2) + x'(t_2)(\bar{t} - t_2) < +\infty$, a contradiction. The same reasoning holds if $x'(t) \to -\infty$. Thus we have proved that $x(t, 0, x_0)$ exists in $[0, +\infty)$.

Let us prove first that $x(t)$ cannot be always nondecreasing for large t. We will prove this by contradiction. Thus suppose that for a nonzero solution $x(t)$ of (8.7.1) we have $x'(t) \geq 0$. Suppose first $x(t) \to +\infty$ as $t \to +\infty$. Then $g(x) \to +\infty$, $|x| > \Delta$ for large t, and then $f(x, x') > 0$ if $x' > \Delta$, $f(x, x') > -M$ otherwise. Thus $f(x, x')x' > -M\Delta$ for large t in either case, and $f(x, x')x' + g(x) \to +\infty$ as $t \to +\infty$. Thus $x'' \to -\infty$, $x' \to -\infty$, $x \to -\infty$, a contradiction.

Let us suppose now that $x(t) \geq 0$, $x(t) \to k$, k finite. Then $x(t) < k$ for all t and $\int^{+\infty} x'(t)\,dt < +\infty$. Thus $\underline{\lim}\, x'(t) = 0$ as $t \to +\infty$. Put $a = |g(k)|$, $b = |f(k, 0)| + 1$, $m = \max(a, b) \geq 1$, and let η be a number such that (i) $\eta > 0$, $\eta < 2^{-4}$, $\eta < \delta/2$; (ii) $\eta < a/2^4 b$ if $a \neq 0$; (iii) $|x - k| < \eta$ implies $|g(x) - g(k)| < a/2$ if $a \neq 0$, < 1 if $a = 0$; (iv) $|x - k| < \eta$, $|x'| < 2\eta$ implies $|f(x, x') - f(k, 0)| < b$. Let t_0 be a number such that $0 \leq x'(t_0) < \eta$, and $0 < k - x(t) < \eta^2/2m < \eta$ for all $t \geq t_0$.

Suppose first $k > 0$. Then $g(k) = a > 0$, and $t \geq t_0$ implies $k - \eta < x(t) < k$, $g(x) > a/2$. In addition, in a right neighborhood of t_0 we certainly have (α) $0 \leq x'(t) < 2\eta$ and, hence, as long that (α) holds, $|f(x, x')| < |f(k, 0)| + b < 2b$. Thus

$x''=-g(x)-f(x,x')x'<-a/2+2b\cdot 2\eta=(-a/2)(1-8b\eta/a)<-a/4$. Thus $x'(t)$ will continuously decrease from the value $x'(t_0)$, $0\leq x'(t_0)<2\eta$, to zero in a finite time and become negative, a contradiction.

Suppose now $k\leq 0$. If $k<0$, then $g(k)=-a<0$, and $t\geq t_0$ implies $k-\eta<x(t)$, $k-\eta^2/2m<x(t)$, $x(t)<k$, $-3a/2<g(x)<-a/2$. In addition, in a right neighborhood of t_0 inequality (α) holds, and hence, $|f(x,x')|<2b$ as before, and

$$x''>a/2-4b\eta=(a/2)(1-8b\eta/a)>a/4>0,$$

$$x''<3a/2+4b\eta=(a/2)(3+8b\eta/a)<2a\leq 2m.$$

If $k=0$, then $a=0$, and $t\geq t_0$ implies $-\eta<x(t)<0$, $-1<g(x)<0$. In addition in a right neighborhood of t_0 inequality (α) holds, and hence $-2b<f(x,x')<0$, and

$$x''>-g(x)-f(x,x')x'>0,$$

$$x''<1+2b\cdot 2\eta<2b\leq 2m.$$

In either case, $x'(t)$ increases as long as (α) holds and hence $x'(t)>c$ for some $c>0$. If (α) holds for all $t\geq t_0$, then $x(t)\to+\infty$, $x(t)>k$ for some $t>t_0$, a contradiction. If $x'(t)$ takes the value η at $t=t_1$ and the value 2η at $t=t_2$, $t_0<t_1<t_2$, then $x''<2m$ implies $t_2-t_1>\eta/2m$. On the other hand $x'(t)\geq\eta$ in $[t_1,t_2]$ and hence $x(t)$ increases from $x(t_1)>k-\eta^2/2m$ to $x(t_2)>x(t_1)+(t_2-t_1)\eta>k-\eta^2/2m+\eta(\eta/2m)=k$, a contradiction.

All this proves that $x(t)$ cannot be always nondecreasing for large t. An analogous argument shows that $x(t)$ cannot be always nonincreasing.

Since $x(t)$ is a nonzero solution of (8.7.1) and a uniqueness theorem holds, we conclude that x, x' are never zero together. Thus $x'=0$ implies $x\neq 0$, and, from (8.7.1), $x''\neq 0$, $-x''$ of the same sign of x at the same points. Thus all points where $x'(t)=0$, are points of maximum if $x(t)>0$, are points of minimum if $x(t)<0$, and thus the minima t_n' and the maxima t_n'' of $x(t)$ are alternate, say $t_1'<t_1''<t_2'<t_2''<\ldots$, and $x_{1n}<0<x_{2n}$, where $x_{1n}=x(t_n')$, $x_{2n}=x(t_n'')$, $n=1,2,\ldots$. In addition $x'(t)>0$ for $t_n'<t<t_n''$, $x'(t)<0$ for $t_n''<t<t_{n+1}'$.

Let $G(x)$ denote the function $G(x)=2\int_0^x g(x)\,dx$. Then $G(0)=0$, $G(x)>0$ for all $x\neq 0$, and $G(x)$ is increasing for $x>0$, decreasing for $x<0$. By multiplying (8.7.1) by $2x'$ and integrating we have

$$(d/dt)\,[x'^2+G(x)]=-2f(x,x')\,x'^2. \tag{8.8.1}$$

Let us consider any two consecutive maxima of a nonzero solution $x(t)$ of (8.7.1), say $t_1=t_n''$, $t_2=t_{n+1}''$, and put $x_1=x(t_1)$, $x_2=x(t_2)$. Thus $x_1>0$, $x_2>0$. Let us prove that (A) there exists a number $\sigma>0$ such that, for any solution $x(t)$ of (8.7.1), if $x_1<\sigma$, then $x_1<x_2$.

Let $\xi_1<0<\xi_2$ be the points where $G(\xi_1)=\delta^2=G(\xi_2)$. Let $\lambda=\min\,[G(\delta), G(-\delta)]$ and $-\delta\leq\zeta_1<0<\zeta_2\leq\delta$ be the points where $G(\zeta_1)=\lambda=G(\zeta_2)$. Put $\sigma=\min\,[-\xi_1,\xi_2,-\zeta_1,\zeta_2]$. Thus $G(\sigma)\leq\delta^2$ and, if any number $\mu>0$ satisfies $\mu<G(\sigma)$, then $\mu<\delta^2$. If $G(x)<G(\sigma)$, then $|x|<\delta$. Suppose $0<x_1<\sigma$, and suppose, if possible $0<x_2<x_1$. Since $x_2'=x'(t_2)=0$, there is a left neighborhood of t_2 where $|x|<\delta$, $|x'|<\delta$, hence $|x|$, $|x'|<\delta$, and, therefore, $f(x,x')<0$. We want to prove that such a neighborhood should cover $[t_1,t_2]$. Indeed, in the contrary case, we shall denote by $[\bar t,t_2]$ the maximal left neighborhood mentioned above, $t_1<\bar t<t_2$, and then either $|\bar x|=|x(\bar t)|=\delta$, or $|\bar x'|=|x'(\bar t)|=\delta$, while $|x(t)|<\delta$, $|x'(t)|<\delta$ for $\bar t<t\leq t_2$. Thus $f(x,x')<0$ in $(\bar t,t_2)$, $-2fx'>0$, and, by virtue of (8.8.1), $x'^2+G(x)$ is an increasing function of t.

Finally

$$\bar{x}'^2 < G(x_2) < G(\sigma) \leq \delta^2,$$

$$G(\bar{x}) \leq \bar{x}'^2 + G(\bar{x}) \leq G(x_2) < G(\sigma)$$

and, hence, $|\bar{x}'| < \delta$, $|\bar{x}| < \sigma < \delta$, a contradiction. Thus $|x| < \delta$, $|x'| < \delta$ in the whole interval $[t_1, t_2]$, and $x'^2 + G(x)$ is an increasing function of t in $[t_1, t_2]$. Since $x_1' = x_2' = 0$, we have $G(x_1) < G(x_2)$ and finally $x_1 < x_2$, a contradiction. This proves the statement (A). From here it follows that $\overline{\lim}\, x_{2n} \geq \sigma > 0$. Indeed even if some x_{2n} may be smaller than σ then the successive $x_{2,n+1}, x_{2,n+2}, \ldots$ will form an increasing sequence till one is larger than σ. Analogously we can prove that $\underline{\lim}\, x_{1n} \leq -\sigma < 0$.

8.9. Existence of a periodic solution of equation (8.7.1). We shall consider again equation (8.7.1) under the same hypotheses of (8.8. i) and, in addition, the following one:

(H) For each of the four quadrants Q_r, $r = 1, 2, 3, 4$, of the (x, x')-plane, there are three constants $a_r > 0$, $b_r > 0$, $m_r > 0$, such that (H_1) $f(x, x') \geq m_r > 0$ for all $(x, x') \in Q_r$, $|x| \geq a_r$, $|x'| \geq b_r$; (H_2) If $f(x, x')$ is not always ≥ 0 for $(x, x') \in Q_r$, $|x| \geq a_r$, $|x'| \leq b_r$, then $f(x, x')\, x' + g(x) \geq c_r > 0$ $[\leq -c_r < 0]$ for all $(x, x') \in Q_r$ with $|x| \geq a_r$, $|x'| \leq b_r$, $f(x, x') \geq 0$, $r = 1, 4$ $[r = 2, 3]$, where $c_r > 0$ is a constant; (H_3) There are two constants $k_1 > 0$, $k_2 > 0$, such that

$$k_2 (s_2^2 - s_1^2) > G(s_2) - G(s_1) > k_1 (s_2^2 - s_1^2) \geq k_1 (s_2 - s_1)^2$$

for all real numbers s_1, s_2 of the same sign with $|s_2| > |s_1| > 0$.

Condition (H_2) is identically satisfied if $f(x, x') \geq 0$ for $(x, x') \in Q_r$, $|x| \geq a_r$, $|x'| \leq b_r$. Condition H_3 is certainly satisfied if, for instance $k_1 |s| \leq |g(s)| \leq k_2 |s|$. All conditions of (8.8. i) and (H) are verified by the van der Pol equation $x'' + \varepsilon (x^2 - 1)\, x' + x = 0$ with $\varepsilon > 0$, as well as by the Rayleigh equation $x'' + \varepsilon (a + b x'^2)\, x' + x = 0$ with $a < 0$, $b > 0$, $\varepsilon > 0$.

(8.9. i) Existence theorem for a cycle. Under the conditions of (8.8. i) and (H) there is at least one periodic solution for equation (8.7.1) (D. GRAFFI [3]).

We omit the proof which is based on the same elementary considerations as in (8.8). We will resume the discussion of the Liénard equation in § 9 where, by the use of both analytic and topological considerations, we will state and prove (9.4; 9.5) another sufficient condition for a cycle of equation (8.7.1). Also, we shall give there a bibliography on the subject.

8.10. Nonlinear free oscillations. We shall consider the equation

$$x'' + f(x) = 0, \tag{8.10.1}$$

where, for the sake of simplicity, we suppose $f(0) = 0$ and $f(x)$ odd. The function $F(x) = \int_0^x f(u)\, du$ is then even with $F(0) = 0$. By integration of (8.10.1) after multiplication by $2x'$ we have

$$x'^2 = C - 2F(x), \qquad dt = \pm [C - 2F(x)]^{-\frac{1}{2}} dx, \tag{8.10.2}$$

where C is an arbitrary constant. We shall suppose that (8.10.1) satisfies a uniqueness theorem. Then for a nonzero solution of (8.10.1) assuming the value zero at $t=0$, we must have $x' \neq 0$ and thus $C>0$.

I Case. There exists some number $c>0$ for which $C=2F(c)$ and $C-2F(x)>0$ for all $0<x<c$, and, if we suppose $x(0)=0$, $x'(0)>0$, then the sign $+$ holds in (8.10.2) in a right neighborhood of $t=0$. Thus $x'(t)>0$, $x(t)$ increases, and this occurs as long as x remains less than c. Thus $x'(t)>0$, $x(t)$ increasing for all $0<t<T/4$, where $T/4$ (finite, or $+\infty$) is given by

$$T/4 = \int_0^c [C - 2F(x)]^{-\frac{1}{2}} dx. \tag{8.10.3}$$

If $T/4=+\infty$, then $x(t)\to c$, as $t\to+\infty$, and $x(t)$ is nonoscillatory. If $T/4<+\infty$, then $x(t)\to c$ as $t\to T/4$, $x(T/4)=c$, $x'(T/4)=0$. Since $x(t)\to c$ increasing, and $2F(x)<C$ also increases at the left of $x=c$, we conclude that $2f(c)\geq 0$, $x''(T/4)=-f(c)\leq 0$. If $f(c)=0$, then (8.10.1) would have, besides $x(t)$, the solution $x(t)=c=$ constant, both solutions satisfying the condition $x(T/4)=c$, $x'(T/4)=0$. Thus the uniqueness theorem implies $f(c)>0$, $x''(T/4)<0$, and $x'(t)$ must be negative in a right neighborhood of $T/4$. Thus the sign $-$ holds in (8.10.2), $x'(t)<0$, and $x(t)$ decreases in the same neighborhood, more precisely, for all $T/4<t<T/2$, and $x(T/2)=0$, $x'(T/2)<0$. By repeating this reasoning we prove that $x(t)$ is an odd periodic function, oscillating between $+c$ and $-c$, and of period T as given by (8.10.3). If we put

$$T = 4\int_0^c [C - 2F(x)]^{-\frac{1}{2}} dx = 2\pi p^{-1}, \tag{8.10.4}$$

we may develop $x(t)$ in Fourier series of only sines

$$x(t) = \sum_{i=1}^{\infty} c_i \sin i p t \tag{8.10.5}$$

where p depends on the arbitrary constant C.

II Case. $C-2F(x)>0$ for all $x>0$. We shall suppose now $f(x)$ (odd and) periodic of some period $a>0$ of the variable x. Then $f(x)$ has mean value zero, and $F(x)$ is even and periodic of the same period a. Thus $F(x)$ has a maximum M^2 and the hypothesis implies $C>2M^2$. Thus $x'^2\geq C-2M^2>0$ and $x(0)=0$, $x'(0)>0$ implies $x'(t)>0$ for all $t>0$. Also,

$$t = \int_0^x [C - 2F(x)]^{-\frac{1}{2}} dx, \tag{8.10.6}$$

where $[C-2F(x)]^{-\frac{1}{2}}$ is an even, periodic, always positive function of x, whose mean value $\mu = n^{-1}$ is given by

$$\mu = n^{-1} = a^{-1} \int_0^a [C - 2F(x)]^{-\frac{1}{2}} dx,$$

and μ is a decreasing function of C. As $C \to 2M^2 > 0$ with $C > 2M^2$, then μ approaches a number $\mu_0 > 0$, finite, or $+\infty$. Thus $\mu_0 > \mu > 0$, and $\mu_0^{-1} < n < +\infty$. Either of the constants C, μ, n, with $C > 2M^2$, $\mu < \mu_0$, $\mu_0^{-1} < n$ may be taken as an arbitrary constant. If we put $[C - 2F(x)]^{-\frac{1}{2}} = n^{-1} + \varphi(x)$, then $\varphi(x)$ is an even periodic function of x, and then $\Phi(x) = \int_0^x \varphi(\alpha)\, d\alpha$ is an odd periodic function of x of mean value zero. Thus (8.10.6) becomes

$$t = \Phi(x) + n^{-1} x, \tag{8.10.7}$$

where the right hand member has derivative > 0 for all x. Thus (8.10.7) defines $x(t)$ as an implicit function of t, say $x(t) = nt + \Psi(t)$ and $\Psi(t)$ is necessarily odd and periodic of period a/n in t. Thus by developing $\Psi(t)$ in Fourier series we have

$$x = n t + \Sigma c_i \sin i n t, \tag{8.10.8}$$

where n is an arbitrary constant and the coefficients c_i are functions of n. Since (8.10.1) is autonomous we may well replace t by $t + \text{const}$, in either (8.10.5) and (8.10.8). We conclude:

Under the hypotheses mentioned above, equation (8.10.1) may have (I) a two parameter family of solutions of the type

$$x(t) = \sum_{i=1}^{\infty} c_i \sin i(p t + \varepsilon), \tag{8.10.9}$$

where ε is an arbitrary constant, where all p, c_i are functions of a second arbitrary constant C and p is given by (8.10.4); (II) a two parameter family of solutions of the type

$$x(t) = n t + \sum_{i=1}^{\infty} c_i \sin i(n t + \varepsilon), \tag{8.10.10}$$

where n, ε are arbitrary constants and c_i are functions of n.

The limit case $c = 2M^2$ is between the two previous ones, and $x(t)$ may be oscillatory or not. In any case, the two situations (I) and (II) are widely separated, corresponding to a completely different behavior of the solutions of the equation. If $x'^2(0) = c \cong 2M^2$ and the error in the evaluation of c does not allow to know with certainty whether $c > 2M^2$, or $c = 2M^2$, or $c < 2M^2$, then it is not possible to conclude as to whether $x(t)$ will present the behavior (8.10.9) or (8.10.10) as $t \to +\infty$. The situation is analogous to the behavior of a system around a point of instable equilibrium, where slight variations may produce the most different asymptotic behavior.

The example of the exact pendulum

$$x'' + k^2 \sin x = 0, \qquad k^2 = g/l, \tag{8.10.11}$$

shows both types of solutions. Here $f(x) = k^2 \sin x$,

$$F(x) = k^2(1 - \cos x), \qquad M^2 = 2k^2.$$

If $0 < C < 4k^2$, then (8.10.11) has periodic solutions of the type (8.10.9), if $C < 4K^2$ then (8.10.11) has solutions of the type (8.10.10). The former correspond to oscillations around the stable equilibrium position, the latter to complete revolutions around the axis. For the solutions of the type (8.10.9), the amplitudes c, the period $T = 2\pi p^{-1}$, and the constant C are related by the relations $C = 2k^2 (1 - \cos c) = 4k^2 \sin^2(c/2)$,

$$T = 2\pi p^{-1} = 4\int_0^c [2k^2 \cos x - 2k^2 \cos c]^{-\frac{1}{2}} dx = 4(2k)^{-1} \int_0^c [\sin^2(c/2) - \sin^2(x/2)]^{-\frac{1}{2}} dx.$$

By the substitution $\sin(x/2) = \sin(c/2) \sin\psi$, we have

$$T = 2\pi p^{-1} = 4k^{-1} \int_0^{\pi/2} [1 - \sin^2(c/2) \sin^2\psi]^{-\frac{1}{2}} d\psi = 2\pi k^{-1} [1 + 4^{-1} \sin^2(c/2) + \cdots]$$

or, for small c,

$$T = 2\pi k^{-1}(1 + 16^{-1} c^2 + \cdots), \qquad p = k(1 - 16^{-1} c^2 + \cdots). \tag{8.10.12}$$

The coefficients c_i can be determined by substitution of (8.10.9) in (8.10.11) and development in power series of k. We obtain

$$x = c_1 \sin(pt + \varepsilon) + (192)^{-1} c_1^3 \sin 3(pt + \varepsilon) + \cdots, \tag{8.10.9*}$$

with

$$c = c_1 - (192)^{-1} c_1^3 + \cdots,$$

where c and ε can now be assumed as arbitrary constants and p is given by (8.10.12). For the solutions of the type (8.10.11) an analogous procedure gives

$$x = nt + \varepsilon + k^2 n^{-2} \sin(nt + \varepsilon) + 8^{-1} n^{-4} k^4 \sin 2(nt + \varepsilon) + \cdots, \tag{8.10.10*}$$

where n, ε are arbitrary constants. It shall be pointed out that in (8.10.10*) the development is a power series of k^2, while in (8.10.9*) x depends upon k in a quite different way. Both (8.10.9*) and (8.10.10*) are of the form

$$x = \psi(c, u), \qquad u = nt + \varepsilon, \qquad n = n(c), \tag{8.10.13}$$

where ε, c are arbitrary constants and n a constant depending on c. Indeed we have $n = c$ in formula (8.10.10*) and $n = p = p(c)$ in formula (8.10.9*).

8.11. Invariant surfaces. We shall consider now an autonomous nonlinear system

$$x' = F(x), \tag{8.11.1}$$

$x = (x_1, \ldots, x_n)$, $F = (F_1, \ldots, F_n)$, presenting a periodic solution $x = f(t)$, $-\infty < t < +\infty$, whose period we may take as equal to 2π. We shall suppose, with N. LEVINSON [17], that $F(x)$ has continuous first partial derivatives of the first three orders [and hence $f(t)$ is continuous with its derivatives of the first four orders]. We shall suppose that the solution $x=f(t)$, $f=(f_1, \ldots, f_n)$, is stable in the sense that its linear variational system (1.7; 6.4) has $n-1$ characteristic exponents with negative real parts [beside the characteristic exponent which is equal to zero according to (6.4)].

Obviously, system (8.11.1) admits of the one-parameter family of solutions $x=f(t+c)$, c an arbitrary constant, and these solutions represent in the x-space E_n the same orbit say C_0: $x=f(u)$, $0 \leqq u \leqq 2\pi$, while in the (t, x)-space E_{n+1}, they are distinct trajectories filling the cylinder shell Γ_0 of all points (t, x) with $-\infty < t < +\infty$, and $x=f(u)$, $0 \leqq u \leqq 2\pi$. If we consider any arbitrary number $T_0 > 0$, the curve C_0': $x=f(u)$, $0 \leqq u \leqq 2\pi$, $t=T_0$, is identical to the curve C_0: $x=f(u)$, $0 \leqq u \leqq 2\pi$, $t=0$, and we may identify C_0' to C_0, and all points $(t+mT_0, x)$, $m=0, \pm 1, \pm 2, \ldots$, in E_{n+1}, to (t, x). This operation reduces the cylinder Γ_0 above to a torus S_0. We will determine below a convenient value for the number T_0.

Let us consider now a perturbation of system (8.11.1), namely a periodic system

$$x' = F(x) + \varepsilon G(t, x, \varepsilon), \tag{8.11.2}$$

where ε is a real small parameter, G is continuous in (t, x, ε), periodic with respect to t of a given period T, and G has continuous partial derivatives of the first three orders. A solution of this system starting at $t=0$ from a point of C_0 does not necessarily lie on Γ_0. Under the hypotheses listed above, N. LEVINSON [17] has proved the following statement: For all $|\varepsilon|$ sufficiently small there is a fixed integer N and a closed curve, depending on ε, $C=C(\varepsilon)$: $x=\varphi(u)$, $0 \leqq u \leqq 2\pi$, $\varphi(0)=\varphi(2\pi)$, such that the solutions of (8.11.2), say $x=x\big(t; \varphi(u), \varepsilon\big)$, starting at $t=0$ from the points of the curve C, i.e., at $[x=\varphi(u),\ t=0]$, pass at $t=NT$, through the points of a curve C': $x=x\big(NT; \varphi(u), \varepsilon\big)$, $t=NT$ which, by a convenient parametrization, is the same curve C, say C': $[x=\varphi(v),\ t=NT]$. By identifying in E_{n+1}, C' to C, and all points $(t+mNT, x)$ $m=0, \pm 1, \pm 2, \ldots$, to (t, x), we may conclude that the same solutions $x=x\big(t; \varphi(u), \varepsilon\big)$, $-\infty < t < +\infty$, with $0 \leqq u \leqq 2\pi$, fill a torus $S=S(\varepsilon)$, depending on ε. The curve C is close to the curve C_0 in the sense that if $\eta(\varepsilon)=\max \|\varphi(u)-f(u)\|$, $0 \leqq u \leqq 2\pi$, we have $\eta(\varepsilon) \to 0$ as $\varepsilon \to 0$, and $C(0)=C_0$.

If we assume T_0 above to be $T_0=NT$, we may interpret this result by saying that the torus $S(\varepsilon)$ associated with the perturbed periodic system (8.11.2) for all $|\varepsilon|$ sufficiently small reduces, for $\varepsilon=0$, to the torus $S_0=S(0)$ (with $T_0=NT$) associated with the autonomous system (8.11.1). The torus $S(\varepsilon)$ is said to be an invariant surface for the system (8.11.2).

The solutions $x = x(t, \varphi(u), 0)$, thought of as orbits in $S(\varepsilon)$, behave, on $S(\varepsilon)$, as was mentioned in (6.9), and thus give rise on $S(\varepsilon)$ either to the periodic case, or to the ergodic case discussed in (6.9).

Results of the same type have been obtained by F. G. FRIEDLANDER [3] for the system of order 2,

$$x_1' = \omega x_2 + \varepsilon f(t, x_1, x_2), \qquad x_2' = -\omega x_1 + \varepsilon g(t, x_1, x_2),$$

where f, g are periodic functions of t. For other recent results concerning invariant surfaces see S. P. DILIBERTO [1, 2, 3], S. P. DILIBERTO and G. HUFFORD [1], N. KRYLOV and N. BOGOLYUBOV [13], G. HUFFORD [1]. In connection with this subject see also K. O. FRIEDRICHS [4], I. G. MALKIN [6], G. REEB [3], A. N. KOLMOGOROV [1, 2].

8.12. Bibliographical notes. Concerning the existence of periodic solutions of nonlinear systems by the use of the methods of POINCARÉ, LYAPUNOV, VAN DER POL and KRYLOV and BOGOLYUBOV, the following further references are of interest. I. G. MALKIN [11] has studied general systems of the form $x' = Ax + \varepsilon f(t, x)$, where $x = (x_1, \ldots, x_n)$, A is a constant matrix whose characteristic roots have negative real parts, $f = (f_1, \ldots, f_n)$ periodic in t with $\partial f_j/\partial x_h$ Lipschitzian, and has discussed the existence of periodic solutions. S. N. SIMANOV [1] has proved that the condition concerning the partial derivatives is unessential. G. I. BIRYUK [1] has discussed the existence of quasi periodic solutions when f is quasi periodic. For general theorems concerning boundedness of solutions of systems of the form $x'' + A(t)x = F(t, x, x')$ see G. BORG [4]. For systems of the form $\Sigma f_{jk}(D) x_k = \Phi_j(x_1, \ldots, x_n, t)$ see B. V. BULGAKOV [8] who has discussed both the questions of the existence of periodic and almost periodic solutions. For more particular systems see also L. I. MANDELSTAM [1] and L. I. MANDELSTAM and N. PAPALEXI [1]. For a necessary condition for the existence of periodic solutions of a system of the form $x' = Ax + \varepsilon f(x)$ see N. G. BULGAKOV [1]. For systems of the form $x' = Ax + \varepsilon F(t)x + G(x)$, A a constant matrix, ε a small parameter, F quasi harmonic, see N. P. ERUGIN [2] who has improved previous results of J. Z. SHTOKALO [10]. For discontinuous control systems (on and off) see I. FLÜGGE-LOTZ [1]. For periodic and almost periodic solutions in the complex field see E. A. CODDINGTON [1]. See also P. HARTMAN and A. WINTNER [12], A. WINTNER [24]. For periodic solutions of nonlinear systems containing various parameters see C. OBI [1—7].

On general questions of non linear mechanics see the expository papers of D. GRAFFI [16], and of K. O. FRIEDRICHS and J. J. STOKER [2], besides the books mentioned in (1.10).

For studies on the equations $x'' + x + \sin x = 0$ and other analogous interesting the stability of synchronous motors see F. TRICOMI [1], L. AMERIO [2, 3, 4], C. BÖHM [1], YU. N. BAKAEV [1], G. SEIFERT [1], L. N. BELYUSTINA [2], N. V. BUTENIN [3], W. D. HAYES [1]. On questions of sinchronization see particularly the extensive research of J. HAAG [1—24].

For existence theorems for periodic solutions (closed orbits) of dynamic systems based on the direct method of the calculus of variations see L. TONELLI [1, 2, 3] and A. RAZMADÉ [1]. For an expository paper on the Ritz-Galerkin variational method see K. KLOTTER [1].

For a heuristic method for the determination of periodic solutions of non linear systems and their stability see N. MINORSKI [29—31].

8.13. Nonlinear resonance. In this and in the next section we refer briefly to a series of experimental phenomena whose theory shows their inherent nonlinear character. General references should be made, e.g., to N. W. MCLACHLAN [5] and J. J. STOKER [1]. Other references are given below. In § 2 we have discussed the phenomenon of linear resonance which occurs when a physical system able to oscillate is subjected to forced oscillations due to an external force whose period is close to the period of the free oscillations of the system. As we have pointed out (2.7) no physical system reflects the picture we have given there and this is due to inherent nonlinearity of all physical systems for actual (noninfinitesimal) displacements.

To illustrate the nonlinear phenomenon let us consider the piano wires W_0 and W, of which W_0 is kept oscillating at its own frequency, while W is being tuned so as to give the same pitch, or a pitch close to the one of W_0. As the pitch of W approaches the one of W_0, the typical well known "beats" appear. Apparently the two notes have the same pitch but the volume of sound rises and falls at regular intervals. This interval is the beat frequency. As a matter of fact the periods are not exactly the same, but the ear cannot distinguish the difference. On the other hand, when the two vibrations have the same phase, the impulses are added, when the phases are in opposition the impulses are subtracted and the volume of sound diminishes. The beat frequency is nothing else than the differences of the frequencies of W_0 and W.

As the tuning proceeds the beat frequency becomes smaller and it may occur that it is not heard at all. Then the wire W oscillates exactly at the same frequency of the wire W_0, even if a small amount of tuning is being done; that is, for a small amount of tuning, the frequency of W is "locked" to the frequency of W_0, and the oscillations have no tendency to become large as the linear theory suggests. Nevertheless a quite different phenomenon occurs which is called the "libration". A light additional sound may be heard whose frequency is neither the one of W_0, nor the one of the proper oscillations of W_0, nor the frequency of the beat, but completely different. Its frequency depends mainly on the construction of the two wires and on the magnitude and types of the blows which are given to them to keep them vibrating. The phenomenon of the libration is well known also in astronomy and other fields.

This behavior has been essentially justified by a penetrating analysis, where the two types of solutions determined in (8.10) for the pendulum have an essential role. We refer to E. W. BROWN [1], for the details and many applications. See also N. MINORSKY [6].

8.14. Prime movers. We mention here briefly some of the ideas discussed in an article by P. LE CORBEILLER [2]. On the same subject see Lord RAYLEIGH [3], R. VON MISES [2], B. VAN DER POL [2], A. LIÉNARD [1], R.L. WEGEL [1], P. LE CORBEILLER [3, 4], and the books of N.W. MCLACHLAN [5] and J. J. STOKER [1]. A physical system regulated by the equation $m x'' + k x = 0$ is generally called a harmonic oscillator. It may be interpreted as a material point of mass m moving along the x-axis under the influence of an elastic restoring force $-k x$, $k > 0$. The oscillations $x(t)$ of the system are characterized by their constant total energy $\frac{1}{2} m x'^2 + \frac{1}{2} k x^2 = C$. The addition in the system of a "viscous resistance" of the type $-h x'$, $h > 0$, leads to the equation $m x'' + h x' + k x = 0$ whose solutions are oscillatory in character with amplitudes approaching zero as $t \to +\infty$. A "resistance" of opposite sign $h x'$ would create undamped oscillations with amplitudes approaching $+\infty$ as $t + \infty$. Thus it has been thought that any "resistance" of the type $\varphi(x, x')$ with sign opposite to x' for x, or x' large, and of the same sign of x', for x, or x', small, could create oscillations which die down if too large, tend to become large if too small, and thus there may be one (or more) periodic oscillations toward which all (or nearby) solutions approach if $t \to +\infty$. This was indeed the explanation of Lord RAYLEIGH for the sustained periodic oscillations which he had experimentally observed. The idea was developed later by VAN DER POL and successively by many others [cf. the discussion of the Liénard equation in (8.7), (8.8), (9.4), (9.5)].

A system regulated by an equation $m x'' + \varphi(x, x') + x = 0$ having periodic oscillations is often called a *prime mover*. This terminology refers to the fact that it represents a machine capable of transforming a force independent of time but depending on the position x and velocity x' of the system, into a periodic motion. Most of the "machines" using a "natural" source of energy are of this type: solar heat, wind, falling water, chemical reactions, thermal energy in a steam engine, electrical feedback circuits, clocks, mechanical toys, musical instruments. We shall analyse below some of these systems, which are all necessarily nonlinear.

a) *Violin string.* A typical property of dry friction is involved. A body of weight W at rest will not start moving until the driving force F has the value

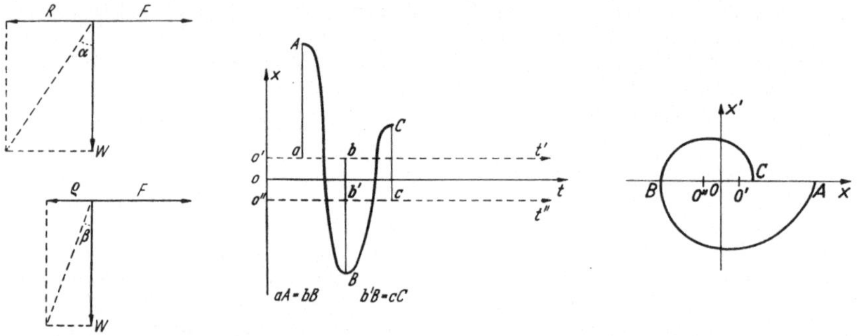

$R = W \tan\alpha$. Afterwards, the resisting force has a constant value $\varrho = W \tan\beta < R$ where the angle β is much smaller than α. Thus a body of mass m subjected to the restoring force $-kx$ and to dry friction is regulated by the equation

$$m x'' \pm \varrho + k x = 0 \tag{8.14.1}$$

where the middle term has always the same sign as x'. A translation $y = x \mp \varrho/k$, changes the equation above into $m y'' + k y = 0$. Hence the solutions x are made up of arcs of sinusoids AB of axis $0't$ and BC of axis $0''t$. The solutions are obviously damped and the movement stops at the first peak A, or B, or $C, \ldots$, for which $-Kx$ is smaller than $W \tan\alpha$. In the phase plane (x, x') the trajectories are made up of half circumferences AB, $BC, \ldots$ with centers at the points $0' = (0, \varrho/k)$, $0'' = (0, -\varrho/k)$.

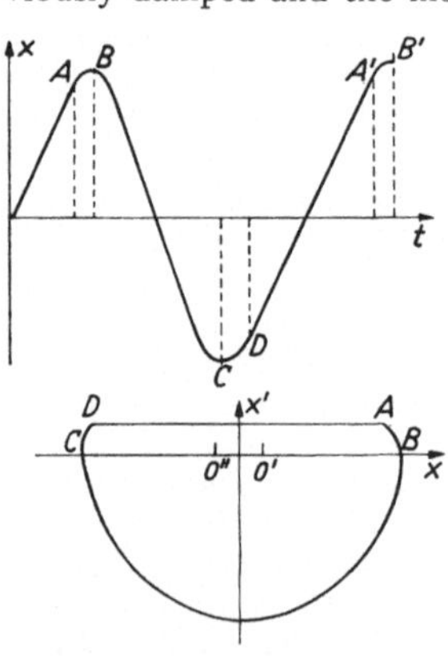

In the case of a violin string, there is dry friction between the string and the bow. Let us start at a moment 0 when the bow pulls the string away from its position of equilibrium with the constant velocity $V > 0$. At a certain point A the restoring force of the string reaches the value $P \tan\alpha$, P being the pressure of the bow on the string. Then the string begins to glide under the bow and, with regard to the violin, will perform oscillations regulated by the equation (8.14.1) where the second term has the sign $+$, or $-$, according as $V - x' > 0$, or < 0. Thus $x(t)$ describes an arc of the sinusoid of axis $0't$; i.e., the point (x, x') describes an arc of circumference with center $0'$. At a certain instant D the cord has exactly velocity $x' = V$, i.e., velocity zero with respect to the bow and then it is locked with the bow and moves at the speed V

with the bow, until an instant A' is reached analogous to A. The movement is obviously periodic.

A variant of the reasoning above occurs when in equation (8.14.1) the second term is $\pm\varrho$ according $x' > 0$ or $x' < 0$. Then the arc $ABCD$ in the xx'-plane is made up of the arcs AB, CD of center $0'$ and of the arc BC of center $0''$.

Phenomena analogous to the violin cord are the following ones: oscillation of a body pulled along a rough road by a string attached to the block through a spring which is made to travel at a constant speed; whirling of a shaft caused by solid friction in a loose bearing; the screeking of a knife cutting hard wood; the screeking of a shoe-brake pressing on a wheel.

b) Steam engine. The double acting steam engine consists essentially of a piston P, a cylinder C with end valves v_1 and v_2, and a slide valve V. Steam entering one end of the cylinder (say end 1) pushes the piston toward the other

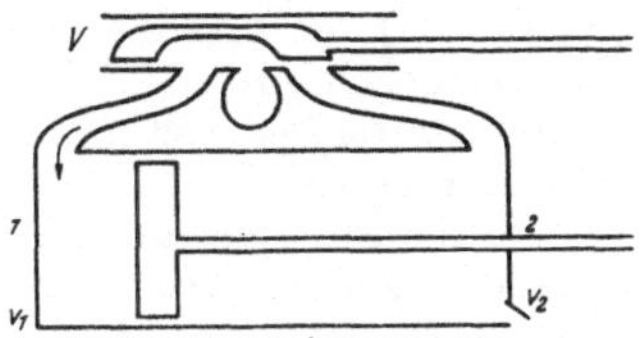

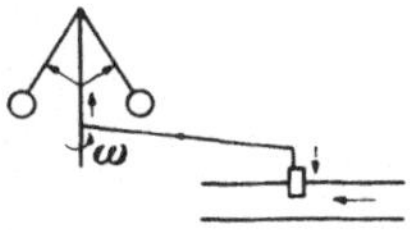

end (2). Then the slide valve V works so that the steam entering 2 pushes the piston back. The valves v_1, v_2 are exhaust valves and are open alternately.

An analogous situation occurs with clocks and watches. The analogue of the slide valve V is the escapement mechanism.

c) Watt governor. A parameter x increases by increasing velocity and closes down a valve which regulates the arrival of steam. The mathematical discussion leads to a nonlinear system of differential equations. By supposing the oscillations small the system can be linearized and then conditions of stability can be written by using Hurwitz criteria. When these conditions are satisfied as usual in well built governors, the deviations from equilibrium are aperiodic and die down as $t \to +\infty$. If the same conditions are not satisfied then the linearized system has oscillations which diverge as $t \to +\infty$, while the nonlinear system has periodic undamped oscillations which produce the unwanted phenomenon of "hunting" of the governor. A quite analogous situation is presented by the vocal cords in the larynx. The following schematization is due to R. L. Wegel. A stream of air of constant velocity comes from the lungs and passes between the vocal cords and, in the position of equilibrium, the restoring force due to the elasticity of the cords balances the lateral pressure of the air flow. If this equilibrium is unstable, then the mathematical discussion leads to a system of differential equations, quite similar to the one for the Watt governor. The (nonlinear) system has then a periodic solution, which corresponds to the situation in which the cords vibrate.

An analogous situation occurs in the reed musical instruments. More complicated situations are presented by steam and water reaction turbines, flutes, windmills, and organ pipes.

d) The Rayleigh equation. We consider here the Rayleigh equation

$$m x'' - (a - b x'^2) x' + k x = 0, \qquad a > 0,\ b > 0.$$

By changing time t and unknown function x we can reduce it to its dimensionless form

$$x'' - \varepsilon (x' - 1/3\, x'^3) + x = 0, \qquad \varepsilon > 0. \tag{8.14.2}$$

By putting $v = x'$, $F(v) = v - 1/3v^3$, we have the system

$$v\,(dv/dx) - \varepsilon F(v) + x = 0, \quad \text{or} \quad dv/dx = [\varepsilon F(v) - x]/v. \tag{8.14.3}$$

If Ox, Ov, $O\varphi$ are axes as in the illustration the graph of $\varphi = -\varepsilon F(v)$ is shown. Equations (8.14.3) represent a field of directions which can be used for a graphical approximate solution of the given equation. A geometrical interpretation is given as follows. For any point $P = (x, v)$ of the xv-plane determine the point M where a parallel through P to the x-axis crosses $\varphi = F(v)$, and the point N which is the projection of M on the x-axis. Then $M = (\varepsilon F(v), v)$ and $N = (\varepsilon F(v), 0)$. The slope of NP is $v/(x - \varepsilon F(v))$ and thus the direction (arrow at P) orthogonal to NP, is the direction of the trajectory passing through P. This construction may be used for a graphical solution of the Rayleigh equation. For $\varepsilon = 0$ the point N coincides with O and then the trajectories are all circles of center O. For ε small it has a periodic solution (cycle) which is approximately a circle C of center O and radius 2, while all other solutions are spirals which approach C as $t \to +\infty$. The period of the periodic solution is 2π (for ε small). Thus (8.14.2) has, for ε small, a sustained self-excited periodic solution of period close to the one of the pendulum $x'' + x = 0$, and a fixed large amplitude close to 2. The van der Pol equation $x'' - \varepsilon(1 - x^2)\,x' + x = 0$ presents an analogous situation. Indeed by taking $y = x'$ in (8.14.2) as a new unknown and differentiating, we obtain the van der Pol equation.

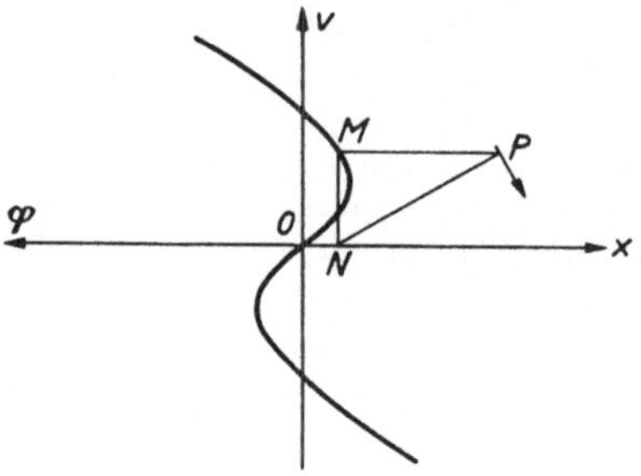

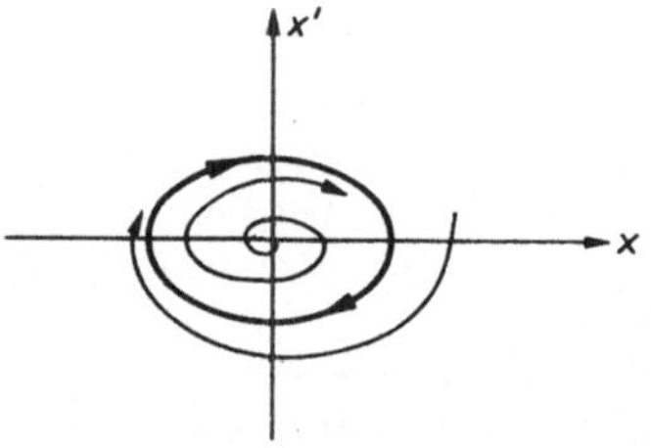

e) The feedback electrical circuit with triode. A triode has a plate (anode), a grid, and a filament (cathode). The anode has a potential V_a and the grid a potential V_g with respect to the filament. The working of the triode depends, with good approximation, upon a linear combination $u = V_g + D\,V_a$ of the two potentials, with $0 < D < 1$, a constant depending on the triode. The filament is heated by a small battery and emits electrons. If these reach the plate (anode), then their flow constitutes a current i_p (plate current). It takes a minimum potential $u = u_0 < 0$ to stop all electrons from reaching the anode. Thus $i_p = 0$ for $u \le u_0$. As u increases above u_0, more and more electrons may reach the plate and thus i_p increases. As a potential $u = u_1$ is reached, practically all electrons reach the anod and i_p reaches a value $i_1 > 0$ which does not increase by further increasing u. Thus $i_p = i$, for all $u \ge u_1$. The function $i_p = f(u)$ is a characteristic of the triode (see illustration). Its graph has a flex point at $u = E_0$, and, usually, the triode is kept working at a potential u oscillating in a small neighborhood of E_0. The first two nonzero terms of the Taylor expansion of $f(u)$ about E_0 constitute a sufficient approximation:

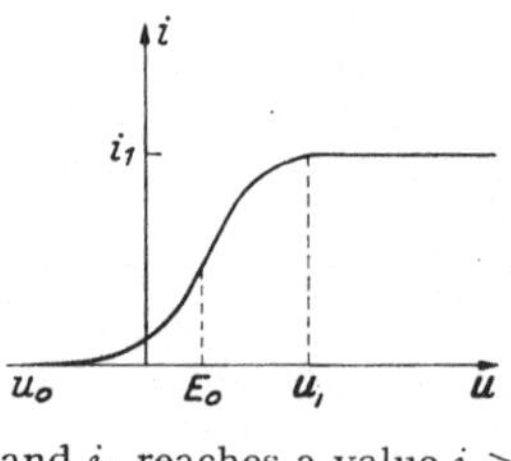

$$f(u) = a\,(u - E_0) - b\,(u - E_0)^3/3, \qquad a > 0, \qquad b > 0.$$

A typical self oscillatory (feedback) circuit is given in the illustration, to which we refer for the designations. The equations regulating the circuit are $i_p = i_L + i_c + i_R$, $V_g = M(di_L/dt)$, and

$$L(di_L/dt) = q_c/C = R\,i_R = E - V_a.$$

The first equation concerns the electricity balance at the node connected to the plate. The second concerns the inductive coupling of the plate and grid sections of the circuit (feedback) of mutual inductance M. We shall require M to be sufficiently large. The last equations assure that the drop of potential through L, C, and R are the same and equal to $E - V_a$. There q_c is the quantity of electricity which is at each instant on one system of plates of the condenser, and hence $dq_c/dt = i_c$. By differentiation and manipulation we have

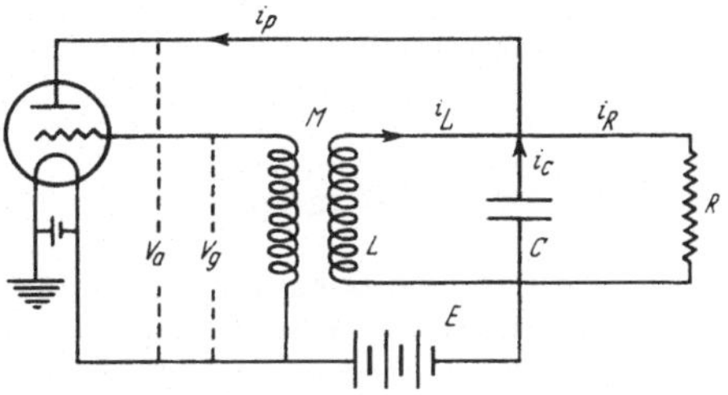

$$LC(d^2 i_L/dt^2) = i_c = i_p - i_L - i_R = f(u) - i_L - LR^{-1}(di_L/dt),$$

with $u = V_g + D V_a$. Hence, we have

$$LC(d^2 i_L/dt^2) + LR^{-1}(di_L/dt) + i_L = f[DE + (M - LD)(di_L/dt)].$$

We will regulate the system in such a way that $DE = E_0$. By putting $(M - LD) \cdot di_L/dt = V$, the expression in brackets is $E_0 + V$. On the other hand, by differentiation and multiplication by $M - LD$ we have

$$LC(d^2 V/dt^2) + LR^{-1}(dV/dt) + V = (M - LD)(a - b V^2)(dV/dt).$$

By the change of independent variable $t = (LC)^{\frac{1}{2}}\tau$, and by supposing M sufficiently large, the two constants $B = (LC)^{-\frac{1}{2}}(M - LD)b$, and $\varepsilon = (LC)^{-\frac{1}{2}}(M - LD - LR^{-1})$ are positive and the equation becomes

$$(d^2 V/d\tau^2) + V - (\varepsilon - B V^2)\, dV/d\tau) = 0.$$

By taking $V = (\varepsilon B^{-1})^{\frac{1}{2}} x$, we obtain the van der Pol equation

$$x'' - \varepsilon(1 - x^2)\,x' + x = 0, \qquad \varepsilon > 0.$$

By a different elimination we could have obtained the Rayleigh equation.

Many other examples of stable selfsustained oscillations could be associated with differential equations similar to the Raleigh and van der Pol equations: the selfexcited oscillations in electrical transmission lines due to the action of the wind; the disaster of the Tacoma bridge under sustained large oscillations due to constant lateral wind; the fluttering of airplanes wings.

8.15. Relaxation oscillation. In (8.14. d) we have considered the equation $x'' - \varepsilon(x' - x'^3/3) + x = 0$ with ε small. If ε is large and we use the same graphical scheme used there, we see that the graph of $\varphi = -\varepsilon F(v)$ is a large horizontal zig zag, and the graphical method shows that the cycle follows closely the outer arcs of $\varphi = \varepsilon F(v)$ and its two vertical tangents. (The picture represents computation performed with $\varepsilon = 10$. The oscillations $x = x(t)$ are made up of two equal arcs connected by a sudden reversal. These oscillations have been called by van der Pol *relaxation oscillations*. The period of these oscillation is close to $1.6\, a/k$ for ε large. The oscillations appear to become somewhat jerky very rapidly for ε large, and they can be thought of as a periodic sequence of nonoscillatory phenomena.

A hydraulic model due to VAN DER POL for these oscillations is represented in the illustration. Two containers are resting on a shaft which oscillates between two fixed positions about a central fulcrum. Each of the containers has two openings, one on the top, and the other on a side. In the upper container water flows in from a faucet conveniently located. When the water reaches a certain level in the uppei container, it will cause

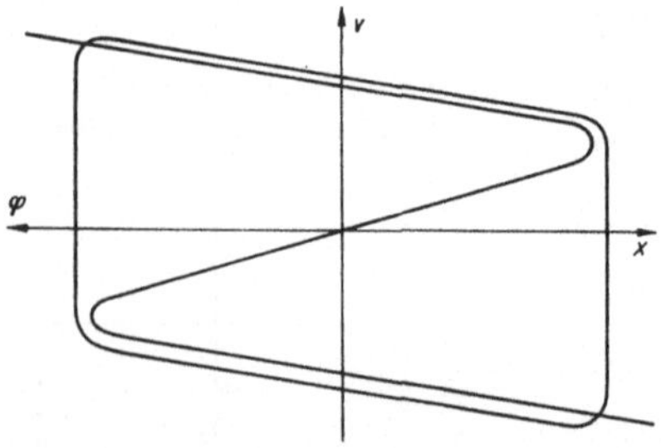

the system to invert its position. Then water will flow out of the first container, while the other is being filled. The system will then alternately assume the two fixed positions. For research on relaxation oscillations see, e.g., VAN DER POL and L. VAN DER MARK [1], D. GRAFFI [10, 11], L. CAPRIOLI [2], P. LE CORBEILLER [1], N. MINORSKY [6], J. LA SALLE [2].

§ 9. Analytical-topological methods

9.1. Poincaré theory of the critical points. Consider an autonomous system $(n=2)$

$$x_1' = P(x_1, x_2), \quad x_2' = Q(x_1, x_2), \quad ('= d/dt) \tag{9.1.1}$$

where P, Q are real continuous functions of x_1, x_2 in some open set D of the (x_1, x_2)-plane E_2. Any solution $x_1 = x_1(t)$, $x_2 = x_2(t)$ of (9.1.1) in D, existing in some maximal interval $\alpha < t < \beta$, defines an orbit, or trajectory Γ in D. Since (9.1.1) is autonomous, also $x_1 = x_1(t+d)$, $x_2 = x_2(t+d)$ is a solution of (9.1.1), where d is an arbitrary constant, and defines the same orbit Γ.

If $P \neq 0$ along an arc γ of Γ, then P has always the same sign on γ, $x_1(t)$ is then an increasing, or decreasing function of t, and the inverse function $t = t(x_1)$ exists. Thus $x_2(t)$ can be thought of as a function of x_1, say $x_2 = X_2(x_1) = x_2[t(x_1)]$, the arc γ has the nonparametric representation $x_2 = X_2(x_1)$, and the function $X_2(x_1)$ is a solution of the reduced time-independent equation

$$dx_2/dx_1 = Q/P. \tag{9.1.2}$$

Conversely, if at a point $(x_{10}, x_{20}) \in D$ we have $P \neq 0$, then P has always the same sign in a neighborhood U of (x_{10}, x_{20}), and we may consider any solution $X_2(x_1)$ of (9.1.2) in a linear neighborhood of x_{10}, with $X_2(x_{10}) = x_{20}$, representing an arc γ in U. Then γ has also the parametric representation $x_1 = \varphi(t)$, $x_2 = X_2[\varphi(t)]$, where $\varphi(t)$ is any (increasing, or decreasing) solution of the equation $\varphi' = P[\varphi, X_2(\varphi)]$, with $\varphi(t_0) = x_{10}$ where P has always the same sign. Thus γ is an arc of a trajectory Γ in D relative to system (9.1.1). Analogous considerations hold if $Q \neq 0$ by exchanging x_1 and x_2. No requirement has been made assuring uniqueness.

Every point (x_1, x_2) at which either P, or Q, or both, are $\neq 0$, is called a *regular point* of (9.1.1), and system (9.1.1) defines there a direction which is the direction of any orbit passing through it. We shall denote by $\theta(p)$ the direction associated by (9.1.1) to any regular point p. By virtue of (1.2) we know that mild conditions on P, Q, say P, Q Lipschitzian in a neighborhood of (x_1, x_2), assure the local uniqueness of the orbit Γ through (x_1, x_2).

From the considerations above it is clear that the only points where an exceptional behavior of the trajectories Γ of (9.1.1) in D can be expected are the points where both P and Q are zero. These points are said to be the *critical* points of (9.1.1) (POINCARÉ). As we will see, the number, position and type of the critical points give an extraordinary insight into the behavior of the solutions of (9.1.1) as a whole, just as the singular points of an analytical function have an exceptional bearing on its properties.

We shall now study the critical points. If (x_{10}, x_{20}) is such a point, we may move it to the origin by a translation in E_2. We shall suppose thus that (0, 0) is a critical point; i.e., $P(0, 0) = Q(0, 0) = 0$.

Main assumption. In the following it will be always assumed that system (9.1.1) can be written in the form

$$x_1' = a\,x_1 + b\,x_2 + f(x_1, x_2), \qquad x_2' = c\,x_1 + d\,x_2 + g(x_1, x_2), \tag{9.1.3}$$

where $ad - bc \neq 0$, a, b, c, d, f, g real, and $|f|, |g| \leq \psi(r)$, $r = (x_1^2 + x_2^2)^{\frac{1}{2}}$, $\psi(r) = o(r)$ as $r \to 0$. All this implies, in particular, that in a convinient neighborhood U of the origin the second members of (9.1.3) are both zero only at the origin; that is, (0, 0) is an isolated critical point.

POINCARÉ'S *main result.* The behavior of the solutions of (9.1.3) near the origin is essentially the same as the behavior of the solutions of the linear system:

$$x_1' = a\,x_1 + b\,x_2, \qquad x_2' = c\,x_1 + d\,x_2. \tag{9.1.4}$$

We have already studied the latter in (2.5). The following proof is based on A. WINTNER [11], and E. A. CODDINGTON and N. LEVINSON [3].

In the following use will be made of polar coordinates (r, ω), $r > 0$, $x_1 = r\cos\omega$, $x_2 = r\sin\omega$. Thus an arc γ of a trajectory Γ not passing through the origin has a representation $r = r(t)$, $\omega = \omega(t)$, where $\omega(t)$ is any one of the infinitely many continuous functions with $x_2(t) = r(t) \cdot \cos\omega(t)$, $x_1(t) = r(t)\sin\omega(t)$.

The origin is said to be an *attractor* (for $t \to +\infty$, or $t \to -\infty$) provided there is a neighborhood U of (0, 0) such that each solution $x_1(t)$, $x_2(t)$ with $[x_1(0), x_2(0)] \in U$ exists in $[0, +\infty)$ or $(-\infty, 0]$, and $r(t) \to 0$ as

$t \to +\infty$, or $t \to -\infty$. It is said to be *a spiral point* provided $r(t) > 0$, $r(t) \to 0$, $\omega(t) \to \infty$ as $t \to +\infty$, or $t \to -\infty$; a *node* if $r(t) > 0$, $r(t) \to 0$, $\omega(t) \to c$, c some constant; a *proper node* if, in addition, for every constant c there is a solution with $\omega(t) \to c$. The same points will be said to be *stable*, or *unstable* provided the behaviors above occur as $t \to +\infty$, or $t \to -\infty$. The origin is said to be a *center* if there are at least countably many closed trajectories C_n in U encircling the origin with diam $(C_n) \to 0$ as $n \to +\infty$. The concept of *saddle point* will be considered later.

The use of all these terms in (2.5) for linear systems (9.1.4) clearly agrees with the definitions above. In case of a center all trajectories were closed. We will now divide POINCARÉ's main statement into a series of simple theorems. No local theorem of uniqueness is really required. Nevertheless in the proofs which follow, for the sake of simplicity, we will suppose that a uniqueness theorem holds at every noncritical point.

(9.1. i) If the origin is an attractor for (9.1.4) it is an attractor also for (9.1.3).

Proof. Suppose the origin is an attractor for (9.1.4) as $t \to +\infty$. Then, according to (2.5), the origin is a stable node, or a stable spiral point and both characteristic roots have negative real parts. By applying (6.2. i) we conclude that $x_1(t)$, $x_2(t) \to 0$ as $t \to +\infty$; that is, $r(t) \to 0$ and the origin is an attractor for (9.1.3). If the origin is an attractor for (9.1.4) as $t \to -\infty$, then it is an unstable node, or an unstable spiral point, and we may repeat the reasoning above by just changing first t into $-t$.

Remark. In the case both f, g are analytic then (9.1. i) can be proved as a consequence of LYAPUNOV's theorems (§ 6). The same holds for all the statements below (9.1. ii—vi).

(9.1. ii) If the origin is a spiral point for (9.1.4), then it is a spiral point also for (9.1.3).

Proof. Suppose that the origin is a stable spiral point for (9.1.4). Then system (9.1.4) has the canonical form

$$u' = \varrho u, \quad v' = \bar{\varrho} v, \quad \text{where} \quad v = \bar{u}, \quad \varrho = \alpha + i\beta, \quad \bar{\varrho} = \alpha - i\beta, \quad \alpha < 0, \quad \beta \neq 0.$$

By taking $y_1 = 2^{-1}(u+v)$, $y_2 = (2i)^{-1}(u-v)$, we have also the canonical form

$$y_1' = \alpha y_1 + \beta y_2, \qquad y_2' = -\beta y_1 + \alpha y_2,$$

and x_1, x_2 and y_1, y_2 are related by a linear transformation with real coefficients. Thus (9.1.3) has the canonical form

$$y_1' = \alpha y_1 + \beta y_2 + p(y_1, y_2), \qquad y_2' = -\beta y_1 + \alpha y_2 + q(y_1, y_2), \tag{9.1.5}$$

with $p, q = o(r)$. In polar coordinates (r, ω), $y_1 = r\cos\omega$, $y_2 = r\sin\omega$, we have

$$r^2 \omega' = y_1 y_2' - y_2 y_1' = -\beta r^2 + o(r^2)$$

or $r \to 0$. By (9.1. i) we know already that every solution starting sufficiently close to the origin has $r \to 0$ as $t \to +\infty$. Thus we have $\omega' = -\beta + o(1)$ as $t \to +\infty$, and hence $\omega(t) = -\beta t + o(t)$, i.e., $\omega(t) \to \pm\infty$ according as $\beta > 0$ or $\beta < 0$, and $t \to +\infty$. This proves that the origin is a stable spiral point for (9.1.5) and thus for (9.13). Analogous reasoning holds if the origin is an unstable spiral point for (9.1.4).

(9.1. iii) If the origin is a proper node for (9.1.4), if $|f|, |g| \leq \psi(r)$, $\psi(r) = o(r)$ as $r \to 0$, and $\int_{+0} r^{-2}\psi(r)\,dr < +\infty$, then the origin is a proper node also for (9.1.3). In particular this holds if $f, g = O(r^{1+\varepsilon})$, $\varepsilon > 0$.

Proof. Suppose that the origin is a stable proper node for (9.1.4). Then by (2.5), (9.1.4) has the canonical form $u' = \alpha u$, $v' = \alpha v$, where $\varrho = \alpha < 0$, and (9.1.3) has the canonical form

$$u' = \alpha u + p(u, v), \qquad v' = \alpha v + q(u, v),$$

where $|p|$, $|q| < K_1\psi(r)$, $K_1 > 0$ a constant. In polar coordinates $u = r\cos\omega$, $v = r\sin\omega$, we have

$$r r' = \alpha r^2 + \alpha r F_1(r, \omega), \qquad r^2\omega' = r^2 F_2(r, \omega), \tag{9.1.6}$$

$$F_1(r, \omega) = \alpha^{-1}\cos\omega\, p(r\cos\omega, r\sin\omega) + \alpha^{-1}\sin\omega\, q(r\cos\omega, r\sin\omega),$$

$$F_2(r, \omega) = r^{-1}\cos\omega\, q(r\cos\omega, r\sin\omega) - r^{-1}\sin\omega\, p(r\cos\omega, r\sin\omega),$$

and hence, in particular

$$r r' = \alpha r^2 + o(r^2) \quad \text{as} \quad r \to 0. \tag{9.1.7}$$

By (9.1. i) we know already that for any solution $r = r(t)$, $\omega = \omega(t)$, starting sufficiently close to the origin, say at (r_0, ω_0), $r_0 = r(0)$, $\omega_0 = \omega(0)$, we have $r \to 0$ as $t \to +\infty$. By (9.1.7) we deduce that $r' < 0$ for all r sufficiently small, i.e., $r(t)$ is a decreasing function of t. We may suppose $r_0 > 0$ so small that $r(t)$, $0 \leq t < +\infty$, is always decreasing. Then the inverse function $t = t(r)$, exists and the solution above defines a function $\omega = \Omega(r) = \omega[t(r)]$. The function $\Omega(r)$ satisfies the reduced time-independent equation which follows from (9.1.6) by division

$$d\omega/dr = \Phi(\omega, r), \qquad \Phi = F_2\alpha^{-1}(r + F_1)^{-1},$$

and $|F_2| \leq K_2 r^{-1}\psi(r)$, $|F_1| \leq K_3\psi(r)$, and finally $|\Phi| \leq K_4 r^{-2}\psi(r)$, where K_2, K_3, K_4 are constants. By integration we have $\omega_0 - \omega(r) = \int_r^{r_0} \Phi\,dr$. Since $|\Phi| \leq K_4 r^{-2}\psi(r)$ and thus $\int_{+0}^{r_0} |\Phi|\,dr \leq K_5 < +\infty$, we conclude that the limit $\omega(0+)$ exists, is finite, and

$$\omega_0 - \omega(0+) = \int_{+0}^{r_0} \Phi\,dr. \tag{9.1.8}$$

Usual continuity considerations assure that, by keeping r_0 constant, the number $\omega(0+)$ can be thought of as a continuous function of the initial value ω_0. On the other hand, from (9.1.8) it follows that $\omega(0+) \to \pm\infty$ as $\omega_0 \to \pm\infty$. Hence $\omega(0+)$ takes all possible values, and (0, 0) is a stable proper node for (9.1.3).

Remark. The conditions of (9.1. iii) are certainly satisfied if f, g are analytic, since then $f, g = O(r^2)$.

On the other hand the conclusion of (9.1. iii) does not hold necessarily under the only assumption $\psi(r) = o(r)$, as the following example shows:

$$x_1' = -x_1 - x_2\log^{-1} r, \qquad x_2' = -x_2 + x_1\log^{-1} 2, \qquad r = (x_1^2 + x_2^2)^{\frac{1}{2}}.$$

The polar form of this system is

$$r' = -r, \qquad \omega' = \log^{-1} r,$$

and hence

$$r = c\,e^{-t}, \qquad \omega(t) = k - \log(t - \log c),$$

where c, k are arbitrary constants. Thus $r \to 0$, $\omega(t) \to -\infty$ as $t \to +\infty$, and the origin is a stable spiral point. For the linear system $x_1' = -x_1$, $x_2' = -x_2$, the origin is a proper node instead.

(9.1. iv) If the origin is a center for (9.1.4), then it is either a center or a spiral point for (9.1.3).

Proof. Suppose that the origin is a center for system (9.1.4). Then system (9.1.4) has the canonical form $u = i\beta u$, $v' = -i\beta v$, $v = \bar{u}$, $\beta \neq 0$, and, by taking $y_1 = 2^{-1}(u+v)$, $y_2 = (2i)^{-1}(u-v)$, also the real canonical form $y_1 = \beta y_2$, $y_2 = -\beta y_1$. Thus system (9.1.3) has the canonical form

$$y_1' = \beta y_2 + p(y_1, y_2), \qquad y_2' = -\beta y_1 + q(y_1, y_2).$$

In polar form we have

$$r' = F_1(r, \omega), \qquad \omega' = -\beta + F_2(r, \omega) =$$

$$F_1(r, \omega) = \cos\omega\, p(r\cos\omega, r\sin\omega) + \sin\omega\, q(r\cos\omega, r\sin\omega) = o(r),$$

$$F_2(r, \omega) = r^{-1}\cos\omega\, q(r\cos\omega, r\sin\omega) - r^{-1}\sin\omega\, p(r\cos\omega, r\sin\omega) = o(1).$$

Thus $r' = o(r)$, $\omega' = -\beta + o(1)$ as $r \to 0$. This implies that, given $\varepsilon > 0$, there exists $r_0 > 0$ such that $r' > -\varepsilon r$ for all $r \geq r_0$, and thus, if $r(0) = r_0$, then also $r(t) \geq r_0 e^{-\varepsilon t}$ for all $t \geq 0$, and hence $r(t) > 0$. Thus $\omega(t)$ is defined, and $\omega' = -\beta + o(1)$ implies that $\omega(t)$ is a monotone function of t, provided r_0 is sufficiently small, and $\omega(t) \to \pm\infty$ as $t \to +\infty$, according as $\beta \lesseqgtr 0$. Thus the inverse function $t = t(\omega)$ exists and the function $r = \Omega(\omega) = r[t(\omega)]$ exists and satisfies the equation

$$dr/d\omega = F(r, \omega), \quad F(r, \omega) = F_1(r, \omega)[-\beta + F_1(r, \omega)]^{-1}. \tag{9.1.9}$$

The function F is continuous in some circle $0 \leq r \leq r_1$, $r_1 > 0$, and $F(r, \omega + 2\pi) = F(r, \omega)$, $F(r, \omega) = o(r)$ as $r \to 0$ uniformly in ω. Let $M = \max F$ for $0 \leq r \leq r_2$, for any $0 < r_2 \leq r_1$. Then $M = o(r_2)$ as $r_2 \to 0$. For any (r_0, ω_0) with $0 < r_0 < r_2/2$, the solution $r = r(\omega)$ of (9.1.9) with $r(\omega_0) = r_0$, exists at least in the interval $|\omega - \omega_0| < r_2/2M$ and is in the circle $0 < r < r_2$. Since $M = o(r_2)$ we may suppose r_2 so small that $r(\omega)$ exists in some interval larger than $[\omega_0 - 2\pi, \omega_0 + 2\pi]$. Thus we can consider the difference $r(\omega_0) - r(\omega_0 + 2\pi)$ or $r(\omega_0) - r(\omega_0 - 2\pi)$. If $r(\omega_0) - r(\omega_0 + 2\pi) = 0$, then the solution is periodic. If this occurs for countably many values of $r_0 \to 0$ (and a given ω_0) then the origin is a center. If this is not the case then we may suppose r_2 so small that there are no periodic solutions in the circle $0 < r < r_2$. Then the function $r(\omega) - r(\omega + 2\pi)$ has always the some sign since, otherwise, by continiuity considerations, there would be a w_0 with $r(\omega_0) - r(\omega_0 + 2\pi) = 0$, and hence a periodic solution. It is not restrictive to suppose that the function above is positive.

Since $|dr/d\omega| \leq M$ we conclude that the functions $r_k(\omega)$ are equicontinuous and thus the convergence $r_k(\omega) \to \bar{r}(\omega)$ is uniform. By the remark of (1.1) we conclude that $\bar{r}(\omega)$, $0 \leq \omega \leq 2\pi$, is a solution of (9.1.9 and even a periodic solution. This is not contradictory only if $\bar{r}(\omega) = 0$. Thus $r_k(\omega) \to 0$ uniformly and hence $r(\omega) \to 0$ as $\omega \to +\infty$; that is, the origin is a spiral point.

Remark. The following example shows that (9.1. iv) cannot be improved. The system

$$x_1' = -x_2 - r x_1, \qquad x_2' = x_1 - r x_2, \qquad r = (x_1^2 + x_2^2)^{\frac{1}{2}} \geq 0,$$

has the polar form $r' = -r^2$, $\omega' = 1$. Hence the solution starting at any point (r_0, ω_0), $r_0 > 0$, has the expression $r(t) = (t + r_0^{-1})^{-1}$, $\omega(t) = t + \omega_0$, $t \geq 0$, and $r \to 0$, $\omega \to +\infty$ as $t \to +\infty$. Thus the origin is a spiral point. Instead the linear system

$x_1' = -x_2$, $x_2' = x_1$ has for solutions all the circles $r = r_0$, $\omega = t + \omega_0$, and hence the origin is a center.

The following example shows that even in case the origin is a center for the nonlinear system (9.1.3), not all solutions in a neighborhood of the origin need be periodic. Consider the system

$$x_1' = -x_2 + x_1 r \sin(\pi r^{-1}), \qquad x_2' = x_1 + x_2 r \sin(\pi r^{-1}),$$

whose polar form is $r' = r^3 \sin(\pi r^{-1})$, $\omega' = 1$. The circles C_n: $r = n^{-1}$, $n = 1, 2, \ldots$, are all periodic solutions. We have now $r' > 0$ if $r > 1$, and if $(2m+1)^{-1} < r < (2m)^{-1}$, $r' < 0$ if $(2m)^{-1} < r < (2m-1)^{-1}$, $m = 1, 2, \ldots$. Thus the circles C_n are the only periodic solutions. All other solutions spiral outwards outside C_1, spiral inwards between C_{2m-1} and C_{2m}, spiral outwards between C_{2m} and C_{2m+1}.

(9.1. v) If the origin is a node for (9.1.4), then it is a node for (9.1.3) also.

Proof. Suppose that the origin is a stable (improper) node for (9.1.4). Then either of the two situations described in (2.5a), or (2.5b) occurs. Suppose that the former occurs. Then (9.1.4) has the canonical form $u' = \lambda u$, $v' = \mu v$, with μ, λ real, $\mu < \lambda < 0$, and (9.1.3) has the canonical form

$$u' = \lambda u + p(u, v), \qquad v' = \mu v + q(u, v), \qquad p, q = o(r). \tag{9.1.10}$$

By using the polar form we deduce

$$\omega' = 2^{-1}(\mu - \lambda) \sin 2\omega + o(1), \qquad r' = \lambda \cos^2 \omega + \mu \sin^2 \omega + o(1) \quad \text{as} \quad r \to 0,$$

while, by (9.1. i), we known already that $r \to 0$ as $t \to +\infty$ for every solution starting sufficiently close to the origin. We may even suppose $r(t)$ a decreasing function of t, since $r' \leq \lambda + o(1)$.

Given $\varepsilon > 0$ let us consider the four sectors T_r: $|\omega - (r-1)\pi/2| < \varepsilon$, $r = 1, 2, 3, 4$, and observe that, on the boundary lines and for $r > 0$ sufficiently small, the sign of ω' shows that the orbits enter T_1 and T_3 while they leave T_2 and T_4. In addition any orbit already sufficiently close to the origin and not in any of the four sectors must enter either T_1 or T_3. Indeed as long as they are there, ω' has constant sign, (r, ω) moves toward T_1 or T_3, and $|\omega'| \geq 2^{-1}(\lambda - \mu) \sin 2\varepsilon$. Thus in a finite interval of time ω' must enter either T_1 or T_3.

Since $\varepsilon > 0$ is arbitrary, we conclude that for all orbits around the origin, the argument ω approaches one of the four numbers $0, \pi/2, \pi, 3\pi/2 \pmod{2\pi}$ as $t \to +\infty$, and that for all orbits starting from any point (r_0, ω_0), with $|\omega - \pi/2| \geq \varepsilon$, or $|\omega - 3\pi/2| \geq \varepsilon$, and r_0 sufficiently small, we have either $\omega \to 0$, or $\omega \to \pi$. This assures that the origin is a stable improper node.

It is easy to prove that for at least two orbits we have $\omega \to \pi/2$, $\omega \to 3\pi/2$ respectively. Indeed, if we consider the two radii OA, OB with $\omega = \pi/2 \pm \varepsilon$, and the arc AB of radius r sufficiently small, all points of AB are starting points of orbits approaching the origin. Those sufficiently close to $A[B]$ are starting points of orbits leaving T_2 at some point of the radius $OA[OB]$. If C is the point of AB separating the points for which the orbit leaves T_2 through OA, and those points for which this does not occur, then the orbit γ starting at C cannot leave T_2 either through OA, or through OB, and hence must approach O within T_2. Now if we take any sector T_2' analogous to T_2 relative to a number ε', $0 < \varepsilon' < \varepsilon$, it is clear that γ must remain within T_2' because, otherwise, γ would approach O with $\omega \to 0$, $\omega \to \pi/2$, and γ would leave T_2, a contradiction. Thus γ approaches O with $\omega \to \pi/2$. Analogous reasoning holds for the sector T_4.

The case (2.5b) can be discussed analogously.

Remark. If we suppose that the functions $f(x_1, x_2)$, $g(x_1, x_2)$ have continuous first partial derivatives in a neighborhood of the origin, and thus $p(u, v)$, $q(u, v)$ have the same property, then there are only two orbits approaching the origin with $\omega \to \pi/2$ and $\omega \to 3\pi/2$ respectively.

To prove this observe first that $\omega \to \pi/2$, implies $u/v \to 0$ as $t \to +\infty$, and, by (9.1.10), $v'/v = \mu + o(1)$, $\mu < 0$. Hence $v > 0$ implies $v' < 0$ for large t, and thus $v(t)$ is decreasing and the inverse function $t = t(v)$ exists. Then the function $u = U(v) = u[t(v)]$ is a solution of the reduced equation which can be derived from (9.1.10) by division, say

$$du/dv = (\lambda u + p)(\mu v + q)^{-1}.$$

Suppose now that there are two solutions as above, say $\psi_1(v)$, $\psi_2(v)$. Then both relations hold

$$d\psi_i/dv = [\lambda \psi_i(v)\, p(\psi_i(v), v)]\, [\mu v + q(\psi_i(v), v)]^{-1} \equiv A_i B_i^{-1},$$

$i = 1, 2$. If $\psi = \psi_1 - \psi_2$, by difference and manipulation we have also

$$d\psi/dv = [\lambda\psi(v) + \Delta_1] B_1^{-1} + [\lambda\psi_2 + p(\psi_2(v), v)]\, \Delta_2 B_1^{-1} B_2^{-1}, \qquad (9.1.11)$$

where $\Delta_1 = p(\psi_2(v), v) - p(\psi_1(v), v)$ and Δ_2 has analogous expression in terms of q. By the uniqueness theorem we have $\psi \neq 0$, and thus we may suppose $\psi > 0$. By the mean value theorem we have also $\Delta_1 = \psi(v)\, \bar{p}_u$, $\Delta_2 = \psi(v)\, \bar{q}_u$ where the partial derivatives p_u, q_u are taken at convenient points (ξ, v), $\psi_1 < \xi < \psi_2$. Since p_u, $q_u = 0$ at the origin, we conclude that $\bar{p}_u \to 0$, $\bar{q}_u \to 0$ as $r \to 0$, and thus, by (9.1.11) we derive $\psi'(v) = \lambda\psi(v)\mu^{-1}v\,(1 + o(1))$ as $v \to 0$, or $v\psi'/\psi < \gamma < 1$ for any fixed γ, $\lambda\mu^{-1} < \gamma < 1$, and v sufficiently small. Then $\psi(v)/v < Cv^{\gamma - 1}$ where C is a constant, and thus $\psi(v)/v \to \infty$ as $v \to 0$, a contradiction, since $\psi_i(v)/v \to 0$ and hence $\psi(v)/v \to 0$ as $v \to 0$. This proves that there is only one orbit approaching the origin with $\omega \to \pi/2$. An analogous argument holds for $\omega \to 3\pi/2$.

A saddle point is not easy to define. We shall say that the origin is a saddle point if there are two orbits approaching it along opposite directions and all other orbits starting sufficiently near to either of them and to the origin tend away from them as $t \to +\infty$. Then the statement holds:

(9.1. vi) If the origin is a saddle point for (9.1.4) and f, g have continuous partial derivatives in a neighborhood of the origin, then the origin is a saddle point for (9.1.3) also.

The proof is based on the same arguments of the remark above and of the proof of (9.1. v).

As an application of the previous results let us consider the Liénard equation $x'' + f(x)\, x' + g(x) = 0$ under the same hypotheses of (8.7. i). Then the corresponding system (8.7.4) has the form $x' = X(x, y)$, $y' = Y(x, y)$ with $X = y - F(x)$, $Y = -g(x)$. Hence $X = Y = 0$ only at $(0, 0)$. Thus the origin is the only critical point. If we suppose $f(0) \neq 0$, $g'(0) \neq 0$, then the conditions of (8.7. i) imply $f(0) < 0$, $g'(0) > 0$, and we have $X = -x f(0) + y + o(x)$, $Y = -g'(0)\, x + o(x)$ as $x \to 0$. Thus $a = -f(0)$, $b = 1$, $c = -g(0)$, $d = 0$, $ad - bc = g'(0) \neq 0$, and the characteristic equation for the linear system is $\varrho^2 + \varrho f(0) + g'(0) = 0$. Hence the origin is an unstable spiral point, or an unstable node for the linear system and for the non linear system (8.7.4) as well. For instance, for the van der Pol equation $x'' + \varepsilon(x^2 - 1)\, x' + x = 0$ with $\varepsilon > 0$ the origin is an unstable spiral point if $0 < \varepsilon < 2$, and an unstable node if $\varepsilon \gtrsim 2$.

On the behavior of the solutions around a critical point, see, besides the original papers of Poincaré and the paper of A. Wintner already quoted, also P. Hartman and A. Wintner [19], S. Lefschetz [1, 4, 5], E. R. Lonn [1], I. G. Petrovsky [1], W. Hurewicz [1], H. Dulac [1—3]. A necessary and sufficient condition for a center is given by N. A. SAHARNIKOV [1—4] (cf. also M. FROMMER [1], and N. N. BAUTIN [7]).

The singular points for systems of the form $x' = A[a - x - f(y)]$, $y' = B \cdot [b - y - g(x)]$ are discussed by W. J. CUNNINGHAM [1]; for systems derived from a differential equation $y'' = R(x, y, y')$, R rational in y' and algebraic in (x, y), see N. P. ERUGIN [9]; for systems of the form $x\,dy = (P/Q)\,dx$, or $x\,dy = yF(x, y)\,dx$, see T. KATO [1, 2]; for systems of the form $x^{a+1}\,dy = (P/Q)\,dx$, see J. KIMURA [1, 2] and previous results of G. MALMQUIST [1]; for systems of the form $yy' = F(x, y)$, see I. S. KOUKLESS [1—3]; for systems of the form $x' = f(x) + ay$, $y' = bx + cy$, see V. A. PLISS [1, 2]; for systems of the form $dy = f(y/x)\,dx$, see G. E. SILOV [1]. For singular points with $ad - bc = 0$, a, b, c, d not all zero, see K. A. KEIL [1]; for the behavior of the solutions in the case P, Q are discontinuous along a given curve, see YU. K. SOLNCEV [1].

For systems of the form $dy/dx = (\pm x + M)/(y + N)$, M, N polynomials of degree 2 in x, y, see L. N. BELYUSTINA [1], K. S. SIBIRSKII [1], and previous results of W. KAPTEYN [1], H. DULAC [1], N. N. BAUTIN [7], N. A. SAHARNIKOV [4]. For systems of the form $dy/dx = M/N$ see L. S. LYAGINA [1], and I. G. PETROVSKY and E. M. LANDIS [1]. For extensions of the theory of the singular points to n-th order systems $x_j' = \Sigma_k a_{jk} x_k + f_j(x, t)$, $j = 1, \ldots, n$, $x = (x_1, \ldots, x_n)$, see J. HAAG [19], where f_j are Lipschitzian functions. For the case where f is analytic see J. CHAZY [1, 2]. Cf. also A. WINTNER [11]. For studies on $n = 3$ see L. AMERIO [5, 6, 7, 9], and L. V. REIZIN [1], A. A. SESTAKOV [1—5].

9.2. Poincaré-Bendixson theory. We still consider, as in (9.1), the autonomous system

$$x_1' = f(x_1, x_2), \qquad x_2' = g(x_1, x_2), \tag{9.2.1}$$

where f, g are continuous functions of (x_1, x_2) in an open set D of the (x_1, x_2)-plane E_2. The critical points (ξ, η) of (9.2.1) are those points of D where $f(\xi, \eta) = g(\xi, \eta) = 0$. Since $x_1 = \xi$, $x_2 = \eta$ is then a solution of (9.2.1) we conclude that each of the critical points constitutes a degenerate orbit. We shall suppose, in addition to the main assumption of (9.1), that through every point of D there passes one and only one orbit. Thus the critical points will be no exception.

If $x_1 = x_1(t)$, $x_2 = x_2(t)$ is a solution of (9.2.1) existing in the whole interval $t_0 \leq t < +\infty$, or $-\infty < t \leq t_0$, or $-\infty < t < +\infty$, we shall say that it represents an orbit, or trajectory C^+, C^-, C respectively, i.e., a positive semiorbit, a negative semiorbit, or a full orbit. A point $(\xi, \eta) \in E_2$ is said to be a *limit point* of C^+, if there is a sequence $[t_n]$ with $t_n \to +\infty$, $x_1(t_n) \to \xi$, $x_2(t_n) \to \eta$ as $n \to +\infty$. We shall denote by $L(C^+)$, or $L^+(C^+)$, the set of the limit points of C^+ in E_2. Analogous definitions have $L(C^-)$, or $L^-(C^-)$. Finally, for a full orbit C, the sets $L^+(C)$ and $L^-(C)$ are obviously defined. Periodic solutions define a closed orbit C, which is a full orbit and $L^+(C) = L^-(C) = C$. The degenerate trajectories (critical points) fall into this category. In the following we shall always denote by K any given compact subset of D, i.e., bounded and closed. We shall also denote by $\bar{A}$, A^0, A^* the closure, the set of the interior points, and the boundary of any given set A. We give below the important theorems (9.2. vii) and (9.2. x) due to H. POINCARÉ [5] and I. BENDIXSON [1].

We shall need a few lemmas. The exposition is based on E. A. CODDINGTON and N. LEVINSON [3].

(9.2. i) If $C^+ \subset K \subset D$, then $L(C^+)$ is a nonempty, compact, connected set (i.e., a continuum), and $L(C^+) \subset K$.

Proof. As usual in topology, given a sequence $[A_n]$ of sets $A_n \subset K$, we denote by lim sup A_n (lim inf A_n] as $n \to +\infty$ the set of all points p such that in every neighborhood U of p there are points of infinitely many sets A_n [of all but finitely many sets A_n]. If the sets A_n are continua and lim inf A_n is not empty, then a ZORETTI's theorem states that lim sup A_n is a continuum [G. T. WHYBURN, Analytic Topology, Chap. I (9.1)]. If A_n denotes the arc $x_1 = x_1(t)$, $x_2 = x_2(t)$, $n \leq t \leq n+1$, and p_0 any point of accumulation of the first end points of the arcs A_n, then lim inf A_n contains p_0 and is not empty, and then $C^+ = \lim\sup A_n$ is a continuum.

(9.2. ii) If $C^+ \subset K \subset D$ and $L(C^+)$ contains a regular point p_0, then the orbit Γ through p_0 is a full orbit and $\Gamma \subset L(C^+) \subset K$.

Proof. By definition there is a sequence $[t_n]$, with $[x_1(t_n), x_2(t_n)] \to p_0$, $t_n \to +\infty$ as $n \to +\infty$. Since (9.2.1) is autonomous also $x_1 = X_{1n}(t) = x_1(t_n + t)$, $x_2 = X_{2n}(t) = x_2(t_n + t)$, is a solution and $X_{1n}(0)$, $X_{2n}(0) \to p_0$ as $n \to +\infty$. By the remark of (1.1) the sequence γ_n: $x_1 = X_{1n}(t)$, $x_2 = X_{2n}(t)$, $0 \leq t \leq 1$, converges uniformly toward a solution γ: $x_1 = X_1(t)$, $x_2 = X_2(t)$, $0 \leq t \leq 1$, through p_0, $\gamma \subset L(C^+)$, and the limit, say p_1 of the second end points of γ_n, is certainly a regular point. By repetition of the same argument we deduce that γ exists in $0 \leq t \leq 2$. By indefinite repetition of the same argument we conclude that γ exists in $[0, +\infty)$. The same argument applies to $(-\infty, 0]$ also. Thus Γ is a full orbit, and $\Gamma \subset L(C^+)$.

A segment l in D is said to be a *transversal* if all its points p are regular and the direction $\theta(p)$ associated by (9.2.1) to p is different from that of l (9.1).

(9.2. iii) (a) Every regular point $p \in D$ is an interior point of some transversal l which may have any direction different from $\theta(p)$. (b) Every orbit which meets a transversal crosses it, and all such orbits cross it in the same direction (i.e., all from left to right, or all from right to left with respect to any orientation of l). (c) If p_0 is any interior point of l, and $\varepsilon > 0$ any given number, then there is a circle σ of center p_0 such that every orbit passing through any point $p \in \sigma$ at $t = 0$ crosses l at some time t with $|t| < \varepsilon$. (d) If a finite closed arc γ [of a solution of (9.2.1)] crosses l, it does so in a finite number of points, whose order on γ is the same as the order on l. If γ is periodic, then it meets l in only one point.

Proof. (a) is obvious, and (b) is a consequence of the continuity with respect to the initial conditions. To prove (c), denote by $a x_1 + b x_2 + c$ the straight line containing l, by $x_1 = x_1(t, \xi, \eta)$, $x_2 = x_2(t; \xi, \eta)$ the solution of (9.2.1) passing through (ξ, η) at $t = 0$, and by $L = L(t; \xi, \eta) = a x_1(t, \xi, \eta) + b x_2(t; \xi, \eta) + c$. Then $L(0, x_{10}, x_{20}) = 0$, $\partial L(0, x_{10}, x_{20})/\partial t \neq 0$, where $p_0 = (x_{10}, x_{20})$. Thus (c) is a consequence of the implicit function theorem. To prove (d), let us denote by $\lambda_0 = \widehat{p_1 p_2}$ the arc of γ between two successive distinct intersections p_1, p_2 with l, and by l_0 the segment $p_1 p_2$ of l. Then $\lambda_0 + l_0$ is the boundary of a Jordan region R. By virtue of (b), of the two complementary arcs λ_1, λ_2 of γ, one is inside and one is outside R, while of the two complementary

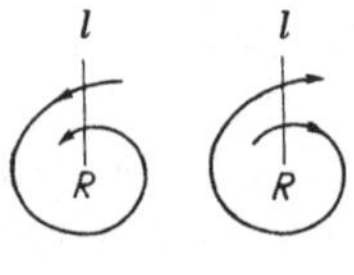

segments l_1, l_2 of l, also one is inside and one is outside R. (The two possible cases are given in the illustration.) Thus γ is periodic if and only if $p_1 = p_2$.

(9.2. iv) If C^+ and $L(C^+)$ have a point in common, then C^+ is a periodic orbit.

Proof. Let p_1 be a common point of C^+ and $L(C^+)$. If $p_1 = (x_{10}, x_{20})$ is a critical point then $C^+ = L(C^+) = p_1$ and C^+ is a (degenerate) periodic orbit. Otherwise we have $x_{10} = x_1(t_1)$, $x_{20} = x_2(t_1)$ for some t_1, and p_1 is a regular point; hence, by (a) above, there is a transversal l through p_1. Let σ be the circle of center p_1 relative to $\varepsilon = 1$ and (c) above. Since $p_1 \in L(C^+)$, there are points $\bar{p}_n \in \sigma$, and we may suppose $\bar{p}_n = [x_1(t_n), x_2(t_n)]$, $t_n > t_{n-1} + 2$, $n = 2, 3, \ldots$. Thus C^+ crosses l at times between t_{n-1} and t_{n+1}, and thus infinitely many times, at points p_n, $n = 1, 2, \ldots$, If say $p_2 \neq p_1$, then, by (d) above, the points p_n tend away from p_1, and p_1 is not on $L(C^+)$, a contradiction. Thus $p_2 = p_1$ and C^+ is periodic.

(9.2. v) A transversal l cannot meet $L(C^+)$ in more than one point.

The same argument above holds.

(9.2. vi) If $L(C^+)$ contains a periodic orbit Γ, then $L(C^+) = \Gamma$.

Proof. Suppose $L(C^+) - \Gamma \neq 0$. Since $L(C^+)$ is connected, Γ contains a point of accumulation p_0 of $L(C^+) - \Gamma$. Let l be a transversal through p_0. For every circle σ of center p_0, there are points $p \in L(C^+) - \Gamma$ and, if p is close enough to p_0, certainly the orbit C_p through p crosses l, by virtue of (9.2. iii). Then by (9.2. ii), C_p is a full orbit and, also, C_p is distinct from Γ, since $C_p \subset L(C^+) - \Gamma$. This implies that l contains two distinct points of $L(C^+)$, a contradiction, because of (9.2. v). Thus $L(C^+) = \Gamma$.

(9.2. vii) (Poincaré-Bendixson theorem). If $C^+ \subset K \subset D$, and $L(C^+)$ consists of regular points only, then either $C^+ = L(C^+)$ is a periodic orbit, or $L(C^+)$ is a periodic orbit.

Proof. If C^+ is periodic, then $C^+ = L(C^+)$. If C^+ is not periodic, then by (9.2. ii) there exists a full orbit $\Gamma \subset L(C^+) \subset K$. Take a point $p_0 \in L^+(\Gamma)$. Then $p_0 \in K$ and, by (9.2. ii), also $p_0 \in L(C^+)$. Hence p_0 is regular and we may take a transversal l through p_0 (9.2. iii). Obviously p_0 and Γ^+ are contained in $L(C^+)$, and, by (9.2. v), l can meet $L(C^+)$ in no point but p_0. On the other hand p_0 is a limit of Γ^+, hence l must meet Γ^+ in some point which is then necessarily p_0. By (9.2. iv), Γ is periodic and this implies, by (9.2. vi) that $\Gamma = L(\Gamma^+) = L(C^+)$.

(9.2. viii) If $C^+ \subset K \subset D$, if C_0 is an orbit contained in $L(C^+)$, then C_0 cannot have a regular limit point unless C_0 is periodic.

The same argument holds as for (9.2. vii).

(9.2. ix) If $C^+ \subset K \subset D$ and there are no critical points in K, then K contains a periodic orbit.

This statement is a corollary of (9.2. vii). Note that the periodic orbit (or solution) is $L(C^+)$ and is often called a cycle or a limit cycle.

A particularly well known application of these theorems is the following one. Suppose that an annular region K, or any Jordan region of connectivity one, in the (x_1, x_2)-plane can be defined such that (a) no singular point of (9.2.1) is in K or on the boundary K^* of K; (b) the solutions through the points of K^* enter K. Then for each semiorbit C^+ starting at any point of K^* we have $C^+ \subset K$ and thus K contains a periodic orbit. Note that both (a) and (b) can be verified by simple inspection of system (9.2.1) and of the direction field $\theta(p)$ defined by it. Also, note that if the periodic orbit in K is known to be unique, then the same orbit, as a consequence of (9.2. vii), is stable (1.8).

As an application of these results let us consider the Liénard equation $x''+f(x)x'+g(x)=0$ under the hypotheses of (8.7. i). Then the corresponding system (8.7.4) is of the form $x'=X(x, y)$, $y'=Y(x, y)$, with $X=y-F(x)$, $Y=-g(x)$, and (0, 0) is the only singular point. Any two integral curves as ABA_1 and $A'B'A_2$ [as mentioned in the proof of (8.7. i)] together with the segments AA_1, A_2A' on the positive y-axis constitute the boundary R^* of an annular region R containing the unique periodic solution and no critical point (ABA_1 spirals out and $A'B'A_2$ spirals in). All solutions crossing R^* enter R. Thus by the remark above the unique periodic solution stated in (8.7. i) is asymptotically orbitally stable (1.8). This situation occurs for instance for the van der Pol equation mentioned in (9.1) with $\varepsilon>0$.

As another application of the principle of the annulus let us consider a system $x'=X(x, y, \lambda)$, $y'=Y(x, y, \lambda)$, $\lambda_1\leqq\lambda\leqq\lambda_2$, depending continuously on a parameter λ, for which the origin is the only critical point in a simple closed region R, and precisely a stable spiral point or node for $\lambda<\lambda_0$, and an unstable spiral point or node for $\lambda>\lambda_0$, where $\lambda_1<\lambda_0<\lambda_2$. Suppose that all solutions through the boundary R^* of R (a simple closed curve) enter R for every $\lambda_1<\lambda<\lambda_2$. Suppose that for $\lambda<\lambda_0$ there are no periodic orbits in R. Then for $\lambda>\lambda_0$ there must be a periodic orbit C in R encircling the origin, and the orbit C is stable from the inside (if C is the unique nontrivial periodic orbit in R, then C is stable at both sides). Thus, as λ passes from $\lambda<\lambda_0$ to $\lambda>\lambda_0$, the origin becomes unstable and a periodic orbit (at least) is created in R. This phenomenon (bifurcation, in the terminology of Poincaré) was first observed under analyticity conditions as a consequence of Poincaré's periodicity conditions (8.3) [in a form different from (8.3.5) and then called bifurcation equation] (cf., e.g., N. Minorsky [6, pp. 87—94]).

(9.2. x) If $C^+\subset K\subset D$ and D has only finitely many critical points, then either (a) $L(C^+)$ consists of a single critical point and C^+ tends to it as $t\to+\infty$; or (b) $L(C^+)$ is a periodic orbit; or (c) $L(C^+)$ consists of finitely many critical points and of a set of full orbits Γ joining them in the sense that for each of these orbits Γ, both Γ^+ and Γ^- are single critical points.

Proof. If $L(C^+)$ consists only of critical points, then $L(C^+)$ is a single point, since $L(C^+)$ is connected and there are only finitely many critical points. If $L(C^+)$ has regular points, then $L(C^+)$ is the union of critical points and orbits $\Gamma\subset L(C^+)$ (9.2. ii). If Γ is a limit orbit then it can have regular limit points only if it is periodic (9.2. viii). Thus either Γ is periodic and $L(C^+)=\Gamma$ (9.2. vi), or has no regular limit points, all orbits in $L(C^+)$ are not periodic, and have critical points as limit points. Thus if Γ is such an orbit, and thus a full orbit (9.2. ii), then both $L^+(\Gamma)$ and $L^-(\Gamma)$ are single critical points (distinct, or coincident).

We recall here a few more statements of which we omit the proof for the sake of brevity.

(9.2. xi) A necessary and sufficient condition in order that a periodic orbit C be stable (orbitally stable as $t\to+\infty$) for both the interior and exterior of C, is that, either (a) an orbit Γ approaches C (at each side) as $t \to +\infty$; or (b) there exist periodic orbits in every neighborhood of C (at each side).

If two periodic orbits C_1, C_2, one, say C_2, is interior to C_1, and in the annular region between C_1 and C_2 there are no critical points nor other periodic orbits, then C_1 and C_2 are said to be *adjacent.*

(9.2. xii) Two adjacent orbits cannot be both stable (as $t \to +\infty$) on the sides facing one another.

Remark. Let $H = (\partial f/\partial x_1) + (\partial g/\partial x_2)$ be the divergence of the vector field (f, g) associated with system (9.2.1). Let C: $x_1 = x_1(t)$, $x_2 = x_2(t)$, be a periodic solution of period T (closed orbit) of (9.2.1), and R be the simple Jordan region whose boundary is C. Then by Gauss-Green theorem and (9.2.1) we have

$$\iint_R H(x_1, x_2)\, dx_1\, dx_2 = \int_C (f\, dx_2 - g\, dx_1) = \int_0^T (x_1' x_2' - x_2' x_1')\, dt = 0.$$

From this simple remark it follows that in a simple region R where H has a constant sign there cannot be periodic solutions of (9.2.1) (BENDIXSON's negative criterion).

9.3. Indices. We will suppose that all general hypotheses of (9.2) hold. Let γ be a counterclockwise oriented Jordan curve in D not passing through critical points. Then the total change of the angle $\theta(p)$ along γ divided by 2π is the *index* $i(\gamma)$ relative to γ. More precisely, if we fix arbitrarily a point p_0 on γ and one, say θ_1, of the infinitely many determinations (mod 2π) of $\theta(p)$ at $p = p_0$, then we may think of $\theta(p)$ as a continuous function of p on γ. As p describes completely γ and returns to the point p_0, $\theta(t)$ takes a value θ_2 which may be different from θ_1. The difference $\theta_2 - \theta_1$ is an integral multiple of 2π and does not depend on the choices of p_0 on γ and of θ_1. Thus the integer $i(\gamma)$ is given by

$$i(\gamma) = (\theta_2 - \theta_1)/2\pi = \Delta\theta/2\pi \gtreqless 0. \qquad (9.3.1)$$

If we observe that $\theta(p)$ is the argument of a vector (u, v) of components $P(x_1, x_2)$, $Q(x_1, x_2)$ never both zero on γ, then, for a rectifiable curve γ, the index $i(\gamma)$ is given by

$$i(\gamma) = (2\pi)^{-1} \int_\gamma (P^2 + Q^2)^{-1} (Q\, dP - P\, dQ). \qquad (9.3.2)$$

If we divide the Jordan region J of which γ is the boundary into finitely many Jordan regions of which we denote by $\gamma_1, \gamma_2, \ldots, \gamma_n$ the oriented boundary curves and none of these curves pass through a critical point, then

(9.3. i) $i(\gamma) = i(\gamma_1) + \cdots + i(\gamma_N)$.

The proof is the same as the usual one in complex variable theory.

(9.3. ii) If the Jordan region J whose boundary is γ contains no critical point, then $i(\gamma) = 0$. More generally if the annular region R between two Jordan curves γ_1, γ_2 contains no critical point (and γ_1, γ_2 do not pass through critical points) then $i(\gamma_1) = i(\gamma_2)$.

For the proof we have only to divide R into Jordan regions q of so small a diameter that $\theta(p)$ has an oscillation smaller than 2π in each of them. Then the index of each of those regions q is zero, and by (9.3. i) it follows that $i(\gamma_1) = i(\gamma_2)$.

(9.3. iii) If C is a periodic orbit, then $i(C) = +1$.

Proof. The known elementary property that the sum of the external angles of a convex simple polygonal region is 2π has the following formulation for simple polygonal regions $\pi = A_1 A_2 \ldots A_n A_1$, convex or not. Let us orient each side $A_i A_{i+1}$ in such a way to leave π to the left, and, for each i, let us denote by α_i, or external angle of π at A_i, the angle ($\gtreqless 0$) in which the positive direction $A_{i-1}A_i$ has to rotate to become parallel to $A_i A_{i+1}$, $-\pi < \alpha_i < \pi$, $i = 1, \ldots, n$. Then $\alpha_1 + \cdots + \alpha_n = 2\pi$. Now C is a Jordan curve with continuously turning tangent t. Thus we can divide C into consecutive arcs $A_i A_{i+1}$ in such a way that $\theta(p)$ has an oscillation on each of them which is $< \pi/4$. Also we can take so fine a subdivision that the polygonal line $A_1 A_2 \ldots A_n A_1$ is simple and the corresponding polygon π surrounds a fixed point O arbitrarily taken inside C. Now we may replace each side $A_i A_{i+1}$ by the polygonal line $l_i = A_i B_i C_i A_{i+1}$, made up by two segments $A_i B_i$ and $C_i A_{i+1}$ of the oriented tangents t_i and t_{i+1} to C at A_i and A_{i+1}, and the segment $B_i C_i$ joining B_i and C_i. By taking the segments $A_i B_i$, $C_i A_{i+1}$ sufficiently small we can assure that also the polygonal line $l_1 + \cdots + l_n$ is simple and that the corresponding polygonal region π surrounds O. Now the angle $\theta_{i+1} - \theta_i$ of the tangents t_{i+1} and t_i is equal to the sum of the external angles β_i and γ_i of π' at B_i and C_i. Thus the angle $\Delta\vartheta$ relative to C is equal to the sum of the external angles of π' which is 2π. Thus $i(C) = 1$.

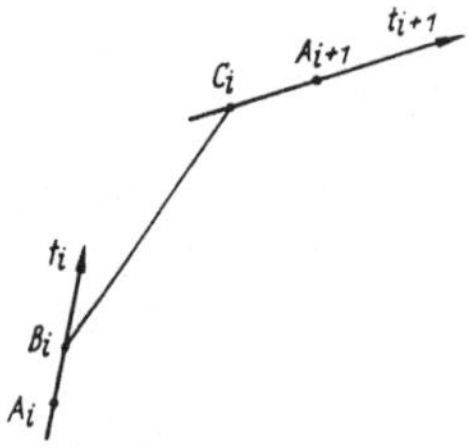

If D contains only finitely many critical points, say $p_1, \ldots, p_k$, and we consider the oriented Jordan curves $\gamma \subset D$ enclosing only one of these points, say p_1, then $i(\gamma)$ does not depend upon γ, and, therefore, is an integer depending on p_1 only, or the *index* $j(p_1)$ of the critical point p_1 with respect to system (9.2.1).

(9.3. iv) If γ is a Jordan curve surrounding the (finitely many) critical points $p_1, \ldots, p_k$, then $i(\gamma) = j(p_1) + \cdots + j(p_k)$.

We have only to divide the Jordan region J whose boundary is γ into finitely many Jordan regions whose boundary curves do not pass through critical points and which contain each only one critical point. Then (9.3. iv) is a consequence of (9.3. i).

(9.3. v) If C is a periodic orbit and the Jordan region R defined by C is inside D, then there is at least one critical point in R. If there are more than one critical point, then the sum of their indices is $+1$.

This statement is a consequence of (9.3. iii) and (9.3. iv).

Let us suppose now that the origin is an isolated critical point and that system (9.2.1) can be written in the form (9.1.3). Then we shall consider, as in (9.1), the two systems

$$\left.\begin{aligned} x_1' &= a\,x_1 + b\,x_2 + f(x_1, x_2), \\ x_2' &= c\,x_1 + d\,x_2 + g(x_1, x_2), \end{aligned}\right\} \qquad (9.3.3)$$

$$\left.\begin{aligned} x_1' &= a\,x_1 + b\,x_2, \\ x_2' &= c\,x_1 + d\,x_2, \end{aligned}\right\} \qquad (9.3.4)$$

where $ad - bc \neq 0$, $f, g = o(r)$.

(9.3. vi) The origin has the same index with respect to systems (9.3.3) and (9.3.4).

Proof. We shall determine the index of the origin with respect to either system by using the circle C_r: $x_1 = r\cos\omega$, $x_2 = r\sin\omega$ and r sufficiently small. Let us observe first that the function $(a\cos\omega + b\sin\omega)^2 + (c\cos\omega + d\sin\omega)^2$ is periodic and continuous and may be zero only if $ac - bd = 0$. Thus $ac - bd \neq 0$ implies that the minimum A and the maximum B of the same function above are positive, $0 < A \leq B$. Let $P(x_1, x_2, \varepsilon)$, $Q(x_1, x_2, \varepsilon)$ denote the expressions $ax_1 + bx_2 + \varepsilon f$, $cx_1 + dx_2 + \varepsilon g$, where $0 \leq \varepsilon \leq 1$, and P_0, Q_0 the same expressions as $\varepsilon = 0$. Then we have $Ar^2 \leq P_0^2 + Q_0^2 \leq Br^2$, and $f, g = o(r)$ on the circle C_r. From

$$P^2 + Q^2 = P_0^2 + Q_0^2 + 2\varepsilon P_0 f + 2\varepsilon Q_0 g + f^2 + g^2$$

we conclude that on the circle C_r we have $(P^2 + Q^2) - (P_0^2 + Q_0^2) = o(r^2)$, uniformly in $0 \leq \varepsilon \leq 1$. Thus $P^2 + Q^2$ is certainly between $Ar^2/2$ and $2Br^2$ on C_r for all r sufficiently small and $0 \leq \varepsilon \leq 1$, $0 \leq \omega \leq 2\pi$. Thus

$$\nu(\varepsilon) = (2\pi)^{-1} \int_{C_r} (P^2 + Q^2)^{-1} (P\,dQ - Q\,dP),$$

is certainly a continuous function of ε for $r > 0$ sufficiently small. Since $\nu(\varepsilon)$ is an integer we conclude that $\nu(0) = \nu(1)$; i.e., the index of the origin is the same for systems (9.3.3) and (9.3.4).

(9.3. vii) The origin has index 1, or -1 according as $ad - bc > 0$, or $ad - bc < 0$; thus it has index -1 if the origin is a saddle point, has index 1 in all other cases.

Proof. We shall consider system (9.3.3) and apply (9.3.2) to the circle C_r as above. Then the index, say j, is given by

$$j = (2\pi)^{-1} (ad - bc) \int_0^{2\pi} [(a\cos + b\sin\omega)^2 + (c\cos\omega + d\sin\omega)^2]^{-1}\,d\omega,$$

and thus j is a continuous function of a, b, c, d provided $ad - bc \neq 0$. Thus j remains constant since j is necessarily an integer. If $ad - bc > 0$, $ad > 0$, $bc < ad$, we may vary the product bc in such a way that it varies monotonically from its value to zero. We can then vary d so that it varies monotonically from its value to a. In all these transformations $ad - bc$ is always $\neq 0$. Thus we may suppose $a = d \neq 0$, $b = c = 0$, and then j is given by $j = (2\pi)^{-1} \int_0^{2\pi} d\theta = +1$. If $ad - bc > 0$, $ad < 0$, then first we vary ad monotonically into a value $a'd' > 0$, $a'd' > bc$, and then we repeat the same process above. If $ad - bc < 0$, then the same argument shows that $j = -1$. Now an inspection of the cases (2.5) shows that $ad - bc$ is < 0 for a saddle point, and > 0 in all other cases.

For extensions of the Poincaré-Bendixson theory to n-th order differential systems see S. A. STEBAKOV [1], T. URA [1], S. K. ZAREMBA [1], G. SOLUTZEV [1], L. L. RAUCH [1], R. E. GOMORY [1], R. E. GOMORY and F. HAAS [1], S. A. ZEVAKIN [1]. It should be mentioned here that the idea to generalize to space the principle of the annulus which holds in the plane was long sought, but not achieved. In 1952 F. B. FULLER [1] proved by means of an example that there is a fourth order differential system which presents a four dimensional torus without singular points, nor periodic solutions, and such that all solutions pass from the exterior into the interior of the torus.

For general structure theorems see L. MARKUS [2—5], N. OTROKOV [1, 2], W. KAPLAN [1]. N. N. BAUTIN [3] had given an example of a system $dy/dx = M/N$, M, N polynomial of degree 2 in x, y, with three cycles. I. G. PETROVSKY and E. M. LANDIS [1] have proved that 3 is the maximum of the number of the cycles of such systems. For other results see T. UNO [1, 2]. On the number of closed orbits of a differential system see also G. SANSONE and R. CONTI [1], L. A. MACCOLL [2], L. E. ELSGOLL [2].

For applications of the Poincaré-Bendixson theory to equations of the form $x'' + x'|x'| \pm (q x' + x) - p^2 x^3 = 0$, see J. CECCONI [1, 2], and L. N. MODONA [1]. For analogous equations see G. SESTINI [1] (cf. previous results of G. H. HARDY and R. H. FOWLER [3]). See also for existence and stability of periodic solutions of Liénard type equations, S. P. DILIBERTO [1]. Finally let us mention that the equation $x''' + [af(x) + b]x'' + af''(x)x'^2 + a(bf(x) - \alpha)x' + \beta f(x) = 0$, treated heuristically by N. LEVINSON [23], has been discussed rigorously by G. COLOMBO [3].

Let us consider a system $x' = F(t, x)$, $x = (x_1, \ldots, x_n) \in E_n$, $t \geq 0$, with $F(t, 0) = 0$, for which existence in the large and usual uniqueness hold for the solutions starting in a sufficiently small neighborhood of the origin. The solution $x = 0$ is said to be *totally stable* (J. L. MASSERA [5]) if the following holds: given $\varepsilon > 0$ there exists $\delta > 0$ such that for every continuous function $F_1(t, x)$ with $\|F_1 - F\| \leq \delta$ for $\|x\| \leq \varepsilon$ and $t \geq 0$, any solution $x_1(t)$ of $x' = F_1(t, x)$ with $\|x_1(0)\| \leq \delta$ exists in $[0, +\infty)$ and $\|x_1(t)\| \leq \varepsilon$ for all $t \geq 0$. The concept agrees with another one discussed by I. G. MALKIN [4]. Conditions for total stability have been given by I. G. MALKIN [4] and J. L. MASSERA [5]. The same concept is connected with the concept of structural stability considered by A. ANDRONOV and L. S. PONTRYAGIN [1]. A theorem for $n = 2$ stated by these authors has been proved recently by H. F. DE BAGGIS [1]. For further results concerning structural stability see M. M. PEIXOTO [1] and M. M. PEIXOTO and M. C. PEIXOTO [1].

9.4. A configuration concerning LIÉNARD'S equation. The method of LEVINSON and SMITH (9.5) for the determination of cycles of the (autonomous) Liénard equation, and the method of the fixed point (9.6) for the determination of periodic solutions of Liénard equations with a forcing periodic term, rest on different topological considerations, but the same preliminary geometrical construction. Namely under convenient assumptions some closed curve C in the (x, x')-plane must be defined, enclosing the origin and some conveniently large circle, having the property that every solution $x = x(t)$, $x' = x'(t)$ crossing C passes from the exterior to the interior of C. The devices on which such curves C are built differ widely from author to author and in the many different situations in which such processes have been applied. The general exposition will be, therefore, greatly simplified if we cover only once the construction of a curve C, and we apply the result both in (9.5) and (9.6). For this purpose we refer below to a fairly general statement concerning the equation

$$x'' + f(x, x')x' + g(x) = e(t), \tag{9.4.1}$$

or the system

$$dx/dt = v, \qquad dv/dt = -f(x, v)v - g(x) + e(t). \tag{9.4.2}$$

Thus in (9.5) we shall suppose $e(t) \equiv 0$, and in (9.6) we shall suppose $e(t)$ nonzero and periodic of a given period T. The following theorem and its proof are due to N. LEVINSON [3].

(9.4. i) Suppose that $e(t)$ is a bounded continuous function, and $g(x)$, $f(x, v)$ possess first order derivatives. Suppose that there are constants a, m, $M > 0$ such that $f(x, v) > m$ if both $|x| \geq a$, $|v| \geq a$; $f(x, v) \geq -M$ for all x and v; $xg(x) > 0$ for $|x| \geq a$, and $|g(x)| \to +\infty$ as $|x| \to +\infty$. Moreover, if $G(x) = \int_0^x g(u)\,du$, suppose that $g(x)/G(x) \to 0$ as $|x| \to +\infty$. Then (A) there is a closed curve C in the (x, v)-plane enclosing the square $|x|, |v| \leq a$, such that each solution $[x(t), v(t)]$ of (9.4.2) crossing C at any time t, passes from the exterior into the interior of C; (B) for every point (x, v) sufficiently remote from the origin there passes a curve C.

Proof of (9.4. i) As it was done in (8.7) and (8.8) we make use of the "energy function" associated with the "motion" described by (9.4.2), namely the function

$$\lambda(x, v) = 2^{-1} v^2 + G(x). \tag{9.4.3}$$

For large $|x|$, $G(x)$ is positive, increasing, and $G(x) \to +\infty$ as $x \to \pm\infty$. Thus the curves $\lambda(x, v) = c$, c a constant, are closed and enclose the origin for large values of $c > 0$. Also it is clear that the curve $\lambda(x, v) = c_2$ encloses the curve $\lambda(x, v) = c_1$ if $c_2 > c_1$. From (9.4.2) and (9.4.3) it follows that

$$d\lambda/dt = -f(x, v)\,v^2 + e(t)\,v \tag{9.4.4}$$

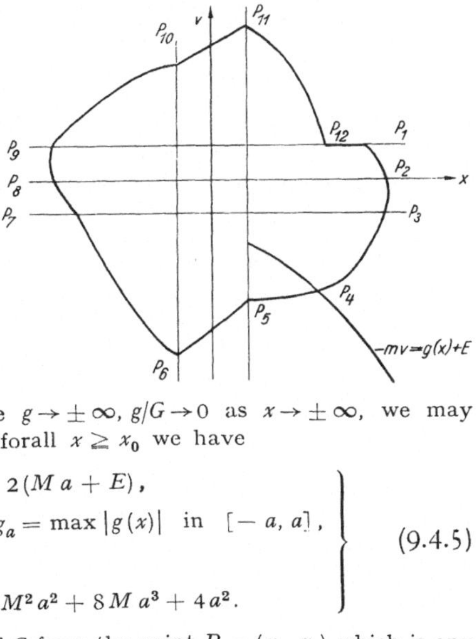

along every solutions S of (9.4.2). If E is a bound for $e(t)$, i.e., $|e(t)| \leq E$ for all t, we may always suppose the constant a of (9.4. i) so large that $ma > 2E$.

We shall build the curve $C = P_1 P_2 \ldots P_{12}$ by means of simple arcs as we will describe. Put $P_i = (x_i, v_i)$ and $\lambda_i = \lambda(x_i, v_i)$, $i = 1, \ldots, 12$. Since $g \to \pm\infty$, $g/G \to 0$ as $x \to \pm\infty$, we may take a number $x_0 > a$ so large that, forall $x \geq x_0$ we have

$$\left.\begin{aligned} & g(x) > 2(Ma + E), \quad g(-x) < -2(Ma + E), \\ & g(x) > 2m M^{-1}(g_a + E), \quad \text{where } g_a = \max|g(x)| \text{ in } [-a, a], \\ & G(x) > 16 a M g(x), \\ & G(x) > 2G(a) + 16 M^{m-1} a E + 64 M^2 a^2 + 8 M a^3 + 4a^2. \end{aligned}\right\} \tag{9.4.5}$$

We shall now start the construction of C from the point $P_4 = (x_4, v_4)$ which is any point on the curve $v = -m^{-1}[g(x) + E]$ with $x_4 \geq x_0$ and far away from the origin. Thus

$$x_4 \geq x_0, \qquad -m v_4 = g(x_4) + E. \tag{9.4.6}$$

and relations (9.4.5) hold for $x = x_4$.

The arc $P_4 P_3$ is a portion of the curve $\lambda(x, v) = \lambda_4 = \lambda(x_4, v_4)$ and $P_3 = (x_3, v_3)$ is the point of it with $v_3 = -a$. On $P_4 P_3$ we have $x \geq a$, $v \leq -a$ and, by virtue of (9.4.4) and $ma \geq 2E$, we have

$$\begin{aligned} d\lambda/dt &< -m v^2 + E|v| = \\ & -m v^2 (1 - E m^{-1} |v|^{-1}) \leq -m v^2 (1 - E/ma) < -2^{-1} m v^2 < 0. \end{aligned}$$

Thus any solution S of (9.4.2) crossing $P_4 P_3$ at any time t passes from the exterior to the interior of C. We have also $\lambda_4 - \lambda_3 = 0$.

The arc $P_2 P_3$ is on the straight line $x = x_4$ and $P_2 = (x_2, v_2)$ is the point of it with $v_2 = 0$. Since $dx/dt = v < 0$, any solution of (9.4.2) crossing $P_2 P_3$ passes from the exterior into the interior of C. On the other hand $x_2 = x_3$ and hence $\lambda_3 - \lambda_2 = 2^{-1} a^2$.

The arc $P_1 P_2$ is a portion of the curve

$$2^{-1} v^2 + G(x) - (M a + E) x = G(x_2) - (M a + E) x_2,$$

and $P_1 = (x_1, v_1)$ as the point of it with $v_1 = a$. On $P_1 P_2$ we have

$$dv/dx = v^{-1} [- g(x) + M a + E],$$

while on every solution S of (9.4.2) crossing $P_1 P_2$ we have

$$dv/dx = v^{-1} [- f(x, v) v - g(x) + e(t)] < v^{-1} [- g(x) + M a + E].$$

Thus each solution S crossing $P_1 P_2$ passes from the exterior to the interior of C In addition we have

$$\lambda_2 - \lambda_1 = G(x_2) - G(x_1) - 2^{-1} a^2 = (M a + E)(x_2 - x_1),$$

and hence

$$\int_{x_1}^{x_2} g(x)\, dx = (M a + E)(x_2 - x_1) + 2^{-1} a^2,$$

where $x_1 > x_4 > x_0$, and $g(x) > 2(Ma + E)$ for all $x \geq x_1$. Thus

$$2(M a + E)(x_2 - x_1) < (M a + E)(x_2 - x_1) + 2^{-1} a^2,$$

$$\lambda_2 - \lambda_1 = (M a + E)(x_2 - x_1) < 2^{-1} a^2.$$

Now let us return to the point P_4. The arc $P_4 P_5$ is on the straight line $v = v_4$. and $P_5 = (x_5, v_5)$ is the point of it with $x_5 = a$. Thus $v_5 = v_4$ and, on the arc $P_4 P_5$ we have

$$dv/dt = - f v - g(x) + e(t) > - g(x) - m v - E,$$

and since (x, v) is now below the curve $-mv = g(x) + E$, we have

$$dv/dt > \big(- g(x) - E\big) + \big(g(x) + E\big) = 0.$$

Thus every solution S of (9.4.2) crossing $P_4 P_5$ passes from the exterior into the interior of C. Moreover, since $v_4 = v_5$, we have

$$\lambda_5 - \lambda_4 = G(a) - G(x_4).$$

The arc $P_5 P_6$ is a portion of a straight line of slope $2M$, and $P_6 = (x_6, v_6)$ is the point with $x_6 = -a$. Thus $v_6 = v_5 - 4 M a$, and on $P_5 P_6$ we have $dv/dx = -2M$. On the other hand on each solution S crossing $P_5 P_6$ we have, by virtue of (9.4.2), (9.4.5) and (9.4.6).

$$\begin{aligned} dv/dx &= v^{-1} [- f(x, v) v - g(x) + e(t)] < \\ &< M - v_5^{-1} (g_a + E) \\ &= M - v_4^{-1} (g_a + E) \\ &= M + m (g_a + E) \big(g(x_4) + E\big)^{-1} < \\ &< M + m (M/2m)\, g(x_4) \big(g(x_4) + E\big)^{-1} < \\ &< M + M/2 = 3M/2 < 2M. \end{aligned}$$

Thus any solution S of (9.4.2) crossing $P_5 P_6$ passes from the exterior into the interior of C. Moreover

$$\begin{aligned} \lambda_6 - \lambda_5 &= 2^{-1} v_6^2 + G(-a) - 2^{-1} v_5^2 - G(a) = 2^{-1} (v_5 + v_6)(- 4 M a) + \\ &\quad + G(-a) - G(a) = - 4 M a v_4 + 8 M^2 a^2 + G(-a) - G(a). \end{aligned}$$

The arc $P_6 P_7$ is a portion of the curve $\lambda(x, v) = \lambda_6 = \lambda(x_6, v_6)$ and $P_7 = (x_7, v_7)$ is the point of it with $v_7 = -a$. As for $P_3 P_4$ we may prove that every solution S of (9.4.2) crossing $P_6 P_7$ passes from the exterior into the interior of C. Moreover $\lambda_7 - \lambda_6 = 0$.

The arc $P_7 P_8$ is a portion of the curve

$$2^{-1} v^2 + G(x) + (M a + E) x = 2^{-1} a^2 + G(x_7) + (M a + E) x_7, \qquad (9.4.7)$$

and P_8 is the point of it with $v = 0$. Here we should have $x_7 < -x_0$. If this is not the case, we can always take x_4 so large as to assure that $x_7 < -x_0$. As for $P_1 P_2$ we may prove that every solution S of (9.4.2) crossing $P_7 P_8$ passes from the exterior into the interior of C and that $\lambda_8 - \lambda_7 < 2^{-1} a^2$. From (9.4.1) we have also $\lambda = -(Ma + E)x +$ constant, and hence $d\lambda/dt = -(Ma + E)v > 0$. Thus we have also $\lambda_8 - \lambda_7 > 0$.

The arc $P_8 P_9$ is a segment of the straight line $x = x_8$ and $P_9 = (x_9, v_9)$ is the point of it with $v_9 = a$. The same considerations hold as for $P_2 P_3$, and $\lambda_9 - \lambda_8 = 2^{-1} a^2$.

The arc $P_9 P_{10}$ is a portion of the curve $\lambda(x, v) = \lambda_9 = \lambda(x_9, v_9)$, and $P_{10} = (x_{10}, v_{10})$ is the point of it with $x_{10} = -a$. The same considerations hold as for $P_3 P_4$, and $\lambda_{10} - \lambda_9 = 0$.

The arc $P_{10} P_{11}$ is on a straight line of slope $2M$ and $P_{11} = (x_{11}, v_{11})$ is the point of it with $x_{11} = a$. The same considerations hold as for $P_5 P_6$, and

$$\lambda_{11} - \lambda_{10} = 4 M a v_{10} + 8 a^2 M^2 + G(a) - G(-a).$$

Finally the arc $P_{11} P_{12}$ is a portion of the curve $\lambda(x, v) = \lambda_{11} = \lambda(x_{11}, v_{11})$, and P_{12} is the point of it with $v_{12} = a$. The usual considerations hold and $\lambda_{12} - \lambda_{11} = 0$.

We shall prove that P_{12} is at the left of P_1. Under this condition, on the segment $P_{12} P_1$ we have

$$\begin{aligned} dv/dt &= -f(x, v) - g(x) + e(t) < -m a - g(x) + E < \\ &< -m a + E < -2E + E = -E < 0. \end{aligned}$$

Thus every solution S of (9.4.2) crossing $P_{12} P_1$ passes from the exterior to the interior of C.

Let us prove now that P_{12} is at the left of P_1: that is, that $x_{12} < x_1$. Since $v_{12} = v_1 = a$ we have only to prove that $\lambda_{12} < \lambda_1$. By combining the eleven evaluations of the differences $\lambda_2 - \lambda_1, \ldots, \lambda_{12} - \lambda_{11}$, we have

$$\lambda_{12} - \lambda_1 = 2a^2 + 4 M a (-v_4 + v_{10}) + 16 M^2 a^2 + G(a) - G(x_4). \qquad (9.4.8)$$

Now we have $\lambda_7 - \lambda_6 = 0$, $0 < \lambda_8 - \lambda_7 < 2^{-1} a^2$, $\lambda_9 - \lambda_8 = 2^{-1} a^2$, $\lambda_{10} - \lambda_9 = 0$, and hence $0 < \lambda_{10} - \lambda_6 < a^2$, and, since $x_6 = x_{10}$, also $0 < v_{10}^2 - v_6^2 < 2a^2$. Thus $v_{10} > |v_6|$ and

$$\begin{aligned} v_{10} &= |v_6| + (v_{10} - |v_6|) = |v_6| + 2a^2 (v_{10} + |v_6|)^{-1} < \\ &< |v_6| + a^2 |v_6|^{-1} < |v_4| + 4 M a + a^2 |v_4|^{-1}. \end{aligned}$$

Since we can always suppose $|v_4| \geq 1$, we conclude that $v_{10} < -v_4 + 4 M a + a^2$ and (9.4.3) becomes

$$\lambda_{12} - \lambda_1 < -8 M a v_4 - G(x_4) + G(a) + 32 M^2 a^2 + 4 M a^3 + 2a^2. \qquad (9.4.9)$$

By virtue of (9.4.6) and (9.4.5) we have now

$$\begin{aligned} -8 M a v_4 - G(x_4) &= 8 M m^{-1} a g(x_4) + 8 M m^{-1} a E - G(x_4) < \\ &< 2^{-1} G(x_4) + 8 M m^{-1} a E - G(x_4) \\ &= -2^{-1} G(x_4) + 8 M m^{-1} a E. \end{aligned}$$

Thus (9.4.9) becomes

$$\lambda_{12} - \lambda_1 < -2^{-1}[G(x_4) - 2G(a) - 16Mm^{-1}aE - 64M^2a^2 - 8Ma^3 - 2a^2]$$

and finally $\lambda_{12} - \lambda_1 < 0$ by virtue of (9.4.5). This implies that $x_{12} < x_1$ and the proof of (9.4. i) is completed.

9.5. Another existence theorem for the Liénard equation. Essentially an application of the Poincaré-Bendixson theorem, the following process, developed by Levinson and Smith and exemplified in the proof of the following theorem, has had recently a number of further applications. The following theorem concerns the existence of cycles of the (autonomous) Liénard differential equation

$$x'' + f(x, x')\,x + g(x) = 0, \qquad (9.5.1)$$

or of the system

$$dx/dt = v, \qquad dv/dt = -f(x, v)\,v - g(x). \qquad (9.5.2)$$

We shall suppose that g and f are continuous together with their first order derivatives. By $G(x)$ we denote the function $G(x) = \int_0^x g(u)\,du$.

(9.5. i) If there are positive constants a, m, M such that $f(x, v) > m$ if both $|x| \geqq a$, $|v| \geqq a$, $f(x, v) \geqq -M$ for all x and v, and $f(0, 0) < 0$, if $xg(x) > 0$ for all $x \neq 0$, and $|g(x)| \to +\infty$, $g(x)/G(x) \to 0$ as $x \to \pm\infty$, then there exists at least one (nonconstant) periodic solution of (9.5.1) N. Levinson [3].

Proof of (9.5. i) The relations $v = 0$, $fv + g = 0$ are satisfied only at $(0, 0)$; hence, the origin is the only critical point of system (9.5.2). Here we have $G(0) = 0$, $G(x) > 0$ for all $x \neq 0$. If $\lambda(x, v) = 2^{-1}v^2 + G(x)$, then the curves α: $\lambda(x, v) = c$, $c > 0$ sufficiently small, are closed, enclose the origin, and are completely contained in a neighborhood of the origin where $f(x, v) < 0$. Moreover the curve $\lambda(x, v) = c_2$ encloses $\lambda(x, v) = c_1$ if $c_2 > c_1$. For every trajectory S of (9.5.2) crossing a curve α we have $d\lambda/dt = -v^2 f(x, v) > 0$. Thus S passes from the interior into the exterior of α. By (9.4. i) there is, on the other hand, a closed curve C enclosing α such that every trajectory S of (9.5.2) crossing C passes from the exterior into the interior of C. Thus if R is the annular region between C and α we conclude that every trajectory S of (9.5.2) crossing once the boundary of R, enters R and remains in R and thus is a half trajectory S^+ in R. Since in R there are no critical points, the curve S^+ cannot have finite length. Indeed in such a case $L(S^+)$ would be a single point, necessarily critical, a contradiction. Thus S^+ has infinite length and thus must spiral within R. By the Poincaré-Bendixson theorem, the set $L(S^+)$ is a periodic solution of (9.5.2) (a limit cycle).

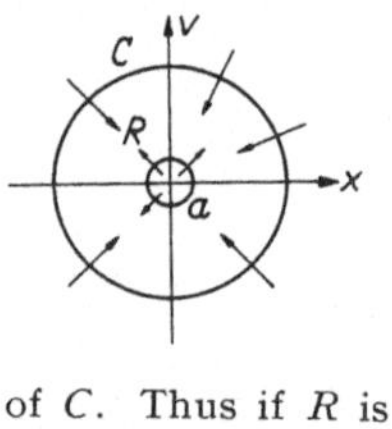

By a discussion analogous to the previous one [(9.4) and the proof above] the following theorem has been proved:

(9.5. ii) If there are positive constants $a, b, M, a < b$, such that $f(x, v) \geqq 0$ for $|x| \geqq a$ and all v, $f(x, v) \geqq -M$ for all x and v, $f(0, 0) < 0$, $xg(x) > 0$ for all $x \neq 0$, $G(x) \to +\infty$ as $x \to \pm\infty$, $\int_a^b f(x, v)\,dx \geqq 10Ma$

for every arbitrary decreasing positive function $v(x)$ in the integration, then (9.5.1) has at least one nonconstant periodic solution (N. LEVINSON and O.K. SMITH [1]).

Two more theorems for the existence of nonconstant periodic solutions of the Liénard equation have been stated in (8.7. i) and (8.9. i). While (8.7. i) assures the uniqueness of the periodic solution, this is not the case for (8.9. i), (9.5. i), and (9.5. ii). A uniqueness theorem is the following one. Let us denote by R_1 and R_2 the parts of the (x, v)-plane where $f<0$, or $f>0$ respectively. Then we shall denote by $R_1(c)$, $R_2(c)$ the arcs of the closed curve $\lambda(x, v)=c$ which respectively lie in R_1 and R_2.

(9.5. iii) Under the conditions of (9.5. ii), if for every c the minimum of $F(x, v)=v^{-2}+v^{-1}f^{-1}\,\partial f/\partial v$ on $R_2(c)$ is positive and exceeds the maximum of $F(x, v)$ on $R_1(c)$, then (9.5.2) possesses a unique nonconstant periodic solution.

The proof of (9.5. iii), which we omit here, consists in proving first that each closed trajectory of (9.5.2) is necessarily asymptotically stable. Then, from the remark of (9.2. xii), we conclude that no two consecutive closed trajectories can be stable. This implies that there is only one closed trajectory for system (9.5.2). Statements analogous to (9.5. ii) are the following ones.

(9.5. iv) If $f(x), g(x)$ are differentiable functions, if there exists $x_1, x_2, -x_1<0<x_2$, such that $f(x)<0$ for $-x_1<x<x_2$, $f(x)\geqq 0$ otherwise, if $x g(x)>0$ for $x\neq 0$, and

$$\int_0^{\pm\infty} g(x)\,dx=\infty, \qquad \int_0^{+\infty} f(x)\,dx=\infty, \qquad \int_0^{-x_1} g(x)\,dx=\int_0^{x_2} g(x)\,dx,$$

then the system

$$dx/dt=v, \qquad dv/dt=-v f(x)-g(x), \tag{9.5.3}$$

has a unique (nonconstant) periodic solution.

(9.5. v) Under conditions (9.5. ii), if for some $x_1, x_2, -x_1<0<x_2$, we have $f(x, v)<0$ for all $-x_1<x<x_2$ and all v, and $f(x, v)\geqq 0$ otherwise; if $\int_0^{-x_1} g(x)\,dx=\int_0^{x_2} g(x)\,dx$, and $v\,\partial f/\partial v\geqq 0$ for all x and v, then system (9.5.2) has a unique periodic solution (N. LEVINSON and O.K. SMITH [1]).

For other applications of the methods above see e.g., Z. MIKOLAJSKA [3], and T. SHIMIZU [1]. For theoretical research on the Liénard equation see D. GRAFFI [3, 7, 8, 13, 15, 17], in particular [8, 13] for bounds of the solutions, and [7, 13, 14] for evaluations of the periods. For other bounds see N. WAX [1]. See also on the same subject V. CAPRA [1], L. CAPRIOLI [1], G. MANARESI [1], G. MALGARINI [1], A. ASCARI [1], A.V. DRAGILEV [1], H. SERBIN [1], G. DUFFING [1], G.F.D. DUFF [1], D.A. FLANDERS and J.J. STOKER [1], R.E. GOMORY and D.E. RICHMOND [1], V.V. KAZAKEVICH [1—3], N.W. McLACHLAN [10], J. SHOHAT [1, 2], S.N. SIMANOV [1],

G. V. SAVINOV [1], J. G. WENDEL [1], K. THEODORCHIK [1—4]. See also the lectures and articles H. J. ECKWEILER [1], K. O. FRIEDRICHS [1, 2], F. JOHN [1], N. LEVINSON [22, 23], K. O. FRIEDRICHS and J. J. STOKER [2], A. N. OBMORSEV [1], E. LEONTOVIC [1], V. A. CECIK [1], A. F. FILIPPOV [1], A. V. DRAGILEV [1], A. DE CASTRO [5], S. P. DILIBERTO [1], E. FISHER [1].

9.6. The method of the fixed point. BROUWER's famous fixed point theorem may have the following formulation. If R: $a_i \leq x_i \leq b_i$, $i=1, 2, \ldots, n$, is a closed interval of the Euclidean x-space E_n, $x=(x_1, \ldots, x_n)$, if T: $y=f(x)$, $x \in R$, is a continuous mapping of R into itself, i.e., if $f=(f_1, \ldots, f_n)$, $f_i=f_i(x_1, \ldots, x_n)$, and $a_i \leq f_i(x_1, \ldots, x_n) \leq b_i$, $i=1, \ldots, n$, for all $x \in R$, then there is at least one point $x_0 \in R$ (fixed point) such that $T x_0 = x_0$, i.e., $f(x_0)=x_0$.

An equivalent formulation is the following: if R is any closed n-cell in E_n and T any continuous mapping from R into itself, i.e., $T(R) \subset R$, then there is at least a point $x_0 \in R$ such that $T x_0 = x_0$.

We give below a typical application of this theorem in the proof of the existence of at least one periodic solution of a Liénard differential equation with a periodic forcing term:

$$x'' + f(x, x')\, x' + g(x) = e(t), \tag{9.6.1}$$

or of the system

$$dx/dt = v, \qquad dv/dt = -f(x, v)\, v - g(x) + e(t). \tag{9.6.2}$$

Here $f(x, x')$, $g(x)$ denote continuous functions with their first order derivatives and $e(t)$ a continuous nonzero periodic function of period T. By $G(x)$ we denote the function $G(x) = \int_0^x g(u)\, du$.

(9.6. i) If there are positive constants a, m, M such that $f(x, v) > m$ if both $|x| \geq a$, $|v| \geq a$, if $f(x, v) \geq -M$ for all x, v, if $x g(x) > 0$ for all $|x| \geq a$, and $|g(x)| \to +\infty$, $g(x)/G(x) \to 0$ as $x \to \pm\infty$, then there exists at least one (non zero) periodic solution of (9.6.1).

Proof. By (9.4. i) there exists a closed curve C, enclosing the origin, such that every solution $S = [x(t), v(t)]$ of (9.6.2), crossing C at any time t, passes from the exterior into the interior of C. Let R be the 2-cell of the (x, v)-plane enclosed by C. For every point $p = (x_0, v_0)$ of R let us consider the solution $[x(t), v(t)]$ of (9.6.2) with $x(0) = x_0$, $v(0) = v_0$, and the point $p' = (x_1, v_1)$ of the (x, v)-plane defined by $x_1 = x(T)$, $v_1 = v(T)$. Then the transformation τ which maps $p \in R$ into p' is defined in R, is continuous in R, and obviously $\tau p \in R$ for every $p \in R$; i.e, $T(R) \subset R$. By Brouwer theorem we conclude that there exists at least one point $(\bar{x}, \bar{v})$ in R, such that, for the corresponding solution $x(t)$, $v(t)$ of (9.6.2) we have

$$x(T) = x(0) = \bar{x}, \qquad v(T) = v(0) = \bar{v}.$$

The solution $[x(t), v(t)]$ is obviously periodic of period T.

Remark. In the theorem (9.6. i) the periodic solution need not be unique. Moreover T need not be the minimum period either of $e(t)$ or of the periodic solutions.

For application of the fixed point theorem to the determination of the periodic solution of the differential equation $x'' + \sigma^2 x = f(x, x', t)$, f periodic, see K. O.

FRIEDRICHS [1]. For other applications see D. GRAFFI [5], A. DE CASTRO [6], G. COLOMBO [7], V. V. NEMYCKII [1—9]. See also L. L. RAUCH [1], A. ANDRONOV and N. BAUTIN [3], G. S. GORELIK [4], F. B. FULLER [2]. For the equation $x''+x|x'|+q(t)x'+x-p(t)x^3=f(t)$, f periodic, see F. STOPPELLI [1]. For a rigorous discussion of the subharmonic solutions of $a(x')x''+bx'+cx=e(t)$, $e(t)$ periodic, see G. COLOMBO [7].

For the study of the Liénard equation containing a forcing term we mention here the papers: G. SANSONE [10—15], C. E. LANGENHOP [1], C. E. LANGENHOP and A. B. FARNELL [1], A. B. FARNELL, C. E. LANGENHOP and N. LEVINSON [1]. See also for analogous results H. A. ANTOSIEWICZ [2, 3, 4] and P. BROCK [1]. Results of R. CACCIOPPOLI and A. GHIZZETTI [1] have been improved by N. LEVINSON [2] by a simpler method.

See D. GRAFFI [16] and G. SANSONE [15] for expository papers. See also A. ASCARI, I. BARBALAT [1], A. DE CASTRO [1—4]. Finally see on the same subject E. and H. CARTAN [1], A. W. GILLIES [1], A. HALANAY [3], R. IGLISH [1—4], A. M. KAC [1, 2], M. URABE [1], R. REISSIG [1], G. E. H. REUTER [1—3]. See also the books by A. ANDRONOV and CHAIKIN [1], J. J. STOKER [1], N. MINORSKI [6], and the recent papers by L. MINOZZI [1], G. MANARESI [1], A. ROSENBLATT [1—5].

9.7. The method of M. L. CARTWRIGHT. This method too, whose main purpose is to determine periodic solutions of differential equations and systems and their properties, is both analytic and topologic. The analytic part consists in theorems of boundedness of the solutions, more precisely, in theorems showing that the solutions will be ultimately in certain fixed regions of the phase-space as $t\to+\infty$. The topological part consists in the application of convenient refinements of BROUWER'S fixed-point theorem. We shall refer below to the main lines of the method in relation to the differential equation

$$x''+f(x)\,x'+g(x)=p(t), \tag{9.7.1}$$

or more precisely to the equation

$$x''+k\,f(x,k)\,x'+g(x,k)=p(t,k), \tag{9.7.2}$$

containing a parameter k. Here f, g, p are continuous functions of x or t, and g satisfies a Lipschitz condition in x so that theorems of uniqueness of solutions and of continuous dependence of the initial conditions hold. The function p is not necessarily periodic in t in the theorem of boundedness.

(9.7. i) (A theorem on boundedness.) Suppose $f(x,k)\geq -b$, for all x, and $f(x,k)\geq b_2>0$ for all $|x|\geq 1$, suppose $g(x,k)\operatorname{sgn} x\geq b_3>0$ for all $|x|\geq 1$, and $|g(x,k)|\leq\gamma(\xi)$, where γ is independent of k, for all $|x|\leq\xi$, and any ξ; suppose $|p(t,k)|<B$, $\left|\int_0^t p(t,k)\,dt\right|<B$ for all t, where b_1, b_2, b_3, B are constants independent of k. Then for every (x_0, y_0) there is a number $t_0=t_0(x_0,y_0)$ such that the solution $x(t)$ of (9.7.2) with $x(0)=x_0$, $x'(0)=y_0$ verifies the relations $|x(t)|<B$, $|x'(t)|<B(k+1)$ for all $t\geq t_0(x_0,y_0)$.

For the sake of brevity we must omit the interesting analytical proof.

(9.7. ii) (A fixed point theorem.) Let T be a $1-1$ continuous transformation of the plane E_2 into itself and let G be an open set and H a simple closed Jordan region with $G \subset H$. Suppose that for every point $P \in H$ there is an integer $n_0(P)$ such that $T^n(P) \in G$ for every $n \geq n_0(P)$. Then there is another simple closed Jordan region J with $G \subset H \subset J \subset E_2$ such that $T(J) \subset J$. Also, there is an integer N such that $T^n(P) \in G$ for every $n \geq N$ and $P \in H$.

Proof. The last statement follows by the Borel covering theorem. Let us prove first that $G\,T(G) \neq 0$. Indeed, if $G\,T(G) = 0$, then $T(P) \in E_2 - G$ for every $P \in G$, which gives a contradiction, namely $T^n(P) \in G$, $T^{n+1}(P) \in E_2 - G$ for $n = n_0(P)$. Now let us consider the set $J = H + T(H) + \cdots + T^N(H)$. Since T is a $1-1$ mapping, all $T^i(H)$, $i = 0, 1, \ldots, N$, are Jordan regions, and since each $H\,T^i(H)$ contains the non-empty set $G\,T(G)$, also the set $H\,T^i(H)$ is also nonempty. By elementary topology it follows that J is a simple Jordan region, and obviously $T(J) \subset J$.

(9.7. iii) Under the same conditions of (9.7. ii) the set $S = \prod_{n=1}^{\infty} T^n(J)$ is a continuum subset of G and $T(S) = S$, i.e., S is invariant under T.

Proof. We have here $T^n(J) \subset J$, $T^{n+1}(J) \subset T^n(J)$, and each set $T^n(J)$ is a closed Jordan region, and hence a continuum. Thus S is not empty, and finally a continuum by virtue of the same Zoretti theorem already mentioned in (9.2. i). Obviously $T(S) = S$.

(9.7. iv) Under the same conditions above there is at least one fixpoint P in S, and the boundary line J^* of J has index number $+1$.

Proof. By Brouwer's theorem (9.6) there is at least one fix point in J and every fixed point of J must lie in S. Now for every $P \in J^*$ the image $T(P)$ is inside J^*. By topologically transforming J into a circle c it is immediately seen that the vector from P to $T(P)$ describes the positive angle 2π when P describes c^*.

(9.7. v) (An existence theorem for periodic solutions.) Under the same conditions of (9.7. i), if $p(t, k)$ is periodic in t of period L, then there is a periodic solution of (9.7.2) of period L.

Proof. Equation (9.7.2) is equivalent to the system

$$dx/dt = v, \quad dv/dt = p(t, k) - k\,f(x, k)\,x' - g(x, k).$$

For every (x_0, v_0) let $x(t)$, $v(t)$ denote the solution with $x(0) = x_0$, $v(0) = v_0$, and let T denote the $1-1$ mapping of the (x, v)-plane into itself which maps (x_0, v_0) into $[x(L), v(L)]$. If $G = [|x| < B, \ |v| < B(k+1)]$ and H is any large circle with $G \subset H$, then by (9.7. i) it follows that for any $P \in H$ there is an integer $n_0 = n_0(P)$ with $T^n(P) \in G$ for all $n \geq n_0$. By (9.7. v) there is at least one fixed point (x, v) in G and the solution $x(t)$, $v(t)$ starting at $(\bar{x}, \bar{v})$ at $t = 0$ is obviously periodic of period L.

For further details on the method and a series of applications to the determination of harmonic and subharmonic solutions of differential equations see M.L. Cartwright [1—8], and M.L. Cartwright and J.E. Littlewood [1—5].

For other applications of the Cartwright method see G. Colombo [4], M. H. A. Newman [1], R. A. Smith [1], D. Graffi [5], A. W. Gillies [1]. For boundedness theorems see also G. E. H. Reuter [4].

9.8. The method of T. Ważewski. The method is essentially topological and it is most recent, since it started with a paper of Ważewski [2] of 1947. A number of problems have been framed into the method, by Ważewski and others (see Note at the end of this paragraph). Though the method is still being modified, some main lines of it seem well established and are given below.

The method concerns real differential systems

$$x'(t) = f(t, x), \qquad x = (x_1, \ldots, x_n), \qquad f = (f_1, \ldots, f_n), \tag{9.8.1}$$

where the vector function f is continuous in an open set W of the $(n+1)$-dimensional $(t, x_1, \ldots, x_n)$-Euclidean space E_{n+1}, and the assumption is moreover made that a uniqueness theorem holds everywhere in W and, hence, also a theorem of continuous dependence upon initial values.

Here by x is denoted a point of the Euclidean n-dimensional $(x_1, \ldots, x_n)$-space E_n (a subspace of E_{n+1}), and by t a point of the one-dimensional t-space R (or E_1, also a subspace of E_{n+1}). We shall denote by t_P, x_P the orthogonal projections of $P=(t, x)$ on R and E_n, and hence $P=(t_P, x_P)$.

Now for every point $P \in W$ consider the corresponding solution of (9.8.1) starting at P, and we shall denote it as usual by $x(t, P)$. Thus $x(t_P, P) = x_P$ and we shall always consider $x(t, P)$ in the maximal (open) interval $D(P)$ in which $x(t, P)$ exists, hence $t_P \in D(P)$.

For each $t \in D(P)$, the pair $[t, x(t, P)]$ is then a point of W and we shall denote it by $Y(t, P)$. Thus $Y(t, P) \equiv [t, x(t, P)]$. Also we shall denote by $L^+(P)$ the set of all points $Q = Y(t, P)$ with $t_P \leqq t$, $t \in D(P)$. Thus $L^+(P)$ is a set of points of W and will be denoted as the semi-integral curve starting at P. Analogously we can define $L^-(P)$ and $L(P)$.

We shall now denote any open subset T of W as a *quill* (fr. tuyau), by $F_r(T, W)$ the boundary of T in W, i.e., $F_r(T, W) = T^* W$, and by $\overline{T}$ the set $\overline{T} = T + F_r(T, W)$, $\overline{T} \subset W$. A semi-integral curve $L^+(P)$, or solution $x(t, P)$, $t_P \leqq t$, $t \in D(P)$, is said to be *asymptotic in* T (or with respect to T), if $L^+(P) \subset T$.

If $L(P)$ is not asymptotic in T then $L^+(P)$ has a first point, say $Q = C(P)$, which is in $F_r(T, W)$, while the arc of $L^+(P)$ with $t_P \leqq t < t_Q$ is in T. The point $Q = C(P) \in F_r(T, W)$ is said to be the *consequent* of P (Poincaré's notation), and a *point of egress* on $F_r(T, W)$. The set S of all points Q of $F_r(T, W)$ which are points of egress, i.e., consequents $C(P)$ of some $P \in T$, is said to be the *set of egress* of T. The subset G of T of all points for which $C(P)$ exists (i.e., for which $L^+(P)$ is not asymptotic in T) is said to be the *left shadow* of S. Thus $G \subset T \subset W$,

$S \subset F_r(T, W)$. In the illustration (where it is assumed $n = 1$) T is the hatched set, G is the part of T which is hatched by continuous lines, all lines (continuous and broken) being integral curves. $F_r(T, W)$ is the arc abc, S the subarc ab. A point Q of egress is said to be of *strict egress* if $Y(t, Q) \in W - \bar{T}$ for every $t_Q < t < t_Q + \varepsilon$, for some $\varepsilon > 0$. The set of all such points is said to be the set S^* of *strict egress*. Therefore $S^* \subset S \subset F_r(T, W)$. In the illustration all interior points of the arc ab are of strict egress, the point b is a point of egress but not of strict egress.

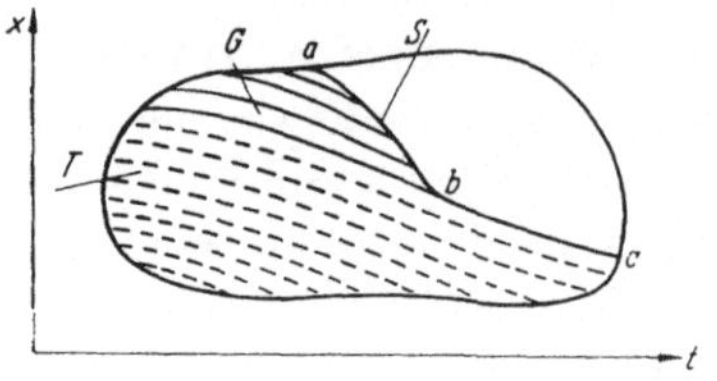

A first problem is to give general conditions [topological in character, and whose verification does not imply any formal solution of system (9.8.1)], assuring that at least one solution $L^+(P)$ is asymptotic in T, that is, lies in T for all $t_P \leq t$, $t \in D(P)$. Such a requirement can be expressed also by saying that $T - G \neq 0$. Another problem consists in establishing that there exist infinitely many solutions $L^+(P)$ which are asymptotic in T, and the number k of parameters on which they depend. If t_0 is some number $t_0 > t_P$, $t_0 \in D(P)$, we may consider the intersection of $T - G$ with the hyperplane $t = t_0$. Then k is the dimension of such an intersection.

If Z is any subset of $\bar{T} = T + F_r(T, W)$, then conditions may be sought in order that $Z(T - G) \neq 0$, i.e., in order that for at least one $P \in Z$ we have $L^+(P) \subset T$.

If $A \subset B$ are any two sets of a given space and $Q = U(P)$ is any continuous mapping from B into A, such that (i) $U(P) \in A$ for every $P \in B$; (ii) $U(P) = P$ for every $P \in A$, then $U(P)$ is said to be a *retraction* from B into A, and A a retract of B (in the sense of Borsuk).

We need first the following lemma:

(9.8. i) If $S = S^*$ and the mapping $Q = K(P)$ is defined by $K(P) = C(P)$ for $P \in G$, $K(P) = P$ if $P \in S$, then $K(P)$ is a retraction from $G + S$ into S.

Proof. By definition $K(P) \in S$ if $P \in G$, $K(P) = P$ if $P \in S$. We have only to prove that $K(P)$ is a continuous mapping. First let P_0 be a point of G, $[P_n]$ any sequence of points of G with $P_n \to P_0$, and let us prove that $C(P_n) \to C(P_0)$.

Let $t_0 = t_{P_0}$, $\tau = t_{Q_0}$ with $C(P_0) = Q_0$. Since $Q_0 \in W$, there is an arc $P''P'''$ of $L^+(P_0)$ containing Q_0 as an interior point, say the arc from $t = \tau - \delta$, to $t = \tau + \delta$, and we may take δ so small that the whole arc $P''P'''$ has diameter $< \varepsilon$. We may take δ so small that the arc $P'P_0$ of $L^-(P_0)$ from $t = t_0 - \delta$ to $t = t_0$ is also in G. Since Q_0 is a point of strict egress we can take δ so small that the point P''' is in the open set $W - \bar{T} = W - T - F_r(T, W)$. On the other hand the whole arc $P'P_0P''$ of $L(P_0)$ is in G. For n sufficiently large certainly we have $t_0 - \delta < t_{P_n} < \tau - \delta$, and the theorem of continuity implies that the whole arc $Q'P_nQ''$ of $L(P_n)$ from $t = t_0 - \delta$ to $t = \tau - \delta$ is in the open set G. On the other hand, for all n sufficiently large the point Q''' of $L^+(P_n)$ with $t=\tau+\delta$ is still in the open set $W-\overline{T}$. In addition the whole arc $Q''Q'''$ of $L^+(P_n)$ from $t=\tau-\delta$ to $t=\tau+\delta$, is a ε-neighborhood of the arc $P''P'''$ of $L^+(P_0)$. Thus $Q''\in G\subset T$, $Q'''\in W-\overline{T}$ and there is a point of $F_r(T, W)$ between Q'' and Q'''. The first of these points is the point $C(P_n)$ and this point is at a distance $<\varepsilon$ from some point of $P''P'''$, while this arc has diameter $<\varepsilon$. Therefore $C(P_n)$ is at a distance $<2\varepsilon$ from $C(P_0)$, and this is true for all n large enough. We have proved that $P_n\to P_0$, $P_0\in G$ implies $C(P_n)\to C(P_0)$.

Suppose now that $P_0\in S=S^*$, $P_n\in G+S$, $P_n\to P_0$, and let us prove that $K(P_n)\to K(P_0)=P_0$. As before we define an arc $P''P'''$ containing P_0 as an interior point, of diameter $<\varepsilon$, made up of all points of $L(P_0)$ from $t=t_0-\delta$, $t=t_0+\delta$. We may suppose $P''\in G$, $P'''\in W-\overline{T}$. Then for all n sufficiently large we have $t_0-\delta<t_{P_n}<t_0+\delta$, and the arc $Q''Q'''$ of $L(P_n)$ from $t=t_0-\delta$ to $t=t_0+\delta$ is in the ε-neighborhood of $P''P'''$. In addition $Q''\in G$ and $Q'''\in W-\overline{T}$. The reasoning is now the same as before, and the lemma is proved.

We ca: now prove the following theorem which is one of the main theorems of WAZEWSKI's theory.

(9.8. ii) If $S=S^*$ and $Z\subset T+S$, if ZS is a retract of S but is no retract of Z, then $Z(T-G)\neq 0$; i.e., there is at least one point $P\in Z$ with $L^+(P)\subset T$.

Proof. By (9.8. i) the mapping $K(P)=C(P)$ if $P\in G$, $=P$ if $P\in S$ is a retraction from $G+S$ into S. Since ZS is a retract of S there is a retraction $Q=V(P)$ from S into ZS. Then $Q=VK(P)$ is a retraction from $G+S$ into ZS. Now suppose, if possible, $Z(T-G)=0$, where $Z\subset T+S$. Then $Z=ZT+ZS=Z(T-G)+ZG+ZS=ZG+ZS$. Thus Z is a subset of $G+S$ and hence $V[K(P)]$ is a retraction of Z into ZS, a contradiction. This implies that $Z(T-G)\neq 0$.

Example. Let us consider the real system of two equations

$$x'=f(t, x, y), \qquad y'=g(t, x, y), \tag{9.8.2}$$

where f, g are continuous functions of t, x, y for all x, y and for $t\geq 0$. Assume a uniqueness theorem holds, and take for W the whole semi-space (t, x, y) with $t>0$. Let $u(t)$, $v(t)$ be any two functions of t, continuous with their first order derivatives, and positive for $t>0$, and let the set T (quill) be defined by $|x|<u(t)$, $|y|<v(t)$, $0<t<+\infty$. Assume that all points (t, x, y) of the set A: $|x|=u(t)$, $|y|<v(t)$, $0<t<+\infty$, are points of strict egress, and that the points of the set B: $|x|\leq u(t)$, $|y|=v(t)$, $0<t<+\infty$, are not points of egress. This is certainly the case if $xf(t, x, y)>uu'$ on A, and $yg(t, x, y)<vv'$ on B. Let Z be the segment $t=t_0$, $|x|\leq u(t_0)$, $y=b$, and suppose $|b|<v(t_0)$. Under these assumptions the conditions of (9.8. ii) are all verified and thus there exists at least one point $P_b\in Z$ such that the solution $[x(t), y(t)]$ of (9.8.2) starting at P_b remains in T for all $t\geq t_0$. By varying b we conclude that the set of such solutions is uncountable. By a more detailed analysis T. WAŻEWSKI has proved that such a set has at least dimension one in the sense of URYSOHN and MENGER.

This example is only meant to give an idea of the interplay of the conditions in (9.8. ii). In reality if a system is given, the choice of a quill T for which (9.8. ii) can be applied successfully may be very difficult. An interesting situation is presented by "perturbed systems". For instance a quill T may be found for a particularly simple system (9.8.2), and then the question occurs whether the same quill T can be used for the perturbed system

$$x'=f(t, x, y)+p(t, x, y), \qquad y'=g(t, x, y)+q(t, x, y)$$

where the functions p, q are sufficiently small. In other words the question may be discussed whether the relations $x(f+p) > u u'$ in A, $y(g+q) < v v'$ in B, are satisfied.

By using Ważewski method, Z. SZMYDTOWNA [1] proved the following theorem which generalizes theorems of PERRON concerning linear systems.

(9.8. iii) Consider the linear differential system

$$dx_i/dt = f_i(t)\, x_i + \sum_{j=1}^{n} g_{ij}(t)\, x_j, \quad i = 1, \ldots, n, \tag{9.8.3}$$

where the coefficients $f_j, g_{ij}, t_0 \leq t < +\infty$, are continuous functions of t. Suppose $n = p + q + \tau$, $0 < q < n$, and

$$R(f_1) \geq \cdots \geq R(f_p) > R(f_{p+1}) \geq \cdots \geq R(f_{p+q}) > R(f_{p+q+1}) \geq \cdots \geq R(f_{p+q+t})$$

$$\int^{+\infty} R\,[f_p(t) - f_{p+1}(t)]\, dt = \infty, \quad \int^{+\infty} R\,[f_{p+q}(t) - f_{p+q+1}(t)]\, dt = \infty,$$

$$g_{ij}[R(f_p - f_{p1})]^{-1} \to 0, \quad g_{ij}[R(f_{p+q} - f_{p+q+1})]^{-1} \to 0 \quad \text{as} \quad t \to +\infty.$$

Then there is a system of $q + r$ linearly independent solutions of (9.8.3) with

$$[|x_1|^2 + \cdots + |x_p|^2 + |x_{p+q+1}|^2 + \cdots + |x_{p+q+2}|^2]\, [|x_{p+1}|^2 + \cdots + |x_{p+q}|^2]^{-1} \to 0$$

as $t \to +\infty$.

The proof of this theorem and also a further extension to nonlinear systems are omitted for the sake of brevity. In application of the Wazewski method and of a theorem of MASSERA [3], recently A. HALANAY [5] has proved the following theorem for nonlinear systems:

(9.8. iv) Consider the real system $dx/dt = Q(x, y) + \varepsilon X(x, y, t)$, $dy/dt = P(x, y) + \varepsilon Y(x, y, t)$, where P, Q are homogeneous polynomials of the same degree and X, Y are analytic functions of x, y, t, periodic in t. Denote by $f(k)$ the function $f(k) = P(1, k)/Q(1, k)$ and suppose that the equation $f(k) = k$ has at least two roots with $f'(k) < 0$. Then for all ε sufficiently small, the system above has periodic solutions.

The requirement concerning the uniqueness of the solution through each point of W is not essential. A. BIELECKI [3] has shown that the concepts above can be modified so as to hold without such a requirement, and then the main theorems can be proved by a modified argument.

For other research on and application of the Wazewski topological method see Z. MIKOLAJSKA [1—4], F. ALBRECHT [1—3], S. LOJASIEWICZ [1, 2], A. PLIS [1, 2], T. WAŻEWSKI [1—15], K. TATARKIEWICZ [1], J. SZARSKI [1—5], Z. SZMYDT [1, 2], Z. SZMYDTOWNA [1], A. HALANAY [4, 5], A. BIELECKI [1—6], I. BARBALAT [1—3].

Chapter IV

Asymptotic developments

§ 10. Asymptotic developments in general

10.1. POINCARÉ'S concept of asymptotic development. Given a function $f(t)$, $t \geq t_0$, of the real variable t, the most immediate information concerning its behavior as $t \to +\infty$ is certainly offered by the limit (if it exists)

$$\lim_{t \to +\infty} f(t) = a_0.$$

Once this limit is known, then $f(t) = a_0 + o(1)$ as $t \to +\infty$, and more precise information on the behavior of $f(t)$ as $t \to +\infty$ is offered by the limit (if it exists)

$$\lim_{t\to+\infty} t[f(t) - a_0] = a_1.$$

Then $f(t) = a_0 + a_1 t^{-1} + o(t^{-1})$ as $t \to +\infty$. It may occur that all the following limits exist

$$\lim_{t\to+\infty} t^n [f(t) - a_0 - a_1 t^{-1} - \cdots - a_{n-1} t^{-n+1}] = a_n, n = 0, 1, 2, \ldots . \quad (10.1.1)$$

Then we have

$$f(t) = a_0 + a_1 t^{-1} + \cdots a_n t^{-n} + R_n, \qquad n = 0, 1, 2, \ldots, \quad (10.1.2)$$

where

$$R_n = o(t^{-n}), \quad \text{as well as} \quad R_n = o(t^{-n-1})$$

as $t \to +\infty$. It is said that the series

$$f(t) \sim \sum_{n=0}^{\infty} a_n t^{-n} \quad (10.1.3)$$

is an *asymptotic expansion* of $f(t)$ as $t \to +\infty$.

The considerations above hold as well if $f(t)$ is a real, or a complex function of the complex variable $t = u + iv$, defined, for instance for all complex numbers t with $u \geqq a$, and where $t \to \infty$ remaining in a sector, say $-\beta < \arg t < \beta$, for some $0 < \beta < \pi/2$, or in a strip $[-\beta < v < \beta, u \geqq a]$. More generally we may suppose that a subset S of the half-plane $u \geqq a$, is given, where S is connected and contains $Z = \infty$, and in all the considerations above it is prescribed that $t \to \infty$ remaining in S and the limit has to be taken uniformly in S [for instance as $u \to +\infty$ uniformly with respect to v for (u, v) in S].

In the theory of functions of a complex variable it is often proved, by successive integration by parts, that the asymptotic expansion holds

$$e^t \int_t^{\infty} s^{-1} e^{-s} ds \sim \sum_{n=0}^{\infty} (-1)^{n+1} n! \, t^{-n-1}.$$

By means of the Laplace transform asymptotic expansions are also often obtained.

It is remarkable that series (10.1.3) need not be convergent for any t, as the series above well shows. On the other hand asymptotic expansions have very simple formal properties, just as under conditions of convergence:

(10.1.i) $f \sim \Sigma a_n t^{-n}$, $g \sim \Sigma b_n t^{-n}$ implies

$f + g \sim \Sigma (a_n + b_n) t^{-n}$,

$fg \sim \Sigma c_n t^{-n}$, where $c_n = a_0 b_n, \ldots, a_n b_0$,

$f' \sim \Sigma d_n t^{-n}$, where $d_0 = d_1 = 0$, $d_n = -(n-1) a_{n-1}$, $n = 2, 3, \ldots$,

$\int_t^{\infty} f \, dt \sim \Sigma e_n t^{-n}$, where $e_0 = 0$, $e_n = n^{-1} a_{n+1}$, $n = 1, 2, \ldots$, the last one provided $a_0 = a_1 = 0$, and where always Σ ranges for $n = 0, 1, 2, \ldots$.

10.2. Ordinary, regular and irregular singular points. It is convenient to summarize first a few points of the theory of linear differential equations for functions of a complex variable z.

(a) The point $z = 0$ is said to be an *ordinary point* for a n-th order linear homogeneous differential equation if the equation can be written in the form

$$y^{(n)} + a_1(z)\, y^{(n-1)} + \cdots + a_n(z)\, y = 0, \tag{10.2.1}$$

where the coefficients $a_i(z)$, $i = 1, \ldots, n$, are single-valued analytic functions in a neighborhood of $z = 0$, regular at $z = 0$. Then there are power series

$$\sum_{m=0}^{\infty} c_m z^m, \tag{10.2.2}$$

which formally satisfy equations (10.2.1), namely all power series which can be obtained by actual substitution in (10.2.1) where the coefficients are replaced by their Taylor expansions, and by solving the ensuing recurrent equations. This leaves exactly the first n coefficients $c_0, \ldots, c_{n-1}$ undetermined. It is then proved that all series so obtained (formal series) actually are convergent in a neighborhood Ω of $z = 0$ [namely the maximum circular neighborhood where all the coefficients $a_i(z)$ are analytic and regular] and define there an n-dimensional manifold of solutions of (10.2.1) all regular in Ω. Thus, in particular, there are n independent solutions, say $y_1, y_2, \ldots, y_n$, of (10.2.1) of this type.

Analogously the point $z = 0$ is said to be an ordinary point for a system of n first order linear homogeneous differential equations, if the system can be written in the form

$$x_i' = \sum_{j=1}^{n} a_{ij}(z)\, x_j, \quad \text{or} \quad x' = A(z)\, x, \tag{10.2.3}$$

where the coefficients $a_{ij}(z)$ are single-valued analytic in a neighborhood of $z = 0$ and regular at $z = 0$. The same statements as above hold for the solutions. Thus, in particular, we can determine a fundamental matrix solution $X(z)$ whose components are all convergent formal series (10.2.2) and therefore analytic functions regular at $z = 0$.

Obviously equation (10.2.1) is reduced to a system (10.2.3) by the usual substitution $x_1 = y$, $x_2 = y'$, $\ldots$, $x_n = y^{(n-1)}$.

(b) The point $z = 0$ is said to be a *regular singular point* for an n-th order linear homogeneous differential equation if the latter can be written in the form (10.2.1) where each coefficient $a_i(z)$ is single-valued analytic in a neighborhood of $z = 0$, and is regular there or has a pole of order at most i at $z = 0$ [and at least one $a_i(z)$ has a pole at $z = 0$]; in other words if the equation can be written in the form

$$z^n y^{(n)} + z^{n-1} b_1(z)\, y^{(n-1)} + \cdots + b_n(z)\, y = 0, \tag{10.2.4}$$

where the coefficients $b_i(z)$ are regular at $z = 0$. Then the solutions are series of the form

$$\sum_{m=0}^{\infty} c_m z^{m+\lambda}, \tag{10.2.5}$$

where λ is a convenient real or complex number, or series of the form

$$\sum_{m=0}^{\infty} \left(c_{m0} + c_{m1} \log z + \cdots + c_{mk} (\log z)^k\right) z^{m+\lambda}. \tag{10.2.6}$$

All these series formally satisfy equation (10.2.4), where the coefficients $b_i(z)$ are replaced by their Taylor expansions. Actually all these series are convergent in a neighborhood of $z=0$ (no matter which determination is used for $\log z$ and for z^λ if λ is not an integer), and define analytic functions of $z \neq 0$ (not necessarily single-valued) in the maximal circular neighborhood Ω of $z=0$ where all the coefficients $b_i(z)$ are regular. As above, a fundamental system $y_1, \ldots, y_n$ can be chosen of such solutions.

Analogously the point $z=0$ is said to be a regular singular point for a system as in (a) if the latter can be written in the form

$$x_i' = z^{-1} \sum_{j=1}^{n} a_{ij}(z)\, x_j, \quad \text{or} \quad x' = z^{-1} A(z)\, x, \tag{10.2.7}$$

where the coefficients $a_{ij}(z)$ are single-valued analytic in a neighborhood of $z=0$, regular at $z=0$, and not all zero at $z=0$, i.e., $A(0) \neq 0$. The same statements as above hold for the solutions, and a fundamental matrix $X(z)$ can be chosen whose components are functions of the types (10.2.5) and (10.2.6).

It may well occur that the solutions are all of the type (10.2.2), or (10.2.5) with λ a nonnegative integer. An example is given by

$$x_1' = z^{-1} x_1, \quad x_2' = z^{-1} x_2,$$

whose solutions are all of the forms $x_1 = az$, $x_2 = bz$, a, b constants. It is then said that $z=0$ is an *apparent singularity*.

In connection with all this the following statement holds: If $B(z)$ is a $n \times n$ matrix whose elements are single-valued analytic in a neighborhood of $z=0$, but at least one is singular there, and $\Phi(z)$ is a fundamental matrix solution of $x' = B(z)\, x$, and the components of Φ are all single-valued analytic and regular at $z=0$, then $\det \Phi(0) = 0$. Indeed $\Phi' = B\Phi$ and, if $\det \Phi(0) \neq 0$, then the matrix $\Phi^{-1}(z)$ has components which are single valued analytic regular at $z=0$. Thus $\Phi' \Phi^{-1}$ is regular at $z=0$, while $\Phi' \Phi^{-1} = B$, which is a contradiction.

Obviously equation (10.2.4) can be reduced to system (10.2.7) by means of the substitution $x_1 = y$, $x_2 = z y'$, $\ldots$, $x_n = z^{n-1} y^{(n-1)}$.

(c) In all other cases the point $z=0$ is said to be an *irregular singular point*. We shall say that $z=0$ is an *irregular singular point of finite type* provided the differential equation can be written in the form (10.2.1) where the coefficients $a_i(z)$ are all single-valued analytic in a neighborhood of $z=0$, are regular or have a pole at $z=0$, and at least one coefficient $a_i(z)$ has at $z=0$ a pole of order $>i$. Analogously the point $z=0$ is said to be an irregular singular point of finite type for a system if the latter can be written in the form

$$x_i' = z^{-1-\mu} \sum_{j=1}^{n} a_{ij}(z)\, x_j, \quad \text{or} \quad x' = z^{-1-\mu} A(z)\, x, \tag{10.2.8}$$

where the coefficients $a_{ij}(z)$ are all single-valued, analytic in a neighborhood of $z=0$, regular at $z=0$, and not all zero at $z=0$, i.e. $A(0) \neq 0$, and $\mu \geq 1$ is an integer.

It may occur here that a series satisfies formally the given differential equation, or system, but is not convergent in any neighborhood of $z=0$. An example is given by

$$z^2 y'' + (3z - 1)\, y' + y = 0 \tag{10.2.9}$$

for which a formal solution (formal series) is

$$\sum_{k=0}^{\infty} k!\, z^k,$$

which is not convergent for all $z \neq 0$. The irregular case (even of finite type) is essentially more difficult than the ordinary and the regular singular cases, and the example above emphasizes one of the many new questions, namely the fact that formal series solutions may not be convergent in any neighborhood of the singular point.

(d) A point $t=\bar{t}$, or the point $t=\infty$, can be reduced to the point $z=0$ by the substitution $z=t-\bar{t}$, or $z=1/t$. Thus a point $t=\bar{t}$, or $t=\infty$, is said to be an ordinary point, or a regular singular point, or an irregular singular point of finite type for a differential equation, or system, provided the point $z=0$ has the same property for the transformed equation or system. For the point $t=\infty$, these three cases correspond to systems written respectively in the forms:

$$x'=t^{-2}A(t)\,x, \qquad t=\infty \quad \text{an ordinary point,}$$

$$x'=t^{-1}A(t)\,x, \qquad t=\infty \quad \text{a regular singular point,}$$

$$x'=t^{\nu}A(t)\,x, \qquad t=\infty \quad \text{an irregular singular point of finite type,}$$

where the elements $a_{ij}(t)$ of the matrix A are all single-valued analytic functions in a neighborhood of $t=\infty$, regular at $t=\infty$ with $A(\infty)\neq 0$, and $\nu\geq 0$ an integer. Indeed the substitution $z=1/t$ reduces them to the systems (10.2.3), (10.2.7), (10.2.8). For instance equation (10.2.9) after the substitution $z=1/t$ and $x_1=y$, $x_2=t\,y'$, becomes the system

$$x_1'=t^{-1}x_2, \quad x_2'=-t^{-1}x_1+(-1+2t^{-1})\,x_2, \quad \text{or} \quad x'=A(t)\,x, \qquad (10.2.10)$$

with $x=(x_1, x_2)$, $v=0$, $A(\infty)\neq 0$, of which a formal solution is

$$\sum_{k=0}^{\infty} k!\,t^{-k}, \qquad -\sum_{k=1}^{\infty} k\,(k!)\,t^{-k}.$$

These series do not converge in any neighborhood of $t=\infty$, but they are asymptotic expansions of an actual solution in a neighborhood of $t=\infty$.

Divergent formal series had been known for centuries and even used (and are still in use) for extremely refined numerical computations (e.g., in astronomy), before their true nature was recognized. It was Poincaré who introduced the concept of asymptotic expansion as in (10.1) and demonstrated in various cases that formal solutions were asymptotic expansions of actual solutions.

10.3. Asymptotic expansions for an irregular singular point of finite type. We may suppose $t=\infty$ to be the irregular singular point in question and the system to be written in the form

$$x'=t^{\nu}A(t)\,x, \qquad (10.3.1)$$

where $\nu\geq 0$ is an integer, $A(\infty)\neq 0$, and the elements $a_{ij}(t)$ of $A(t)$ are all single-valued analytic functions in a neighborhood of $t=\infty$, regular at $t=\infty$, and hence series of the type

$$\sum_{m=0}^{\infty} b_m\,t^{-m}$$

all convergent in a neighborhood of $t=\infty$.

We shall state the existence of a fundamental matrix solution $X(t)$ whose elements have for asymptotic expansions formal series of the type

$$\sum_{i=0}^{M} e^{\sigma_i(t)} \sum_{m=0}^{\infty} \left[c_{im0}+c_{im1}\log t+\cdots+c_{imN_i}(\log t)^{N_i}\right] t^{-m-\lambda_i} \qquad (10.3.2)$$

where the $\sigma_i(t)$ are polynomials in t, and the λ_i are complex numbers; in particular of the types

$$\sum_{m=0}^{\infty} c_m t^{-m}, \quad \sum_{m=0}^{\infty} c_m t^{-m+\lambda}, \quad \sum_{m=0}^{\infty} c_m t^{-m+\lambda} e^{\sigma t}.$$

A very particular case is the equation $y' = y$ of which a solution is $y = e^t$; another particular case is the system (10.2.10) of which a formal solution was given above. An important example is given by the Bessel equation

$$x'' + t^{-1} x' + (1 - \varrho^2 t^{-2})\, x = 0,$$

whose solution $H_\nu^{(1)}(t)$ for $|t|$ large, $|\arg t| < \pi$ has the formal solution and asymptotic expansion

$$H_\nu^{(1)}(t) \sim (2\pi^{-1} t^{-1})^{\frac{1}{2}} \exp\{i\,|t - 2^{-1}\nu\pi - 4^{-1}\pi)\}\left[1 + \sum_{m=1}^{\infty} (-1)^m (\varrho, m)\,(2it)^{-m}\right]$$

where $(\varrho, m) = 2^{-2m}(m!)^{-1}(4\varrho^2 - 1) \ldots (4\varrho^2 - (2m-1)^2)$. An analogous formula holds for the companion solution $H_\nu^{(2)}(t)$. These series are not convergent, except for $\varrho = m + \frac{1}{2}$ when they reduce to finite sums. A last example is given by $y' = t^r y$ of which one solution is $\exp\left(t^{r+1}/r + 1\right)$, and which shows that polynomials $\sigma_i(t)$ of degree > 1 may actually occur in (10.3.2).

(10.3. i) If $A(\infty)$ has distinct characteristic roots λ_i, $i = 1, \ldots, n$, then a fundamental solution $X(t)$ of (10.3.1) exists whose elements are single valued regular analytic functions in a sector $|\arg t| < a$, $|t| \geq B$, $(a > 0,\ B > 0)$, having asymptotic expansions there of the type (10.3.2) with $N = 0$, and where the polynomials $\sigma_i(t)$ are of the form

$$\sigma_i(t) = \lambda_i (\nu + 1)^{-1} t^{\nu+1} + \lambda_{i1} \nu^{-1} t^\nu + \cdots + \lambda_{i\nu} t,$$

in which $\lambda_{i1}, \ldots, \lambda_{i\nu}$ are complex numbers.

The proof, together with a more complicated statement for the case of multiple roots is omitted.

For asymtotic expansions of the solutions of a linear system of differential equations whose coefficients have known asymptotic expansions as $t \to +\infty$, see S. Bochner [1], M. Hukuhara [1—8], H. Kneser [2], T. Carleman [1]. See also A. Kneser [3], E. A. Coddington and A. Wintner [1], P. Hartman and A. Wintner [12], A. Wintner [9].

10.4. Asymptotic developments deduced from Taylor expansions. Often a solution of a differential equation is of the form

$$\sum_{j=1}^{k} z^{\alpha_j} (\log z)^{\varrho_j} F_j(z),$$

where the functions $F_j(z)$ are single-valued analytic in the whole z-plane but for finitely many singular points $(\neq 0)$, and are given by their Taylor expansions around $z = 0$,

$$F_j(z) = \sum_{n=0}^{\infty} c_n z^n,$$

convergent in a circle $|z| < R$ around the origin. The question arises to determine asymptotic expansions of the functions $F(z)$ at $z = \infty$. The case when the coefficients $c_n = g(n)$, $n = 0, 1, \ldots$, are the values taken by a known function $g(w)$ analytic in an open connected set G of the w-plane containing all the nonnegative integers has been discussed in particular. When asymptotic properties of the functions $g(w)$ in G are known, the asymptotic properties of the functions $F(z)$ may often be inferred. As an illustration consider the following result, essentially due to W. B. FORD [4]:

(10.4. i) For $u \geq -h$, $h \geq 0$, let $g(w) = g(u + iv)$ be a single-valued analytic function with the exception of at most finitely many poles. Suppose that for $u \geq -h$, and $|w|$ sufficiently large

$$|g(u + iv)| \leq k \exp\left(l\pi|u| + m\pi|v|\right)(1 + v^2)^{-p}, \tag{10.4.1}$$

where k, l, m, p are non-negative constants, $m < 1$, $p > \frac{1}{2}$. Consider the function defined by

$$F(z) = \sum_{n=0}^{\infty}{}' g(n) z^n, \quad |z| < e^{-l\pi},$$

where the prime indicates that we omit those terms for which n is a pole of $g(w)$ Then the function $F(z)$ is single-valued and analytic in the entire sector $|\arg(-z)| \leq 1 - m)\pi$. Moreover, in this sector

$$F(z) = -\sum_{0<n\leq h}{}' g(-n) z^{-n} - \sum_s r(s) + O(z^{-h}).$$

where $\sum_s$ is extended over the poles s of $g(w)$ with $R(s) \geq -h$, and $r(s)$ is the residue at s of the function

$$\pi g(w)(-z)^w / \sin \pi w.$$

The following theorem, due to J. H. B. KEMPERMAN [1], yields an asymptotic expression of $\Sigma g(\alpha n + \beta) z^n$ in terms of $\Sigma g(n) z^n$.

(10.4. ii) For $u \geq -h$, let $g(w) = g(u + iv)$ be a single-valued analytic function with no poles for which (10.4.1) holds with $m \leq 1$, $p > \frac{1}{2}$. Let $\alpha = |\alpha| e^{i\theta}$ and β be complex constants with $|\theta| < \pi/2$, and consider the analytic functions $F(z) = \Sigma g(n) z^n$, $G(z) = \Sigma g(\alpha n + \beta) z^n$, where the sums range over all integers n for which $n \geq -h$ or $R(\alpha n + \beta) \geq -h$ respectively. Let g_1 be any sector in the ζ-plane, $\theta_1 < \arg \zeta < \theta_2$, which does not contain any half ray $\arg \zeta = \pm m\pi + k(2\cos\theta)/|\alpha|$, $k = \pm 1, \pm 2, \ldots$, and let g be the closure of g_1. Finally consider the function

$$H(\zeta) = G(\zeta^\alpha) - \alpha^{-1} \Sigma_\mu Z_\mu^{-\beta} F(Z_\mu), \quad \zeta \in g, \tag{10.4.2}$$

with $Z_\mu = \zeta \exp(2\pi\alpha^{-1}\mu i)$, where Σ ranges over all integers μ with $|\arg Z_\mu| < m\pi$ where $\zeta \in g_1$, and observe that $H(\zeta)$ is analytic for $|\zeta|$ sufficiently small and ζ in g. Then $H(\zeta)$ is single valued and analytic in the entire region g_1, continuous in g, and $H(\zeta) = O(\zeta^{-h-\beta})$ for large ζ. Thus (10.4.2) is an asymptotic expression for G with ζ in g.

The following theorem is a condensed and somewhat generalized version, due to J. H. B. KEMPERMAN [1] of a number of results due to E. M. WRIGHT [9].

(10.4. iii) Let h be a real constant and let $c_0, \ldots, c_L, \beta, \sigma$ and $\alpha = |\alpha| e^{i\theta}$ be complex constants with $|\theta| < \pi/2$. For $R(\alpha w + \beta) \geq -h$, let $g(w)$ be a single-valued and analytic function, with the exception of at most a finite number of poles, such that for $|w|$ sufficiently large

$$g(w) = \sigma^w \left[\sum_{j=0}^{L} c_j \Gamma^{-1}(\alpha w + \beta + j) + O\left(\Gamma^{-1}(\alpha w + \beta + L + 1)\right)\right], \tag{10.4.3}$$

where $L > h + \frac{1}{2}$. Consider the function

$$F(z) = \Sigma' g(n) z^n$$

where we sum over those integers n for which $R(\alpha n + \beta) \geq -h$ and for which n is not a pole of $g(w)$. Further, let

$$S(z) = \Sigma_s \operatorname{Res}\,[\pi (\sin \pi w)^{-1} g(w) (z e^{-\pi i})^w]_{w=s}$$

where Σ_s ranges over the poles s of $g(w)$ in the half plane $R(\alpha w + \beta) \geq -h$, and where $\arg z$ is chosen in such a way that

$$|\arg (\sigma z e^{-\pi i})^{1/\alpha}| \leq \pi |\alpha|^{-1} \cos \theta + \varepsilon$$

where ε is a fixed number, $0 < \varepsilon < \pi/2$. Then for all large $z = \xi^\alpha$ we have

$$F(z) = \Sigma_\mu Z_\mu^{1-\beta} e^{Z_\mu} \left[\sum_{j=0}^{L} c_j Z_\mu^{-j} + O(Z_\mu^{-L-\frac{1}{2}}) \right] - S(z) + O(\xi^{-h-\beta_1}), \qquad (10.4.4)$$

where $\beta_1 = R(\beta)$, $Z_\mu = (\sigma z e^{2\pi \mu i})^{1/\alpha}$. Further Σ_μ ranges over all integers μ for which $|\arg Z_\mu| \leq \pi/2 + \delta$, δ denoting a fixed number, $0 < \delta < \pi$.

Here it should be emphasized that in (10.4.4) the summation depends on the particular ζ-sector under consideration. This phenomenon is known as the Stokes phenomenon. For $\arg \zeta =$ constant, usually the series obtained from a particular μ is dominating. Only in the neighborhood of certain critical half rays more than one value μ may be of importance.

Particularly interesting is the case of (10.4. iii) where

$$g(w) = \prod_{j=1}^{n_1} \Gamma(\alpha_j w + \beta_j) \Big/ \prod_{k=1}^{n_2} \Gamma(\gamma_k w + \delta_k),$$

where the α_j, etc. are complex constants. Let $\alpha = \Sigma \gamma_k - \Sigma \alpha_j$. If

$$|\arg \alpha| < \pi/2, \qquad |\arg \alpha_j/\alpha| < \pi/2, \qquad |\arg \gamma_k/\alpha| < \pi/2,$$

then one obtains from STIRLING's formula an expansion of the form (10.4.3) for every positive integer L and each half plane $R(\alpha w + \beta) \geq -h$. Here

$$\beta = \Sigma \delta_k - \Sigma \beta_j + (n_1 - n_2 + 1)/2, \qquad \sigma = \alpha^\alpha \Pi \alpha_j^{\alpha_j} \Pi \gamma_k^{-\gamma_k},$$
$$c_0 = (\sqrt{2\pi})^{n_1 - n_2 - 1} \alpha^{\beta - \frac{1}{2}} \Pi \alpha_j^{\beta_j - \frac{1}{2}} \Pi \gamma_k^{-\delta_k + \frac{1}{2}}.$$

Consequently, theorem (10.4. iii) yields the complete asymptotic development of the resulting generalized hypergeometric function $F(z)$.

Remark. E.M. WRIGHT [9] has required condition (10.4.3) only in a sector of angle less than π, and showed that (10.4.4) still holds for part of the z-plane. E.M. WRIGHT [10] also obtained similar results for the case that not $g(w)$ but $g(w) \exp(-\psi(w)$ has the asymptotic behavior (10.4.3) where $\psi(z) = \sum_{j=1}^{M} \alpha_j w^{b_j}$, $(R(b_j) < 1)$. For further details on the subject see W. B. FORD [1—4], E. M. WRIGHT [1—10], and J. H. B. KEMPERMAN [1].

10.5. Equations containing a large parameter. We shall consider a system of n first order homogeneous linear differential equations

$$x_i' = \lambda^r \sum_{j=1}^{n} a_{ij}(t, \lambda)\, x_j, \quad \text{or} \quad x' = \lambda^r A(t, \lambda)\, x, \qquad (10.5.1)$$

where $r \geq 1$ is an integer, and where the coefficients $a_{ij}(t, \lambda)$ are continuous functions of t for $a \leq t \leq b$, and $|\lambda|$ large, and are analytic in λ

for $|\lambda|$ large so that the series

$$a_{ij}(t,\lambda)=\sum_{k=0}^{\infty} a_{ijk}(t)\,\lambda^{-k}, \qquad i,j=1,\ldots,n, \tag{10.5.2}$$

are convergent for $|\lambda|$ large and the $a_{ijk}(t)$ are continuous in $a\leq t\leq b$.

Requested are asymptotic properties of the solutions of system (10.5.1) for $a\leq t\leq b$, and $|\lambda|$ large. As in (10.2) and (10.3), we expect that there are formal series of some type, analogous to the series (10.5.2), and corresponding solutions of system (10.5.1) of which such series are, for each $a\leq t\leq b$, the asymptotic expansion with respect to the parameter λ for $|\lambda|$ large and $\lambda\to\infty$. The parameter λ may be real, or complex and in this case will be supposed to belong to some connected set of the complex λ-plane containing $\lambda=\infty$. The expected statement mentioned briefly above is essentially true, provided the following condition is satisfied, namely, that the characteristic roots $\lambda_i(t)$, $a\leq t\leq b$, $i=1,2,\ldots,n$, all functions of t in $[a,b]$ of the matrix $(a_{ij0}(t))$ are all distinct in $[a,b]$. A precise statement will be given below (10.5. i). The same expected result is essentially true even if a number of the same characteristic roots, say k, $1\leq k\leq n$, coincide in the whole interval $[a,b]$ while the remaining $n-k$ roots are distinct from these and with each other, but then the formal series are more complicated. If two or more roots $\lambda_i(t)$ coincide at a point $t=t_0$ and are distinct left and right of t_0, then it may occur for instance that some roots λ_i are say real at the right of t_0, complex at the left of t_0, and thus the solutions of system (10.5.1) have different behavior right and left of t_0, and the corresponding formal series may be different. This is the Stokes phenomenon and the point $t=t_0$ is said to be a *turning point*, or *transition point*. We shall discuss turning points in (10.6). Here we will refer to some of the known theorems for the case where there are no turning points in $[a,b]$ and all the roots $\lambda_i(t)$ are distinct in $[a,b]$.

Formal series will be used of the type

$$\sum_{k=-N}^{+\infty} c_k(t)\,\lambda^{-k},$$

where the coefficients $c_k(t)$ are continuous functions of t in $[a,b]$. Such a series will be said differentiable in $[a,b]$ if the functions $c_k(t)$ have continuous first derivatives in $[a,b]$, and then the formal derivative of the series above is

$$\sum_{k=-N}^{+\infty} c_k'(t)\,\lambda^{-k}.$$

Finally we shall consider polynomials $p_i(t,\lambda)$ of the type

$$p_i(t,\lambda)=b_{i0}(t)\,\lambda^r+b_{i1}(t)\,\lambda^{r-1}+\cdots+b_{ir}(t)$$

where $b_{ij}(t)$ are continuous functions of t in $[a, b]$ and $b_{i0}(t)$ is a function of t with $b_{i0}'(t) = \lambda_i(t)$ in $[a, b]$. Here too any differentiability properties refer to the coefficients $\lambda_i(t)$ and $b_{ij}(t)$.

(10.5. i) If the coefficients $a_{ij}(t, \lambda)$ are infinitely differentiable in $a \leq t \leq b$, if the roots $\lambda_i(t)$, $i = 1, \ldots, n$, of the matrix $[a_{ij0}(t)]$ are distinct in $[a, b]$, i.e., $\lambda_i(t) - \lambda_j(t) \neq 0$ for all $a \leq t \leq b$, $i \neq j$, $i, j = 1, \ldots, n$, then (10.5.1) has a fundamental system of formal series solutions of the type

$$\left[x_{ij}(t) \sim e^{p_j(t,\lambda)} \sum_{k=0}^{+\infty} c_{ijk}(t)\, \lambda^{-k}, \quad i = 1, \ldots, n\right], \qquad j = 1, \ldots, n \tag{10.5.3}$$

all infinitely differentiable in $[a, b]$. In addition, if there exist a region S of the complex λ-plane bounded by two arcs tending to ∞, if $R[p_i(t, \lambda) - p_j(t, \lambda)] \geq 0$, or ≤ 0 for all $a \leq t \leq b$ and $\lambda \in S$, and for each pair i, j, $(i \neq j, i, j = 1, \ldots, n)$, then there is also a fundamental system $x_{ij}(t, \lambda)$, $j = 1, \ldots, n$, $i = 1, \ldots, n$, of solutions of system (10.5.1), with $\lambda \in S$, $a \leq t \leq b$, of which the series (10.5.3) are asymptotic series for each $a \leq t \leq b$, and $\lambda \in S$, $|\lambda| \to \infty$.

We may add here that, if k roots $\lambda_i(t)$ coincide in $[a, b]$ while the remaining $n - k$ roots are distinct and distinct from these, then an analogous result holds. The only change is that the formal series (10.5.3) proceed in powers of $\lambda^{1/k}$ both in the proper series and in the polynomials. We omit the statement and all proofs.

For extensive research on the subject of the present paragraph (10.5) see R. E. LANGER [1—7] and W. WASOW [3—5], A. KNESER [3], D. MEKSYN [1—4], H. L. TURRITIN [1—4], G. D. BIRKHOFF [2], H. A. KRAMERS [1], J. L. SYNGE [1].

In connection with asymptotic expansions we mention here the WKB-method. Consider the second order differential equation

$$y'' - [\lambda^2 f(x) + g(x)]\, y = 0, \tag{10.5.4}$$

where λ is a large parameter, $f(x)$, $g(x)$ are continuous functions of x in an interval or a region R, and $f(x) \neq 0$ in R. The usual transformation $y = \exp(\lambda \int u(x)\, dx)$ (cf. 5.3), transforms (10.5.4) into the Riccati equation

$$\lambda u' + \lambda^2 u^2 - \lambda^2 f - g = 0,$$

which can be solved formally by a series of the type

$$u = \sum_{h=0}^{\infty} u_h(x)\, \lambda^{-h},$$

where the coefficients $u_h(x)$ can be obtained recursively by the very simple relations $u_0 = \pm f^{\frac{1}{2}}$, $u_1 = -(2u_0)^{-1} u_0'$, $u_2 = (2u_0)^{-1}(g - u_1' - u_1^2)$, and

$$u_{h+1} = -(2u_0)^{-1}\left(u_h' + \sum_{k=1}^{h} u_k\, u_{h+1-k}\right), \qquad h \geq 2.$$

The hypothesis $f(x) \neq 0$ in R assures that in R there are no turning points, and, on the other hand, makes the formulas above valid in R. The simple considerations above, through their extensions and applications, form the basis of the so-called

WKB-method (WENTZEL, KRAMERS, BRILLOUIN). This method has had wide applications in physics. In particular the WKB-method has been used in the study of the Schrödinger (wave) equation

$$\psi''(t) + 8\pi^2 m h^{-1} [E - V(t)] \psi = 0, \tag{10.5.5}$$

(m mass, V potential energy), and of the asymptotic expansions of its solutions for large values of the parameter h. The eigenvalues E_n of (10.5.5) have been determined for nonzero solutions bounded in $(-\infty, +\infty)$. For important remarks on the WKB-method see the already quoted papers of R. E. LANGER [3—6]. See also, e.g., S. GOLDSTEIN [3], H. JEFFREYS [1], G. WENTZEL [1], H. A. KRAMERS [1].

10.6. Turning points and the theory of R. E. LANGER. Extending slightly the concept mentioned in (10.5), by a turning point t_0, $a<t_0<b$, of a system (10.5.1) we mean a point at which two or more characteristic roots $\lambda_i(t)$ coincide but such that all characteristic roots are distinct to the right and left of t_0. In the presence of a turning point at t_0 it occurs that the asymptotic expansions as $|\lambda|\to\infty$ of solution of (10.5.1) are essential different to the right and left of t_0 (or in sectors around t_0 if a complex neighborhood of t_0 is considered). This constitutes the Stokes phenomenon at the turning point t_0. If one allows approximations to solutions of (10.5.1) involving multiple-valued, instead of single-valued, functions, the Stokes phenomenon does not appear.

It is convenient to limit ourselves to the second order equation

$$w'' + \lambda P_1(t, \lambda) w' + \lambda^2 P_2(t, \lambda) w = 0, \qquad (') = d/dt, \tag{10.6.1}$$

for which the theory is most highly developed. First approximation connection formulas due to R.E. LANGER [8] involving Bessel functions and valid in a full neighborhood of a turning point, are classic; but these were only first approximation formulas. More recently in 1949 R. E. LANGER [9] gave a general method for the construction of complete asymptotic expansions of the solutions of (10.6) valid in a full neighborhood of a simple turning point (see also the successive paper of T. M. CHERRY [1]). Thus the Stokes phenomenon was circumvented and important applications followed.

We may assume without loss of generality that $t_0=0$. By the usual transformation $w = u \exp[-2^{-1}\lambda \int P_1(t, \lambda)\, dt]$ we obtain an equation of the form

$$u'' + \lambda^2 Q(t, \lambda) u = 0. \tag{10.6.2}$$

We assume $Q(t, \lambda) = \Sigma q_j(t) \lambda^{-j}$, where $|\lambda|$ is sufficiently large, and the $q_j(t)$, $j=0, 1, \ldots,$ are analytic in a complex neighborhood R of $t=0$, no longer restricting ourselves to an interval. The characteristic roots $\lambda_j(t)$ of (10.6.2), $j=1, 2$, according to (10.5) are the roots of the algebraic equation $\varrho^2 + q_0(t) = 0$, and hence the existence of a turning point at $t=0$ means that $q_0(t)$ has an isolated zero at $t=0$. We shall present an algorithm based upon the work of R. E. LANGER [9] for the determination of formal solutions of (10.6.2) when $q_0(t) = t^\nu q_0^*(t)$, $\nu = 1$, 0, or -1, where $q_0^*(t)$ is analytic and bounded away from zero in R. Thus $t = 0$ is a turning point for $\nu = 1$. We suppose that $\int_0^t q_0^{\frac{1}{2}}(x)\,dx$ is nonzero in R except at $t = 0$ and that $q_0^*(0) = 1$.

A first approximation. Let $\mu = (\nu + 2)^{-1}$ and define $\varphi(t) = q_0^{\frac{1}{2}}(t)$, $\Phi(t) = \int_0^t \varphi(x)\,dx$, $\xi(t, \lambda) = \lambda\,\Phi(t)$, and $\Psi(t) = \varphi^{-\frac{1}{2}}(t)\,\Phi^{\frac{1}{2}-\mu}(t)$, where $q_0^{\frac{1}{2}}$ is either one of the two square roots of q_0 and Ψ is defined so as to be continuous at the origin and hence analytic in R. For instance, for $\nu = 1$, we have $\mu = \frac{1}{3}$, $\frac{1}{2} - \mu = \frac{1}{6}$, and, if $q_0 = a_0 t + a_1 t^2 + \cdots$, $a_0 \neq 0$, we have $\varphi(t) = t^{\frac{1}{2}}(b_0 + \cdots)$, $b_0 \neq 0$, $\Phi(t) = t^{\frac{3}{2}}(2b_0/3 + \cdots)$, and $\Psi(t) = \varphi^{-\frac{1}{2}}\Phi^{\frac{1}{6}} = c_0 + c_1 t + \cdots\; c_0 \neq 0$. Note that $q_0^*(0) = 1$ implies $a_0 = b_0 = 1$. It may be seen that if $C_\mu(\xi)$ is any cylindrical (Bessel) function of order μ, i.e., any solution of the equation $\xi^2 y'' + \xi y + (\xi^2 - \mu^2)\,y = 0$, then the function

$$V(t) = \Psi(t)\,\xi^\mu\,C_\mu(\xi) = \lambda^\mu\,\Phi^{\frac{1}{2}}(t)\,C_\mu[\lambda\,\Phi(t)]\,\varphi^{-\frac{1}{2}}(t),$$

is a solution of the differential equation

$$y'' + [\lambda^2 q_0 + \theta(t)]\,y = 0, \qquad \theta(t) = -\Psi''/\Psi. \tag{10.6.3}$$

If $q_1(t) \equiv 0$, it may be proved that the solutions of this equation approximate the solutions of (10.6.2) up to terms involving λ^{-1}.

The second approximation. If $q_1(t)$ is not identically zero, the solutions of (10.6.3) bear little resemblance to those of (10.6.2). We obtain a closer approximation to (10.6.2) as follows. With $v_j(t, \lambda)$, $j = 1, 2$, denote any pair of linearly independent solutions of (10.6.3) and with tentatively undetermined coefficients $\mu_0(t)$ and $\mu_1(t)$, we define functions $\xi_j(t, \lambda)$ by the formulas

$$\xi_j(t, \lambda) = \mu_0(t)\,v_j(t, \lambda) + \mu_1(t)\,v_j'(t, \lambda)\,\lambda^{-1}, \qquad j = 1, 2.$$

If we differentiate $\xi_j(t, \lambda)$ twice with respect to t, always replacing v_j'' by its equivalent as given by (10.6.3), we find that

$$\left.\begin{aligned} \xi_j'' + [\lambda^2 q_0 + \lambda q_1]\,\xi_j = {} & [-\lambda(2q_0\mu_1' + q_0'\mu_1 - q_1\mu_0) + g_1]\,v_j + \\ & + [\lambda(2\mu_0' + q_1\mu_1) + g_2]\,v_j'\lambda^{-1}, \quad j = 1, 2, \end{aligned}\right\} \tag{10.6.4}$$

where

$$g_1(t, \lambda) = \mu_0'' - \theta\mu_0 - \lambda^{-1}(2\theta\mu_1' + \theta'\mu_1), \qquad g_2(t) = \mu_1'' - \theta\mu_1.$$

We now impose the conditions

$$2q_0\mu_1' + q_0'\mu_1 - q_1\mu_0 = 0, \qquad 2\mu_0' + q_1\mu_1 = 0,$$

in order to make the coefficients of the highest powers of λ in the right member of equation (10.6.4) vanish. A particular solution of this differential system for μ_0 and μ_1 is

$$\mu_0(t) = \cos\omega(t), \qquad \mu_1 = \varphi^{-1}(t)\sin\omega(t),$$

where

$$\omega(t) = \int_0^t (2\varphi(x))^{-1} q_1(x)\,dx.$$

Both μ_0 and μ_1 are analytic in R provided μ_1 is defined to be continuous at $t=0$ when the above formula is indeterminate there. We observe that $\mu_0^2 + q_0\mu_1^2 = 1$. Having chosen μ_0 and μ_1, we may now find that the differential equation satisfied by the functions ξ_j is

$$\xi'' + H_0\xi' + [\lambda^2 q_0 + \lambda q_1 + K_0]\,\xi = 0,$$

in which

$$\begin{aligned} H_0 &= (\lambda D_0)^{-1}(g_1\mu_1 - g_2\mu_0), \\ K_0 &= -D_0^{-1}[g_1(\mu_0 + \lambda^{-1}\mu_1') + g_2(q_0\mu_1 - \lambda^{-1}\mu_0') + \lambda^{-2}\theta\mu_1], \\ D_0 &= 1 + \lambda^{-1}(\mu_0\mu_1' - \mu_0'\mu_1) + \lambda^{-2}\theta\mu_1^2. \end{aligned}$$

The function D_0 is bounded from zero for t in R and $|\lambda|$ sufficiently large. Now $H_0 = -D_0' D_0^{-1}$, and thus, if $z_j = D_0^{-\frac{1}{2}}\xi_j$, the functions z_j solve

$$z'' + [\lambda^2 q_0 + \lambda q_1 + K]\,z = 0,$$

where $K = K_0 + D_0''(2D_0)^{-1} - 3[D_0'(2D_0)^{-1}]^2$. This is the second approximating equation to (10.6.2). It may be seen that K is an analytic function of t and λ for t in R and $|\lambda|$ sufficiently large. The second approximating equation is explicitly soluble, and its solutions may be shown to approximate solutions of (10.6.2) up to terms involving λ^{-1}.

Approximations to any desired degree of accuracy, say to terms involving λ^{-m-1} may be obtained by setting $\eta_j = A z_j + B z_j' \lambda^{-2}$, where $A(t,\lambda) = \sum_0^{m-1} \alpha_j(t)\,\lambda^{-j}$, $B(t,\lambda) = \sum_0^{m-1} \beta_j(t)\,\lambda^{-j}$, and determining the functions α_j and β_j so that the η_j satisfy a differential equation of the form

$$\eta'' + [\lambda^2 Q(t,\lambda) + \lambda^{-m}\Omega(t,\lambda)]\,\eta = 0,$$

where $\Omega(t,\lambda)$ is a bounded analytic function of t and λ for t in R and $|\lambda|$ sufficiently large. We omit the proofs.

Bibliographical notes. LANGER's theory, whose initial underlying motivation was in quantum mechanics, has been developed first for second order differential equations and one turning point, and applied to classical problems of analysis (R. LANGER [5—9]). Successively the theory was extended to equations of higher order and multiple turning points (R.E. LANGER [12—16], and applied to quantum mechanics (scattering problem) and hydrodynamics (stability of the laminar flow of a viscous fluid) (R.E. LANGER [10, 11, 16, 17], C.C. LIN [1, 2], W. WASOW [8, 9], D. MEKSYN [1—4]).

The equation $y''+\lambda^2 Q(x, \lambda) y=0$, where $Q(x, \lambda)=q_0(x)+\lambda^{-1} q_1(x)+\cdots$, $q_0=x^\nu q_0^*(x)$, $q_0^*(x)\neq 0$ in a neighborhood of $x=0$, and $|\lambda|$ large, has an extensive literature. For $\nu=0$ see G. D. BIRKHOFF [2] besides earlier classical papers; for $\nu=1$, where a turning point is present, see R. E. LANGER [9] and T. M. CHERRY [1]; for $\nu=-1$, see N. D. KAZARINOFF [1]; for $\nu=2$, see R. McKELVEY [1]; for $\nu=-2$, see N. D. KAZARINOFF and R. McKELVEY [1] and, for particular cases, E. D. CASHWELL [1]; for ν real, $\nu<-2$, see N. D. KAZARINOFF [3].

The essentially more difficult case of equations containing a large parameter and presenting two simple turning points in a region R has been studied recently by R. E. LANGER [18] who has derived the leading term (first approximation) valid in R and connecting the two turning points, and by N. D. KAZARINOFF [4] who has given the successive approximations. An equation of this type is the Weber equation $y''+(n^2-x^2) y=0$, which was studied by A. ERDÉLYI, M. KENNEDY, and J. McGREGOR [1].

The Whittaker equation $y''+[4^{-1}+k x^{-1}+(4^{-1}-m^2) x^{-2}] y=0$ has been studied by N. D. KAZARINOFF in [1] for m complex and large, and in [4] for both k and m complex and large. For other studies see A. ERDÉLYI and C. A. SWANSON [1], F. W. J. OLVER [1], and R. C. THORNE [1].

10.7. Singular perturbation. Let

$$\varepsilon x_i'=f(t, x_1, \ldots, x_n, \varepsilon), \quad i=1,2, \ldots, n, \quad \text{or} \quad x'=f(t, x, \varepsilon), \tag{10.7.1}$$

be a differential system, containing a parameter ε, where f is a vector function of t, x, ε, whose components f_i are continuous with their partial derivatives $f_{ij}=\partial f_i/\partial x_j$ for all $-\infty<t<+\infty$, $x\in S$, $|\varepsilon|<\varepsilon_0$, for some positive ε_0, and some open set S of the x-space E_n. The situation thus is analogous to the one of (8.6. a), with the difference that the parameter ε multiplies in (10.7.1) the derivatives x_i', $i=1, \ldots, n$. As $\varepsilon\to 0$ system (10.7.1) becomes

$$f(t, x, \varepsilon)=0, \tag{10.7.2}$$

which is not even a differential system.

Suppose now that the vector function $f(t, x, \varepsilon)$ is periodic in T of a given fixed period T, suppose that the algebraic equation (no differential equation) (10.7.2) has a nonzero known periodic solution $p(t)$, $-\infty<t<+\infty$, of period T. We shall now ask, as in (8.8.a) whether for $|\varepsilon|<\varepsilon_1$ and some ε_1 sufficiently small, system (10.7.1) has a solution $x(t, \varepsilon)$, $-\infty<t<+\infty$, periodic in t of the same period T, such that $x(t, \varepsilon) \rightrightarrows p(t)$ uniformly in $(-\infty, +\infty)$ as $\varepsilon\to 0$. If the answer is affirmative we shall say that $x(t, \varepsilon)$ is a *singular perturbation* of $p(t)$ for the system (10.7.1).

Of course the formulation above for systems (10.7.1) of first order differential equations can be repeated for a system of one or more higher order differential equations. The different formulation is not always a question of personal preference. For instance, we may consider one n-th order differential equation of the form

$$\varepsilon^k y^{(n)}=f(y, y', \ldots, y^{(m)}, t, \varepsilon), \tag{10.7.3}$$

where $k, n > m \geqq 0$, are integers, where $-\infty < t < +\infty$, $|\varepsilon| < \varepsilon_0$ for some $\varepsilon_0 > 0$, and f is continuous with all its first order partial derivatives f_i, $i = 0, 1, \ldots, m$, with respect to $y, y', \ldots, y^{(m)}$ for $(y, y', \ldots, y^{(m)})$ belonging to some open set S in E_{m+1}. Then for $\varepsilon = 0$ equation (10.7.3) becomes

$$0 = f(y, y', \ldots, y^{(m)}, t, 0), \tag{10.7.4}$$

which for $m > 1$, is a differential equation of order m. Again we shall suppose that f is periodic in t of a given period T, that (10.7.4) has a known nonzero periodic solution $p(t)$ of period T, and we shall ask whether (10.7.3) has, for $|\varepsilon| < \varepsilon_1$ and some $\varepsilon_1 > 0$, a solution $y(t, \varepsilon)$ periodic of period T with $y^{(\nu)}(t, \varepsilon) \rightrightarrows p^{(\nu)}(t)$ as $\varepsilon \to 0$, in $(-\infty, +\infty)$, $\nu = 0, 1, \ldots, m$.

The formulation (10.7.1), (10.7.2) has been considered by I. M. Volk [3—5], the formulation (10.7.3), (10.7.4) by W. Wasow [3], with results essentially more general than the ones of Volk. We shall refer here to the results of W. Wasow, and briefly afterwards to results of other authors.

Let us suppose, with W. Wasow, that f is an analytic function of all its arguments, real when these arguments are all real, and that, for $-\infty < t < +\infty$, the point $[p(t), \ldots, p^{(m)}(t)]$ belongs to S. Also, if f_i, $i = 0, 1, \ldots, m$, denotes the partial derivatives of f with respect to $y, y', \ldots, y^{(m)}$, respectively, consider the $m+1$ functions

$$q_i(t) = f_i[p(t), \ldots, p^{(m)}(t); t, 0], \qquad i = 0, 1, \ldots, m,$$

and suppose $q_m(t) \neq 0$ for all $-\infty < t < +\infty$. Then we may consider the "full variational equation" and the "reduced variational equation", respectively,

$$\varepsilon^k V^{(m)} = \sum_{i=0}^{m} q_i(t)\, V^{(i)}, \tag{10.7.5}$$

$$0 = \sum_{i=0}^{m} q_i(t)\, v^{(i)}. \tag{10.7.6}$$

Since $q_m(t) \neq 0$ we may solve these equations with respect to the highest derivative. The coefficients of the new linear equations so obtained are then analytic in t.

Both linear equations (10.7.5) have periodic coefficients and then Floquet theory can be applied. We shall suppose explicitely that no one of the characteristic exponents of the reduced variational equation (10.7.6) is zero. According to (8.6) this condition is equivalent to the requirement that the same equation (10.7.6) has no nonzero solution of period T.

A last requirement is that for at least one of the two signs of ε no one of the $(n-m)$-th roots of $\mu = (\operatorname{sign} \varepsilon^k)\,(\operatorname{sign} p(t)) = \pm 1$, has real part zero. For $n-m$ odd this condition is satisfied for either sign of ε. For $n-m$ even, this condition has actually a bearing and some cases may be excluded altogether. We will say that the sign of ε has been restricted.

Under these conditions W. Wasow [3] has proved the following theorem:

(10.7. i) Under the conditions above there is an $\varepsilon_1>0$, such that for $|\varepsilon|<\varepsilon_1$, equation (10.7.3) has a periodic solution $y(t,\varepsilon)$, $-\infty<t<+\infty$, of period T with $y(t,\varepsilon)\rightrightarrows p(t)$ in $-\infty<t<+\infty$, as $\varepsilon\to 0$, where the sign of ε may be restricted.

While the formal procedure used by W. Wasow is the rather standard process of perturbation leading to a system of successive approximations involving the variational equations (10.7.5), (10.7.6), the proof of the convergence of the process is distinguished.

The case where equation (10.7.3) does not depend on t, i.e., the autonomous case, is slightly more complicated than the one above, just as it occurs in the nonsingular perturbation method. Then it occurs that while (10.7.4) may have a periodic solution of a given period T_0, the system (10.7.3) may have for $\varepsilon\neq 0$ small solutions $y(t,\varepsilon)$ periodic of period $T=T(\varepsilon)$ depending on ε, with $T(\varepsilon)\to T_0$ as $\varepsilon\to 0$. Then the requirement $y(t,\varepsilon)\rightrightarrows p(t)$ uniformly in $(-\infty,+\infty)$ cannot be requested. The more convenient requirement $y(t,\varepsilon)\rightrightarrows p(t)$ uniformly in $(0,2T_0)$ is adequate. The theorem is only a repetition of the one above where these modifications are made. As a matter of fact, because of the continuity of all functions involved, these requirements can be formulated simply by requesting that

$$y(t,0)\equiv p(t), \qquad T(0)=T_0.$$

On the same subject see K.O. Friedrichs and W. Wasow [1], I. S. Gradstein [1—11], N. Levinson [8, 13, 18, 19], H. L. Turritin [1—3]. See also J. J. Levin and N. Levinson [1], M. J. Lighthill [1], M. Nagumo [2], A. Tihonov [1—3], W. Wasow [7, 9], G. D. Birkhoff and R. E. Langer [1], A. Besicovitch and J. Tamarkin [1], A. B. Vasileva [1—6], V. M. Volosov [1, 2], M. Volk [1], J. G. Wendel [2].

Bibliography

The italics numbers in parentheses refer to the sections of the book. Star indicates expository papers.

Abelé, J.: [1] Système d'entretien à amplitude autostabilisée (*8*). C. R. Acad. Sci. Paris **214**, 841—842 (1942). — [2] Définition cinématique des oscillations de relaxation (*8*). Ibid. **220**, 511—515 (1945). — [3] Définition cinématique d'oscillations de relaxation discontinues (*8*). Ibid. **221**, 656—658 (1945). — [4] Définition cinématique des oscillations de relaxation (*8*). J. Phys. Radium

(8) **6**, 96—103 (1945). — [5] Les oscillations de relaxation et le problème de leur définition analytique (*8*). Rev. Gén. Sci. Pures Appl. **53**, 68—77 (1946). — [6] Construction d'oscillateurs non linéaires sinusoidaux par la méthode de l'axe mobile (*8*). C. R. Acad. Sci. Paris **225** 1270—1271 (1947).

ADAMOV, N. V.: [1] Le sens géométrique d'une condition de stabilité donné par LYAPUNOV (*1, 4, 5*). Doklady Akad Nauk SSSR, **2**, 361—364 (1935). — [2] Sur quelques conditions de stabilité (*4, 5*). Ibid. **2**, 447—450 (1935). — [3] Sur l'oscillation des intégrales de l'équation du deuxième ordre aux coéfficients périodiques et sur quelques conditions de la stabilité (*4, 5*). Mat. Sbornik, **42** 651—668 (1935). — [4] On the method of successive approximations (*4*). Doklady Akad. Nauk SSSR, (N. S.) **18**, 219—223 (1938). — [5] On finding the periodic solutions of an ordinary differential equation of the first order by the method of successive approximations (*4*). Ibid. **19**, 17—22 (1938). — [6] On certain transformations not changing the integral curves of a differential equation of the first order (*4, 9*). Mat. Sbornik, N. S. **23** (65), 187—228 (1948). Amer. Mat. Soc. Translation No. 31. — [7] Sur quelques propriétés des intégrales d'une équation du second ordre à coefficients périodiques (*4*). C. R. Acad. Sci. Paris **197**, 1280—1282 (1933). — [8] Sur quelques propriétés des transformations qui laissent invariable la courbe intégrale d'une équation du premier ordre (*4*). Doklady Akad. Nauk. SSSR, (N. S.) **29**, 539—543 (1940).

AIZERMAN, M. A.: [1] On the convergence of processes of automatic regulation after large initial deflections (*1, 6, 7, 8*). Avtom. i Telem. **7**, 148—167 (1946). — [2] On taking account of nonlinear functions of several variables in the investigation of the stability of a system of automatic regulation (*1, 6, 7, 8*). Ibid. **8**, 20—29 (1947). — [3] On a problem concerning the stability in the large of dynamical systems (*1, 6, 7, 8*). Uspehi Mat. Nauk, N. S. **4**, 187—188 (1949). — [4] On the determination of the safe and unsafe parts on the boundary of stability (*1, 6, 7, 8*). Prikl. Mat. Meh. SSSR, **14**, 444—448 (1950). — [5] On the construction of resonance graphs for systems with nonlinear feedback (*2*). Akad. Nauk SSSR, Inzen. Sbornik **13**, 151—165 (1952). — [6] A sufficient condition for stability of a class of dynamical systems with periodic coefficients (*4*). Prikl. Mat. Meh. SSSR, **15**, 382—384 (1951). — [7] The theory of automatic control of motion (*). Gostekizdat. Moscow-Leningrad 1952.

—, and F. R. GANTMAHER: [1] Conditions for the existence of a region of stability for a single-contour system of automatic regulation (*2*). Prikl. Mat. Meh. SSSR, **18**, 103—122 (1954).

ALBRECHT, F.: [1] Remarques sur un théorème de T. WAŽEWSKI relatif a l'allure asymptotique des intégrales des équations différentielles (*9*). Bull. Acad. Polon. Sci. **2**, 315—318 (1954). — [2] Points singuliers et solutions périodiques (*9*). Com. Acad. R. P. Romine **5**, 1035—1040 (1955). — [3] Un théorème de comportement asymptotique des solutions des équations et des systèmes d'équations différentielles (*9*). Bull. Acad. Polon. Sci. **4**, 737—739 (1956).

ALEXANDROFF, P., and H. HOPF: [1] Topologie (*). Springer 1935.

AL'MUHAMEDOV, M. I.: [1] On conditions for the existence of stable and unstable centers (*9*). Doklady Akad. Nauk SSSR, (N. S.) **67**, 961—964 (1949). Amer. Math. Soc. Translation no. 34.

AMERIO, L.: [1] Un preliminare teorema di analisi per lo studio dei moti con resistenza passiva (*5, 8*). Rend. Accad. Lincei (7) **3**, 415—426 (1942). — [2] Determinazione delle condizioni di stabilità per gli integrali di un'equazione interessante l'elettrotecnica (*8*). Annali Mat. pura appl. (4) **30**, 75—90 (1949). — [3] Studio asintotico del moto di un punto su una linea chiusa per azione di forze indipendenti dal tempo (*8*). Annali Scuola Norm. Sup. Pisa (3) **3**, 19—57 (1950). — [4] Questioni di stabilità in problemi di meccanica e di elettrotecnica (*8*). Rend. Sem. Mat. Fis. Milano **21**, 82—89 (1950). — [5] Sull'estensione delle nozioni di colle, nodo e fuoco ai sistemi di due eqauzioni differenziali periodiche in tre variabili (*9*). Rend. Accad. Lincei (8) **10**, 206—212, 289—297 (1951). — [6] Analisi delle nozioni di nodo, stella e fuoco estese ai sistemi di due equazioni differenziali in tre variabili (*9*). Riv. Mat. Univ. Parma **3**, 207—231 (1952). — [7] Extension of the notion of saddle point to systems of two differential equations in three variables (*9*). Comm. Pure Appl. Math. **6**, 435—454 (1953). — [8] Sur l'extension de quelques points de la theorie de Poincaré aux systèmes de deux équations différentielles à trois variables (*9*). Coll. Intern. Vibrations, Ile de Porquerolles **1951**, 261—269. — [9] Varietà analitiche chiuse trasformate in sè dai sistemi differenziali periodici (*4*, *9*). Annali Mat. pura appl. (4) **37**, 219—248 (1954).

Andronov, A.: [1] Les cycles limites de Poincaré et la theorie des oscillations autoentretenues (*9*). C. R. Acad. Sci. Paris **189**, 559 (1929).

—, and N. Bautin: [1] Le mouvement d'un avion neutre piloté automatiquement et la theorie des transformations poinctuelles des surfaces (*9*). Doklady Akad. Nauk SSSR, (N. S.) **43**, 189—193 (1944). — [2] Stabilization de route d'un avion neutre par autopilote ayant un servomoteur à vitesse constant et à zone d'insensibilité (*9*). Ibid. **46**, 143—146 (1945). — [3] Sur un cas dégénéré du problème général de régulation directe (*9*). Ibid. **46**, 277—279 (1945). — [4] On the influence of Coulomb friction in the dashpot on processes of indirect regulation (*8*). Izvetiya Akad. Nauk SSSR, **1955**, no. 7, 34—48.

—, N. Bautin and G. S. Gorelik: [1] Autooscillations d'un schema simplifié, contenant une hélice à pas variable automatique (*9*). Doklady Akad. Nauk SSSR, **47**, 263—267 (1945).

—, and C. E. Chaikin: [1] Theory of oscillations (*). Princeton University Press 1949. (Russian orig. publ. in Moscow, 1937.)

— —, L. I. Mandelstam and A. Witt: [1] Exposé des rescherches récentes sur les oscillations non-linéares (*). Zechn. Fiz. SSSR, Leningrad **2**, 81—134 (1935).

—, and E. Leontovic: [1] On oscillations of a system with periodically varying parameters (*4*). Ž. Russ. Fiz., Him. Obsc. Ver. Fiz. **59**, 428—442 (1927). — [2] Generation of limit cycles from a structurally unstable focus or from a center and from a structurally unstable limit cycle (*9*). Doklady Akad. Nauk SSSR, (N. S.) **99**, 885—888 (1954).

—, and A. Mayer: [1] Le problème de Mises dans la théorie de la régulation directe et la théorie des transformations pounctuelles des surfaces (*9*). Doklady Akad. Nauk SSSR, N. S. **43**, 54—58 (1944). — [2] Le problème de Visnegradskii dans la théorie de la régulation directe (*9*). Ibid. **47**, 340—343 (1945). — [3] The simplest linear systems with retardation (*2*). Avtom i Telem. **7**, 95—106 (1946). — [4] Visnegradskii's problem in the theory of direct regulation (*2*). Ibid. **8**, 314—334 (1947).

—, and G. Neumark: [1] Sur les mouvements des modèles idealisés d'horloge ayant deux degrés de liberté. Horloge pregalileene (*8*). Doklady Akad. Nauk SSSR, (N. S.) **50**, 17—20 (1946).

—, and L. S. Pontryagin: [1] Singular points of differential systems (*9*). Doklady Akad. Nauk SSSR, (2) **14**, 247—250 (1937).

—, and A. Witt: [1] On the mathematical theory of the capture of electrons (*8*). J. Appl. Phys. **7**, no. 4 (1930). — [2] Zur Theorie des Mitnehmens (*8*). Arch. Elektrotechn. **24**, 89 (1930). — [3] Sur la théorie mathematiques des autooscillations (*8*). C. R. Acad. Sci. Paris **190**, 256—258 (1930). — [4] Unstetige periodische Bewegungen und die Theorie des Multivibrators von Abraham und Bloch (*9*). Doklady Akad. Nauk. SSSR, **1930**, 189—192. — [5] On stability in the sense of Lyapunov (*8*). J. exp. theoret. Phys. **4**, 606—608 (1933). —

[6] On the mathematical theory of selfoscillating systems with two degrees of freedom (*8*). Ibid. **4**, 249—271 (1933).

Ansov, H. I.: [1] Stability of linear oscillating systems with constant time lag (*2*). J. Appl. Mech. **16**, 158—164 (1949).

—, and J. A. Krumhansl: [1] A general stability criterion for linear oscillating systems with constant time lag (*2*). Quart. Appl. Math. **6**, 337—341 (1948).

Antosiewicz, H. A.: [1] A note on asymptotic stability (*3*). Quart. Appl. Math. **9**, 317—319 (1951). — [2] Forced periodic solutions of systems of differential equations (*8*). Ann. of Math. (2) **57**, 314—317 (1953); **58**, 592 (1953). — [3] On the differential equation $\ddot{x} + k\left(f(x) + g(x)\dot{x}\right)\dot{x} + h(x) = ke(t)$, (*8*). Nat. Bur. Stand., Rep. No. 3412, **1954**. — [4] On non-linear differential equations of the second order with integrable forcing term (*8*). J. London Math. Soc. **30**, 64—67 (1955). — [5] Stable systems of differential equations with integrable perturbation term. Ibid. **31**, 208—212 (1956). — [6] A survey of Lyapunov's second method (*7*). Contributions Theory nonlinear Oscillations, **4**, 141—166, 1958.

—, and M. Abramowitz: [1] A representation for solutions of analytic systems of linear differential equations (*10*). Proc. Acad. Sci. USA, **44**, 382—384 (1954).

—, and P. Davis: [1] Some implications of Lyapunov's conditions for stability (*3, 7*). J. Rational Mech. Anal. **3**, 447—457 (1954).

Arley, N., and V. Borhsenius: [1] On the theory of infinite systems of differential equations and their application to the theory of stochastic processes and the perturbation theory of quantum mechanics (*3, 9*). Acta Math. **76**, 261—322 (1945).

Armellini, G.: [1] Sopra una equazione differenziale della dinamica (*5*). Rend. Accad. Lincei **21**, 111—116 (1935). — [2] Contributo alla dinamica del sistema galattico. Atti Accad. Italia **3**, 73—82 (1941/42). — [3] Sopra l'età dei pianeti e sopra l'incremento dei parametri delle loro orbite a causa del termine cosmogonico. Ibid. **3**, 229—235 (1941/42). — [4] Sopra una classe di equazioni differenziali della meccanica celeste di cui l'integrale generale tende a zero (*5*). Pont. Acad. Sci. Acta **6**, 387—396 (1942). — [5] Sopra l'equazione differenziale del moto centrale newtoniano (*5*). Rend. Accad. Italia (7) **4**, 342—348 (1943).

Artemiev, N. A.: [1] Foundations of the qualitative theory of ordinary differential equations, I (*). Izdat. Leningrad Gos. Univ. **1941**.

Ascari, A.: [1] Studio asintotico di un'equazione relativa alla dinamica del punto (*9*). Ist. Lombardo Sci. Lett. (3) **16**, (85), 278—288 (1952).

Ascoli, G.: Sopra una particolare equazione differenziale del secondo ordine (*3, 5*), I, II. Rend. Ist. Lombardo Sci. Lett. (2) **69**, 167—184, 185—197 (1936). — [2] Sul comportamento asintotico degli integrali delle equazioni differenziali lineari del secondo ordine (*5*). Rend. Accad. Lincei (6) **22**, 234—243 (1935). — [3] Sopra una particolare equazione differenziale del secondo ordine (*5*). Ibid. (6) **22**, 314—319 (1935). — [4] Osservazioni sopra alcune questioni di stabilità (*3*). ibid. (8) **9**, 129—134, 210—213 (1950). — [5] Ricerche asintotiche sopra una classe di equazioni differenziali non lineari (*10*). Annali Scuola Norm. Sup. Pisa (3) **5**, 1—28 (1951). — [6] Questioni asintotiche nal campo delle equazioni differenziali non lineari (*10*). Rend. Sem. Mat. Fis. Milano **22**, 63—73 (1952). — [7] Sul comportamento asintotico delle soluzioni dell'equazione $y'' - (1 + \eta)\, y = 0$, (*5*). Boll. Unione Mat. Ital. (3) **8**, 115—123 (1953). — [8] Sul comportamento asintotico degli integrali dell'equazione $y'' = \left(1 + f(t)\right) y$ in un caso notevole, (*5*). Riv. Mat. Univ. Parma **4**, 11—29 (1953). — [9] Sulla forma asintotica degli integrali dell'equazione $y'' + A(x)\; y = 0$ in un caso notevole di stabilità (*3, 5*). Revista Mat. Fis. Univ. Tucuman **2** (1941). — [10] Sul comportamento asintotico e sulla valutazione approssimata degli integrali delle equazioni differenziali del primo ordine (*3*). Scritti mat. offerti a L. Berzolari **1936**,

617–635. — [11] Sopra un caso di stabilità per l'equazione $y'' + A(x)y = 0$ (*5*). Annali Mat. pura appl. (4) **26**, 199–206 (1947).

ATKINSON, F. V.: [1] On second order linear oscillators (*5*). Riv. Univ. Nac. Tucuman **8**, 71—87 (1951). — [2] Asymptotic properties of a differential equation (*5*). Actas Acad. Ci. Lima **14**, 28—33 (1951). — [3] The asymptotic solution of second order differential equations (*5*). Annali Mat. pura appl. (4) **37**, 347—348 (1954). — [4] On linear perturbation of non-linear differential equations (*8*). Canad. J. Math. **6**, 561—571 (1954). — [5] On second order nonlinear oscillations (*5*). Pacific. J. Math. **5**, 643—647 (1955). — [6] On asymptotically linear second order oscillations (*6*). J. Rational Mech. Anal. **4**, 769—793 (1955).

AVAKUMOVIC, V.: [1] Sur l'équation différentielle de THOMAS-FERMI (*5*). Glas Srpske Akad. Nauka **191**, 163—187 (1948).

AYMERICH, G.: [1] Sulle oscillazioni forzate di due circuiti elettrici non lineari con accoppiamento induttivo e capacitivo (*8*). Atti Sem. Mat. Fis. Univ. Modena **5**, 83—89 (1951). — [2] Oscillazioni forzate periodiche di sistemi non lineari a due gradi di libertà (*8*). Ibid. **5**, 165—177 (1951). — [3] Sulle oscillazioni autosostenute impulsivamente (*8*). Rend. Sem. Fac. Sci. Univ. Cagliari **22**, 34—37, 109—116 (1953). — [4] Oscillazioni periodiche di un sistema di ROCARD a due gradi di libertà (*8*). Ibid. **24**, 51—62 (1954). — [5] Modulazione di ampiezza e di fase nell'oscillatore di ROCARD (*8*). Ibid. **23**, 165—174 (1953).

BAGGIS, H. F. DE: [1] Dynamical systems with stable structures (9). Contributions to the theory of nonlinear oscillations. Ann. of Math. Studies **29**, 37—59 (1952).

BAILEY, H. R.: [1] Existence and stability of periodic solutions of weakly nonlinear differential equations (*8*). To appear.

—, and R. A. GAMBILL: [1] On stability of periodic solutions of weakly nonlinear differential equations (*8*). J. Math. Mech. **6**, 655–668 (1957).

BAKAEV, YU N.: [1] Approximate integration of the differential equation of a pendulum (*8*). Prikl. Mat. Meh. SSSR, **16**, 723—728 (1952).

BAKER, J. G.: [1] Forced vibrations with nonlinear spring constants (*8*). Trans. Amer. Soc. Mech. Engrs. **54** (1932).

BALLIEU, R.: [1] Sur l'unicité de l'intégrale d'une équation différentielle (*1*). Acad. Roy. Belg. Bull. (*5*) **33**, 725—742 (1947).

BARBALAT, I.: [1] Solutions bornées et solutions périodiques pour certaines équations différentielles non linéaires du second ordre. I. (*8*). Acad. Repub. Pop. Romine Bul. Sti. Mat. Fiz. **5**, 393—402, 503—515 (1953). — [2] Une proprieté globale des trajectoires d'un système d'équations différentielles équivalent à l'équation des oscillations non linéaires de LIÉNARD (*8*). Ibid. **6**, 853—860 (1954). — [3] Remarques sur la note "Points singuliérs et solutions périodiques." Ibid. **7**, 325—328 (1955).

—, and A. HALANAY: A criterion of existence of a stable cycle for a nonlinear differential equation (*9*). Acad. R.P. Romine, Stud. Cerc. Math. **7**, 81–95 (1956).

BARBASIN, E. A.: [1] On the theory of general dynamical systems (*9*). Ucenye Zapiske Moskov Gos. Univ. **2**, 110—133 (1948). — [2] Dispersive dynamical systems (*9*). Uspehi Mat. Nauk, (N.S.) **1950**, no. 4 (38), 138—139. — [3] On the stability of the solution of a nonlinear equation of the third order (*8*). Prikl. Mat. Meh. SSSR, **16**, 629—632 (1952). — [4] The method of sections in the theory of dynamical systems (*7, 9*). Mat. Sb, (N.S.) **29** (71), 233—280 (1951).

—, and N. N. KRASOWSKIJ: [1] On stability of motion in the large (*1, 7, 9*). Doklady Akad. Nauk SSSR, N.S. **86**, 453—456 (1952). — [2] On the existence of Lyapunov functions in the case of asymptotic stability in the large (*1, 7, 9*). Prikl. Mat. Meh. SSSR, **18**, 345—350 (1954).

BARBER, N.F., and F. URSELL: [1] The response of a resonating system to a gliding tone (*2*). Phil. Mag. (7) **39**, 345 (1948).

BARBUTI, U.: [1] Sulla stabilità delle soluzioni per la equazione $x''+A(t)x=0$, (*5*). Rend. Accad. Lincei (8) **12**, 170—175 (1952). — [2] Sopra un caso di "risonanza" per la equazione $x''+A(t)x=0$, (*5*). Boll. Unione Mat. Ital. (3) **7**, 154—159 (1952). — [3] Su alcuni teoremi di stabilità (*3, 5*). Annali Scuola Norm. Sup. Pisa (3) **8**, 81—91 (1954). — [4] Contributi al problema della stabilità per i sistemi differenziali lineari ordinari (*3, 5*). Annali Scuola Norm. Sup. Pisa (3) **10**, 185—215 (1957).

BARNES, E.W.: [1] The asymptotic expansions of functions defined by Taylor series (*10*). Phil. Trans. A **206**, 249—297 (1906). — [2] The asymptotic expansions of integral functions defined by generalized hypergeometric series (*10*). Proc. Lond. Math. Soc. (2) **5**, 59—116 (1907); **6**, 141—177 (1908).

BAROCIO, S.: [1] On certain critical points of a differential system in the plane (*9*). Contrib. Theory Nonlinear Oscillations **3**, 127—137 (1956).

BARRETT, J.H.: [1] Behavior of solutions of second order selfadjoint differential equations (*5*). Proc. Amer. Math. Soc. **6**, 247—251 (1955).

BASCH, A.: [1] Über Schwingungen von Systemen mit zwei Freiheitsgraden (*2, 8*). Öster. Ing.-Arch. **8**, 83—86 (1954).

BASOV, V.P.: [1] On solutions of a class of systems of linear differential equations (*3*). Doklady Akad. Nauk SSSR, (N.S.) **80**, 301—304 (1951). — [2] Necessary and sufficient conditions for the stability of solutions of a certain class of systems of linear differential equations in one doubtful case (*3*). Ibid. **81**, 5—8 (1951). — [3] Construction of solutions of a class of systems of linear differential equations (*3*). Prikl. Mat. Meh. SSSR, **18**, 313—328 (1954).

BATEMAN, H.: [1] The control of an elastic fluid (*). Bull. Amer. Math. Soc. **51**, 601—646 (1945).

BAUTIN, N.N.: [1] On the behavior of dynamical systems with small violations of the condition of stability of ROUTH-HURWITZ (*9*). Prikl. Mat. Meh. SSSR. **12**, 613—632 (1948). — [2] Theory of an escapement regulator with spring blude (*9*). Doklady Akad. Nauk SSSR, (N.S.) **61**, 17—20 (1948); **65**, 270—282 (1949); **72**, 19—22 (1950). — [3] On the number of limit cycles appearing with variation of the coefficients from an equilibrium state of the type of a focus or a center (*9*). Mat. Sbornik, (N.S.) **30** (72), 181—195 (1952). Amer. Math. Soc. Translations No. 100. — [4] A dynamical model of a chromometric movement (*9*). Akad. Nauk SSSR, Inzen. Sbornik **12**, 3—22 (1952). — [5] A dynamic model of a watch movement without a characteristic period (*9*). ibid. **16**, 3—12 (1953). — [6] On periodic solutions of a system of differential equations (*9*). Prikl. Mat. Meh. SSSR, **18**, 128 (1954). — [7] Du nombre de cycles limites naissant en cas de variation des coefficients d'un état d'equilibre du type foyer ou centre (*9*). Doklady Akad. Nauk. SSSR, (N.S.) **24**, 669—672 (1939). — [8] Behavior of dynamical systems near the boundary of their region of stability (*, *8*). Gostekizdat. Moscow 1949.

BELLMAN, R.: [1] The stability of solutions of linear differential equations (*3*). Duke Math. J. **10**, 643—647 (1943). — [2] A stability property of solutions of linear differential equations (*3, 5*). Ibid. **11**, 513—516 (1944). — [3] On the stability of systems of differential equations (*3*). Proc. Nat. Acad. Sci. U.S.A. **32**, 190—193 (1946). — [4] The boundedness of solutions of linear differential equations (*3*). Duke Math. J. **14**, 83—97 (1947). — [5] The boundedness of solutions of infinite systems of linear differential equations (*3*). Ibid. **14**, 695 to 706 (1947). — [6] On the boundedness of solutions of nonlinear differential and difference equations (*6, 8*). Trans. Amer. Math. Soc. **62**, 357—386 (1947). — [7] On the boundedness of solutions of nonlinear differential and difference equations (*8*). Ibid. **64**, 374—388 (1948). — [8] On an application of a Banach-Steinhaus theorem to the study of the boundedness of solutions of nonlinear differential and difference equations (*3, 6*). Ann. of Math. (2) **49**, 515—522 (1948). [9] A survey of the theory of the boundedness, stability, and asymptotic behavior of solutions of linear and nonlinear differential and difference equations (*). Off. of Naval Research, Washington, D.C., 1949. — [10] On the existence and boundedness of solutions of nonlinear differential-difference equations (*8*). Ann. of Math. (2) **50**, 347—355 (1949). — [11] On the asymptotic behavior of solutions of $u'' + (1 + f(t))u = 0$, (*5*). Annali Mat. pura appl. (4) **31**, 83—91 (1950). — [12] Stability Theory of differential equations (*). McGraw-Hill 1953. — [13] A survey of the mathematical theory of time-lag, retarded control, and hereditary processes (*, *8*), pp. 107, Rand Corporation 1954.

BELYUSTINA, L. N.: [1] On conditions for the existence of a center (*9*). Prikl. Mat. Meh. SSSR, **18**, 511 (1954). — [2] On the stability of the operating regime of a salient-pole synchronous motor (*8*). Izvetiya Akad. Nauk SSSR, **10**, 131—140 (1954).

BENDIXSON, I.: [1] Sur les courbes définies par des équations différentielles (*9*). Acta math. **24**, 1—88 (1901).

BERNSTEIN, I., and E. IKONIKOV: [1] The mathematical theory of forced vibrations in selfoscillating systems with two degrees of freedom (*8*). Zechn. Fiz. Leningrad **4**, 172—190 (1934).

BERTOLINI, F.: [1] Sugli integrali di una equazione differenziale ordinaria (*3*). Rend. Accad. Lincei (8) **8**, 285—292 (1950). — [2] Sulle soluzioni di un sistema di equazioni differenziali ordinarie (*1*). Rend. Mat. Univ. Roma (5) **9**, 354—366 (1950).

BESICOVITCH, A., and J. TAMARKIN: [1] Über die asymptotischen Ausdrücke für die Integrale eines Systems linearer Differentialgleichungen, die von einem Parameter abhängen (*10*). Math. Z. **21**, 119—125 (1924).

BIEBERBACH, L.: [1] Theorie der Differentialgleichungen (*). Springer 1930.

BIELECKI, A.: [1] Sur une équation différentielle binome du 2-me ordre (*5*). Ann. Univ. Mariae Curie-Sklodowska, Sect. A **4**, 13—17 (1950). — [2] Remarkes sur la méthode de T. WAZEWSKI sur l'étude qualitative des équations différentielles ordinaires (*9*). Bull. Acad. Polon. Sci. **4**, 493—495 (1956). — [3] Sur une méthode de régularization des équations différentielles ordinaires dont les intégrales ne remplissent pas la condition d'unicité (*9*). Ibid. **4**, 497—501 (1956). — [4] Certaine propriétés topologiques des intégrales des equations différentielles ordinaires (*9*). Ibid. **4**, 503—506 (1956). — [5] Extension de la méthode de T. WAZEWSKI aux equations au paratingent (*9*). Ann. Univ. Mariae Curie-Sklodowska, Sect. A **9**, 37—61 (1957). — [6] Certaine propriétés topologiques des solutions des equations au paratingent (*9*). Ibid. **9**, 63—79 (1957).

BIERNACKI, M.: [1] Sur l'equation $x'' + A(t)x = 0$, (*5*). Prace Mat.-Fiz. **40**, 163—171 (1932). — [2] Sur les zéros des intégrales de l'équation $x^{(5)}(t) + A(t)x = 0$, (*5*). Ann. Soc. Polon. Math. **21**, 26—37 (1948). — [3] Sur un théorème dans la théorie des équations différentielles (*1*). Prace Mat.-Fiz. **47**, 129—141 (1949). — [4] Sur la dérivée logarithmique des intégrales des équations différentielles linéaires (*5*). Ann. Univ. Mariae Curie-Sklodowska, Sect. A **6**, 55—64 (1954). — [5] Sur l'équation différentielle $y^{(4)} + A(x)y = 0$, (*5*). Ibid. **6**, 65—78 (1954). — [6] Sur le nombre minimum des zéros des intégrales de l'équation $y^{(n)} + A(x)y = 0$, (*5*). Ibid. **7**, 15—18 (1953).

BIHARI, I.: [1] A generalization of a lemma of BELLMAN and its application to uniqueness problems of differential equations (*3, 6*). Acta Math. Acad. Sci. Hung. **7**, 71—94 (1956). — [2] Researches on boundedness and stability of the solutions of nonlinear differential equations (*3, 6*). Ibid. **8**, 261—278 (1957).

Bilharz, H., and S. Schottlaender: [1] Periodische Lösungen einer geregelten Bewegung (*8*). Arch. Math. **5, 479—491** (1954).

Birkhoff, G. D.: [1] Boundary value and expansion problems of ordinary linear differential equations (*10*). Trans. Amer. Math. Soc. **9**, 373—395 (1908). — [2] On the asymptotic character of the solutions of certain linear differential equations containing a parameter (*10*). Ibid. **9**, 219—231 (1908). — [3] Existence and oscillation theorems for a certain boundary value problem (*5*). Ibid. **10**, 259—270 (1909). — [4] Singular points of ordinary linear differential equations (*10*). Ibid. **10**, 436—470 (1909). — [5] A simplified treatment of the regular singular point (*10*). Ibid. **11**, 199—202 (1910). — [6] On the solutions of ordinary linear homogeneous differential equations of the third order (*5*). Ann. of Math. **12**, 103—127 (1911). — [7] Quelques théorèmes sur le mouvement des systèmes dynamiques (*1, 6*). Bull. Soc. Math. France **40**, 305—323 (1912). — [8] Equivalent singular points of ordinary linear differential equations (*10*). Math. Ann. **74**, 134—139 (1913). — [9] On a simple type of irregular singular point (*10*). Trans. Amer. Math. Soc. **14**, 426—476 (1913). — [10] Proof of Poincaré's geometric theorem (*9*). Ibid. **14**, 14—22 (1913). — [11] Theorem concerning the singular points of ordinary linear differential equations (*10*). Proc. Nat. Acad. Sci. U.S.A. **1**, 578—581 (1915). — [12] The restricted problem of three bodies (*9*). Rend. Circ. Mat. Palermo **39**, 1—70 (1915). — [13] Dynamical systems with two degrees of freedom (*9*). Proc. Nat. Acad. Sci. U.S.A. **3**, 314—316 (1917). — [14] Dynamical systems with two degrees of freedom (*6*). Trans. Amer. Math. Soc. **18**, 199—300 (1917). — [15] Sur la démonstration directe du dernier théorème de H. Poincaré par M. Dantzig (*9*). Bull. Sci. Math. **42**, 41—43 (1918). — [16] Surface transformations and their dynamical applications (*9*). Acta math. **43**, 1—119 (1920). — [17] Celestial mechanics (*6*). Bull. Nat. Res. Council **4**, 1—22 (1922). — [18] Note on the expansion problems of ordinary linear differential equations (*10*). Rend. Circ. Mat. Palermo 2. sem. **36**, 1—12 (1923). — [19] An extension of Poincaré's last geometric theorem (*6*). Acta math. **47**, 297—311 (1925). — [20] Stabilità e periodicità nella dinamica (*1, 6*). Periodico di Mat. (4) **6**, 262—271 (1926). — [21] Sur la signification des équations canoniques de la dynamique (*6*). C. R. Acad. Sci. Paris **183**, 516 (1926). — [22] Über gewisse Zentralbewegungen dynamischer Systeme (*6*). Nachr. Ges. Wiss. Göttingen **1926**, 81—92. — [23] Stability and the equations of dynamics (*1, 6*). Amer. J. Math. **49**, 1—38 (1927). — [24] On the periodic motions of dynamical systems (*1, 9*). Acta math. **50**, 359—379 (1927). — [25] Dynamical systems (*). Amer. Math. Soc. Coll. Publ., New York **1927**. — [26] A remark on the dynamical role of Poincaré's last geometrical theorem (*9*). Acta Sci. Math. Szeged **4**, 6—11 (1928). — [27] A new criterion of stability (*8*). Atti Congr. Internat. Mat. Bologna **5**, 5—13 (1928). — [28] Einige Probleme der Dynamik (*8*). Jber. dtsch. Math. Ver. **38**, 1—16 (1929). — [29] Divergente Reihen und singuläre Punkte gewöhnlicher Differentialgleichungen (*10*). Preuß. Akad. Wiss. **11**, 171—183 (1929). — [30] Une généralisation à *n* dimensions du dernier théorème de géométrie de Poincaré (*6*). C. R. Acad. Sci. Paris **192**, 196 (1931). — [31] Proof of a recurrence theorem for strongly transitive systems (*6*). Proc. Nat. Acad. Sci. U.S.A. **17**, 650—660 (1931). — [32] Sur quelques courbes fermées remarquable (*6*). Bull. Soc. Math. France **60**, 1—26 (1932). — [33] Sur l'existence de regions d'instabilité en dynamique (*6*). Ann. Inst. H. Poincaré **2**, 369—386 (1932). — [34] Recent contributions to the ergodic theory (*6*). Proc. Nat. Acad. Sci. U.S.A. **18**, 279—282 (1932). — [35] Nouvelles recherches sur les systèmes dynamiques (*6*). Mem. Pont. Accad. Nuovi Lincei (3) **1**, 85—216 (1935). — [36] Note sur la stabilité en dynamique (*1, 6*). J. de Math. (9) **15**, 339—344 (1936). — [37] Sur le problème restreint des trois corps (*6*). Annali Scuola Norm. Sup. Pisa (2) **4**, 267—306 (1935); **5**, 1—42 (1936).

Birkhoff, G. D., and F. R. Bamfork: [1] Divergent series and singular points of ordinary differential equations (*10*). Trans. Amer. Math. Soc. **32**, 114—146 (1930).

—, and O. D. Kellog: [1] Invariant points in function space (*9*). Trans. Amer. Math. Soc. **23**, 96—115 (1922).

—, and R. E. Langer: [1] The boundary problems and developments associated with a system of ordinary linear differential equations of first order (*10*). Proc. Amer. Acad. Arts Sci. **58**, 51—128 (1923).

—, and D. C. Lewis: [1] On the periodic motions near a given periodic motion of a dynamical system (*1, 9*). Annali Mat. pura appl. (4) **12**, 117—133 (1933).

Biryuk, G. I.: [1] On an existence theorem for almost periodic solutions of certain systems of nonlinear differential equations with a small parameter (*8*). Doklady Akad. Nauk SSSR, (N. S.) **96**, 5—7 (1954); **97**, 577—579 (1954).

Blaguière, A.: [1] Les oscillateurs non linéaires et le diagramme de Nyquist (*8*). J. Phys. Radium (8) **13**, 527—540 (1952).

Bloh, Z. S.: [1] Some estimates of the quality of regulation from the frequency characteristics (*2*). Avtom. i Telem. **16**, 258—268 (1955).

Boas, M., R. P. Boas and N. Levinson: [1] The growth of solutions of a differential equation (*3, 5*). Duke Math. J. **9**, 847—853 (1942).

Bocher, M.: [1] On regular singular points of linear differential equations of the second order whose coefficients are not necessarily analytic (*3*). Trans. Amer. Math. Soc. **1**, 40—52 (1900). — [2] Leçons sur les méthodes de Sturm (*5, **). Paris 1917.

Bochner, S.: [1] Allgemeine lineare Differenzgleichungen mit asymptotisch konstanten Koeffizienten (*10*). Math. Z. **33**, 426—450 (1931). — [2] Homogeneous systems of differential equations with almost periodic coefficients (*3, 4*). J. London Math. Soc. **8**, 283—288.

—, and D. V. Widder: [1] A homogeneous differential system of infinite order with nonvanishing solution (*3*). Bull. Amer. Math. Soc. **54**, 409—415 (1948).

Bogdanov, Yu. S.: [1] On normal systems of Lyapunov (*3, 5*). Doklady Akad. Nauk SSSR, (N. S.) **57**, 215—217 (1947).

Bogolyubov, N.: [1] On some statistical methods in mathematical physics (*6*). Akad. Nauk Ukrain SSSR, **1945**.

—, and N. M. Krylov: [1] Quelques exemples d'oscillations non linéaires (*8*). C. R. Acad. Sci Paris **194**, 957—960 (1932). — [2] Sur le phenomène de l'entrainement en radiotechnique (*8*). Ibid. **194**, 1064—1066 (1932). — [3] Fundamental problems of the nonlinear mechanics (*8*). Congres Internat. Math., Zürich 1932, 2, 270—272. — [4] Recherches sur la stabilité statique et la stabilité dynamique des machines synchrones (*8*). Congrès Internat. d'Électricité, Paris 1932, 4, 179—205. — [5] Les phénomène de demultiplication de frequence en radiotechnique (*8*). C. R. Acad. Sci. Paris **194**, 1119—1122 (1932). — [6] Recherches sur la stabilité longitudinale des avions (*8*), pp. 60, Kiev 1932. — [7] Sur qualques propriétés générale des resonances dans la mécanique non linéaire (*8*). C. R. Acad. Sci. Paris **197**, 908—910 (1933). — [8] Problèmes fondamentaux de la mécanique non linéaire (*8*). Bull. Acad. Sci. URSS, **1933**, 475—498. Rev. gén. Sci. **44**, 9—19 (1933) (translation). — [9] Méthodes nouvelles de la mécanique non linéaire dans leur application à l'étude de l'oscillateur a lampe, pp. 242, Moscow 1934. — [10] Über einige Methoden der nicht linearen Mechanik in ihren Anwendungen zur Theorie der nichtlinearen Resonanz. Schweiz. Bauzeitung (8) **103**, 255—257, 267—270 (1934). — [11] L'application des méthodes de la mécanique nonlinéaire à la théorie des perturbations des systèmes canoniques (*8*). Ukrain. Akad. Nauk Kiev **1934**, no. 4, pp. 56. — [12] Sur quelques développments formals en séries dans la mécanique non linéaire (*8*). Ibid. **1934**, no. 5, pp. 99. — [13] Méthodes de la mécanique non linéaire appliquées à l'étude des oscillations stationnaires (*8*). Ibid. **1934**, no. 8, pp. 112. — [14] Sur les solutions quasi-periodiques des équations de la mécanique non linéaire (*8*). C. R. Acad. Sci. Paris **199**, 1592—1593 (1934). — [15] Méthode symbolique de la méchanique non linéaire dans leur applications à l'étude de résonance dans l'oscillateur (*8*). Bull. Acad. Sci. USSR, **1934**, 7—34. — [16] Sur l'étude du cas de resonance dans les problemès de la mécanique non linéaire (*8*). Ibid. **200**, 113—115 (1935). — [17] Méthods de la mécanique non linéaire appliquées à la théorie des oscillations stationaires (*8*). Casopis pro. pest matem. **64**, 107—115 (1935). — [18] Méthodes approchées de la mécanique non linéaire dans leur application des mouvements périodiques et de divers phénomènes de résonance s'y rapportant (*8*). Ukrain. Akad. Nauk Kiev **1935**, no. 14, pp. 114. — [19] Calculation by methods of nonlinear mechanics of the vibrations in lattice girders with consideration of the normal forces (*8*). Ukrain. Sci. Res. Inst. Armament Recueil, Kiev **1935**, 25—42. — [20] Sur quelques théorèmes de la théorie générale de la mesure (*6*). C. R. Acad. Sci. Paris **201**, 1002—1003 (1935). — [21] Les mesures invariantes et la transitivité (*6*). Ibid. **201**, 1454—1456 (1935). — [22] Influence of resonance in transverse vibrations of rods caused by periodic normal forces at one end (*6*). Ukrain. Sci. Res. Inst. Armament Recueil, Kiev **1935**. — [23] Upon some new results in the domain of nonlinear mechanics (*6*). Indian Acad. Sci. Proc. A **3**, 523—526 (1936). — [24] Application de la mécanique non linéaire à quelques problèmes de la radiotechnique moderne (*8*). Onde Électrique **15**, 508—531 (1936). — [25] Les mouvements stationaires généraux dans les systèmes dynamiques de la mécanique non linéaire (*8*). C. R. Acad. Sci. Paris **202**, 200—201 (1926). — [26] Sur les proprietés ergodiques de l'équation de Snoluchovsky (*6*). Bull. Soc. Math. France **64**, 49—56 (1936). — [27] Les mesures invariantes et transitives dans la mécanique non linéaire (*6*). Mat. Sbornik (N.S.) **1**, 707—710 (1936). — [28] Mécanique non linéaire (*8, 9*). Ukrain. Akad. Nauk. Kiev **3**, 29—53 (1937). — [29] La théorie générale de la mesure dans la mécanique non linéaire (*6*). Ibid. **3**, 55—112 (1957) (French resumé 113—118). — [30] La théorie générale de la mesure dans son application a l'étude des systèmes dynamique non linéaire (*6*). Ann. of Math. **38**, 65—113 (1937). — [31] Sur les itérations répétées avec les paramètres variables (*8*). Ukrain. Akad. Nauk Kiev **3**, 201—211 (1937); in Ukrainian, 191—200. — [32] L'effet de la variation statistique des paramètres sur le mouvement des systèmes dynamiques conservatifs pendant des periods de temps suffisament longues (*9*). Ibid. **3**, 136—153, 172—190 (1937); in Ukrainian 119—136, 154—171. — [33] Introduction to nonlinear mechanics (*, *8, 9*). Ibid. **1—2**, pp. 365 (1937). — [34] Introduction to nonlinear mechanics (a free translation by S. Lefschetz) (*). Ann. of Math. Studies **11**, pp. 106 (1947).

Bogolyubov, N., and Yu. A. Mitropolsky: [1] Asymptotic methods in the theory of nonlinear oscillations (*). Moscow 1955; second ed. 1958. Engl. tr.: Hindustan Publ. Corp., Delhi 1961 (Gordon and Breach Sci. Publ., New York).

Bohl, P.: [1] Über die Bewegung eines mechanischen Systems in der Nähe einer Gleichgewichtslage (*8*). J. reine u. angew. Math. **127**, 179—276 (1904). — [2] Über eine Differentialgleichung der Störungstheorie (*4, 8*). J. für Math. **131**, 268—321 (1906). — [3] Sur certaines équations différentielles d'un type général utilisables en mécanique (*4, 8*). Bull. Soc. Math. France **38**, 5—138 (1910). — [4] Über die hinsichtlich der unabhängigen und abhängigen Variablen periodische Differentialgleichung erster Ordnung (*6, 8*). Acta Math. **40**, 321—336 (1916).

Böhm, C.: [1] Nuovi criteri di esistenza di soluzioni periodiche di una nota equazione differenziale non lineare (*8*). Annali Mat. pura appl. (*4*) **35**, 343—353 (1953).

BONDARENKO, G. V.: [1] Hill's equation and its application in the domain of engineering vibrations (*4*). Izdat. Akad. Nauk SSSR, Moscow **1936**.

BORG, G.: [1] Über die Stabilität gewisser Klassen von linearen Differentialgleichungen (*4*). Ark. Mat. Astr. Fys. **31** A, no. 1, 31 pp. (1944). — [2] Eine Umkehrung der Sturm-Liouvilleschen Eigenwertaufgabe. Bestimmung der Differentialgleichung durch die Eigenwerte (*5*). Acta math. **78**, 1—96 (1946). — [3] Inverse problems in the theory of characteristic values of differential systems (*5*). C. R. Dixième Congrès Math. Scandinaves 1946, 172—180. — [4] Bounded solutions of a system of differential equations (*8*). Ark. Mat. Astr. Fys. **34** B, no. 24, 7 pp. (1948). — [6] Two notes concerning stability (*4*). Colloque Internat. Vibrations non linéaires, Ile de Porquerolles, 1951, 21—29. — [5] On a Lyapunov criterion of stability (*4*). Amer. J. Math. **71**, 67—70 (1949).

BORŮVKA, O.: [1] Sur les intégrales oscillatoires des équations différentielles linéaires du second ordre (*5*). Cekoslovack. Mat. Z. **3**, 199—255 (1953).

BOTHWELL, F. E.: [1] Transients in multiply periodic nonlinear system (*8*). Quart. Appl. Math. **8**, 247—254 (1950).

BOTTEMA, O.: [1] On the stability of the equilibrium of a linear mechanical system (*2*). Z. angew. Math. Phys. **6**, 97—104 (1955).

BOUTROUX, P.: [1] Leçons sur les functions définies par les équations différentielles du premier ordre (*). Gauthier-Villars 1908.

BRAINERD, J. G.: [1] Note on modulation (*4*). Proc. Inst. Radio Engrs., N. Y. **28**, 136 (1940). — [2] Stability of oscillations in systems obeying MATHIEU's equation (*4*). J. Franklin Inst. **253**, 135 (1942).

—, and C. N. WEYGANDT: [1] Solutions of MATHIEU's equations (*4*). Phil. Mag. **30**, 458 (1940).

BREIT, G., and G. E. BROWN: [1] Perturbation methods for Dirac radial equations (*8*). Phys. Rev. (*2*) **76**, 1307—1310 (1949).

BRILLOUIN, L.: [1] A practical method for solving HILL's equation (*10*). Quart. Appl. Math. **6**, 167—178 (1948). — [2] The B. W. K. approximation and HILL's equation (*4, 5*). Quart. Appl. Math. **6**, 169 (1949); **7**, 363—380 (1950).

BROCK, P.: [1] The nature of solutions of a Rayleigh-type forced vibration equation with a large coefficient of damping (*8*). J. Appl. Phys. **24**, 1004—1007 (1953).

BRODIN, J.: [1] Stabilité et continuité parametrique d'un servomechanisme linéaire à coefficient dépendant du temp (*3*). C. R. Acad. Sci. Paris **234**, 800—801 (1952).

BROMBERG, P. V.: [1] On the problem of stability of a class of nonlinear systems (*6, 7, 8*). Prikl. Mat. Meh. SSSR. **14**, 561—562 (1950).

BROWN, E. W.: [1] Elements of theory of resonance (*). Rice Inst. Pamphl. **19**, 1—60 (1932).

BROWN, G. S., and D. P. CAMPBELL: [1] Principles of servomechanisms (*, *2*), pp. 400, Wiley 1948.

BROWNELL, F. H.: [1] Nonlinear difference-differential equations and contributions to the theory of nonlinear oscillations (*8*). Ann. of Math. Studies **20**, 89—148 (1950).

BRUIJN, N. G. DE: [1] A note on VAN DER POL's equation (*10*). Philips Res. Rep. **1**, 401—406 (1946). — [2] The asymptotically periodic behavior of the solutions of some linear functional equations (*10*). Amer. J. Math. **71**, 313—330 (1949). — [3] On some linear functional equations (*8*). Publ. Math. Debrecen **1**, 129—134 (1950).

BÜCKNER, H.: [1] Inequalities for solutions of linear differential equations. A contribution to the theory of servomechanisms (*2*). Edinburgh Math. Notes **38**, 13—16 (1952).

BULGAKOV, A. A.: [1] The dynamics of contact synchronization of electrical drives (*9*). Avtom. i Telem. **8**, 108—116 (1947).

BULGAKOV, B. V.: [1] Maintained oscillations of automatically controlled systems (*6, 7, 8*). Doklady Akad. Nauk SSSR, (N. S.) **37**, 250—253 (1942). — [2] Regulating circuits with links having up to several degrees of freedom (*2*). Prikl. Mat. Meh. SSSR, **14**, 619—634 (1950). — [3] The discriminant curve and the domain of aperiodic stability (*2*). Ibid. **14**, 453—458 (1950). — [4] On the application of the method of Poincaré to a free pseudo-linear oscillating system (*8*). Ibid. **6**, no. 4 (1942). — [5] Periodic processes in free pseudo-linear oscillatory systems (*8*). Frankl. Inst. 235, 1943. — [6] Problems of the control theory with nonlinear characteristics (*8*). Prikl. Mat. Meh. SSSR, **10**, 313—332 (1946). — [7] On the accumulation of disturbances in linear oscillatory systems with constant parameters (*2*). Doklady Akad. Nauk SSSR, (N. S.) **51**, 343—345 (1946). — [8] On the method of VAN DER POL and its application to nonlinear control problems (*8*). J. Frankl. Inst. **235**, 595—615 (1943); **241**, 31—54 (1946). — [9] On equivalence and consistency of systems of linear differential equations with constant coefficients (*2*). Prikl. Mat. Mech. SSSR, **16**, 15—22 (1952). — [10] Oscillations (*), Izdat. Moscow-Leningrad 1949, 464 pp.; 2. edit. 1954, 891 pp.

—, and N. T. KUZOVKOV: [1] On the accumulation of disturbances in linear systems with varying parameters (*2*). Prikl. Mat. Meh. SSSR, **14**, 7—12 (1950).

—, and M. Z. LITVIN-SEDOL: [1] On a problem of automatic regulation with nonlinear characteristic (*6, 7, 8*). Avtom. i Telem. **10**, 329—341 (1949).

— N. G.: [1] Oscillations of quasilinear autonomous systems with many degrees of freedom and a nonanalytic characteristic of nonlinearity (*8*). Prikl. Mat. Meh. SSSR, **19**, 265—272 (1955).

BURDINA, V. I.: [1] A criterion of boundedness for solutions of a system of differential equations of 2nd order with periodic coefficients (*4*). Doklady Akad. Nauk SSSR, (N. S.) **90**, 329—332 (1953). — [2] On boundedness of the solutions of a system of differential equation (*4*). Ibid. **93**, 603—606 (1953).

BURGERS, J. M.: [1] Adiabatic invariants of mechanical systems (*8*). Phil. Mag. **1917**, 33.

BURTON, L. P.: [1] Oscillation theorems for the solutions of linear nonhomogeneous, second order differential systems (*5*). Pacific J. Math. **2**, 281, 289 (1952).

BUTENIN, N. V.: [1] On the theory of forced oscillations in a nonlinear mechanical system with two degrees of freedom (*8*). Prikl. Mat. Meh. SSSR, **13**, 337—348 (1949). — [2] On the theory of "resonance" in a mechanical auto-oscillating system with gyroscopic terms (*8*). Ibid. **14**, 45—56 (1950). — [3] On the theory of forced synchronization (*8*). In memory of A. A. ANDRONOV. Izdat. Akad. Nauk SSSR, Moscow **1955**, 187—195.

BUTLEWSKI, Z.: [1] On the real solutions of linear differential equations (*5*). Wiadomosci Mat. **54**, 17—81 (1937). — [2] Sur les integrales bornées des équations différentielles (*5*). Ann. Soc. Polon. Math. **1939**. — [3] Sur les intégrales d'une équation différentielle du second ordre (*5*). Mathematica, Cluj **12**, 36—48 (1936). — [4] Sur les intégrales oscillantes et bornées d'une équation différentielle du second ordre (*5*). Annali Scuola Norm. Sup. Pisa (2) **9**, 187—200 (1940). — [5] Sur les zéros des intégrales réelles des équations différentielles linéaires (*5*). Mathematica, Cluj **17**, 85—110 (1941). — [6] Sur les intégrales bornées des équations différentielles (*5*). Annals Soc. Polon. Math. **18**, 47—54 (1945). — [7] Sur les intégrales d'un système d'équations différentielles linéaires ordinaires (*5*). Studia Math. **10**, 40—47 (1948). — [8] Sur les intégrales d'un système d'équations différentielles (*5*). Ann. Univ. Mariae Curie-Sklodowska, Sect. A **4**, 73—104 (1950). — [9] Sur les intégrales oscillantes d'une équations du second ordre (*5*). Bull. Soc. Amis Sci. Poznan,

Ser. B, **11**, 3—22 (1951). — [10] Un théorème de l'oscillation (*5*). Ann. Soc. Polon. Math. **24**, 95—110 (1951).

BYLOV, B. F.: [1] On an estimate for the characteristic numbers of the solution of almost diagonal systems of linear differential equations (*3*). Prikl. Mat. Meh., SSSR, **14**, 114—116 (1950). — [2] On the characteristic numbers of the solutions of linear differential equations (*3*). Ibid. **14**, 341—352 (1950). — [3] On stability beyond the greatest characteristic exponent (*6*). Dokl. Akad. Nauk SSSR, (N. S.) **103**, 181—184 (1955).

CACCIOPPOLI, R.: [1] Sopra un criterio di stabilità (*3, 5*). Rend. Accad. Lincei (6) **11**, 251—254 (1930).

—, and A. GHIZZETTI: [1] Ricerche asintotiche per una particulare equazione differenziale non lineare (*8*). Rend Accad. Italia (7) **3**, 427—440, 493—501 (1942).

CAFIERO, F.: [1] Un'osservazione sulla continuità rispetto ai valori iniziali degli integrali dell'equazione: $y' = f(x, y)$ (*1*). Rend. Accad. Lincei (8) **3**, 479—482 (1947). — [2] Su due teoremi di confronto relativi ad un'equazione differenziale ordinaria del primo ordine (*5*). Boll. Unione Mat. Ital. (3) **3**, 124—128 (1948).

CAHEN, G.: [1] Systèmes électromécaniques non linéaires (*8*). Rev. gén. Électr. **62**, 277—293 (1953).

CALAMAI, G.: [1] Sul sistema canonico di una classe di equazioni differenziali del secondo ordine a coefficienti periodici (*4, 5*). Rend. Accad. Lincei (6) **19**, 560—566 (1934). — [2] Sulla stabilità delle soluzioni per l'equazione differenziale del secondo ordine a coefficienti periodici (*4, 5*). Rend. Accad. Italia (7) **3**, 183—193 (1942).

CALIGO, D.: [1] Sulla soluzione della equazione lineare omogenes del secondo ordine a coefficienti periodici (*5*). Boll. Unione Mat. Ital. (2) **3**, 370—372 (1941). — [2] Un criterio sufficiente di stabilità per le soluzioni dei sistemi di equazioni integrali lineari e sue applicazioni ai sistemi di equazioni differenziali lineari (*3*). Atti Secondo Congr. Unione Mat. Ital., Bologna, 1940, p. 177—185. — [3] Un criterio sufficiente di stabilità per le soluzioni dei sistemi di equazioni integrali lineari e sue applicazioni (*3*). Rend. Accad. Italia (7) **1**, 497—506 (1940). — [4] Sulle equazioni differenziali lineari del secondo ordine a coefficienti periodici (*4, 5*). Mem. Accad. Italia (7) **13**, 1025—1033 (1943). — [5] Sopra una classe di equazioni differenziali non lineari (*8*). Mem. Accad. Sci. Torino (3) **1**, 24 pp. (1952). — [6] Sulla integrazione delle equazioni differenziali del secondo ordine a riferimento razionale (*8*). Rend. Mat. Univ. Roma (5) **11**, 322—337 (1952). — [7] Comportemento asintotico degli integrali dell'equazione $y'' + A(x)y = 0$ nell'ipotesi $\lim A(x) = 0$ (*3, 5*). Boll. Unione Mat. Ital. (2) **3**, 286—295 (1941).

CAMBI, E.: [1] Una equazione differenziale del secondo ordine a coefficiente periodico reciproco di quello di MATHIEU (*4, 5*). Rend. Accad. Lincei (8) **1**, 1035—1041, 1181—1187 (1946). — [2] Una equazione differenziale del secondo ordine a coefficiente periodico reciproco di quello di MATHIEU (*4, 5*). Ricerca Sci. **17**, 186—190 (1947). — [3] Sulla integrazione delle equazioni differenziali lineari a coefficienti periodici (*4*), Rend. Mat. Univ. Roma (5) **7**, 103—123 (1948). — [4] Trigonometric components of a frequency-modulated wave (*4*). Proc. I. R. E. **36**, 42—49 (1948). — [5] The simplest form of second order linear differential equation with periodic coefficient, having finite singularities (*5*). Proc. Roy. Soc. Edinburgh, Sect. A **63**, 27—51 (1950).

CAMERON, R. H.: [1] Linear differential equation with almost periodic coefficients (*3, 4*). Duke Math. J. **1**, 356—360 (1935). — [2] Linear differential equations with almost periodic coefficients (*3, 4*). Acta math. **69**, 21—56 (1938).

CAMPBELL, R.: [1] Comportement asymptotique des fonctions de Mathieu associées pour des paramètres infiniment grands (*10*). C. R. Acad. Sci. Paris **225**, 371—373 (1947).

Capra, V.: [1] Sulle vibrazioni libere di un sistema meccanico ad un grado di libertà (*8, 9*). Rend. Sem. Mat. Politec. Torino **13**, 307—325 (1954).

Caprioli, L.: [1] Sulle soluzioni periodiche di una equazione fortemente nonlineare (*8, 9*). Boll. Unione Mat. Ital. (3) **9**, 271—280 (1954). — [2] Su un modello meccanico per le oscillazioni di rilassamento (*8, 9*). Rend. Accad. Sci. Ist. Bologna (11) **1**, 152—168 (1954).

Carathéodory, C.: [1] Vorlesungen über reelle Funktionen (*). Teubner 1927, 2nd ed.

Carleman, T.: [1] Dévélopments asymptotiques des solutions d'une classe d'équations différentielles linéaires (*10*). Acta Math. **43**, 319—336 (1922). — [2] Sur les équations intégrales singulières à noyau reel et symétrique (*). Uppsala Univ. Arsk. **3**, p. 228 (1923). — [3] Application de la théorie des équations intégrales lineaires aux systèmes d'équations différentialles non linéaires (*8*). Acta Math. **59**, 63—87 (1932).

Carleson, L.: [1] On infinite differential equations with constant coefficients (*2*). Math. Scand. **1**, 31—38 (1953).

Carrier, G. F.: [1] On the nonlinear vibration problem of the elastic string (*8*). Quart. Appl. Math. **3**, 157—165 (1945).

Cartan, E., and H.: [1] Note sur la génération des oscillations entretenues (*8*). Ann. Postes Télég. Téléph. **14**, 1196 (1925).

Cartwright, M. L.: [1] On nonlinear differential equations of the second order. I. The equation $\ddot{x} - k(1 - x^2)\dot{x} + x = p \cdot k\lambda \cos(\lambda t + \alpha)$, k small and λ near 1 (*9*). Proc. Cambridge Philos. Soc. **45**, 495—501 (1949). — [2] Forced oscillations in nearly sinusoidal systems (*9*). J. Inst. Electr. Engrs. **95**, 88—96 (1948). — [3] Topological aspect of forced oscillations (*, *9*). Research **1**, 601—606 (1948). — [4] Nonlinear vibrations (*9*). Rep. for the Brit. Assm. Advanc. Sci. **6**, no. 21 (1949). — [5] Forced oscillations in nonlinear systems (*9*). Contrib. to the Theory of Nonlinear Oscillations. Ann. of Math. Studies 1950, no. 20, 149—241. — [6] Forced oscillations in nonlinear systems (*9*). J. Res. Nat. Bur. Stand. **45**, 514—518 (1950). — [7] Van der Pol's equation for relaxation oscillations (*9*). Contrib. to the Theory of Nonlinear Oscillations. Ann. of Math. Studies 1952, no. 29, 3—18. — [8] Nonlinear vibrations: a chapter in mathematical history (*9*). Math. Gaz. **36**, 81—88 (1952).

—, E. T. Copson and J. Greig: [1] Nonlinear vibrations (*9*). Advanc. Sci. **6**, no. 21, 12 pp. (1949).

—, and J. E. Littlewood: [1] Addendum to "On nonlinear differential equations of the second order" (*9*). Ann. of Math. (2) **50**, 504—505 (1949). — [2] Some fixed point theorems. With appendix by H. D. Ursell (*9*). ibid. (2) **54**, 1—37 (1951). — [3] On nonlinear differential equations of the second order. The equation $\ddot{y} - k(1 - y^2)\dot{y} + y = b\lambda k \cos(\lambda t + a)$, k large (*9*). J. London Math. Soc. **20**, 180—189 (1945). — [4] On nonlinear differential equations of the second order. The equation $\ddot{y} + kf(y)\dot{y} + g(y, k) = p(t) = p_1(t) + kp_2(t)$; $k > 0$, $f(y) \geq 1$ (*9*). Ann. of Math. (2) **48**, 472—494 (1947). — [5] Errata: On nonlinear differential equations of the second order (*9*). Ibid. (2) **49**, p. 1010 (1948).

Cashwell, E. D.: [1] The asymptotic solutions of an ordinary differential equation in which the coefficient of the parameter is singular (*10*). Pacific J. Math. **1**, 337—352 (1951).

Castro, A. de: [1] On small oscillations of dissipative systems (*8, 9*). Rev. Mat. Hisp-Amer. (4) **10**, 47—50 (1950). — [2] Recurrence formulas for the differential equation $y'' + 2ng(x)y' + r(x)y = 0$ (*5*). Ibid. (4) **11**, 217—221 (1951). — [3] On the asymptotic behavior of integrals of linear differential equations (*3*). Rev. Acad. Ci. Madrid **45**, 351—362 (1951). — [4] Studies on nonlinear mechanics. On the general differential equation of relaxation oscillations (*8*). Rev. Mat. Hisp.-Amer. (4) **12**, 266—280, 317—329 (1952). — [5] Soluzioni periodiche di una equazione differenziale del secondo ordine (*8*). Boll. Unione Mat. Ital. (3) **8**, 26—29 (1953). — [6] Sopra l'equazione differenziale delle oscillazione nonlineari (*8, 9*). Riv. Mat. Univ. Parma **4**, 133—143 (1953). — [7] Sulle oscillazioni non lineari dei sistemi in uno o più gradi di libertà (*8*). Rend. Sem. Mat. Univ. Padova **22**, 294—304 (1953). — [8] Sopra l'equazione differenziale di risposta di un circuito elettrico (*8*). Boll. Unione Mat. Ital. (3) **9**, 167—169 (1954). — [9] Un teorema di confronto per l'equazione differenziale delle oscillazioni di rilassamento (*8*). Ibid. (3) **9**, 280—282 (1954). — [10] Sull'esistenza ed unicità delle soluzioni periodiche dell' equazione $x''+f(x, x')x'+g(x)=0$ (*9*). Boll. Unione Mat. Ital. (3) **9**, 369—372 (1954).

Cecconi, J.: [1] Su di una equazione differenziale di rilassamento (*9*). Rend. Accad. Lincei (8) **9**, 38—44 (1951). — [2] Su di una equazione differenziale non lineare di secondo ordine (*9*). Annali Scuola Norm. Sup. Pisa (3) **4**, 245—278 (1951).

Cecik, V. A.: [1] Necessary and sufficient conditions of semistability of a limit cycle (*9*). Uspehi Mat. Nauk SSSR, (N. S.) **10**, 183—187 (1955).

Cesari, L.: [1] Sulla stabilità delle soluzioni delle equazioni differenziali lineari (*3*). Annali Scuola Norm. Sup. Pisa (2), **3**, 131—148 (1940). — [2] Proprietà asintotiche delle equazioni differenziali lineari ordinarie (*3*). Rend. Sem. Mat. Univ. Roma (4) **3**, 171—193 (1940). — [3] Un nuovo criterio di stabilità per le soluzioni delle equazioni differenziali lineari (*3*). Annali Scuola Norm. Sup. Pisa (2) **9**, 163—186 (1940). — [4] Sulla stabilità delle soluzioni dei sistemi di equazioni differenziali lineari a coefficienti periodici (*4*). Mem. Accad. Italia (6) **11**, 633—695 (1941). — [5] Existence theorem for periodic solutions of nonlinear Lipschitzian differential systems and fixed point theorem (*8*). Contrib. Theory Nonlinear Oscillations **5**, 115—172 (1960). — [6] Functional analysis and periodic solutions of nonlinear differential equations (*8*). Contrib. Theory Differential Equations **1**, (1962). — [7] Periodic solutions of nonlinear differential systems (*8*). Proc. Symp. Polyt. Inst. Brooklyn, **10**, 545—560 (1961). — [8] Existence theorems for periodic solutions of nonlinear differential systems (*8*). Boletin Soc. Mat. Mexicana, **1960**, 24—41. — [9] Problems of asymptotic behavior and stability (*8*). Trans. AIEE, Appl. Ind., **80**, Sept. 1961, 161—166. — [10] Branching of cycles of autonomous nonlinear differential systems (*8*). To appear.

—, and H. R. Bailey: [1] Boundedness of solutions of linear differential systems with periodic coefficients (*4*). Archive Rat. Mech. Anal. **1**, 246—271 (1958).

—, and J. K. Hale: [1] Second order linear differential systems with periodic L-integrable coefficients (*4*). Riv. Mat. Univ. Parma **5**, 55—61 (1954); **6**, p. 159 (1955). — [2] A new sufficient condition for periodic solutions of weakly nonlinear differential systems (*8*). Proc. Amer. Math. Soc. **8**, 757—764 (1957).

Cetaev, N. G.: [1] Un théorème sur l'instabilité (*7*). Doklady Akad. Nauk SSSR, **2**, 529—534 (1934). — [2] Theorem concerning the nonstability of regular systems (*6, 7*). Prikl. Mat. Meh. SSSR, **8**, 323—326 (1944). — [3] The smallest characteristic number (*3, 6, 7*). Ibid. **9**, 193—196 (1945). — [4] Calculation of particular solutions for systems of linear differential equations with constant coefficients (*2, 6, 7*). Ibid. **10**, 291—294 (1946). — [5] Stability of motion (*). Izdat. Akad. Nauk SSSR, Moscow 1946, 204 pp. — [6] Concerning the stability and instability of irregular systems (*6, 7*). Prikl. Mat. Meh. SSSR, **12**, 639—642 (1948). — [7] On the sign of the smallest characteristic number (*6, 7*). Ibid. **12**, 101—102 (1948). — [8] On the choice of parameters of a stable mechanical system (*2, 6, 7*). Ibid. **15**, 371—372 (1951). — [9] On unstable equilibrium in certain cases when the force function is not maximum (*7*). Ibid. **16**, 89—93 (1952).

Chazy, J.: [1] Sur les courbes définies par les équations différentielles du second ordre (*9*). Bull. Sci. Math. (2) **45**, 270—280 (1921). — [2] Sur les multiplicités singulières du problème des trois corps (*9*). Ibid. **56**, 79—104 (1932). — [3] Sur une généralisation du pendule cycloidal D'HUYGENS (*9*). C. R. Acad. Paris **213**, 93—98 (1941). — [4] Sur une équation différentielle de premier order et du second degré (*9*). Ann. Sci. École Norm. Sup. Paris (3) **61**, 45—71 (1944). — [5] Sur la théorie des centres (*9*). C. R. Acad. Sci. Paris **221**, 7—10 (1945). — [6] Sur les courbes définies par les équations différentielles (*9*). ibid. **221**, 457—459 (1945).

CHEPELEFF, W. M.: [1] Zur Frage der Stabilität der Bewegung (*2, 5*). Appl. Math. Mech. **3**, 144—148 (1936).

CHERRY, T. M.: [1] Uniform asymptotic formulae for functions with transition points (*10*). Trans. Amer. Math. Soc. **68**, 224—257 (1950).

CIMINO, M.: [1] Sul comportamento asintotico degli integrali di una equazione della dinamica (*5*). Boll. Unione Mat. Ital. (2) **5**, 78—87 (1943). — [2] Sulle soluzioni dell'equazione generale del potenziale newtoniano di una sfera fluida in equilibrio (*5*). Ibid. (3) **8**, 164—172 (1953).

CODDINGTON, E. A.: [1] The classical existence theorem of nonlinear analytic differential equations (*8*). Proc. Amer. Math. Soc. **1**, 738—743 (1950).

—, and N. LEVINSON: [1] A boundary value problem for a nonlinear differential equation with a small parameter (*8*). Proc. Amer. Math. Soc. **3**, 73—81 (1952). — [2] Perturbations of linear systems with constant coefficients possessing periodic solutions (*8*). Contrib. Theory of Nonlinear Oscillations. Ann. of Math. Studies 1952, no. 29, 19—35. — [3] Theory of ordinary differential equations (*), pp. 429. McGraw-Hill 1955,

—, and A. WINTNER: [1] On the classical existence theorem of analytic differential equations (*8*). Amer. J. Math. **71**, 886—892 (1949).

COHEN, H.: [1] The stability equation with periodic coefficients (*8*). Quart. Appl. Math. **10**, 266—270 (1952). — [2] Subharmonic synchronization for the forced VAN DER POL equation (*8*). Coll. Internat. Vibrations nonlinéaires, Ile de Porquerolles 1951, 169—187. — [3] On subharmonic synchronization of nearly-linear systems (*8*). Quart. Appl. Math. **13**, 102—105 (1955).

COHN, G. I., and B. SALTZBERG: [1] Solution of nonlinear differential equations by the reversion method (*8*). J. Appl. Phys. **24**, 180—186 (1953).

COLLATZ, L.: [1] Über Stabilität von Regelungen mit Nachlaufzeit (*8*). Z. angew. Math. Mech. **25/27**, 60—63 (1947).

COLOMBO, G.: [1] Sulle frequenze e sullo smorzamento delle oscillazioni di una sfera vibrante radialmente in un fluido (*8*). Rend. Sem. Mat. Univ. Padova **17**, 107—114 (1948). — [2] Sull'equazione differenziale non lineare del terzo ordine di un circuito oscillante triodico (*9*). Ibid. **9**, 114—140 (1950). — [3] Sulle oscillazioni non lineari in due gradi di libertà (*9*). Ibid. **19**, 413—441 (1950). — [4] Sopra un sistema non lineare in due gradi di libertà (*9*). Ibid. **21**, 64—98 (1952). — [5] Sopra un fenomeno di isteresi oscillatoria (*9*). Ibid. **21**, 370—382 (1952). — [6] Sopra un caso singolare che si presenta in un problema di stabilità in meccanica non lineare (*9*). Ibid. **22**, 123—133 (1953). — [7] Sulle oscillazioni forzate di un circuito compredente una bobina a nucleo di ferro (*9*). Ibid. **22**, 380—398 (1953); **23**, 407—421 (1954).

CONTE, S. D., and W. C. SANGREN: [1] An asymptotic solution for a pair of first order equations (*10*). Proc. Amer. Math. Soc. **4**, 696—702 (1953).

CONTI, R.: [1] Criteri sufficienti di stabilità per i sistemi di equazioni integrali lineari (*3*). Rend. Accad Lincei (8) **11**, 154—169 (1951). — [2] Un teorema di confronto per le equazioni alle differenze finite, lineari, del 2° ordine (*3*). Boll. Unione Mat. Ital. (3) **6**, 208—213 (1952). — [3] Un criterio sufficiente di stabilità per i sistemi di equazioni differenziali lineari del primo ordine, omogenee (*3*). Ibid. (3) **6**, 288—293 (1951). — [4] Soluzioni periodiche dell'equazione di

LIÉNARD generalizzata. Esistenza ed unicità (8). Ibid. (3) **7**, 111—118 (1952). — [5] Sulla stabilità dei sistemi di equazioni differenziali lineari (3). Riv. Mat. Univ. Parma **6**, 3—35 (1955). — [6] Sulla t-similitudine tra matrici e la stabilità dei sistemi lineari (3). Rend. Accad. Lincei (8) **19**, 247—250 (1955). — [7] Limitazioni in ampiezza delle soluzioni di un sistema di equazioni differenziali e applicazioni (1). Boll. Un. Mat. Ital. (3) **11**, 344—349 (1956).

COOKE, K. L.: [1] The asymptotic behavior of the solutions of linear and nonlinear differential-difference equations (8). Trans. Amer. Math. Soc. **75**, 80—105 (1953). — [2] The rate of increase of real continuous solutions of algebraic differential-difference equations of the first order (8). Pacific J. Math. **4**, 483—501 (1954).

CORBEILLER, P. LE: [1] Les systèmes auto-entretenues et les oscillations de relaxation (*, 8). Paris 1931. — [2] Theory of prime movers, (8) 1—18. Brown Univ. Lecture notes 1942/43. — [3] On the oscillations of triodes and regulators (8). C. R. 3. Congr. Mec. Appl., Stockholm 1930, 3, p. 205. — [4] The nonlinear theory of the maintenance of oscillations (8). J. Inst. Electr. Engin., London **79**, 361 (1936).

CORDUNEANU, C.: [1] Solutions presque-périodiques des équations différentielles non linéaires du second ordre (8). Com. Acad. R. P. Romine **5**, 793—797 (1955).

CORONATO, S.: [1] Teoremi di confronto per equazioni differenziali lineari del secondo ordine (5). Rend. Accad. Sci. Fis. Mat. Napoli (4) **9**, 75—82 (1939).

COTTON, E.: [1] Sur les solutions asymptotiques des équations différentielles (3). Ann. Sci. Ecole Norm. Sup. Paris (3) **28**, 473—521 (1911). — [2] Sur l'instabilité de l'équilibre (6). C. R. Acad. Sci. Paris **1911**, 153. — [3] Approximations successives et les équations différentielles (*). Mem. Act. Sci. 28. Herman 1928.

COURANT, R., and D. HILBERT: [1] Methoden der Mathematischen Physik (*). Springer 1924.

COUWENHOVEN, A. C.: [1] Über die Schüttelerscheinungen elektrischer Lokomotiven mit Kurbelantrieb (4). Forschungsarb. VDI, **1919**, H. 218.

CRAMER, L.: [1] Die Verringerung der Zahl der Stabilitätskriterien bei Voraussetzung positiver Koeffizienten der charakteristischen Gleichung (2). Z. angew. Math. Mech. **33**, 221—227 (1953).

CRANDALL, S. H.: [1] On restoring forces which admit forcing terms of non-critical amplitude (8). Bull. Amer. Math. Soc. **53**, 633—636 (1947).

CRUM, M. M.: [1] Associated Sturm-Liouville systems (5). Quart. J. Math., Oxford Ser. (2) **6**, 121—127 (1955).

CUDOV, L. A.: [1] The inverse Sturm-Liouville problem (5). Mat. Sbornik, (N. S.) **25** (67), 451—456 (1949).

CUÉNOD, M.: [1] Étude des propriétés d'un réglage automatique (2, *). Bull. techn. Suisse Rom. **73**, 105—115, 121—125 (1947).

CUNNINGHAM, W. J.: [1] Simultaneous nonlinear equations of growth (9). Bull. Math. Biophys. **17**, 101—110 (1955).

CYPKIN, YA. Z.: [1] The degree of stability of systems with retarded feedback (2). Avtom. i Telem. **8**, 145—155 (1947).

DALECKII, YU. L.: [1] On the asymptotic solution of a vector differential equation (3). Doklady Akad. Nauk SSSR, (N. S.) **92**, 881—884 (1953).

DAMKÖHLER, W.: [1] Periodische Extremalen von Minimumtyp in Ringbereichen (6). Annali Scuola Normale Sup. Pisa (2) **5**, 127—140 (1936).

DEMIDOVIC, B. P.: [1] Periodic solutions of nonlinear systems of the second order of ordinary differential equations whose second members are periodic relatively to the independent variable (8). Doklady Akad. Nauk SSSR, (N. S.) **61**, 601—603 (1948). — [2] On a critical case of instability in the sense of LYAPUNOV (3). Ibid. **72**, 1005—1008 (1950). — [3] On stability in the sense of LYAPUNOV of a linear system of ordinary differential equations (*3*). Mat. Sbornik, (N. S.) **28**, (70), 659—684 (1951). — [4] On some properties of the characteristic exponents of systems of ordinary linear differential equations with periodic coefficients (*4*). Sci. Rep. Moscow State Univ. **163**, 123—132 (1952). — [5] On a case of almost periodicity of a solution of an ordinary differential equation of the lst order (*8*). Uspehi Mat. Nauk, (N. S.) **8**, no. 6 (58), 103—106 (1953). — [6] On a generalization of N. Bogolyubov's principle of averaging (*8*). Doklady Akad. Nauk SSSR, (N. S.) **96**, 693—694 (1954). — [7] On some averaging theorems for ordinary differential equatione (*). Mat. Sbornik, (N. S.) **35** (77), 73—92 (1954).

Denjoy, A.: [1] Sur les courbes défines par les équations différentielles à la surface du tore (*6*). J. de Math. (9) **11**, 333—375 (1932).

Dieudonné, J.: [1] Sur la méthode du col (*9*). Bol. Soc. Mat., São Paulo **2**, 7—34 (1947).

Diliberto, S. P.: [1] On systems of ordinary differential equations (*8*, *9*). Contrib. to the Theory of nonlinear Oscillations **3**. Ann. of Math. Studies **20**, 1—38 (1950). — [2] An application of periodic surfaces (solution of a small divisor problem) (*8*). Contrib. Theory Nonlinear Oscillations **3**, 257—260 (1956). — [3] Bounds for periods of periodic solutions (*8*). Ibid. **3**, 269—276 (1956).

—, and G. Hufford: [1] Perturbation theorems for nonlinear ordinary differential equations (*8*). Contrib. Theory Nonlinear Oscillations **3**, 207—236 (1956).

—, and M. D. Marcus: [1] A note on the existence of periodic solutions of differential equations (*8*). Contrib. Theory Nonlinear Oscillations **3**, 237—242 (1956).

Dini, U.: [1] Studi sulle equazioni differenziali lineari (*3*). Annali Mat. Pura Appl. (3) **2**, 297—324 (1899); **3**, 125—183 (1900); **11**, 285—335 (1905); **12**, 179—262 (1906).

Dirichlet, L.: [1] Über die Stabilität des Gleichgewichts (*7*). J. reine u. angew. Math. **32** (1846).

Dolgolenko, Yu. V.: [1] Stability and auto-oscillations of a class of relay systems of automatic discontinuous regulation (*8*). Akad. Nauk SSSR, Inzen. Sbornik **13**, 161—176 (1952).

Donskaya, L. I.: [1] On the structure of the solutions of a system of three linear differential equations in the neighborhood of the irregular singular points (*3*). Doklady Akad. Nauk SSSR, (N. S.) **80**, 321—324 (1951).

Dorodnicyn, A. A.: [1] Asymptotic solution of van der Pol's equation (*10*). Prikl. Mat. Mech. SSSR, **11**, 313—328 (1947). Amer. Math. Soc. Translation no. 88.

Doyle, T. C.: [1] Invariant theory of the general ordinary, linear, homogeneous, second order, differential boundary problem (*5*). Duke Math. J. **17**, 249—261 (1950).

Dragilev, A. V.: [1] Periodic solutions of the differential equation of nonlinear oscillations (*8*, *9*). Prikl. Mat. Meh. SSSR, **16**, 85—88 (1952).

Dramba, C.: [1] Sur les singularités de certains systèmes différentiels (*9*). Bull. Math. Soc. Roumaine Sci. **48**, 27—31 (1947).

Dubois-Violette, P.: [1] Sur les points singuliers exceptionells des équations différentielles du premier ordre, considérés comme limites de points singuliers simples (*8*). C. R. Acad. Sci. Paris **217**, 567—569 (1943). — [2] Contribution à l'étude de la stabilité des circuits de régulation et des servomécanismes (*2*). Ibid. **230**, 1380—1383 (1950). — [3] Sur la stabilité des régulateurs automatiques par action intégrale dérivée seconds conjuguées (*2*). Ibid. **230**, 1448—1450 (1950). — [4] Étude de l'influence des temps de propagation sur la stabilité des servomécamismes régulateurs par la méthode de fusion des racines (*2*). Ibid. **230**, 1499—1501 (1950).

Dubosin, G. N.: [1] A stability problem for constantly acting disturbances (*8*). Vestnik Moskov. Univ. **1952**, no. 2, 35—40. — [2] Foundations of the theory of stability of motion (*), pp. 318. Izdat. Moskov. Univ. **1952**. — [3] Some remarks on theorems of the second method of A. LYAPUNOV (*7*). Vestnik Moskov. Univ. **1950**, no. 10, 27—31. — [4] On integration of a system of linear equations of the second order by the method of A. LYAPUNOV (*3*). Moskov. Gos. Univ. Trudy Astr. Inst. **24**, 109—121 (1954).

DUFF, G. F. D.: [1] Limit cycles of systems of the second order (*9*). Proc. Nat. Acad. Sci. U.S.A. **36**, 749—752 (1950). — [2] Limit cycles and rotated vector fields (*9*). Ann. of Math. (2) **57**, 15—31 (1953).

—, and N. LEVINSON: [1] On the non-uniqueness of periodic solutions of an asymmétric Liénard equation (*8*). Quart. Appl. Math. **10**, 86—88 (1952).

DUFFIN, R. J.: [1] Nonlinear networks (*8*). Bull. Amer. Math. Soc. **52**, 833—839 (1946); **53**, 963—971 (1947); **55**, 119—129 (1949).

DUFFING, G.: [1] Erzwungene Schwingungen bei veränderlichen Eigenfrequenz (*8*, *). Braunschweig, Vieweg, 1918.

DULAC, H.: [1] Détermination et intégration d'une certaine classe d'équations différentielles ayant pour point singulier un centre (*9*). Bull. Sci. Math. (2) **32**, 230—252 (1908). — [2] Solutions d'une système d'équations différentielles dans le voisinage à valeurs singulières (*9*). Ibid. **40**, 324—383 (1912). — [3] Sur les cycles limites (*9*). Ibid. **61**, 45—188 (1923).

DUNKEL, O.: [1] Regular singular points of a system of homogeneous linear differential equations of the first order (*3*). Proc. Amer. Acad. Arts. Sci. **38**, 341—370 (1912).

DUVAKIN, A. P., and A. M. LETOV: [1] On the stability of regulating systems with two organs of regulation (*7*). Prikl. Mat. Meh. SSSR, **18**, 163—166 (1954).

DYHMAN, E. I.: [1] On the reduction principle (*7*). Izvestiya Akad. Nauk Kazah. SSSR, **4**, 73—84 (1950). — [2] Some stability theorems (*7*). Ibid. **4**, 85—97 (1950).

ECKWEILER, H. J.: [1] Nonlinear differential equations of the van der Pol type with a variety of periodic solutions (*8*, *9*). Studies in nonlinear Vibration Theory, 4—49, New York Univ., 1946.

EFGRAFOV, M. A.: [1] A new proof of PERRON'S theorem (*3*). Isvestiya Akad. Nauk SSSR, **17**, 77—82 (1953).

EINAUDI, R.: [1] Sugli esponenti caratteristici di una configurazione di equilibrio di un sistema dissipativo (*4*). Rend. Accad. Lincei (6) **21**, 336—343 (1935). — [2] Sulle configurazioni di equilibrio instabile di una piastra sollecitata da sforzi tangenziali pulsanti (*4*). Atti Accad. Gioenia Catania **20**, 1 (1935); mem. 5, 1, 1937. — [3] Sulle vibrazioni quasi armoniche di un sistema dissipativo (*4*). Atti Ist. Veneto Sci. Lett. **95**, 425—444 (1936).

EL'SGOL'L, L. E.: [1] An estimate for the number of singular points of a dynamical system defined on a manifold (*9*). Mat. Sbornik, (N. S.) **26** (68), 215—223 (1950). Amer. Math. Soc. Translation no. 68. — [2] Remark on the estimation of the number of points of rest of dynamical systems with retarded argument (*9*). Moskov. Gos. Univ. Uc. Zap. **165**, II. 221—222 (1954).

EL'SIN, M. I.: [1] On a problem of oscillations linear differential equations of the second order (*4*, *5*). Doklady Acad. Nauk SSSR, **18**, 141—145 (1938). — [2] On finding the phase of the solutions of a differential equation of the second order (*4*, *5*). Sci. Notes Moscow State Univ. **45**, 97—105 (1940). — [3] On the conditions for which the solutions of a second order linear differential equation has two zeros (*4*, *5*). Doklady Acad. Nauk SSSR, **51**, 573—576 (1946). — [4] On linear system with established spherical motion (*5*). Ucenye Zapiski Moskov. Gos. Univ. 135, Matematika **2**, 173—187 (1948). — [5] Qualitative problems on the linear differential equation of the second order (*5*). Doklady Akad. Nauk SSSR, (N.S.) **68**, 221—224 (1949). — [6] The phase method and the classical method of comparison (*5*). Ibid. **68**, 813—816 (1949). — [7] On the decremental estimate of amplitudes (*5*). Ibid. **69**, 7—10 (1949). — [8] Qualitative solution of a linear differential equation of the second order (*5*). Uspehi Mat. Nauk, (N.S.) **5**, no. 2 (36), 155—158 (1950). — [9] The method of comparison in the qualitative theory of an incomplete differential equation of second order (*5*). Mat. Sbornik, (N.S.) **34** (76), 323—330 (1954).

Erdélyi, A.: [1] Über die kleinen Schwingungen eines Pendels mit oszillierendem Aufhängepunkt (*8*). Z. angew. Math. Mech. **14**, 435 (1934). — [2] Asymptotic expansions (*, *10*), Dover, 1955.

—, M. Kennedy, and J. McGregor: [1] Parabolic cylinder functions of large order (*10*). J. Rat. Mech. Anal. **3**, 459—485 (1954).

—, and C. A. Swanson: [1] Asymptotic forms of Whittaker's confluent hypergeometric functions (*10*). Mem. Amer. Math. Soc., no. **25**, 1—49 (1957).

Ermolaev, L.: [1] On uniform stability in the first approximation of denumerable almost linear and nonlinear systems of differential equations (*1, 4, 7*). Izvestiya Akad. Nauk Kazah SSSR, **1** (6), 88—105, 106—114 (1952).

Ersov, B. A.: [1] On stability in the large of a certain system of automatic regulation (*7, 9*). Prikl. Mat. Meh. SSSR, **17**, 61—72 (1953). — [2] A theorem on stability of motion in the large (*8, 9*). Ibid **18**, 381—383 (1954).

Erugin, N. P.: [1] Reducible systems (*3*, *). Trudy Mat. Inst. Steklov **13**, 95pp. (1946). — [2] On asymptotically stable solutions of certain systems of differential equations (*3, 8*). Prikl. Mat. Meh. **12**, 157—164 (1948). — [3] The generalization of a theorem of Lyapunov (*4*). Ibid. **12**, 633—638 (1948). — [4] A closed solution of a parabolic inhomogeneous boundary problem (*3*). Ibid. **14**, 215—217 (1950). — [5] A qualitative investigation of the integral curves of a system of differential equations (*8, 9*). Ibid. **14**, 659—664 (1950). — [6] On certain questions of stability of motion and the qualitative theory of differential equations in the large (*1, 7, 9*). Ibid. **14**, 459—512 (1950). — [7] On the continuation of solutions of differential equations (*1*). Ibid. **15**, 55—58 (1951). — [8] Some general questions of the theory of stability of motion (*11, 9*). Ibid. **15**, 227—236 (1951). — [9] Theorems on instability (*6, 7, 9*). Ibid. **16**, 335—361 (1952). — [10] Analytic theory of nonlinear systems of ordinary differential equations (*8, 9*). Ibid. **16**, 465—486 (1952). — [11] On a problem of the theory of stability of systems of automatic regulation (*8, 9*). Ibid. **16**, 620—628 (1952). — [12] Construction of the whole set of systems of differential equations having a given integral curve (*9*). Ibid. **16**, 659—670 (1952). — [13] The methods of A. Lyapunov and questions of stability in the large (*7, 9*). Ibid. **17**, 389—400 (1953).

Esclangon, E.: [1] Nouvelles recherches sur les fonctions quasipériodiques (*1, 3*). Ann. Obs. Bordeaux **16**, 51—226 (1917).

Esipovic, E. M.: [1] On stability of solutions of a class of differential equations with retarded argument (*2*). Prikl. Mat. Meh. SSSR, **15**, 601—608 (1951).

Evans, R. L.: [1] Solution of linear ordinary differential equations containing a parameter (*10*). Proc. Amer. Math. Soc. **4**, 92—94 (1953). — [2] Asymptotic and convergent factorial series in the solution of linear ordinary differential equations (*10*). Ibid. **5**, 88—92, 1000 (1954). — [3] Errors in asymptotic solutions of linear ordinary differential equations (*10*). Quart. Appl. Math. **12**, 295—300 (1954).

Faedo, S.: [1] Il teorema di Fuchs per le equazioni differenziali lineari a coefficienti non analitici e proprietà asintotiche delle soluzioni (*3*). Annali Mat. pura appl. (4) **25**, 111—133 (1946). — [2] Sulla stabilità delle soluzioni delle equazioni differenziali lineari (*3*). Rend. Accad. Lincei (8) **2**, 564—570, 757—764 (1947); **3**, 37—43, 192—198 (1947). — [3] Proprietà asintotiche delle soluzioni dei sistemi differenziali lineari omogenei (*3*). Ann. Mat. pura appl. (4) **23**, 25—50 (1944); **26**, 207—215 (1947).

FARNELL, A. B., C. E. LANGENHOP and N. LEVINSON: [1] Forced periodic solutions of a stable nonlinear system of differential equations (*6*). J. Math. Phys. **29**, 300—302 (1951).

FATOU, P.: [1] Sur un critère de stabilité (*3*). C. R. Acad. Sci. Paris **189**, 967—969 (1929).

FAURE, R.: [1] Sur certaines solutions périodiques d'équations différentielles non linéaires. Cas des vibrations forcées. Influence de la frequence (*6*). Annali. Mat. pura appl. (4) **42**, 165—188 (1956).

FAVARD, J.: [1] Sur les équations différentielles linéaires à coefficients presque-périodiques (*3, 4*). Acta Math. **51**, 31—81 (1928).

FÉJER, L.: [1] Über Stabilität und Labilität eines materiellen Punktes in understrellenden Mittel (*1*). J. f. Math. **131**, 216—223 (1906).

FEL'DBAUM, A.A.: [1] Integral criteria for the quality of a regulation (*8*). Avtom. i Telem. **9**, 3—19 (1948).

FEŠČENKO, S. F.: [1] Estimate of the error in the asymptotic behavior of integrals of ordinary linear differential equations having a parameter (*3*). Dopovidi Akad. Nauk Ukrain. SSSR, **1951**, 156—162. — [2] Asymptotic solutions of an infinite system of differential equations with slowly varying parameters (*10*). Ibid. **1954**, 82—86. — [3] Estimate of the error in the asymptotic solution of an infinite system of differential equations with slowly varying coefficients (*10*). Ibid. **1955**, 211—216.

FIFER, S.: [1] Studies in nonlinear Vibration Theory (*8*). J. Appl. Phys. **22**, 1421 to 1428 (1951).

FILIPPOV, A. F.: [1] A sufficient condition for the existence of a stable limit cycle for an equation of the second order (*9*). Mat. Sbornik, (N.S.) **30** (72), 171—180 (1951).

FIRSOV, G. A.: [1] On the question of the oscillation of a ship provided with an active stabilizer (*8*). Izvestiya Akad. Nauk SSSR, **1945**, 995—1002. — [2] On the question of the stiffness of a ship under the influence of a squall (*8*). Ibid. **1945**, 648—656.

FISHER, E : [1] The period and amplitude of the van der Pol limit cycle (*8, 9*). J. Appl. Phys. **25**, 273—274 (1954).

FITE, W. B.: [1] Concerning the zeros of the solutions of certain differential equations (*5*). Trans. Amer. Math. Soc. **19**, 341—352 (1917).

FLANDERS, D. A., and J. J. STOKER: [1] The limit case of relaxation oscillations (*8, 9*). Studies in nonlinear Vibration Theory. New York Univ. 1946, 50—64.

FLOQUET, G.: [1] Sur les equation différentielles linéaires à coefficients périodiques (*4*). Ann. École Norm. Sup, Paris (2) **12**, 47—89 (1883).

FLÜGGE-LOTZ, I.: [1] Discontinuous automatic control (*8*). Princeton 1953. 168pp.
— H. F. HODAPP, K. KLOTTER, H. MEISSINGER and K. SCHOLZ: [1] Über Bewegungen eines Schwingers unter dem Einfluß von Schwarz-Weiß-Regelungen (*8*). Zeit. angew. Math. Mech. **25/27**, 97—113 (1947).

FOÀ, E.: [1] Sull'equazione del moto smorzato con parametri variabili e su un caso di instabilità (*5*). Pont. Acta Sci. Acad. **6**, 9—16 (1943).

FOGAGNOLO, B.: [1] Sulle vibrazioni dei sistemi a due gradi di libertà (*2*). Rend. Mat. Univ. Roma (5) **5**, 220—233 (1946).

FORD, W. B.: [1] Sur les équations linéaires aux différences finies (*3*). Annali Mat. pura appl. **13**, 263—328 (1907). — [2] Studies on divergent series and summability (*10*). Macmillan 1916. [3] On the behavior of integral functions in distant protions of the plane (*10*). Bull. Amer. Math. Soc. **34**, 91—106 (1928). — [4] The asymptotic developments of functions defined by MACLAURIN series (*1, 10*). Univ. Michigan Press 1936.

FORSTER, H.: [1] Über das Verhalten der Integralkurven einer gewöhnlichen Differentialgleichung erster Ordnung in der Umgebung eines singularen Punktes (*9*). Math. Z. **43**, 271—320 (1937).

FOWLER, R. H.: [1] The form near infinity of real continuous solutions of a certain differential equation of the second order (*8*). Quart. J. Math. Oxford Ser. **45**, 289—350 (1914). [2] Some results on the form near infinity of real continuous solutions of a certain type of second order differential equation (*8*). Proc. London Math. Soc. (2) **13**, 314—371 (1914). — [3] The solution of EMDEN's and similar differential equations (*8*). Quart. J. Math. **2**, 259—288 (1931).

FRANKLIN, P.: [1] Almost periodic recurrent motions (*6*). Math. Z. **30**, 325—331 (1929).

FRIEDLANDER, F. G.: [1] On the forced vibrations of quasi-linear systems (*8*). Quart. J. Mech. Appl. Math. **3**, 364—376 (1950). — [2] On the asymptotic behavior of the solutions of the differential equation $x'' + x = kf(x, x', t)$ (*8*). Proc. Cambridge Philos. Soc. **46**, 406—418 (1950). — [3] On the recurrent solutions of a class of nonlinear differential equations (*8*). Ibid. **47**, 315—330 (1951). — [4] On the oscillations of a bowed string (*8*). Ibid. **49**, 516—530 (1953).

FRIEDRICHS, K. O.: [1] On nonlinear vibrations of third order (*9*). Studies in nonlinear Vibration Theory. New York Univ. 1946, p. 65—103. — [2] Justification of the perturbation process (*6*). Brown Univ. lecture notes, 1942/43, 1—15. — [3] Combination oscillations (*8, 9*). Ibid. **1**, 16—21. — [4] Fundamentals of POINCARE's theory (*, *8*). Proceedings Symposium on nonlinear Circuit Analysis. New York 1953, 56—67. — [5] Advanced ordinary differential equations. New York Univ. lecture notes, 1950.

—, P. DECORBEILLER, N. LEVINSON and J. J. STOKER: [1] Nonlinear mechanics (*8, 9*). Brown Univ. mimeographed lecture notes 1942/43.

—, and J. J. STOKER: [1] Forced vibrations of systems with nonlinear restoring force (*8*). Quart. J. Appl. Math. **1**, 97—115 (1943). — [2] Introduction to nonlinear mechanics (*8, 9*). Brown Univ. lectures notes, 1942/43, 1—78.

—, and W. WASOW: [1] Singular perturbations of nonlinear oscillations (*10*). Duke Math. J. **13**, 367—381 (1946). — [2] On simple harmonic vibrations of a system with nonlinear characteristics (*8*). Studies in nonlinear vibration theory. New York Univ. 1946, p. 104—192. — [3] On harmonic vibrations out of phase with the exciting force (*8*). Comm. Appl. Math. **1**, 341—359 (1948).

FROMMER, M.: [1] Über das Auftreten von Wirbeln und Strudeln (geschlossener und spiraliger Integralkurven) in der Umgebung rationaler Unbestimmtheitsstellen (*9*). Math. Ann. **109**, 395—424 (1934).

FUBINI, G.: [1] Studi asintotici per alcune equazioni differenziali (*5*). Rend. Accad. Lincei (*6*) **26**, 253—259 (1937).

FULLER, F. B.: [1] Note on trajectories in a solid torus (*9*). Ann. of Math. (2) **56**, 438—439 (1952). — [2] The existence of periodic points. Ibid. **57**, 229—230 (1953).

—, W. R.: [1] Existence of cycles for weakly nonlinear real differential systems containing differences (*8*). To appear.

FURUYA, S.: [1] VAN DER POL's equation with harmonic disturbance (*8*). Comm. Math. Univ. St. Paul. **3**, 7—13 (1954). — [2] Periodic solutions of a nonlinear differential equation (*8*). Ibid. **4**, 47—51 (1955). — [3] Periodic solutions of the VAN DER POL-MATHIEU equation (*8*). Ibid. **3**, 109—113 (1955).

GADSDEN, C. P.: [1] An electrical network with varying parameters (*5, 8*). Quart. Appl. Math. **8**, 199—205 (1950).

GAGLIARDO, E.: [1] Sul comportamento asintotico degli integrali dell'equazione differenziale $y'' + A(x)y = 0$ con $A(x) \geq 0$ (*5*). Boll. Unione Mat. Ital. (3) **8**,

177—185 (1953). — [2] Sul comportamente degli integrali dell'equazione differenziale non lineare $x''+f(x)x'+g(x)=0$ con $g(x)$ crescente e $f(x)$ positiva per $|x|< M$ (*5*). Ibid. (3) **8**, 309—314 (1953). — [3] Sui criteri di oscillazione per gli integrali di un'equazione differenziale lineare del secondo ordine (*5*). Ibid. (3) **9**, 177—189 (1954).

GAMBIER, B.: [1] L'équation différentielle linéaire du second ordre (*5*). Nouv. Ann. de Math. (6) **2**, 2—23 (1927).

GAMBILL, R. A.: [1] Stability criteria for linear differential systems with periodic coefficients (*4*). Riv. Mat. Univ. Parma **5**, 169—181 (1954). — [2] Criteria for parametric instability for linear differential systems with periodic coefficients (*4*). Ibid. **6**, 37—43 (1955). — [3] A fundamental system of real solutions for linear differential systems with periodic coefficients (*4*). Ibid. 311—319 (1956).

—, and J. K. HALE: [1] Subharmonic and ultraharmonic solutions for weakly nonlinear systems (*8*). J. Rational Mech. Anal. **5**, 353—398 (1956).

GAPONOV, V. I.: [1] Theory of entrainment (*9*). J. Tekhn. Fiz. Moscow **5**, 821 (1935).

GAUTHIER, L.: [1] Au sujèt de la recherche des cycles limites (*9*). Actes du Colloque Internat. des Vibrations non linéaires. Ile de Porquerolles **1951**, p. 257—259.

GAVRILOV, N. I.: [1] On stability according to LYAPUNOV of systems of linear differential equations (*3*). Doklady Akad. Nauk SSSR, (N. S.) **84**, 425—428 (1952). — [2] On a method in the theory of stability according to LYAPUNOV (*3*). Ibid. **84**, 657—660 (1952).

GAVURIN, M. K.: [1] On systems of differential equations of the form $y'=Ay^2+2By+C$. Doklady Akad. Nauk SSSR, (N. S.) **84**, 205—208 (1952).

GEJLER, L. B.: [1] Zur Frage der aperiodischen Stabilität linearer Systeme (*3*). Uspehi Mat. Nauk **4**, 2 (30), 206—208 (1949); **5**, 4 (38), 176 (1950).

GEL'FAND, I. M., and B. M. LEVITAN: [1] On the determination of a differential equation from its spectral function (*4*). Izvestiya Akad. Nauk SSSR, Ser. Mat. **15**, 309—366 (1951). Amer. Math. Soc. Translations (2) **1**, 253—304.

—, and V. B. LIDSKII: [1] On the structure of the regions of stability of linear canonical systems of differential equations with periodic coefficients (*4*). Uspehi Mat. Nauk, (N. S.), **10**, 3—40 (1955); Trans. Amer. Math. Soc. (2) **8**, 143—181.

GELFOND, A. O.: [1] Linear differential equations of infinite order with constant coefficients and asymptotic periods of entire functions (*10*). Trudy Mat. Inst. Steklov **38**, 47—67 (1951). — [2] On PERRON's theorem in the theory of difference equations (*3*). Izvestiya Akad. Nauk SSSR, Ser. Mat. **17**, 83—86 (1953).

GENNARO, A. DE: [1] Alcuni criteri di stabilità per le soluzioni di un'equazione differenziale lineare (*4*). Giorn. Mat. Battaglini **2** (78) 42—54 (1948). — [2] Sul comportamento asintotico degli integrali di certe equazioni differenziali lineari ordinarie (*4*). Ibid. **3** (79), 49—62 (1950).

GHIZZETTI, A.: [1] Sul comportamento asintotico degli integrali delle equazione differenziali ordinarie, lineari ed omogenee (*3*). Giorn. Mat. Battaglini (4) **1** (77), 5—27 (1947). — [2] Un teorema sul comportamento asintotico degli integrali delle equazioni differenziali lineari omogenee (*3*). Rend. Mat. Univ. Roma (5) **8**, 28—42 (1949).

GILLIES, A. W.: [1] On the transformations of singularities and limit cycles of the variational equations of VAN DER POL (*9*). Quart. J. Mech. Appl. Math. **7**, 152—167 (1954). — [2] The periodic solutions of a non-linear differential equation of the second order with unsymmetrical non-linear damping, and a forcing term (*8*). Quart. J. Mech. Appl. Math. **8**, 107—128 (1955).

GINSBURG, I. P.: [1] On sufficient conditions for the stability of solutions of the equation $y''+py'+qy=0$ (*5*). Sci. Rep. Leningrad State Univ., Meh. **114**, 200—204 (1949).

GIULIANO, L.: [1] Generalizzazione di un lemma di GRONWALL e di una disegua-glianza di PEANO (*3*). Rend. Accad. Lincei (8) **1**, 1264—1271 (1946). — [2] Sul teorema di confronto di STURM (*5*). Boll. Unione Mat. Ital. (3) **2**, 16—19 (1947).

GOL'DFARB, L. S.: [1] On some nonlinearities in systems of regulation (*8*). Avtom. i Telem. **8**, 349—383 (1947).

GOL'DIN, A. M.: [1] On a criterion of LYAPUNOV (*4, 5*). Prikl. Mat. Meh. SSSR, **15**, 379—384 (1951).

GOLDSTEIN, S.: [1] MATHIEU functions (*4*). Trans. Cambridge Philos. Soc. **23**, 303 (1927). — [2] Characteristic numbers of MATHIEU's equation with imaginary parameter (*4*). Phil. Mag. **8**, 836 (1929). — [3] A note on certain approximate solutions of linear differential equations of the second order, with an application to the Mathieu equation (*4, 10*). Proc. London Math. Soc. (2) **28**, 81—90 (1928); (2) **33**, 246—252 (1932).

—, and N. W. MCLACHLAN: [1] Sound waves of finite amplitude in an exponential loudspeaker horn (*4, 5*). Wireless Engineer **11**, 427 (1934). J. Acoust. Soc. Amer. **6**, 275 (1935).

GOLDWEIZER, A. L., and A. I. LURE: [1] The mathematical theory of the equilibrium of elastic shells (*). Prikl. Mat. Meh. SSSR, **11**, 565—592 (1947). — Amer. Math. Soc. Translation No. 9.

GOLOMB, M.: [1] Bounds for solutions of nonlinear differential systems (*6*). Archive Rat. Mech. Anal. **1**, 272—282 (1958). — [2] Solutions of certain nonautonomous systems by series of exponential functions (*8*). Illinois J. Math. **3** (1958). — [3] Expansion and boundedness theorems for solutions of linear differential systems with periodic or almost periodic coefficients (*4*). Archive Rat. Mech. Anal. **2**, 284—308 (1958).

—, and E. USDIN: [1] A theory of multi-dimensional servo-systems (*2*). J. Franklin Inst. **253**, 29—57 (1952).

GOMORY, R. E.: [1] Trajectories tending to a critical point in 3-space (*9*). Ann. of Math. (2) **61**, 140—153 (1955).

—, and F. HAAS: [1] A study of trajectories which tend to a limit cycle in three-space (*9*). Ann. of Math. (2) **62**, 152—161 (1955).

—, and D. E. RICHMOND: [1] Boundaries for the limit cycle of VAN DER POL's equation (*8, 9*). Quart. Appl. Math. **9**, 205—209 (1951).

GONZALEZ BAZ, E.: [1] Relation between the parameter and the dimensions of the periodic solution of the VAN DER POL equation (*8*). Comisión Impulsora y Coordinadora, Investigación Científica (Mex.) 1949, p. 87—95.

GORBUNOV, A. D.: [1] On a method for obtaining estimates of the solution of a system of ordinary linear homogeneous differential equations (*7*). Vestnik Moskov. Univ. **1950**, no. 10, 19—26. — [2] Some questions of the qualitative theory of ordinary linear homogeneous differential equations with variable coefficients (*7*). Moskov. Gos. Univ. Ucenye Zapiski 165, Mat. **7**, 39—78 (1954). — [3] On estimates of the coordinates of solutions of systems of ordinary linear differential equations (*7*). Vestnik Moskov. Univ. **9**, no. 5, 27—31 (1954). — [4] On conditions of asymptotic stability of the zero solution of a system of ordinary linear homogeneous differential equations (*7*). Ibid. **8**, no. 9, 49—55 (1953). — [5] On some properties of the solutions of systems of ordinary linear homogeneous differential equations (*3*). Vestnik Moskov. Univ. **6**, 3—15 (1951).

GORELIK, G. S.: [1] Resonance appearing in linear systems with periodically varying parameters (*4*), Z. Tehn. Fiz. **4**, 1783—1817 (1934); **5**, 195—215 (1935). — [2] Retroaction retardée (*4*). Akad. Sci. SSSR, Ž. Fiz. **1**, 465—470 (1939). — [3] On the theory of retarded feedback (*4*). Ž. Tehn. Fiz. **9**, 450—454 (1939).

GORELIK, G. S. and G. HINTZ: [1] Resonanzerscheinungen in linearen Systemen mit periodisch veränderlichen Parametern (*4*). Z. Tehn. Fiz. **4**, 1783 (1934); **5**, 195, 489 (1935).

GORSIN, S.: [1] On stability of motion with constantly acting disturbances (*6, 7*). Izvestiya Akad. Nauk Kazan SSSR, **2**, 46—73 (1948). — [2] Critical cases (*6, 7*). Ibid. **2**, 74—101 (1948). — [3] On the stability of the solutions of a denumerable system of differential equations with constantly acting disturbances (*6, 7*). Ibid. **2**, 32—38 (1948). — [4] On LYAPUNOV's second method (*7, 9*). Ibid. **4**, 42—50 (1950). — [5] Some criteria of stability with constant disturbances (*6, 7*). Ibid. **4**, 51—56 (1950).

GOSSE, R.: [1] Sur une équation de LANGMUIR généralisée (*5, 8*). Ann. Inst. Fourier Grenoble **1949**, 1; **1950**, 5—11.

GOT, T.: [1] Détermination des solutions périodiques stables de certaines équations différentielles quasi harmoniques (*9*). C. R. Acad. Sci. Paris **230**, 612—614 (1950).

GOTTLIEB, M. J.: [1] Oscillation theorems for self-adjoint boundary value problems (*5*). Duke Math. J. **15**, 1073—1091 (1948).

GOTTSCHALK, W. H., and G. A. HEDLUND: [1] Topological Dynamics (*). Amer. Math. Soc. Coll. Publ. **36** (1955).

GOURSAT, E.: [1] Cours d'analyse mathématique (*). 4. edit. Paris 1922—1927.

GRABAR, M. I.: [1] On strong ergodicity of dynamical systems (*9*). Doklady Akad. Nauk SSSR, (N. S.) **95**, 9—12 (1954).

GRADSTEIN, I. S.: [1] On behaviour of solutions of systems of linear differential equations degenerating in the limit (*2, 8, 10*). Doklady Akad. Nauk SSSR, (N. S.) **53**, 391—394 (1946). — [2] The solution of systems of linear equations by L. L. GUTENMAKER's electrical models (*2, 8, 10*). Izvestiya Akad. Nauk SSSR, **1947**, 529—584. — [3] Linear differential equations with small coefficients for the higher derivatives (*3, 8, 10*). Doklady Akad. Nauk SSSR, (N. S.) **59**, 841—843 (1948). — [4] On a class of nonlinear differential equations with small ceofficients for certain derivatives (*8, 10*). Ibid. **64**, 441—443 (1949). — [5] Differential equations with small coefficients for the derivatives and LYAPUNOV's theory of stability (*7, 8, 10*). Ibid. **65**, 789—792 (1949). — [6] Nonlinear differential equations with small coefficients for certain derivatives (*7, 8, 10*). Ibid. **66**, 789 to 792 (1949). — [7] On the behavior of the solutions of systems of linear differential equations with constant coefficients, degenerating in the limit (*7, 8, 10*). Izvestiya Akad. Nauk SSSR, **13**, 253—280 (1949). — [8] Linear equations with variable coefficients and small parameters in the highest derivatives (*7, 8, 10*). Mat. Sbornik, (N. S.) **27** (*69*), 47—68 (1950). — [9] Application of A. LYAPUNOV's stability theory to the theory of differential equations with small coefficients of the derivatives (*7, 8, 10*). Doklady Akad. Nauk SSSR, (N. S.) **81**, 985—986 (1951). — [10] Differential equations in which various powers of a small parameter appear as coefficients of the derivatives (*10*). Ibid. **82**, 5—8 (1952). — [11] On continuous dependence of asymptotic stability upon a parameter (*10*). Uspehi Mat. Nauk, (N. S.) **9** (62), no. 4, 163—166 (1954).

GRAFFI, D.: [1] Gli invarianti adiabatici come metodo di integrazione approssimata di equazioni differenziali (*5, 8*). Rend. Accad. Lincei (6) **9**, 15 (1932). — [2] Sul calcolo degli autovalori per una corda non omogenea (*5*). Attisec. Congr. Unione Mat. Ital., Bologna **2**, 353—359 (1940). — [3] Sopra alcune equazioni differenziali non lineari della fisica-matematica (*8, 9*). Mem. Accad. Sci. Bologna (9) **7**, 121—129 (1940). — [4] Sopra alcune equazioni differenziali della radiotecnica (*8, 9*). Ibid. (9) **9**, 145—153 (1942). — [5] Forced oscillations for several nonlinear circuits (*9*). Ann. of Math. (2) **54**, 262—271 (1951). — [6] Equazioni delle oscillazioni non lineari in relazione alle applicazioni (*, *8*). Atti Quarto

Congr. Unione Mat. Ital., Taormina **1**, 218—231 (1951). — [7] Sul periodo delle oscillazioni dei sistemi non lineari a più gradi di libertà (*8*, *9*). Colloque Internat. Vibrations non linéaires, Ile de Porquerolles 1951, p. 189—193. — [8] Sulle oscillazioni forzate nei sistemi non lineari a due gradi di libertà (*8*). Rend. Accad. Lincei (8) **16**, 176—180 (1954). — [9] Sulla teoria delle oscillazioni elastiche con ereditarietà (*3*). Nuovo Cim. **5**, 310—317 (1928). — [10] Le oscillazioni di rilassamento (*8*). Alta Frequenza **11**, 80—98 (1942). — [11] Sul modello di Rocard per le oscillazioni di rilassamento (*8*). Accad. Scienze Ferrara **1942**. — [12] Sulla teoria delle oscillazioni libere in un sistema soggetto a forze elastiche con ereditarietà (*3*). Rend. Sem. Mat. Modena **3**, 1—23 (1949). — [13] Sulle oscillazioni forzate nella meccanica non lineare (*8*, *9*). Riv. Mat. Univ. Parma **3**, 317—326 (1952). — [14] Sul periodo delle oscillazioni nei sistemi non lineari a due gradi di libertà (*8*). Mem. Accad. Sci. Bologna (10) **9**, 17—22 (1952). — [15] Sull'espressione delle soluzioni periodiche nei sistemi non lineari a due gradi di libertà (*8*, *9*). Atti Accad. Sci. Bologna **1953**. — [16] Oscillazioni non lineari (*, *8*). Confer. Sem. Mat. Univ. Bari no. 1, 18 pp., 1954. — [17] Su alcune equazioni differenziali non lineari (*8*, *9*). Rend. Accad. Sci. Ist. Bologna (10) **1**, 57—64 (1954).

GRAVES, W. M. H.: [1] On a certain family of periodic solutions of differential equations with an application to the triode oscillations (*8*). Proc. Roy. Soc. Lond., Ser. A **103**, 516—524 (1923).

GREGUS, M.: [1] Application of dispersions to boundary problems of the second order (*5*). Mat.-Fyz. Casopis Slovensk. Akad. Vied. **4**, 27—37 (1954).

GROBMAN, D. M.: [1] On characteristic exponents of systems near to a linear one (*6*). Doklady Akad. Nauk SSSR, (N. S.) **75**, 157—160 (1950). — [2] Systems of differential equations analogous to linear ones (*6*). Ibid. **86**, 19—22 (1952). — [3] Characteristic exponents of systems near to linear ones (*6*). Mat. Sbornik, (N. S.) **30** (72), 121—166 (1952).

GRONWALL, T. H.: [1] Note on the derivatives with respect to a parameter of the solution of a system of differential equations (*3*). Ann. of Math. **20**, 292—296 (1918).

GROSS, W.: [1] Zur Theorie der Differentialgleichungen mit festen kritischen Punkten (*9*, *10*). Math. Ann. **78**, 332—342 (1918).

GRÜNWALD, E.: [1] Lösungsverfahren der Laplace-Transformation für Ausgleichvorgänge in linearen Netzen, angewandt auf selbsttätige Regelungen (*2*). Arch. Elektrotechn. **35**, 379—400 (1941).

GUBAR, N. A.: [1] A characteristic of composite singular points of a system of two differential equations by means of simple singular points of neighboring systems (*9*). Doklady Akad. Nauk SSSR, (N. S.) **95**, 435—438 (1954).

GUSAROV, L. A.: [1] On the boundedness of the solutions of a linear equation of the second order (*3*, *5*). Doklady Akad. Nauk. SSSR, (N.S.) **68**, 217—220 (1949). — [2] On the approach to zero of the solutions of a linear differential equation of the second order (*5*). Ibid. **71**, 9—12 (1950). — [3] On some properties of solutions of a linear differential equation of second order (*5*). Moskov. Gos. Univ. Uc. Zap. **1954**, 223—237.

GUSAROVA, R. S.: [1] On the criteria of boundedness of the solutions of linear equations of the second order (*4*). Uspehi Mat. Nauk (N.S.) **4**, 132—133 (1949). — [2] On bounded solutions of a linear differential equation with periodic coefficients (*4*). Prikl. Mat. Meh. SSSR, **13**, 241—246 (1949); **14**, 313—314 (1950).

HAACKE, W.: [1] Über die Stabilität eines Systems von gewöhnlichen linearen Differentialgleichungen zweiter Ordnung mit periodischen Koeffizienten, die von Parametern abhängen (*4*). Math. Z. **56**, 65—79 (1952); **57**, 34—45 (1952). — [3] Über die nichtlineare Mechanik (*, *8*). Phys. Bl. **9**, 398—405 (1953).

HAAG, J.: [1] Étude asymptotique des oscillations de relaxation (*8, 10*). Ann. Sci. École Norm. Sup., Paris (3) **60**, 35—64, 65—111 (1943). — [2] Exemples concrets d'étude asymptotique d'oscillations de relaxation (*8, 10*). Ibid. (3) **61**, 73—117 (1944). — [3] Sur la théorie de la synchronisation (*8*). C. R. Acad. Sci. Paris **221**, 682—684 (1945). — [4] Sur l'amortissement et l'entretien des oscillations à *n* degrés de liberté (*8*). Ibid. **221**, 734—736 (1945). — [5] Sur la régime transitoire précédant la synchronisation (*8*). Ibid. **222**, 314—316 (1946). [6] Sur la stabilité des solutions de certains systèmes d'équations différentielles (*8*). Ibid. **222**, 623—624 (1946). — [7] Sur certaines systèmes différentiels à solutions périodiques (*8*). Ibid. **223**, 446—449 (1946). — [8] Sur la synchronisation sous-harmonique (*8*). Ibid. **223**, 525—527 (1946). — [9] Sur la synchronisation d'un système à plusiers degrés de liberté (*8*). Ibid. **223**, 877—879 (1946). — [10] Sur la stabilité des solutions de certains systèmes d'équations différentielles (*8*). Bull. Sci. Math. (2) **70**, 21—36 (1946). — [11] Sur certains systèmes d'équations différentielles à solutions périodiques (*8*). Ibid. (2) **70**, 155—172 (1946). — [12] Sur certaines systèmes d'équations différentielles définies par des fonctions périodiques et discontinues (*8*). Ibid. (2) **71**, 205—219 (1947). — [13] Sur la synchronisation des systèmes à plusiers degrés de liberté(*8*). Ann. Sci. École Norm. Sup. Paris (3) **64**, 285—338 (1947). — [14] Sur l'existence et la stabilité des solutions périodiques de certains systèms différentiels (*8*). Ibid. (3) **65**, 299—335 (1948). — [15] Sur les oscillateurs à amplitude stabilisée (*8*). C. R. Acad. Sci. Paris **226**, 1567—1568 (1948). — [16] Sur la synchronisation des systèmes oscillants non linéaires (*8*). Ibid. **227**, 649—651 (1948). — [17] Sur certains systèmes différentiels à solution périodique lentement variable (*8*). Ibid. **230**, 1229—1231 (1950). — [18] Les oscillations non linéaires en chronométrie (*, *8*). Colloque Internat. Vibrations non linéaires, Ile de Porquerolles 1951, 1—16. — [19] Cols, noeuds et foyers (*9*). Bull. Sci. Math. (2) **74**, 167—192 (1950). — [20] Sur certains systèmes différentiels à solution périodique lentement variable (*8*). Ibid. (2) **75**, 15—21 (1951). — [21] Sur la synchronisation des systèmes oscillants non linéaires (*8*). Ann. Sci. École Norm. Sup. Paris (3) **67**, 321—392 (1950). — [22] À propos de l'équation de MATHIEU (*8*). C. R. Acad. Sci. Paris **232**, 661—663 (1951). — [23] Sur la synchronisation d'un oscillateur par une force sinusoidale indépendante de la vitesse (*8*). Ibid. **233**, 117—118 (1951). — [24] Les mouvements vibratoires, 2 vols (*). Presses Univ. de France 1952, 268 pp.

HAAS, F.: [1] A theorem about characteristics of differential equations on closed manifolds (*9*). Proc. Acad. Sci. U.S.A. **38**, 1044—1047 (1952). — [2] On the global behavior of differential equations on two-dimensional manifolds (*9*). Proc. Amer. Math. Soc. **4**, 630—635 (1953). — [3] On the total number of singular points and limit cycles of a differential equation. Contrib. Theory Nonlinear Oscillations **3**, 137—172 (1956).

—, V. B.: [1] On a nonlinear differential equation containing a small parameter. Contrib. Theory Nonlinear Oscillations **3**, 57—84 (1956).

HADAMARD, J.: [1] Sur certaines propriété des trajectoires en dynamique (*9*). J. de Liouville (5) **3** (1897). — [2] Sur l'itération et les solutions asymptotiques des équations différentielles (*9*). Bull. Sci. Math. **26** (1901).

HALANAY, A.: [1] On a linear differential equation with an almost periodic coefficient (*4*). Doklady Akad, Nauk SSSR, (N. S.) **88**, 419—422 (1953). — [2] Solutions presque-périodiques de l'équation de RICCATI (*5*). Acad. Repub. Pop. Romane, Stud. Cerc. Mat. **4**, 345—354 (1953). — [3] Nouveaux criteriums d'existence des solutions périodiques pour l'équation des oscillations non linéaires forcées (*8*). Acad. R. P. Romine, Bul. Sti. Mat. Fiz. **5**, 373—391 (1953). — [4] Relativement à la methode du petit paramètre (*9*). Ibid. **6**, 483 to 488 (1954). — [5] Points singulière et solutions périodiques (*9*). Ibid. **7**, 319—324 (1955). — [6] Solutions presque périodique des systèmes d'équations différentielles nonlineaires (*9*). Ibid. **7**, 13—17 (1956). — [7] Almost periodic solutions of certain nonlinear systems (*9*). Gaz. Mat. Fiz. (a) **6**, 396—399 (1955).

HALE, J. K.: [1] Evaluations concerning products of exponential and periodic functions (*4*). Riv. Mat. Univ. Parma **5**, 63—81 (1954). — [2] On boundedness of the solution of linear differential systems with periodic coefficients (*4*). Ibid. **5**, 137—167 (1954). — [3] Periodic solutions of nonlinear systems of differential equations (*8*). Ibid. **5**, 281—311 (1954). — [4] On a class of linear differential equations with periodic coefficients (*4*). Illinois J. Math. **1**, 98—104 (1957). — [5] Linear systems of first and second order differential equations with periodic coefficients (*4*). Ibid. **2**, 586—591 (1958). — [6] Sufficient conditions for the existence of periodic solutions of systems of weakly nonlinear first and second order differential equations (*8*). Journ. Math. Mech. **7**, 163—172 (1958). — [7] A short proof of a boundedness theorem for linear differential systems with periodic coefficients (*4*). Archive Rat. Mech. Anal. **2**, 429—434 (1959). — [8] On the behavior of the solutions of linear periodic differential systems near resonance points (*8*). Contrib. Theory nonlinear Oscillations **5**, 55—90 (1960). — [9] On the stability of periodic solutions of weakly nonlinear periodic and autonomous differential systems (*8*). Ibid., **5**, 91—114 (1960). — [10] On differential equations containing a small parameter (*8*). Contrib. Theory Differential Equations, **1**, 1962. — [11] Nonlinear oscillations (*, *8*). McGraw-Hill 1963. — [12] The asymptotic behavior of the solutions of systems of differential equations (*8*). Purdue University Thesis, 1953.

HAHN, W.: [1] Theorie und Anwendung der direkten Methode von LJAPUNOV (*7*), pp. 142. Ergebn. Math. u. ih. Grenz. (N.F) **22**, Springer 1959.

HAMEL, G.: [1] Über die Instabilität der Gleichgewichtslage eines Systems von zwei Freiheitsgraden (*8*). Math. Ann. **57**, 541—553 (1903). — [2] Über lineare homogene Differentialgleichungen zweiter Ordnung mit periodischen Koeffizienten (*4*). Ibid. **73**, 371—412 (1913).

HARASAHAL, V.: [1] On fundamental solutions of denumerable systems of differential equations (*3, 6*). Izvestiya Akad. Nauk Kazah. SSSR, Sez. Mat. Meh. **4**, 98—108 (1950). — [2] Stability in the first approximation of solutions of denumerable systems of differential equations (*3, 6*). Ibid. **5**, 136—141 (1951). — [3] On stability in the first approximation of the solutions of denumerable systems of differential equations (*3, 6, 7*). Izvestiya Akad. Nauk Kazah. SSSR, Sez. Mat. Meh. **3**, 77—84 (1949).

HARTMAN, P.: [1] On the solutions of an ordinary differential equation near a singular point (*5*). Amer. J. Math. **68**, 495—504 (1946). — [2] The L^2-solutions of linear differential equations of second order (*5*). Duke Math. J. **14**, 323—326 (1947). — [3] Differential equations with nonoscillatory eigenfunctions (*5*). Ibid. **15**, 697—709 (1948). — [4] Unrestricted solution fields of almost-separable differential equations (*5*). Trans. Amer. Math. Soc. **63**, 560—580 (1948). — [5] On a theorem of MILLOUX (*3, 5*). Amer. J. Math. **70**, 395—399 (1948). — [6] On the linear logarithmic exponential differential equation of the second order (*5*). Ibid. **70**, 764—779 (1948). — [7] A characterization of the spectra of one dimensional wave equations (*5*). Ibid. **71**, 915—920 (1949). — [8] On bounded GREEN's kernel for second order linear ordinary differential equations (*5*). Ibid. **73**, 646—656 (1951). — [9] On the eigenvalues of differential equations (*5*). Ibid. **73**, 657—662 (1951). — [10] On non-oscillatory linear differential equations of second order (*5*). Ibid. **74**, 389—400 (1952). — [11] On the derivatives of solutions of linear, second order, ordinary differential equations (*5*). Ibid. **75**, 173—177 (1953). — [12] On linear second order differential equations with small coefficients (*5*). Ibid. **73**, 955—962 (1951). — [13] On the essential spectra of ordinary differential operators (*5*). Ibid. **76**, 831—838 (1954). — [14] On the zeros of the solutions of second order linear differential equations (*5*). J. London Math. Soc. **27**, 492—496 (1952).

HARTMAN, P. and C. R. PUTNAM: [1] The least cluster point of the spectrum of boundary value problems (*5*). Amer. J. Math. **70**, 849—855 (1948). — [2] The gaps in the essential spectra of wave equations (*3, 5*). Ibid. **72**, 849—862 (1950). — [3] The essential spectrum and averages of the potential (*5*). Duke Math. J. **23**, 83—92 (1956). — [4] The essential spectra belonging to bounded and half-bounded potentials (*5*). Ibid. **24**, 561—570 (1956).

—, and A. WINTNER: [1] Asymptotic distributions and the ergodic theorem (*7*). Amer. J. Math. **61**, 977—984 (1939). — [2] On the asymptotic behavior of the solutions of a nonlinear differential equation (*5*). Ibid. **68**, 301—308 (1946). — [3] An oscillation theorem for continuous spectra (*5*). Proc. Nat. Acad. Sci. U.S.A. **33**, 376—379 (1947). — [4] The asymptotic arcus variation of solutions of real linear differential equations of second order (*5*). Amer. J. Math. **70**, 1—10 (1948). — [5] Criteria of non-degeneracy for the wave equation (*5*). Ibid. **70**, 295—308 (1948). — [6] On the orientation of unilateral spectra (*5*). Ibid. **70**, 309—316 (1948). — [7] On the asymptotic problems of the zeros in wave mechanics (*5*). Ibid. **70**, 461—480 (1948). — [8] On non-conservative linear oscillators of low frequency (*5*). Ibid. **70**, 529—539 (1948). — [9] On the Laplace-Fourier transcendents (*10*). Ibid. **71**, 367—372 (1949). — [10] Oscillatory and nonoscillatory linear differential equations (*5*). Ibid. **71**, 627—649 (1949). — [11] A separation theorem for continuous spectra (*5*). Ibid. **71**, 650—662 (1949). — [12] On the classical existence theorem of linear differential equations (*3*). Ibid. **71**, 859—864 (1949). — [13] On the derivatives of the solutions of onedimensional wave equations (*5*). Ibid. **72**, 148—156 (1950). — [14] On nonlinear differential equations of first order (*8*). Ibid. **72**, 347—358 (1950). — [15] On the essential spectra of singular eigenvalue problems (*5*). Ibid. **72**, 545—552 (1950). — [16] On the classical transcendents of mathematical physics (*8*). Ibid. **73**, 381—389 (1951). — [17] On the non-increasing solutions of $y''=f(x, y, y')$ (*8*). Ibid. **73**, 390—404 (1951). — [18] An inequality for the amplitudes and areas in vibration diagrams of time dependent frequency (*5*). Quart. Appl. Math. **10**, 175—176 (1952). — [19] On the behavior of the solutions of real binary differential systems at singular points (*9*). Amer. J. Math. **75**, 117—176 (1953). — [20] On nonoscillatory linear differential equations (*5*). Ibid. **75**, 717—730 (1953). — [21] Linear differential and difference equations with monotone solutions (*5*). Ibid. **75**, 731—743 (1953). — [22] On an oscillation criterion of LYAPUNOV (*5*). Ibid. **73**, 885—890 (1951). — [23] Linear differential equations with completely monotone solutions (*5*). Ibid. **76**, 199—206 (1954). — [24] On nonoscillatory linear differential equations with monotone coefficients (*5*). Ibid. **76**, 207—219 (1954). — [25] On monotone solutions of systems of nonlinear differential equations (*5, 8*). Ibid. **76**, 860—866 (1954). — [26] Integrability in the large and dynamical stability (*6*). Ibid. **65**, 273—278 (1943). — [27] Asymptotic integrations of linear differential equations (*3*). Amer. J. Math. **77**, 45—86 (1955). — [28] On the assignment of asymptotic values for the solutions of linear differential equations of second order (*5*). Ibid. **77**, 475—483 (1955). — [29] On linear, second order differential equations in the unit circle (*5*). Trans. Amer. Math. Soc. **78**, 492—500 (1955).

HARTOG, J. P. DEN: [1] Mechanical Vibrations (*), 2nd edit. McGraw-Hill, 1940.

HARTREE, D. R., A. PORTER, A. CALLENDER and A. B. STEVENSON: [1] Time lag in a control system (*2*). Proc. Roy. Soc. Lond., Ser. A **161**, 460—476 (1937).

HAUPT, O.: [1] Über das asymptotische Verhalten der Lösungen gewisser linearer gewöhnlicher Differentialgleichungen (*3, 5*). Math. Z. **48**, 289—292 (1913). — [2] Über lineare homogene Differentialgleichungen zweiter Ordnung mit periodischen Koeffizienten (*4, 5*). Math. Ann. **79**, 278—285 (1919). — [3] Über Lösungen linearer Differentialgleichungen mit Asymptoten (*3, 5*). Math. Z. **48**, 212—220 (1942). — [4] Über das asymptotische Verhalten der Lösungen gewisser linearer gewöhnlicher Differentialgleichungen (*3, 4, 5*). Math. Ann. **48**, 289—292 (1942).

HAYASHI, C.: [1] Forced oscillations with nonlinear restoring force (*8*). Mem. Fac. Eng. Kyoto Univ. **13**, 180—197, 187—207 (1951). — [2] Stability investigation of the nonlinear periodic oscillations (*8*). Ibid. **14**, 92—102 (1952). — J. Appl. Phys. **24**, 344—348 (1953). — [3] Forced oscillations in nonlinear systems (*, *8*). Osaka, 1953. 164 pp.

HAYES, N. D.: [1] Roots of the transcendental equation associated with a certain difference-differential equation (*2*). J. London Math. Soc. **25**, 226—232 (1950).

— W. D.: [1] On the equation for a damped pendulum under constant torque (*8*). Z. angew. Math. Phys. **4**, 398—401 (1953).

·HAZEBROEK, P., and B. L. VAN DER WAERDEN: [1] Theoretical considerations on the optimum adjustment of regulators (*2*). Trans. Amer. Math. Soc. **72**, 309—315 (1950). — [2] The optimum adjustment of regulators (*2*). Ibid. **72**, 317—322 (1950).

HILB, O.: [1] Zur Theorie der linearen funktionalen Differentialgleichungen (*2, 3*). Math. Annal. **78**, 137—170 (1917).

HILL, G. W.: [1] Mean motion of the lunar perigee (*8*). Acta math. **8**, 1—36 (1886).

HILLE, E.: [1] Nonoscillation theorems (*5*). Trans. Amer. Math. Soc. **64**, 234—252 (1948). — [2] Behavior of solutions of linear second order differential equations (*5*). Ark. Mat. **2**, 25—41 (1952).

HOHLOV, R. V.: [1] On the theory of entrainement for small amplitude of the external force (*8*). Doklady Akad. Nauk SSSR, (N.S.) **97**, 411—414 (1954).

HOK, G.: [1] Response of linear resonant systems to excitation of a frequency varying linearly with time (*2*). J. Appl. Phys. **19**, 242—250, 623 (1948).

HOPF, E.: [1] Abzweigung einer periodischen Lösung von einer stationären Lösung eines Differentialsystems (*8*). Ber. Sächs. Akad. Wiss., Leipzig **95**, 3—22 (1943). — [2] Ergodentheorie (*9*). Ergebn. Math. u. i. Grenzgeb. 5, 1937, Chelsea 1948.

HORN, J.: [1] Bewegungen in der Nähe einer Gleichgewichtslage (*8*). J. f. Math. **131**, 224—245 (1906).

HORT, W.: [1] Die Differentialgleichungen des Ingenieurs (*), 2. Aufl. Berlin 1925. — [2] Technische Schwingungslehre (*), 2. Aufl., Springer 1942.

HORVAY, G.: [1] Unstable solutions of a class of Hill differential equations (*4*). Quart. Appl. Math. **4**, 385—396 (1947).

—, and S. W. YUAN: [1] Stability of a rotor blade flapping motion when the hinges are tilted. Generalization of the "rectangular ripple" method of solution (*4*). J. Aeronaut. Sci. **14**, 583—593 (1947).

HUANG, T. C.: [1] Harmonic oscillations of nonlinear two degrees of freedom systems (*8*). J. Appl. Mech. **22**, 107—110 (1955).

HUBER, A.: [1] Das Verhalten der Integrale der Gibbs-Duhem-Marguelsschen Gleichung für binäre Gemische in der Umgebung ihrer festen singulären Stellen (*9*). Sitzgsber. Österr. Akad. Wiss. **1951**, 181—197.

HUFFORD, G.: [1] Banach spaces and the perturbation of ordinary differential equations (*8*). Contrib. Theory Nonlinear Oscillations **3**, 173—196 (1956).

HUKUHARA, M.: [1] Sur un système de deux équations différentielles ordinaires non linéaires a coefficients réels (*10*). J. Fac. Sci. Univ. Tokyo, Sect. 1 **6**, 295—317 (1917). — [2] Sur les points singuliers des équation différentielles linéaires (*3, 10*). J. Fac. Sci. Univ. Hokkaido (1) Math. **2**, 13—81 (1934—1936). —

[3] Sur les points singuliers des équations différentielles linéaires (*10*). Mem. Fac. Sci. Kyusyu Imp. Univ. A **2**, 1—25 (1941); **2**, 125—137 (1942); **3**, 67—73 (1944). — [4] Sur les propriétés asymptotiques des solutions d'un systéme d'équations différentielles linéaires contenant un paramètre (*3, 10*). Mem. Fac. Eng. Kyushyu Imp. Univ. **8**, 249—284 (1937). — [5] On the expansion of the solution of differential equations in the neighborhood of their singular point (*10*). ibid. **4**, 1—7 (1949). — [6] On singular points of the ordinary differential equation of the first order (*10*). Ibid. **4**, 9—21 (1949). — [7] Sur la généralisation des théorèmes de M. M. MALMQUIST (*10*). J. Fac. Sci. Univ. Tokyo, Sect. 1 **6**, 77—84 (1949). — [8] Sur un theorem de KNESER (*9*). ibid. **6**, 329—344 (1953).

HUKUHARA, M., and M. NAGUMO: [1] On a condition of stability for a differential equation (*3, 5*). Proc. Imp. Akad. Tokyo **6**, 131—132 (1930).

HULL, T. E., and C. FROESE: [1] Asymptotic behavior of the inverse of a Laplace transform (*2*). Canad. J. Math. **7**, 116—125 (1955).

HUREWICZ, W.: [1] Ordinary differential equations (*). Brown Univ. 1943.

HURWITZ, A.: [1] Über die Bedingungen unter welchen eine Gleichung nur Wurzeln mit negativen reellen Teilen besitzt (*2*). Math. Ann. **46**, 273—284 (1895).

IBRASEV, H. I.: [1] Some cases of stability of solutions of a denumerable system of differential equations (*6, 7*). Izvestiya Akad. Nauk Kazah. SSSR. **5**, 119—135 (1951).

IELCHIN, M.: [1] Sur le problème d'oscillation pour l'équation différentielle linéaire du deuxième ordre (*4*). Doklady Akad. Nauk SSSR, (N. S.) **18**, 141—145 (1938).

IGLISH, R.: [1] Zur Theorie der Schwingungen (*8*). Mh. Math. u. Phys. **37**, 325—342 (1930); **39**, 173—220 (1932); **42**, 7—36 (1935). — [2] Über die Lösungen des Duffingschen Schwingungsproblems bei großen Parameterwerten (*8*). Math. Ann. **111**, 568—581 (1935). — [3] Die erste Resonanzkurve beim Duffingschen Schwingungsproblem (*8*). Math. Ann. **112**, 221—245 (1936). — [4] Über den Resonanzbegriff bei nicht linearen Schwingungen (*8*). Z. angew. Math. Mech. **17**, 249—258 (1937).

IMAZ, C.: [1] Sobre ecuaciones differenciales lineales periodicas con un parametro pequeno (*4*). Bol. Soc. Matem. Mexicana **1961**, 19—51.

INCE, E. L.: [1] General solution of HILL's equation (*4*). Monthly Not. Roy. Astr. Soc. **75**, 436 (1915). — [2] Research into the characteristic numbers of the Mathieu equation (*4*). Proc. Roy. Soc. Edinburgh **46**, 20—29 (1925). — [3] Periodic solutions of linear differential equations with periodic coefficients (*4*). Proc. Cambridge Phil. Soc. **23**, 44—46 (1926). — [4] The real zeros of solutions of a linear differential equation with periodic coefficients (*4*). Proc. Lond. Math. Soc. **25**, 53—58 (1926). — [5] Ordinary differential equations (*). London, Longmans, Green 1927.

JACOBSTHAL, W.: [1] Asymptotische Darstellungen von Lösungen gewöhnlicher linearer Differentialgleichungen (*10*). Math. Ann. **56**, 129—154 (1902).

JAMES, H. M., N. B. NICHOLS and R. S. PHILLIPS (editors): [1] Theory of Servomechanisms (*, *2*). McGraw-Hill 1947, 375pp.

JEFFREYS, H.: [1] On certain approximate solutions of linear differential equations of the second order (*10*). Proc. London Math. Soc. **23**, 428—436 (1925).

JOHN, F.: [1] On simple harmonic vibrations of a system with nonlinear characteristics (*8, 9*). Studies in nonlinear Vibration Theory, 104—192. New York Univ. 1946. — [2] On harmonic vibrations out of phase with the exciting force (*8*). Comm. on Appl. Math. **1**, 341—359 (1948).

JONES, C. W.: [1] On reducible nonlinear differential equations occurring in mechanics (*9*). Proc. Roy. Soc. Lond., Ser. A **217**, 327—343 (1953).

JOUNIN, H.: [1] Sur le calcul des fréquences propres des systèmes non linéaires (*8*). C. R. Acad. Sci. Paris **222**, 1203—1205 (1946).

KAC, A. M.: [1] On the approximate solution of nonlinear differential equations of the second order (*8*). Prikl. Mat. Meh. SSSR, **14**, 111—113 (1950). — [2] Biharmonic oscillations of a dissipative nonlinear system which are induced or sustained by a harmonic disturbing force (*8*). Ibid. **18**, 425—444 (1954). — [3] Forced oscillations of nonlinear systems with one degree of freedom and near to conservative ones (*8*). Ibid. **19**, 13—32 (1955).

KALININ, S. V.: [1] On the stability of periodic motions in the case when one root is equal to zero (*7*). Prikl. Mat. Meh. SSSR, **12**, 671—672 (1948). — [2] On the stability of periodic motions in the case when one of the roots is zero (*7*). Ibid. **13**, 247—252 (1949).

KALLMAN, H., and M. PÄSLER: [1] Zur Integration der gestörten zeitabhängigen Schrödinger-Gleichung (*8*). Z. Physik **126**, 749—759 (1949).

KAMENKOV, G. V.: [1] On stability of motion over a finite interval of time (*7*). Prikl. Mat. Meh. SSSR, **17**, 529—540 (1953).

— and A. A. LEBEDEV: [1] Remarks to a paper on stability over a finite interval of time (*7*). Prikl. Mat. Meh. SSSR, **18**, 512 (1954).

KAMKE, E.: [1] Differentialgleichungen reeller Funktionen (*). Leipzig 1930. — [2] A new proof of STURM's comparison theorem (*5*). Ann. Math. Monthly **46**, 417—421 (1939). — [3] Über STURMS Vergleichungssätze für homogene lineare Differentialgleichungen zweiter Ordnung und Systeme von zwei Differentialgleichungen erster Ordnung (*5*). Math. Z. **47**, 788—795 (1942). — [4] Differentialgleichungen. Lösungsmethoden und Lösungen (*), Bd. 2. Leipzig 1943, 1948. — [5] Differentialgleichungen (*). Edwards 1945.

KAPLAN, W.: [1] Regular curve-families filling the plane (*9*). Duke Math. J. **7**, 154—185 (1940); **8**, 11—46 (1941). — [2] Some methods for analysis of the flow in phase space (*1, 9*, *). Proc. Symposium on Nonlinear Circuit Analysis, New York 1953, 99—106. — [3] Stability theory (*). Ibid. 1956, 3—21.

KAPTEYN, W.: [1] On the centers of integral curves of differential equations of first order and first degree (*9*). Nederl. Akad. Wetensch. **19**, 1446—1457 (1911); **20**, 354—1365 (1911); **21**, 27—33 (1912).

KARIMOV, D. H.: [1] Sur les solutions périodiques des équations différentielles nonlinéaires du type parabolique (*8*). Doklady Akad. SSSR, (N. S.) **25**, 3—6 (1939); **28**, 403—406 (1940); **46**, 175—178 (1945); **54**, 293—295 (1946). — [2] On periodic solutions of nonlinear equations of the fourth order (*8*). Ibid. **57**, 651—653 (1947); **1949**, 3—7.

KÁRMÁN, TH. v.: [1] The engineer grapples with nonlinear problems (*). Bull. Amer. Math. Soc. **46** (1940).

KARREMAN, G.: [1] Some types of relaxation oscillations as models of all-or-none phenomena (*8*). Bull. Math. Biophys. **11**, 311—318 (1949).

KARTVELISVILI, N. A.: [1] On conditions for the oscillation of an automatic regulator (*2*). Doklady Akad. Nauk. SSSR, (N. S.) **61**, 21—23 (1948).

KASAHARA, S.: [1] On the existence of periodic solutions for certain differential equations (*8*). Proc. Japan Acad. **29**, 544—547 (1953).

KATÖ, T.: [1] Sur les points singuliers des équations différentielles ordinaires du premier ordre (*9*). Nat. Sci. Rep. Ochanimizu Univ. **2**, 13—17 (1951); **4**, 36—39 (1953; **5**, 1—4 (1954). — [2] On singular points of ordinary differential equations of the first order (*9*). Ibid. **1**, 17—21 (1951). — [3] On the least eigenvalue of the Hill equation (*4*). Quart. Appl. Math. **10**, 292—294 (1952).

KATZ, A. M.: [1] Constrained oscillations within the domain of resonance (*2*). Akad. Nauk. SSSR., Inzen. Sbornik **3**, no. 2, 100—125 (1947).

KAUDERER, H.: [1] Zur Analyse der Dämpfung freier Schwingungen (*8*). Ing.-Arch. **22**, 251—257 (1954).

KAZAKEVICH, V. V.: [1] On approximate integration of VAN DER POL'S, equation (*8, 9*). Doklady Akad. Sci. SSSR, (N. S.) **49**, 414—417 (1945). — [2] Sur l'intégration approximative des systèmes oscillatoires à un degré de liberté (*8, 9*). Ibid. **51**, 107—110 (1946). — [3] Sur le processus d'établissement de systèmes d'os-cillation à un degré de liberté (*8, 9*). Ibid. **49**, 486—489 (1945). — [4] Multiply valued systems and the simplest dynamical models of clocks (*8*). Ibid. **74**, 665—668 (1950).

KAZARINOFF, N. D.: [1] Asymptotic forms for the Whittaker functions with both parameters large (*10*). J. Math. Mech. **6**, 341—360 (1957). — [2] Asymptotic solution with respect to a parameter of a differential equation having an irregular singular point (*10*). Proc. Amer. Math. Soc. **7**, 62—69 (1956). — [3] Asymptotic expansions for the Whittaker functions of large complex order *m* (*10*). Trans. Amer. Math. Soc. **78**, 305—328 (1955). — [4] Asymptotic theory of second order differential equations with two turning points (*10*). Archive Rat. Mech. Anal., to appear. — [5] Asymptotic solutions with respect to a parameter of ordinary differential equations having a regular singular point (*10*). Mich. Math. J. **4**, 207—220 (1957).

—, and R. MCKELVEY: [1] Asymptotic solution of differential equations in a domain containing a regular singular point (*10*). Canad. J. Math. **8**, 97—104 (1956).

KEIL, K. A.: [1] Das qualitative Verhalten der Integralkurven einer gewöhnlichen Differentialgleichung erster Ordnung in der Umgebung eines singulären Punktes (*9*). Jber. dtsch. Math. Ver. **57**, 111—132 (1955).

KELLER, J. B.: [1] Bowing of violin strings (*8*). Comm. Pure Appl. Math. **6**, 483—495 (1953).

KEMPERMAN, J. H. B.: [1] The analytic continuation and asymptotic expansion of a function defined by its Taylor series. To appear.

KESTIN, J., and S. K. ZAREMBA: [1] Geometrical methods in the analysis of ordinary differential equations. Introduction to nonlinear mechanics (*, *9*). Appl. Sci. Res. **3**, 149—189 (1953).

KIMURA, T.: [1] Sur une généralisation d'un théorème de MALMQUIST (*9*). Comm. Math. Univ. St. Paul **2**, 23—28 (1953); **3**, 97—107 (1955). — [2] Sur les points singuliers des équations différentielles ordinaires du premier ordre (*9*). Ibid. **2**, 47—49 (1954).

KITAMURA, T.: [1] Some inequalities on a system of solutions of linear simultaneous differential equations (*3*). Tohoku Math. J. **49**, 308—311 (1943).

KITO, F.: [1] On a property of linear ordinary differential equations which relates to "end effect" in theory of curved shells (*3*). Proc. Fac. Engr. Keiogjuku Univ. **3**, no. 8, 16—21 (1950).

KLOTTER, K.: [1] Das Ausschlag-Zeit-Diagramm einer „einfachen Schwebung" (*2*). Z. angew. Math. Mech. **30**, 190 (1950). — [2] Schwingungen mit endlich vielen Freiheitsgraden (*). Naturforschung und Medizin in Deutschland 1939—1946, Bd. 4, Teil II, 67—85. Weinheim: Verlag Chemie 1953. — [3] Steady state vibrations in systems having arbitrary restoring and arbitrary damping forces (*, *8*). Proc. Symposium on Nonlinear Circuit Analysis, New York, 1953, 234—257.

—, and G. KOTOWSKI: [1] Über die Stabilität der Lösungen Hillscher Differentialgleichungen mit drei unabhängigen Parametern (*4*). Z. angew. Math. Mech. **23**, 213 (1943).

—, and E. PINNEY: [1] A comprehensive stability criterion for forced vibrations in nonlinear systems (*, *8*). J. Appl. Mech. **20**, 9—12 (1953); a translation in Rev. gen. Électr. **63**, 559—562 (1954).

Kneser, A.: [1] Untersuchungen über die reellen Nullstellen der Integrale linearer Differentialgleichungen (*5*). Math. Ann. **42**, 409—435 (1893). — [2] Studien über die Bewegungsvorgänge in der Umgebung instabiler Gleichgewichtslagen(*9*). J. reine angew. Math. **115**, (1895); **118** (1897). — [3] Untersuchungen und asymptotische Darstellung der Integrale gewisser Differentialgleichungen bei großen reellen Werten des Arguments (*5, 10*). Ibid. **116**, 178—212 (1896); **117**, 72—103 (1897); **120**, 267—275 (1899).

—, H.: [1] Reguläre Kurvenscharen auf den Ringflächen (*10*). Math. Annalen **91**, 135—154 (1924). — [2] Die Reihenentwicklung bei schwach singulären Stellen linearer Differentialgleichungen (*6, 10*). Ark. Math. **2**, 413—419 (1951) (1949/50).

Kocin, N. E.: [1] On the oscillations of a crankshaft (*2*). Prike. Mat. Meh. SSSR, **2**, 3—28 (1934).

Kolmogorov, A. N.: [1] On dynamical systems with an integral invariant on the torus (*6, 8*). Doklady Akad. Nauk SSSR, (N. S.) **93**, 763—766 (1953). — [2] On conservation of conditionally periodic motions for a small change in Hamilton's function (*8*). ibid. **98**, 527—530 (1954).

Kononenko, V. O.: [1] On a case of coupled oscillations (*8*). Dopovidi Akad. Nauk SSSR, **1951**, 74—81.

Koosis, P.: [1] One-dimensional repeating curves in the nondegenerate case. Contrib. Theory Nonlinear Oscillations **3**, 277—285 (1956).

Košlyakov, N. S.: [1] Investigation of a class of differential equations with doubly periodic coefficients (*5*). Izvestiya Akad. Nauk SSSR, **16**, 537—562 (1952).

Koukless, I. S.: [1] Sur les conditions nécessaires et suffisante pour l'éxistence d'un centre (*9*). Doklady Akad. Nauk SSSR, (N. S.) **42**, 160—163 (1944). — [2] Sur quelques cas de distinction entre un foyer et un centre (*9*). Ibid. **42**, 208—211 (1944). — [3] Sur deux groupes fondamentaux de points singuliers (*9*). Ibid. **42**, 253—255 (1944).

Kovalenko, K. R. and M. G. Krein: [1] On some investigations of A. Lyapunov on differential equations with periodic coefficients (*4*). Doklady Akad. Nauk SSSR., N. S. **75**, 495—498 (1950).

Krall, G.: [1] Qualche complemento alla teoria degli invarianti adiabatici secondo Levi Civita (*8*). Rend. Accad. Lincei (*6*) **9**, 13 (1931). — [2] Problemi intorno alla stabilitá dell'equilibrio elastico (*8*). Ricerca Scientifica **1940**. — [3] Meccanica tecnica delle vibrazioni (*), 2 vols. Zanichelli 1940.

Kramers, H. A.: [1] Das Eigenwertproblem im eindimensionalen periodischen Kraftfelde (*4, 10*). Phys. **2**, 483—490 (1935).

Krasnoselskii, M. A., and M. G. Krein: [1] On the principle of averaging in nonlinear mechanics (*8*). Uspehi Mat. Nauk, (N. S.) **10**, no. 3 (65), 147—152 (1955). — [2] Nonlocal existence theorems and uniqueness theorems for systems of ordinary differential equations (*1*). Doklady Akad. Nauk SSSR, (N. S.) **102**, 13—16 (1955).

Krasnuškin, P. E.: [1] On the asymptotic representation of solutions of the wave equation (*10*). Vestnik Moskov. Univ. **3**, no. 6, 73—76 (1948).

Krasovskiĭ, A. A.: [1] On a vibrational method for linearizing some nonlinear systems (*8*). Avtom i Telem. **9**, 20—29 (1948).

—, N. N.: [1] Theorems on stability of motions determined by a system of two equations (*7, 9*). Prikl. Mat. Meh. SSSR **16**, 547—554 (1952). — [2] On a problem of stability of motion in the large (*7, 9*). Doklady Akad. Nauk SSSR, (N. S.) **88**, 401—404 (1953). — [3] On stability of solutions of a system of two differential equations (*7, 9*). Prikl. Mat. Meh. SSSR, **17**, 651—672 (1953). — [4] On stability of solutions of a system of second order in critical cases (*7, 9*). Doklady Akad. Nauk SSSR, (N. S.) **93**, 965—967 (1953). — [5] On stability of motion in the large for constantly acting disturbances (*7, 9*). Prikl. Mat. Meh. SSSR, **18**, 95—102 (1954). — [6] On the behavior in the large of the integral curves of a system of two differential equations (*7, 9*). Ibid. **18**, 149—154 (1954). — [7] On the inversion of theorems of A. Lyapunov and N. G. Cetaev on instability for stationary systems of differential equations (*7*). ibid. **18**, 513—532 (1954). — [8] Sufficient conditions for stability of solutions of a system of nonlinear differential equations (*7, 9*). Doklady Akad. Nauk SSSR, N. S. **98**, 901—904 (1954). — [9] On stability in the large of the solution of a nonlinear system of differential equations (*7, 9*). Prikl. Mat. Meh. SSSR. **18**, 735—737 (1954). — [10] On stability of motion in the critical case of a single zero root (*7*). Mat. Sbornik. **37** (79), 83—88 (1955). — [11] On inversion of K. P. Persidskii's theorem on uniform stability (*3, 7*). Prikl. Mat. Meh. SSSR. **19**, 273—278 (1955). — [12] On conditions of inversion of A. Lyapunov's theorems on instability for stationary systems of differential equations (*7*). Dokl. Akad. Nauk SSSR, (N. S.) **101**, 17—20 (1955). — [13] On the stability in the presence of arbitrary initial perturbations of the solutions of a system of three equations (*8*). Prikl. Mat. Meh. SSSR. 339—350 (1953).

Krein, M. G.: [1] A generalization of some investigations of A. Lyapunov on linear differential equations with periodic coefficients (*4*). Doklady Akad. Nauk SSSR, (N. S.) **73**, 445—448 (1950). — [2] On certain problems on the maximum and minimum of characteristic values and on the Lyapunov zones of stability (*4, 5*). Prikl. Mat. Meh. SSSR, **15**, 323—348 (1951); Amer. Math. Soc. Translations (2) **1**, 163—188. — [3] On the theory of entire matrix function of exponential type (*3, 4*). Ukrain. Mat. Ž. **3**, 164—173 (1951). — [4] On an application of an algebraic process to the theory of entire matrices (*3, 4*). Uspehi Mat. Nauk **6**, 171—177 (1951). — [5] On inverse problems of the theory of filters and λ-zones of stability (*4, 5*). Doklady Akad. Nauk SSSR, (N. S.) **93**, 767—770 (1953). — [6] On the theory of λ-zones of stability of canonical systems of linear differential equations with periodic coefficients (*4*). Collected memoirs dedicated to A. Andronov. Izvestiya Akad. Nauk SSSR, pp. 413—498, Moscow 1955. — [7] On criteria of stable boundedness of solutions of periodic canonical systems (*4*). Prikl. Mat. Meh. SSSR. **19**, 641—680 (1955).

Kruming, A. A.: [1] Estimate of the radius of convergence of power series in a small parameter which represent periodic solutions of systems of differential equations (*8*). Ukrain. Mat. Z. **5**, 434—438 (1953).

Krzyzański, M.: [1] Sur les solutions de l'équation linéaire du type parabolique déterminées par les conditions initiales (*3*). Ann. Soc. Polon. Math. **18**, 145—156 (1945).

Kučer, D. L.: [1] On some criteria for the boundedness of the solution of a system of differential equations (*3*). Doklady Akad. Nauk SSSR, (N. S.) **69**, 603—606 (1949).

Kulebakin, V. S.: [1] On the behavior of continuously perturbed automatized linear systems (*2*). Ibid. **68**, 855—858 (1949).

Kulikov, N. K.: [1] On the determination of the limit of the general solution of a non linear differential equation of the first order (*8*). Prikl. Mat. Meh. SSSR, **16**, 729—734 (1952).

Kumar, S. A.: [1] The transmission factors of potential barriers (*10*). Proc. Nat. Inst. Sci. India **10**, 373—385 (1944).

Kunin, L. A.: [1] Determination of a finite region of initial deviations for which the motions remain asymptotically stable for a system of two equations of the first order (*7*). Prikl. Mat. Meh. SSSR, **16**, 539—546 (1952).

Kuntzmann, J., J. Daniel, and M. Min-Yuan: [1] Stabilité des systèmes de réglage. Méthodes d'étude (*, *8*). Rev. gén. Électr. **52**, 149—152 (1952).

KURZWEIL, J.: [1] On the reversiblility of the first theorem of LYAPUNOV concerning the stability of motion. Czech. Math. J. **5** (80), 382—398 (1955). — [2] On the inversion of the second theorem of LYAPUNOV on stability of motion. Ibid. **6** (81), 217—259 (1956).

KUWAGAKI, A.: [1] Sur quelques équations functionelles et leurs solutions caractéristiques (*3*). Mem. Coll. Sci. Univ. Kyoto, **26**, 271—277 (1951).

KUZMIN, P. A.: [1] On the theory of stability of motion (*7*). Prikl. Mat. Meh. SSSR, **18**, 125—127 (1954).

KYNER, W. T.: [1] A fixed point theorem (*9*). Contrib. Theory Nonlinear Oscillations **3**, 197—206 (1956).

LAGRANGE, J. L.: [1] Mécanique Analytique (*), pp. 512, Paris 1788.

LAHAYE, E.: [1] Les développements des intégrales des équations $dy/dx = P(x, y)/Q(x, y)$ dans le domaine des valeurs qui annullent simultanément P et Q (*9*). Mem. Acad. Roy. Belg. (2) **20**, no. 5, 123 pp. (1945).

LAMPARIELLO, G.: [1] Sulla natura analitica delle soluzioni delle equazioni differenziali lineari a coefficienti periodici (*4*). Rend. Accad. Lincei (6) **21**, 637—644 (1935).

LANCASTER, D. E.: [1] Some results concerning the behavior at infinity of real continuous solutions of algebraic differential equations (*2*). Bull. Amer. Math. Soc. **46**, 169—177 (1940).

LANDAU, E.: [1] Über einen Satz von Herrn ESCLANGON (*1, 3*). Math. Ann. **102**, 177—188 (1930).

LANGENHOP, C. E.: [1] Note on LEVINSON's existence theorem for forced periodic solutions of a second order differential equations (*9*). J. Math. Phys. **30**, 36—39 (1951).

—, and A. B. FARNELL: [1] The existence of forced periodic solutions of second order differential equations near certain equilibrium points of the unforced equation (*9*). Contrib. to the Theory of nonlinear Oscillations. Ann. of Math. Studies **20**, 291—312.

LANGER, R. E.: [1] The asymptotic location of the roots of a certain transcendental equation (*10*). Trans. Amer. Math. Soc. **31**, 837—844 (1929). — [2] On the zeros of exponential sums and integrals (*10*). Bull. Amer. Math. Soc. **37**, 213—239 (1931). — [3] On the asymptotic solutions of ordinary differential equations, with an application to the Bessel functions of large order (*10*). Trans. Amer. Math. Soc. **33**, 23—64 (1931). — [4] On the asymptotic solutions of differential equations, with an application to the Bessel functions of large complex order (*10*). Ibid. **34**, 447—480 (1932). — [5] The asymptotic solutions of ordinary linear differential equations of the second order, with special reference to the Stokes phenomenon (*10*). Bull. Amer. Math. Soc. **40**, 545—582 (1934). — [6] The asymptotic solutions of certain linear ordinary differential equations of the second order. Trans. Amer. Math. Soc. **36**, 90—106 (1934). — [7] The solutions of the Mathieu equation with a complex variable and at least one parameter large. Ibid. **36**, 637—695 (1934). — [8] On the asymptotic solution of ordinary differential equations with reference to the Stokes phenomenon about a singular point (*10*). Ibid. **37**, 397—416 (1935). — [9] The asymptotic solutions of ordinary linear differential equations of the second order, with special reference to a turning point (*10*). Ibid. **67**, 461—490 (1949). — [10] On the wave equation with small quantum numbers (*10*). Phys. Rev. (2) **75**, 1573—1578 (1949). — [11] Asymptotic solutions of a differential equation in the theory of microwave propagation (*10*). Comm. Pure Appl. Math. **3**, 427—438 (1950). — [12] The solutions of the differential equation $v''' + \lambda^2 z v' + 3\mu\lambda^2 v = 0$ (*10*). Duke Math. J. **22**, 525—542 (1955). — [13] On the asymptotic forms of the solutions of ordinary linear differential equations of the third order in a region containing a turning point. Trans. Amer. Math. Soc. **80**, 93—123 (1955). — [14] The solutions of a class of ordinary linear differential equations of the third order in a region containing a multiple turning point (*10*). Duke Math. J. **23**, 93—110 (1956). — [15] On the construction of related differential equations (*10*). Trans. Amer. Math. Soc. **81**, 394—410 (1956). — [16] On the asymptotic solutions of a class of ordinary differential equations of the fourth order, with special reference to an equation of hydrodynamics (*10*). Ibid. **84**, 144—191 (1957). — [17] On the stability of the laminar flow of a viscous fluid (*10*). Bull. Amer. Math. Soc. **46**, 257—263 (1940). — [18] The asymptotic solutions of a linear differential equation of the second order with two turning points (*10*). Trans. Amer. Math. Soc., to appear.

LaSalle, J.: [1] Uniqueness theorems and successive approximations (*1*). Ann. of Math. (2) **50**, 722—730 (1949). — [2] Relaxation oscillations (*8*). Quart. Appl. Math. **7**, 1—19 (1949).

Lauer, H., R. Lesnick and L. E. Hatson: [1] Servomechanisms fundamentals (*, *2*). McGraw-Hill, 1947.

Lawden, D. F.: [1] Mathematics of Engineering Systems (linear and nonlinear) (*, *2*), pp. 380. Wiley, 1954.

Leavitt, W. G.: [1] On systems of linear differential equations (*10*). Amer. J. Math. **73**, 690—696 (1951).

Lebedev, A. A.: [1] On the problem of stability of motion over a finite interval of time (*7*). Prikl. Mat. Meh. SSSR **18**, 75—94 (1954). — [2] On stability of motion during a given interval of time (*7*). Ibid. **18**, 139—148 (1954).

Ledinegg, E.: [1] Über ein dem klassischen Minimumproblem homogener Differentialgleichungen vom Sturm-Liouvilleschen Typus zugeordnetes Variationsprinzip (*5*). Acta physica Austriaca **3**, 273—276 (1949).

Lefévre, P.: [1] L'étude de la stabilité des systemes linéaires par la méthode du diagramme de phase généralisé (*2*). Mem. Artillerie Franç. **26**, 503—588 (1952).

Lefschetz, S.: [1] Existence of periodic solutions of certain differential equations (*8*, *9*). Proc. Mat. Acad. Sci. U.S.A. **29**, 90 (1943). — [2] Lectures on Differential Equations (*). Ann. of Math. Studies 1948, no. 14, 210 pp. — [3] (editor) Contributions to the Theory of nonlinear Oscillations (*). Ann. of Math. Studies **1**, 1950, no. 20, 350 pp.; **2**, 1952, no. 29, 116 pp.; **3**, 1956, no. 36, 285 pp.; **4**, 1958, no. 41, 211 pp. — [4] Notes no differential equations (*9*). Contrib. Theory Nonlinear Oscillations **2**, 61—73 (1952). — [5] On Liénard's differential equation (*, *9*). Proc. Sympos. Appl. Math. **5**, 149—153 (1954). — [6] Complete families of periodic solutions of differential equations (*8*). Comm. Math. Helv. **28**, 341—345 (1954). — [7] Differential equations: Geometric theory (*), pp. 360, Interscience, 1957.

Lehmann, N. J.: [1] Die Stabilitätsfrage bei rückgekoppelten Verstärkern (*2*). Z. angew. Math. Mech. **28**, 23—29, 59—64 (1948).

Leighton, W.: [1] Bounds for the solutions of a second order linear differential equation (*5*). Proc. Nat. Acad. Sci. U.S.A. **35**, 190—191, 422 (1949). — [2] Principal quadratic functionals and self-adjoint second-order differential equations (*5*). Ibid. **35**, 192—193 (1949). — [3] On self-adjoint differential equations of second order (*5*). Ibid. **35**, 656—657 (1949). — [4] A substitute for the Picone formula (*5*). Bull. Amer. Math. Soc. **55**, 325—328 (1949). — [5] The detection of the oscillation of solutions of a second order linear differential equation (*5*). Duke Math. J. **17**, 57—61 (1950). — [6] On self-adjoint differential equations of second order (*5*). J. London Math. Soc. **27**, 37—47 (1952).

Leonhard, A.: [1] Stabilitätskriterium insbesondere von Regelkreisen bei vorgeschriebener Stabilitätsgüte (*2*). Arch. Elektrotechn. **39**, 100—107 (1948).

Leonov, M. Ya.: [1] On quasiharmonic oscillations (*4*, *5*). Prikl. Mat. Meh. SSSR, **10**, 575—580 (1946). — [2] The parametric representation of quasiharmonic oscillations (*4*, *5*). Doklady Akad. Nauk SSSR, (N. S.) **62**, 161—162 (1948). — [3] Certain criteria of dynamical stability (*4*, *5*). Prikl. Mat. Meh. SSSR, **12**,

737—748 (1948). — [4] On the theory of quasiharmonic oscillations (*4, 5*). Dopovidi Akad. Nauk Ukraine SSSR, **1948**, 1—57. — [5] The stability of quasiharmonic oscillations (*4, 5*). Doklady Akad. Nauk SSSR, (N. S.) **64**, 645 to 648 (1949). — [6] An approximate method for investigating quasiharmonic oscillations (*4, 5*). Akad. Nauk Ukraine **2**, 5—8 (1953).

LEONTIEV, A. E.: [1] Differential-difference equations (*3*). Mat. Sbornik **24** (66), no. 3, 347—374 (1949). — Amer. Math. Soc. Trans. **1952**, no. 78.

LEONTOVIC, E.: [1] On the generation of limit cycles from separatrices (*9*). Doklady Akad. Nauk SSSR, (N. S.) **78**, 641—644 (1951).

—, and A. MAYER: [1] On a scheme determining the topological structure of the separation of trajectories (*9*). Doklady Akad. Nauk SSSR, (N. S.) **103**, 557—560 (1955).

LETOV, A. M.: [1] The regulation of the stationary state of a system subjected to constant perturbing forces (*8*). Prikl. Mat. Meh. SSSR, **12**, 149—156 (1948). — [2] On the theory of an isodromic regulator (*8*). Ibid. **12**, 363—368 (1948). — [3] Strictly unstable regulating systems (*8*). Ibid. **14**, 183—192 (1950). — [4] Bounds for the smallest characteristic value of a class of regulating systems (*8*). Ibid. **15**, 591—600 (1951). — [5] Stability of control systems with two regulating organs (*8*). Ibid. **17**, 401—410 (1953). — [6] Stability of unsteady motions of control systems (*8*). Ibid. **19**, 257—264 (1955). — [7] Stability of nonlinear regulatory systems (*). Gostekizdat, Moscow 1955.

LETTENMAYER, F.: [1] Über das asymptotische Verhalten der Lösungen von Differentialgleichungen und Differentialgleichungssystemen (*3*). Sitzgs.ber. bayer. Akad. Wiss., Münch., Math.-naturwiss. Kl. **1929**, 201—252.

LEVENSON, M. E.: [1] Harmonic and subharmonic response for the Duffing equation $x''+\alpha x+\beta x^3=F\cos\omega t(\alpha>0)$ (*8*). J. Appl. Phys. **20**, 1045—1051 (1949).

LEVI, B., and J. L. MASSERA: [1] Study in the large of a differential equation of the second order (*9*). Math. Notae **7**, 91—155 (1947).

—, E.: [1] Sul comportamento asintotico delle soluzioni dei sistemi die equazioni differenziali lineari omogenee (*3*). Rend. Accad. Lincei (8) **8**, 465—470 (1950); **9**, 26—31 (1950).

LEVI CIVITA, T.: [1] Sur l'instabilité de certain substitions (*8*). C. R. Acad. Sci. Paris **130**, 103—106 (1900). — [2] Sur l'instabilité de certain solutions périodiques (*8*). Ibid. **130**, 170—173 (1900). — [3] Sur le problème restreint des trois corps (*8*). Ibid. **131**, 236—239 (1900). — [4] Sopra alcuni criteri di instabilità (*8*). Annali Mat. pura appl. (3) **5**, 221—308 (1901). — [5] Sur la recherche des solutions particulières des systèmes différentielles et sur les mouvements stationaires (*8, 9*). Prace Mat.-Fiz. **17**, 1—40 (1906). — [6] Sur les équations linéaires à coefficients périodiques et sur le moyen mouvement du noed lunaire (*4*). Ann. Sci. École Norm. Sup. Paris (2) **28**, 325—376 (1911). — [7] Sur la regularisation du probleme des trois corps (*9*). Acta Math. **42**, 99—144 (1919). — [8] Sugli invarianti adiabatici (*8*). Atti Congr. Internat. Fis. 1927.

—, and U. AMALDI: [1] Lezioni di meccanica razionale (*). Vols. 2. Zanichelli, 1927.

LEVIN, B. YA.: [1] On the growth of the solutions of the Sturm-Liouville equation (*5*). Odessa State Univ. Mat. **2**, 39—43 (1938).

J. J., and N. LEVINSON: [1] Singular perturbations of nonlinear systems of differential equations and an associated boundary layer equation (*8, 10*). J. Rational Mech. Anal. **3**, 247—270 (1954).

LEVINSON, N.: [1] On certain nonlinear differential equations of the second order (*8*). Proc. Nat. Acad. Sci. U.S.A. **29**, 222—223 (1943). — [2] Correction to "On certain nonlinear differential equations of the second order" (*8*). Ibid. **29**, 281 (1943). — [3] On the existence of periodic solutions for second order differential equations with a forcing term (*8, 9*). J. Math. Phys. Mass. Inst. Techn.

22, 41—48 (1943). — [4] On a nonlinear differential equation of the second order. Ibid. (8) **22**, 181—187 (1943). — [5] Transformation theory of nonlinear differential equations of the second order (*6*). Ann. of Math. (2) **45**, 723—737 (1944). — [6] The asymptotic behavior of a system of linear differential equations (*3*). Amer. J. Math. **68**, 1—6 (1946). — [7] Perturbations of discontinuous solutions of nonlinear systems of differential equations (*8*). Proc. Nat. Acad. Sci. U.S.A. **33**, 214—218 (1947). — [8] A simple second order differential equation with singular motions (*8*, *10*). Ibid. **34**, 13—15 (1948). — [9] The asymptotic nature of solutions of linear systems of differential equations (*3*). Duke Math. J. **15**, 111—126 (1948). — [10] On stability of nonlinear systems of differential equations (*6*). Colloq. Math. **2**, 40—45 (1949). — [11] Criteria for the limit-point case for second order linear differential operators (*5*). Casopis Pest. Mat. Fys. **74**, 17—20 (1949). — [12] The inverse Sturm-Liouville problem (*5*). Mat. Tidsskr. B **1949**, 25—30. — [13] A second order differential equation with singular solutions (*8*, *10*). Ann. of Math. (2) **50**, 127—153 (1949). — [14] Determination of the potential from the asymptotic phase (*5*). Phys. Rev. (2) **75**, 1445 (1949). — [15] On the uniqueness of the potential in a Schrödinger equation for a given asymptotic phase (*5*). Danske Vid. Selsk. Mat.-Fys. Medd. **25**, no. 9, pp. 29 (1949). — [16] The first boundary value problem for $\varepsilon Du + A(x, y) u_x + B(x, y) u_y + C(x, y) u = D(x, y)$ for small ε (*5*). Ann. of Math. **51**, 428—445 (1949). — [17] Small periodic perturbations of an autonomous system with a stable orbit (*8*). Ibid. (*2*) **52**, 727—738 (1950). — [18] An ordinary differential equation with an interval of stability, a separation point, and an interval of instability (*8*, *10*). J. Math. Phys. **28**, 215—222 (1950). — [19] Perturbations of discontinuous solutions of nonlinear systems of differential equations (*8*, *10*). Acta math. **82**, 71—106 (1950). — [20] A simplified proof of the expansion theorem for singular second order linear differential equations (*5*). Duke Math. J. **18**, 57—71 (1951). — [21] Certain explicit relationships between phase shift and scattering potential (*5*). Phys. Rev. (2) **89**, 755—757 (1953). — [22] Liénard's Method (*8*, *9*). Brown. Univ. lecture notes **49**, 1—6 (1942/43). — [23] Mathematical treatment of the third order differential equation of a triod oscillator (*8*). Ibid. 1—9 (1942/43).

Levinson, N., and O. K. Smith: [1] A general equation for relaxation oscillations (*8*, *9*). Duke Math. J. **9**, 382—403 (1942).

Lewis, D.C.: [1] On the role of first integrals in the perturbation of periodic solutions (*8*). Annals of Math. **63**, 535—548 (1936). — [2] Metric properties of differential equations (*8*). Amer. J. Math. **71**, 294—312 (1949). — [3] Differential equations referred to a variable metric (*8*). Ibid. **73**, 48—58 (1951). — [4] On the perturbation of a periodic solution when the variational system has nontrivial periodic solutions (*8*). J. Rat. Mech. and Anal. **4**, 795—815 (1955). — [5] Families of periodic solutions of systems having relatively invariant line integrals (*8*). Proc. Amer. Math. Soc. **6**, 181—185 (1955). — [6] Periodic solutions of differential equations containing a parameter (*8*). Duke Math. J. **22**, 39—56 (1955).

Lidskii, V. B.: [1] On the number of solutions with integrable square of the system of differential equations $y'' + P(t) y = \lambda y$ (*5*). Doklady Akad. Nauk SSSR, (N. S.) **95**, 217—220 (1954).

—, and M. G. Neygaus: [1] On the boundedness of the solutions of the linear systems with periodic coefficients (*5*). Doklady Akad. Nauk SSSR, (N. S.) **77**, 189—192 (1951).

Liénard, A.: [1] Étude des oscillations entretenues (*8*). Rev. gén. Électr. **23**, 901—946 (1928). — [2] Oscillations auto-entretenues (*8*). Proc. III. Congr. Appl. Mech., Stockholm **3**, 173 (1931).

Lighthill, M. J.: [1] A technique for rendering approximate solutions to physical problems uniformly valid (*10*). Phil. Mag. (7) **40**, 1179—1201 (1949).

Lin, C.C.: [1] On the stability of two dimensional parallel flows (*10*). Proc. Nat. Acad. Sci. USA. **30**, 316—323 (1944). — [2] On the stability of two dimensional parallel flows. I. General theory (*10*). Quart. Appl. Math. **3**, 119—146 (1945). II. Stability in an inviscid fluid (*10*). Ibid. **3**, 218—234 (1945). III. Stability in a viscous fluid. Ibid. **3**, 277—301 (1946). — [3] The theory of hydrodynamic stability (*), pp. 155. Cambridge Univ. Press 1955.

Lindelöf, E.: [1] Sur la croissance des intégrales des équations différentielles algebraiques du premier ordre (*3*). Bull. Soc. Math. France **27**, 205—215 (1899).

Lindstedt, A.: [1] Differentialgleichungen der Störungstheorie (*8*). Mém. Sci. St. Petersbourgh **31** (1883).

Litzman, O.: [1] The Peano function and orbital stability of a differential equation of the first order (*9*). Publ. Fac. Sci. Univ. Masaryk **1953**, 65—90.

Loicyanskii, L. G.: [1] Free and forced vibrations with resistance laws which are quadratic and intermediate between linear and quadratic (*8*). Akad. Nauk SSSR., Inzen. Sbornik **18**, 139—148 (1954).

Lojasiewicz, S.: [1] Sur l'allure asymptotique des integrales du système d'équations différentielles au voisinage de point singulier (*9*). Ann. Polon. Math. **1**, 34—72 (1954). — [2] Sur un théorème de Kneser (*9*). Ann. Soc. Polon. Math. **24**, no. 2, 148—152 (1951).

Lonn, E. R.: [1] Über singuläre Punkte gewöhnlicher Differentialgleichungen (*9*). Math. Z. **44**, 507—530 (1938).

Los, F. S.: [1] On the principle of averaging for differential equations in Hilbert space (*8*). Ukrain. Mat. Z. **2**, no. 3, 87—93 (1950).

Love, C.: [1] On linear differentia' and difference equations (*2*). Amer. J. Math. **38**, 57—80 (1916).

Lubkin, S., and J. J. Stoker: [1] Stability of columns and strings under periodically varying forces (*4*). Quart. Appl. Math. **1**, 215 (1943).

Ludeke, C. A.: [1] A method of equivalent linearization for nonlinear oscillatory systems with large nonlinearity (*8*). J. Appl. Phys. **20**, 694—699 (1949).

Lure, A. I.: [1] Stability of one type of systems under control (*8*). Prikl. Mat. Meh. SSSR, **9**, 353—367 (1945). — [2] Investigation of the stability of motion of a dynamic system (*8*). Ibid. **11**, 445—448 (1947). — [3] On auto-oscillations in some regulating systems (*8*). Avtom i Telem. **8**, 335—348 (1947). — [4] On the stability of the auto-oscillations of regulating systems (*8*). Ibid. **9**, 361—362 (1948). — [5] On periodic solutions of systems of linear equations with constant coefficients (*2*). Prikl. Mat. Meh. SSSR, **12**, 353—362 (1948). — [6] On a canonical form of the equations of the theory of automatic regulation (*6, 7, 8*). Ibid. **12**, 651—666 (1948). — [7] On the character of the bounds of the region of stability of regulating systems (*1, 6, 7, 8*). Ibid. **14**, 371—382 (1950). — [8] Some nonlinear problems of the theory of automatic regulation (*, *8*). Izdat. Moscow-Leningrad 1951, 215 pp. — [9] On the problem of the stability of regulating systems (*6, 7, 8*). Prikl. Mat. Meh. SSSR, **15**, 67—74 (1951). — [10] On strictly unstable regulating systems (*6, 7, 8*). Ibid. **15**, 251—254 (1951).

—, and G. M. Fialko: [1] On the stability of regulation in the presence of retardation in the measuring organ of the regulator (*7, 8*). Akad. Nauk SSSR, Inzen. Sbornik **4**, no. 2, 109—112 (1948).

—, and V. N. Postnikov: [1] Conserning the theory of stability of regulating systems (*8*). Prikl. Mat. Meh. SSSR, **8**, 246—248 (1941).

Lusin, U., and P. Kouznetzoff: [1] Sur l'invariabilité absolue et l'invariabilité à ε-près dans la théorie des équations différentielles (*3*). Doklady Akad. Nauk SSSR, (N. S.) **51**, 251—253 (1946); **80**, 325—327, 335—337 (1951).

LUSTERNIK, L.: [1] Sur un problème limite dans la théorie des équations différentielles non linéaires (*9*). Doklady Akad. Nauk SSSR, (N. S.) **33**, 5—8 (1941).

LUTWINISZYN, J.: [1] A certain boundary problem of a vibrating string (*8*). Arch. Mec. Appl., Gdansk. **2**, 75—88 (1950).

LYAGINA, L. S.: [1] The integral curves of the equation $y' = \frac{a x^2 + b x y + c y^2}{d x^2 + e x y + f y^2}$ (*9*). Uspehi Mat. Nauk, (N. S.) **6**, no. 2 (42), 171—183 (1951).

LYAPUNOV, A.: [1] Sur les mouvements hélicoidaux permanents d'un corps solide (*6*). Comm. Soc. Math. Kharkov (2) **1** (1888). — [2] On the stability of motion in a particular case of the problem of three bodies (*6*) Ibid. (2) **2**, 1—94 (1889). — [3] Problème général de la stabilité du mouvement (*, *6*, *7*). Ibid. 1892; **3**, 265—272 (1893). Ann. Fac. Sci. Toulouse (2) **9**, 204—474 (1907). Ann. of Math. Studies **17** (1949). ONTI, 1935; GIT, Moscow-Leningrad 1950. — [4] Sur un série relative à la théorie des équations différentielles linéaires à coefficients périodiques (*4*). C. R. Acad. Sci. Paris **123**, 1248 to 1252 (1896). — [5] Sur l'instabilité de l'équilibre dans certains cas on la fonction de force n'est pas maximum (*7*). J. Liouville **5**, 81—94 (1897). — [6] Sur une équation différentielle linéaire du second ordre (*4*). C. R. Acad. Sci. Paris **128**, 910—913 (1899). — [7] Sur une équation transcendante et les équations différentielles linéaires du second ordre à coefficients périodiques (*4*). Ibid. **128**, 1085—1088 (1899). — [8] Sur un série relative à la théorie d'une équation différentielle linéaire du second ordre (*5*). Ibid. **131**, 1185—1188 (1900). — [9] Sur une série dans la théorie des équations différentielles linéaires du second ordre à coefficients périodiques (*5*). Zad. Akad. Nauk Fiz.-Mat. (8) **13**, 1—70 (1902). — [10] On a problem of the theory of linear second order differential equations with periodic coefficients (*4*). Proc. Kharkov Math. Soc. (2) **5**, 190—192 (1896); 193—254 (1897).

LYASCENKO, N. YA.: [1] On a separation theorem for a system of linear differential equations (*3*). Doklady Akad. Nauk SSSR, (N. S.) **97**, 965—967 (1954). — [2] On asymptotic stability of solutions of a system of differential equations (*3*). Ibid. **96**, 237—239 (1954); **104**, 177—179 (1955). — [3] On a separation theorem for a linear system of differential equations with almost periodic coefficients (*3*). Ukrain. Mat. Ž. **7**, 47—55 (1955).

MACCOLL, L. A.: [1] Fundamental Theory of Servomechanisms (*), pp. 130. Van Nostrand, 1945. — [2] Pseudo closed trajectories in the family of trajectories defined by a system of differential equations (*9*). Quart. Appl. Math. **8**, 255 to 263 (1950).

MACMILLAN, W. D.: [1] An existence theorem for periodic solutions (*8*). Trans. Amer. Math. Soc. **13**, 146—158 (1912). — [2] On the reduction of certain differential equations of the second order (*8*). Ibid. **19**, 205—222 (1918).

MAGNUS, K.: [1] Erzwungene Schwingungen des linearen Schwinger bei nichtharmonischer Erregung (*2*). Z. angew. Math. Mech. **31**, 324—329 (1951).

MAILLET, E.: [1] Sur les équations différentielles et les systemes de reservoirs (*8*). J. École Polyt. (2) **13**, 27—56 (1909). — Bull. Soc. Math. France **33**, 129—145 (1905). — J. de Math. **5**, 225—262 (1909).

MAĬZEL, A. D.: [1] On stability in the first approximation (*7*). Prikl. Mat. Meh. SSSR. **14**, 171—182 (1950).

MAKAROV, I. P.: [1] New criteria for stability according to LYAPUNOV in the case of an infinite triangular matrix (*7*). Doklady Akad. Nauk SSSR, (N. S.) **62**, 289—292 (1948). — [2] Conditions for the approach to zero of the solutions of an inhomogeneous infinite system of differential equations (*7*). Ibid. **68**, 225—228 (1949). — [3] New criteria of stability according to LYAPUNOV in the case of an infinite triangular matrix (*7*). Mat. Sbornik, (N. S.) **30** (72) 53—58 (1952).

MAKAROV, S. M.: [1] Investigations on the characteristic equation of a linear system of two equations of the first order with periodic coefficients (*4*). Trudy Kubish. Aviat. Inst. **1**, 24—29 (1952). — Prikl. Mat. Meh. SSSR, **15**, 373—378 (1951).

MALGARINI, G.: [1] Studio asintotico del moto d'un oscillatore elastico, con resistenza di tipo subviscoso (*8, 9*). Rend. Ist. Lombardo Sci. Lett. (3) **17**, (86) 258—280 (1953).

MALKIN, I. G.: [1] Certain questions on the theory of the stability of motion in the sense of LYAPUNOV (*7*). Sbornik Nancnyh Trudov Kazanskogo Aviacionnogo Inst. P. A. Baranova, no. 7, 1937. Amer. Math. Soc. Translation No. 20. — [2] On the stability of motion in the sense of LYAPUNOV (*7*). Mat. Sbornik, (N. S.) **3** (45), 47—100 (1938). Amer. Math. Soc. Translation no. 41. — [3] Some basic theorem of the theory of stability of motion in critical cases (*7*). Prikl. Mat. Meh. SSSR, **6**, 411—448 (1942). Amer. Math. Soc. Translation no. 38. — [4] Stability in the case of constantly acting disturbances (*7, 8*). Prikl. Mat. Meh. SSSR, **8**, 241—245 (1944). Amer. Math. Soc. Translation no. 8. — [5] Stability of periodic motions of dynamic systems (*7, 8*). Ibid. **8**, 327—331 (1944). — [6] Oscillations of systems with one degree of freedom close to systems of Lyapunov (*7, 8*). Ibid. **12**, 561—596 (1948). Amer. Math. Soc. Translation no. 22. — [7] Oscillations of systems with several degrees of freedom. Prikl. Mat. Meh. SSSR. **12**, 673—690 (1948). Amer. Math. Soc. Translation no. 21. — [8] On POINCARÉ's theory of periodic solutions (*8*). Prikl. Mat. Meh. SSSR, **13**, 633—646 (1949). — [9] The methods of LYAPUNOV and POINCARÉ in the theory of nonlinear oscillations, OGIZ. Moscow-Leningrad 1949. 244 pp. — [10] Oscillations of quasilinear systems with a nonanalytic characteristic of nonlinearity (*8*). Prikl. Mat. Meh. SSSR, **14**, 13—22 (1950). — [11] On the theory of oscillations of quasilinear systems with many degrees of freedom (*8*). Ibid. **14**, 353—370 (1950). — [12] On the theory of stability of regulating systems (*8*). Ibid. **15**, 59—66 (1951). — [13] On the solution of a stability problem in the case of two purely imaginary roots (*8*). Ibid. **15**, 255—257 (1951). — [14] On a method of solution of the problem of stability in the critical case of a pair of purely imaginary roots (*8*). Ibid. **15**, 473—484 (1951). — [15] Solution of some critical cases of the problem of stability of motion (*8*). Ibid. **15**, 575—590 (1951). — [16] A theorem on stability in the first approximation (*8*). Doklady Akad. Nauk SSSR, (N. S.) **76**, 783—784 (1951). — [17] On the characteristic values of linear differential equations (*6*). Prikl. Mat. Meh. SSSR, **16**, 3—14 (1952). — [18] On the construction of Lyapunov functions for systems of linear equations (*7*). Ibid. **16**, 239—242 (1952). — [19] On a problem of the theory of stability of systems of automatic regulation (*9*). Ibid. **16**, 365—368 (1952). — [20] On the stability of systems of automatic regulation (*9*). Ibid. **16**, 495—499 (1952). — [21] On a theorem concerning stability of motion (*7*). Doklady Akad. Nauk SSSR, (N. S.) **84**, 877 to 878 (1952). — [22] Theory of stability of motion (*, *7*). Izdat. Gos. Moscow-Leningrad 1952, 432 pp. — [23] On the reversibility of LYAPUNOV's theorem on asymptotic stability (*7*). Prikl. Mat. Meh. SSSR, **18**, 129—138 (1954). — [24] On resonance in quasiharmonic systems (*8*). Ibid. **18**, 459—463 (1954). — [25] On almost periodic oscillations of nonlinear nonautonomous systems (*8*). Prikl. Mat. Meh. SSSR, **18**, 681—704 (1954).

MALMQUIST, J.: [1] Sur les functions à un nombre fini de branches satisfaisant à une equation différentielle du premier ordre (*9*). Acta Math. **36**, 297—343 (1913); **74**, 175—196 (1941).

MAMBRIANI, A.: [1] Su un teorema relativo alle equazioni differenziali del secondo ordine (*5*). Rend. Acad. Lincei (6) **9**, 620—622 (1929). — [2] Osservazioni su il classico teorema di confronto di Sturm (*5*). Riv. Mat. Univ. Parma **2**, 111—114 (1951).

MAMMANA, G.: [1] Sopra un nuovo metodo di studio delle equazioni differenziali lineari (*5*). Math. Z. **25**, 734—748 (1926). — [2] Sistemi differenziali, autovalori e autofunzioni (*, *5*). Napoli 1938.

MANACORDA, T.: [1] Soluzioni periodiche di una equazione differenziale non lineare (*8*). Rend. Accad. Lincei (8) **1**, 1046—1050 (1946). — [2] Sul comportamento asintotico degli integrali dell'equazione; $y''(x) + p(x)y'(x) + q(x)y(x) = 0$ quando $\lim q(x) = +\infty$ (*5*). Ibid. (8) **2**, 537—541, 752—757 (1947). — [3] Sopra un'equazione differenziale non lineare della dinamica del punto (*8*). Rend. Ist. Lombardo Sci. Lett. (3) **11** (80) (1947); **1949**, 85—98. — [4] Estensione alle equazioni differenziali lineari del secondo ordine omogenee complete di una formula di HARTMANN e WINTNER per la valuatazione asintotica del numero degli zeri di un integrale (*5*). Boll. Unione Mat. Ital. (*3*) **3**, 205—210 (1948). — [5] Sul comportamento asintotico di una classe di equazioni differenziali lineari non omogenee (*3*). Ibid. (3) **6**, 304—311 (1951); (3) **7**, 281—284 (1952). — [6] Sul comportamento asintotico degli integrali di una classe di equazioni differenziali non lineari (*8*). Ibid. (3) **7**, 137—142 (1952). — [7] Studio di un circuito non lineare col metodo stroboscopico di N. MINORSKY (*8*). Ibid. (3) **8**, 281—285 (1953). — [8] Vibrazioni forzate in un particolare sistema oscillante non lineare. Rend. Accad. Lincei (8) **4**, 557—561 (1948).

MANARESI, G.: [1] Sull'equazione di Liénard generalizzata (*8*, *9*). Bull. Unione Mat. Ital. (3) **8**, 59—64 (1953). — [2] Su alcuni teoremi di media nella meccanica non lineare (*8*). Atti Sem. Mat. Fis. Univ. Modena **6** (1951/52); **1953**, 78—86. — [3] Sulle soluzioni sottoarmoniche semplici nei sistemi non lineari a due gradi di libertà (*8*). Boll. Unione Mat. Ital. (3) **9**, 412—417 (1954).

MANDELSTAM, L. I.: [1] Systems with periodical coefficients with many degrees of freedom and small nonlinearity (*8*). Akad. Nauk SSSR, Eksper. Teoret. Fiz. **15**, 605—612 (1945).

MANDELSTAM, L. I., and N. PAPALEXI: [1] Über Resonanzschwingungen bei Frequenzteilung (*2*, *8*). Z. Physik **73**, 223 (1932). — [2] On the phenomena of resonance of the *n*th order (*2*). J. Techn. Phys. **2**, no. 7—8 (1932).

MANGERON, D. I.: [1] Mécanique non-linéaire sur les systèmes oscillatoires non linéaires (*8*). Bull. École Polytech. Jassy **1946**, 62—66.

MARACKOV, M.: [1] Über einen Liapounoffschen Satz (*7*). Bull. Soc. Phys.-Math. Kazan (3) **12**, 171—174 (1940).

MARCHENTE, E.: [1] Teoremi di confronto per problemi al contorno relativi a sistemi di due equazioni differenziali del primo ordine (*5*). Rend. Sem. Mat. Padova **12**, 81—88 (1941).

MARCUS, M.: [1] Some results on the asymptotic behavior of linear systems (*3*). Canad. J. Math. **7**, 531—538 (1955).

—, M. D.: [1] An invariant surface theorem for a nondegenerate system. Contrib. Theory nonlinear Oscillations **3**, 243—256 (1956). — [2] Repeating solutions for a degenerate system. Ibid. **3**, 261—268 (1956).

MARDEN, M.: [1] The geometry of the zeros of a polynomial (*2*). Math. Surveys **3**, Amer. Math. Soc. (1949).

MARKOV, A.: [1] Stabilität im Lyapunovschen Sinne und Fastperiodizität (*6*). Math. Z. **36**, 708—738 (1933).

MARKOVIC, Z.: [1] Sur les solutions de l'équation différentielle linéaire du second ordre à coefficient périodique (*5*). Proc. Lond. Math. Soc. (2) **31**, 417—438 (1930).

MARKUS, L.: [1] Escape times for ordinary differential equations (*8*). Rend. Sem. Mat. Politecnico Torino **11**, 271—277 (1952). — [2] On completeness of invariant measure defined by differential equations (*9*). J. Math. pure appl. (9) **31**, 341—353 (1952). — [3] Invariant measures defined by differential equations (*9*). Proc. Amer. Math. Soc. **4**, 89—91 (1953). — [4] A topological theory for ordinary differential equations in the plane (*9*), Colloque de topologie et géometrie différentielle, pp. 8. Strasbourg 1952. no. 9. — [5] Global structure of ordinary differential equations in the plane (*9*). Trans. Amer. Math. Soc. **76**, 127—148 (1954). — [6] Continuous matrices and the stability of differential systems (*3*). Math. Z. **62**, 310—319 (1955). — [7] Oscillation and disconjugacy for linear differential equations with almost periodic coefficients (with R. A. MOORE) (*5*). Acta Math. **95**, 99—123 (1956). — [8] Asymptotically autonomous differential systems (*9*). Contrib. Theory nonlinear Oscillations **3**, 17—29 (1956).

MARTIENSSEN, O.: [1] Über neue Resonanzerscheinungen in Wechselstromkreisen (*8*). Phys. Z. **11** (1910).

MARTIN, M. H.: [1] Real asymptotic solutions of real differential equations (*8*). Bull. Amer. Math. Soc. **45**, 475—481 (1940).

— W. T.: [1] Linear difference equations with arbitrary real span (*2*). Acta Math. **69**, 57—98 (1938).

MASSERA, J. L.: [1] The number of subharmonic solutions of nonlinear differential equations of the second order (*6*). Ann. of Math. (2) **50**, 118—126 (1949). — [2] On LYAPUNOV's conditions of stability (*6, 7*). Ibid. (2) **50**, 705—721 (1949). — [3] The existence of periodic solutions of systems of differential equations (*6*). Duke Math. J. **17**, 457—475 (1950). — [4] Remarks on the periodic solutions of differential equations (*9*). Bol. Fac. Ingen. Montevideo **2**, 43—53 (1950). — [5] Total stability and approximately periodic vibrations (*8*). Ibid. **2**, 135—145 (1954). — [6] Sur un théorème de G. SANSONE sur l'équation de LIÉNARD (*8*). Boll. Union Mat. Ital. (3) **9**, 367—369 (1954). — [7] Contribution to stability theory (*9*). Annals of Math. **64**, 182—206 (1956).

MATHIEU, E.: [1] Memoire sur le mouvement vibratoire d' une membrane de forme elliptique (*4*). J. Math. Pures Appl. **13**, 137 (1868).

McCARTHY, J.: [1] A method for the calculation of limit cycles by successive approximations (*9*). Contrib. Theory nonlinear Oscillation, Ann. of Math. Ser. no. 29, 75—79 (1952).

McHARG, E. A.: [1] A differential equation (*8*). J. London Math. Soc. **32**, 83—85 (1947).

McKELVEY, R. W.: [1] The solutions of second order linear ordinary differential equations about a turning point of order two (*10*). Trans. Amer. Math. Soc. **79**, 103—123 (1955).

McLACHLAN, N. W.: [1] HILL's differential equation (*4*). Math. Gaz. **29**, 68 (1945). — [2] Mathieu functions and their classification (*4*). J. Math. Phys. Mass. Inst. Techn. **25**, 209 (1946). — [3] Theory and application of Mathieu functions (*). Oxford, Clarendon Press, 1947. 401 pp. — [4] Periodic solution of a certain nonlinear differential equation (*8*). Math. Gaz. **32**, 64—66 (1948). — [5] Ordinary Nonlinear Differential Equations in Engineering and Physical Sciences (*), pp. 201. Oxford, Clarendon Press, 1950. — [6] Theory of vibrations (*). Dover, 1951. — [7] Nonlinear differential equations having a periodic coefficient (*8*). Math. Gaz. **35**, 32—36 (1951). — [8] Application of MATHIEU's equation to stability of nonlinear oscillator (*4, 8*). Ibid. **35**, 105—107 (1951). — [9] On a nonlinear differential equation in hydraulics (*, *8*). Proc. Sympos. Appl. Math. **5**, 49—61 (1954). — [10] Two theorems on ordinary nonlinear differential equations (*8, 9*). Math. Gaz. **39**, 200—202 (1955).

MEEROV, M. V.: [1] On systems of autoregulation stabilized for an arbitrarily large coefficient of amplification (*2*). Avtom. i Telem. **8**, 225—242 (1947).

MEISSNER, E. [1] Über Schüttelschwingungen in Systemen mit periodisch veränderlicher Elastizität (*4*). Schweiz. Bauzeitg. **72**, 95—98 (1918). — [2] Resonanz bei konstanter Dampfung (*2*). ZAMM **1935**, 63—70.

MEKSYN, D.: [1] Solutions of OSEEN's equation for an elliptical cylinder in a viscous fluid (*10*). Proc. Roy. London Math. Soc. (A) **163**, p. 232 (1937). — [2] Fluid motion between parallel planes. Dynamical stability (*10*). Ibid. **186**, 391—409 (1946). — [3] Stability of viscous flow between rotating cylinders (*10*). Ibid. **187**, 480—504 (1946). — [4] Asymptotic integrals of a fourth order differential equation containing a large parameter (*10*). Proc. London Math. Soc. (2) **49**, 436—457 (1947).

METTLER, E.: [1] Allgemeine Theorie der Stabilität erzwungener Schwingungen elastischer Körper (*4*). Ing.-Arch. **17**, 418—449 (1949).

MIHLIN, S. G.: [1] Singular integral equations (*10*). Uspehi Mat. Nauk, (N. S.) **3**, no. 3 (25), 29—112 (1948). Amer. Math. Soc. Translations no. 24.

MIKOLAJSKA, Z.: [1] Sur les mouvements asymptotiques d'un point matérial mobile dans le champ des forces repoussantes (*9*). Bull. Acad. Polon. Sci. **1**, 11—13 (1953). — [2] Sur une propriété asymptotique des intégrales d'une équation différentielle du second ordre (*9*). Ibid. **2**, 113—116 (1954). — [3] Sur l'équation généralisée des oscillations entretenues (*9*). Ibid. **2**, 309—313 (1954). — [4] Sur l'allure asymptotique des intégrales des systèmes d'équations différentielles au voisinage d'un point asymptotiquement singulier (*9*). Ann. Pol. Math. **1**, 277—305 (1955).

MIKUSINSKI, J. G.: [1] Sur les intégrales de quelques équations différentielles linéaires (*5*). Ann. Univ. Mariae Curie-Sklodowska, Sect. A **1**, 23—34 (1946). — [2] Sur l'équation différentielle $y^{(6)} + y = 0$ (*5*). Ibid. **1**, 35—40 (1946). — [3] On FITE's oscillation theorems (*5*). Colloqu. Math. **2**, 34—39 (1949).

MILLER, J. C. P.: [1] On a criterion for oscillatory solutions of a linear differential equation of the second order (*5*). Proc. Cambridge Phil. Soc. **36**, 283—287 (1940).

— K. S.: [1] A remark on stability (*3*). J. Appl. Phys. **22**, 1054—1057 (1951); **25**, 407—408 (1954).

MILLOUX, H.: [1] Sur l'équation différentielle $\ddot{x} + A(t)x = 0$ (*5*). Prace Mat.-Fiz. **41**, 39—54 (1934).

MILNE, W. E.: [1] A theorem of oscillation (*5*). Bull. Amer. Math. Soc. **28**, 102—104 (1922).

MINKEVIC, M. I.: [1] Theory of integral funnels in dynamical systems without uniqueness (*9*). Ucenye Zapiski Moskov. Gos. Univ. **2**, 134—151 (1948). — Doklady Akad. Nauk SSSR, (N. S.) **59**, 1049—1052 (1948); **60**, 341—343 (1948).

MINORSKY, N.: [1] Self-excited oscillations in dynamical systems possessing retarded actions (*8*). J. Appl. Phys. **9**, 65—71 (1942). — [2] Mechanical self-excited oscillations (*8*). Proc. Nat. Acad. Sci. U.S.A. **30**, 308 (1944). — [3] Introduction to nonlinear Mechanics. (I. Topological Methods of nonlinear Mechanics; II. Analytical Methods of nonlinear Mech.; III. Nonlinear Resonance) (*, *8*, *9*). D. Taylor Model Basin, Rep. 534, 546, 558. — [4] On parametric excitation (*4*). J. Franklin Inst. **240**, 25—46 (1945). — [5] On nonlinear phenomenon of self-rolling (*8*). Proc. Nat. Acad. Sci. U.S.A. **31**, 346—349 (1945). — [6] Introduction to nonlinear Mechanics. Topological Methods. Analytical Methods. Nonlinear Resonance. Relaxation Oscillations (*, *8*, *9*). Edwards 1947. 464 pp. — [7] Modern trends in nonlinear mechanics (*, *8*, *9*). Advances in Applied Mechanics, edited by R. V. v. MISES and

T. v. Karman, pp. 41—103. New York, Academic Press, 1948. — [8] Self-excited mechanical oscillations (*8*). J. Appl. Phys. **19**, 332—338 (1948). — [9] Sur une classe d'oscillations autoentretenues (*8*). C. R. Acad. Sci. Paris **226**, 1122—1124 (1948). — [10] Self-excited oscillations in systems possessing retarded actions (*8*). Proc. Seventh Int. Congr. Appl. Math. **4**, 43—51 (1948). — [11] On certain applications of difference-differential equations (*8*). Dept. of Eng. and Math., Stanford Univ., 1948, 38pp. — [12] Energy fluctuations in a van der Pol oscillator (*8*). J. Franklin Inst. **248**, 205—223 (1949). — [13] Meccanica non lineare (*8*). Boll. Unione Mat. Ital. (3) **5**, 313—330 (1950). — [14] Sur l'excitation paramétrique (*8*). C. R. Acad. Sci. Paris **231**, 1417—1419 (1950). — [15] Sur une équation différentielle de la physique (*8*). Ibid. **232**, 1060—1062 (1951). — [16] Sur l'oscillateur non linéaire de Mathieu (*8*). Ibid. **232**, 2179—2180 (1951). — [17] Sur le pendule entretenu par un courant alternatif (*8*). Ibid. **233**, 728—729 (1951). — [18] Parametric excitation (*4*, *8*). J. Appl. Phys. **22**, 49—54 (1951). — [19] Stationary solutions of certain nonlinear differential equations (*8*). J. Franklin Inst. **254**, 21—42 (1952). — [20] Sur l'interaction des oscillations non linéaires (*8*). C. R. Acad. Sci. Paris **234**, 292—294 (1952). — [21] Sur les systems à l'action retardée (*8*). Ibid. **234**, 1945—1947 (1952). — [22] Sur l'excitation asynchrone (*8*). Ibid. **237**, 643—645 (1953). — [23] Sur l'excitation asynchrone (*8*). Ibid. **237**, 964 to 966 (1953). — [24] Sur le phénomène Béthenod (*8*). Colloque Intern. Vibrations non linéaires, Ile de Porquerolles, 1951, 223—234. — [25] On interaction of nonlinear oscillations (*8*). J. Franklin Inst. **256**, 147—165 (1953). — [26] Sur les systèms non linéaires à deux degrés de liberté (*8*). C. R. Acad. Sci. Paris **238**, 646—647 (1954). — [27] La methode stroboscopique et ses applications (*8*). Bull. Soc. Franc. Méc. **4**, no. 13, 15—26 (1954). — [28] On the stroboscopic method (*8*). Studies in math. and mech. presented to R. von Mises, 192—199. New York, Academic Press, 1954. — [29] On asynchronous action (*8*). J. Franklin Inst. **259**, 209—219 (1955). — [30] Sur la méthode stroboscopique (*8*). Mem. Accad. Sci. Bologna (10) **9**, 23—29 (1952). — [31] Oscillatory systems containing inertial parameters (*8*). Proc. Symposium on nonlinear Circuit Analysis, 154—160. New York, 1953. — [32] Sur les systèmes non-linéaires à deux degrés de liberté (*8*). Rend. Sem. Mat. Politec. Torino **13**, 59—70 (1954). — [33] Sur l'éspace paramétrique de l'équation de M. Liénard (*9*). C. R. Acad. Sci. Paris **240**, 1508—1509 (1955). — [34] Sur la résonance non linéaire (*8*). Ibid. **240**, 2482—2484 (1955).

Minozzi, L.: [1] Sulle soluzioni sottoarmonische dell'equazione di Liénard (*8*). Boll. Unione Mat. Ital. (3) **9**, 196—198 (1954).

Mirolyubov, A. A.: [1] Solution of differential difference equations with linear coefficients (*3*). Doklady Akad. Nauk. SSSR, (N. S.) **85**, 1209—1210 (1952).

Miscenko, E. F., and L. S. Pontryagin: [1] Periodic solutions of systems of differential equations near to discontinuous ones (*9*). Dokl. Akad. Nauk SSSR, (N. S.) **102**, 889—891 (1955).

Mises, R. v.: [1] Die Grenzschichte in der Theorie der gewöhnlichen Differentialgleichungen (*8*). Acta Sci. Math. Szeged **12**, 29—34 (1950). — [2] Dynamical problems in the theory of machines, (B) Governors (*8*). Enzykl. der Mathem. Wiss., vol. 4/2, 254. 1911.

Mitropolskii, Yu, A.: [1] Investigation of oscillations in nonlinear systems with many degrees of freedom and slowly varying parameters (*8*). Ukrain. Mat. Ž. **1**, no. 2, 85—98 (1949). — [2] Slow processes in nonlinear oscillating systems with many degrees of freedom (*8*). Prikl. Math. Meh. SSSR, **14**, 139—170 (1950). — [3] On oscillations in gyroscopic systems while passing through resonance (*8*). Ukrain. Mat. Ž. **5**, 333—349 (1953). — [4] Forced oscillations in nonlinear systems while passing through resonance (*8*). Inzen. Sb. **15**, 89—98 (1953). — [5] On unsteady oscillations in systems with many degrees of freedom (*8*). Ukrain, Mat. Ž. **6**, 176—189 (1954). — [6] On the effect on a nonlinear oscillator of a "sinusoidal" force with modulated frequency (*8*). ibid. **6**, 442—447 (1954). — [7] On passage through a resonance of second order (*8*). Ibid. **7**, 121—123 (1955). — [8] Transient processes in nonlinear oscillatory systems (*). Izdat. Akad. Nauk. Kiev, 1955.

MIZOHATA, S.: [1] On the existence of systems of periodic solutions for several non linear circuits (*8*). Mem. Coll. Sci. Univ. Kyoto, Ser. A Math. **27**, 115—121 (1952). — [2] Sur les phénomènes de sauts dans certains systèmes non linéaires (*8*). Ibid. 203—221 (1953). — [3] Sur certaines équations différentielles régissant quelques phénomènes héréditaires (*8*). Math. Japonicae **3**, 1—5 (1953).

—, and M. YAMAGUTI: [1] On the existence of periodic solutions of the nonlinear differential equation $\ddot{x} + a(x)\dot{x} + \Phi(x) = p(t)$ (*8*). Mem. Coll. Sci. Univ. Kyoto, Ser. A Math. **27**, 109—113 (1952).

MODONA, L. N.: [1] Su di una equazione differenziale non lineare del secondo ordine (*8*). Boll. Unione Mat. Ital. (3) **8**, 428—441 (1953).

MOORE, R. A.: [1] The behavior of solutions of a linear differential equation of second order (*5*). Pacific. J. Math. **5**, 125—145 (1955).

MORAWETZ, C. S.: [1] Asymptotic solutions of the stability equations of a compressible fluid (*10*). J. Math. Phys. **33**, 1—26 (1954).

MORDUCHOW, M., and L. GALOWIN: [1] On double-pulse stability criteria with damping (*8*). Quart. Appl. Math. **10**, 17—23 (1952).

MORRIS, G. R.: [1] A differential equation for undamped forced nonlinear oscillations. I. (*8*). Proc. Cambridge Philos. Soc. **51**, 297—312 (1955).

MORSE, M.: [1] The calculus of variations in the large (*). Amer. Math. Soc. Coll. Publ. **18** (1934).

MOSER, J.: [1] Störungstheorie des kontinuierlichen Spektrums für gewöhnliche Differentialgleichungen zweiter Ordnung (*5*). Math. Ann. **125**, 366—393 (1953). — [2] Periodische Lösungen des restringierten Dreikörperproblems, die sich erst nach vielen Umläufen schließen (*8*, *9*). Ibid. **126**, 325—335 (1953). — [3] Über periodische Lösungen kanonischer Differentialgleichungssysteme (*8*). Nachr. Akad. Wiss. Göttingen **1953**, 23—48. — [4] Stabilitätsverhalten kanonischer Differentialgleichungssysteme (*8*, *9*). Ibid. **1955**, 88—120. — [5] Nonexistence of integrals for canonical systems of differential equations (*8*). Comm. Pure on Appl. Math. **8**, 409—436 (1955). — [6] The analytic invariants of an area preserving mapping near a hyperbolic fixed point (*9*). Ibid. **9**, 673—692 (1956). — [7] New aspects in the theory of stability of Hamiltonian systems. Comm. Pure on Appl. Math. **9**, 81—114 (1958).

MÜLLER, K. E.: [1] Über die Schüttelschwingungen des Kuppelstangenantriebes (*4*), p. 174. Schweiz. Baus., **1919**.

— M.: [1] Über die Existenz periodischer Lösungen bei gewissen Systemen gewöhnlicher Differentialgleichungen erster Ordnung (*8*). Math. Z. **48**, 128—135 (1912).

MÜLLER-STROBEL, J.: [1] Störungstheorie und Stabilität. Anwendung der Störungstheorie zur näherungsweisen Bestimmung der statischen Stabilitätsgrenzen von Synchronmaschinen in vermaschten Netzen (*2*). Arch. Elektrotechn. **37**, 555—587 (1943).

MUNUDATA, K.: [1] Some exact solutions in nonlinear oscillations (*8*). J. Phys. Soc. Japan **7**, 383—391 (1952).

MYSKIS, A. D.: [1] The general theory of differential equations with a retarded argument (*2*). Uspehi Mat. Nauk, (N. S.) **4**, no. 5, 99—141 (1949). Amer. Math. Soc. Translation no. 55, 1951. — [2] Linear homogeneous differential equations of the first order with retarded arguments (*2*). Ibid. **5**, no. 2 (36), 150—162 (1950). — [3] Investigation of a class of differential equations with retarded arguments by means of a generalized Fibonacci series (*2*). Doklady Akad. Nauk SSSR, (N. S.) **71**, 13—169 (1950). — [4] On the solutions of linear homogeneous differential equations of the first order of unstable type with a retarded argument (*2*). Ibid. **70**, 953—956 (1950). — [5] Supplementary bibliographical material to the paper "General theory of differential equations with retarded arguments" (*2*). Uspehi Mat. Nauk, (N.S.) **5**, no. 2 (36), 148—154 (1950). — [6] General theory of differential equations with retarded arguments (*2*). Ibid. **4**, no. 5 (33), 99—141 (1949). Amer. Math. Soc. Translation no. 55. — [7] On solutions of linear homogeneous differential equations of the second order of periodic type with retarded argument (*4*). Mat. Sbornik, (N.S.) **28**, 15—54 (1951). — [8] On solutions of linear homogeneous differential equations of the first order of stable type with a retarded argument (*4*). Ibid. **28** (70), 641—658 (1951). — [9] Linear differential equations with retarded argument (*, *2*, *4*), 255pp. Izdat. Moscow-Leningrad 1951.

NAGUMO, M.: [1] Über die Lage der Integralkurven gewöhnlicher Differentialgleichungen (*9*, *10*). Proc. Phys.-Math. Soc. Japan (3) **24**, 551—559 (1942). — [2] Über das Verhalten des Integrals von $\lambda y'' + f(x, y, y', \lambda) = 0$ für $\lambda \to 0$ (*10*). Ibid. **21**, 529—534 (1939).

— Z.: [1] On a forced discontinuous oscillation (*8*). Proc. Fac. Eng. Keio Univ. **7**, 36—43 (1954).

NAIMARK, H. A.: [1] Linear differential operators (*), pp. 351. Izdat. Gos. Moscow 1954.

NARDINI, R.: [1] Sulle vibrazioni quasi armoniche di un sistema dissipativo con elasticità periodica (*4*). Boll. Unione Mat. Ital. (3) **4**, 370—373 (1949). — [2] Sul comportamento asintotico degli integrali di un'equazione differenziale della dinamica (*5*). Rend. Accad. Lincei (8) **7**, 47—61 (1949). — [3] Sulla stabilità delle vibrazioni quasi armoniche di un sistema dissipativo (*4*). Ibid. (8) **6**, 603—608 (1949). — [4] Su un sistema dissipativo ad n gradi di libertà (*3*). Rend. Accad. Lincei (8) **7**, 224—227 (1949). — [5] Sul comportamento asintotico della soluzione di un problema della magneto-idrodinamica (*8*). Ibid. (8) **16**, 225—231, 341—348.

NEHARI, Z.: [1] On the zeros of solutions of second-order linear differential equations (*5*). Amer. J. Math. **76**, 689—697 (1954).

NEMYCKII, V. V.: [1] The method of fixed-points in analysis (*9*). Uspehi Mat. Nauk, (N. S.) **1**, 141—174 (1936). — [2] Intégration qualitative du système $dx/dt = Q(x, y)$; $dy/dt = P(x, y)$ en première approximation (*9*). Doklady Akad. Nauk SSSR, (N. S.) **38**, 190—192 (1943). — [3] Intégration qualitative du système d'équations différentielles $dx/dt = Q(x, y)$; $dy/dt = P(x, y)$ (*9*). Mat. Sbornik, (N.S.) **16** (58), 307—344 (1945). — [4] Intégration qualitative du système $dx/dt = Q(x, y)$, $dy/dt = P(x, y)$ au moyen de réseaux universels de lignes polygonales (*9*). Ucenye Zapiski Moskov. Mat. **2**, 34—52 (1946). — [5] Topological problems of the theory of dynamical systems (*, *9*). Uspehi Mat. Nauk, (N.S.) **4**, no. 6 (34), 91—153 (1949). — [6] Problems of the qualitative theory of differential equations (*9*). Vestnik Moskov. Univ. **8**, 19—39 (1952). — [7] LYAPUNOV's method of rotating functions for finding oscillatory regimes (*9*). Doklady Akad. Nauk SSSR, (N. S.) **97**, 33—36 (1954). — [8] Some problems of the qualitative theory of differential equations (*9*). Uspehi Mat. Nauk, (N. S.) **9**, no. 3 (61), 39—56 (1954). — [9] Estimate of the regions of asymptotic stability of nonlinear systems (*7*, *9*). Dokl. Akad. Nauk SSSR, (N. S.) **101**, 803—804 (1955).

NEMYCKII, V. V., and V. V. STEPANOV: [1] Qualitative theory of differential equations (*, *9*). GITL, Moscow, 2nd edit., 1949.

NEWMAN, M. H. A.: [1] On the ultimate boundedness of the solutions of certain differential equations (*9*). Composito Math. **8**, 142—156 (1950).

NIJENHUIS, W.: [1] A note on a generalized van der Pol equation (*8*). Phillips Res. Rep. **4**, 401—406 (1949).

NIKITIN, V. P., V. K. TURKIN and N. P. KUNICKII: [1] On the stability of operation of an amplidyne electric drive (*2*). Izvestiya Akad. Nauk SSSR, **1946**, 1567 to 1580. — [2] Stability diagrams for systems of the fifth order (*8*). Doklady Akad. Nauk SSSR, (N. S.) **58**, 591—594 (1947). — [3] On diagrams exhibiting to what extent damping of a transient process differs from the damping according to a simple exponential law (*8*). Ibid. **59**, 1097—1099 (1948).

NIKOLENKO, L. D.: [1] On oscillation of solutions of the differential equation $y'' + p(x) y = 0$ (*5*). Ukrain. Mat. Ž. **7**, 124—127 (1955).

NIKOL'SKIĬ, G. N.: [1] On a problem of indirect regulation (*8*). Akad. Nauk SSSR, Inzen. Sbornik **4**, no. 2, 113—132 (1948).

NISHIMO, K.: [1] Some notes on the subharmonic resonance in the nonlinear mechanical vibratory system (*8*). J. Jap. Soc. Appl. Mech. **3**, 121—126 (1950).

NOHEL, J. A.: [1] Stability of perturbed periodic motions (*8*). J. Reine Angew. Math. **203**, 64—79 (1960).

NOUGMANOVA, CH.: [1] Sur la stabilité des mouvements périodiques (*4*, *7*). Doklady Akad. Nauk SSSR, (N. S.) **42**, 202—204 (1944).

OBI, C.: [1] Subharmonic solutions of nonlinear differential equations of the second order (*8*). J. London Math. Soc. **25**, 217—226 (1950). — [2] Periodic solutions of nonlinear differential equations of the second order (*8*). Proc. Cambridge Phil. Soc. **47**, 741—751 (1951). — [3] Periodic solutions of nonlinear differential equations of the second order (*8*). Ibid. **47**, 752—755 (1951). — [4] A nonlinear differential equation of the second order with periodic solutions whose associated limit cycles are algebraic curves (*8*). J. London Math. Soc. **28**, 356—360 (1953). — [5] Researches on the equation $\ddot{x} + (\varepsilon_1 + \varepsilon_2 x)\dot{x} + x + \varepsilon_3 x^2 = 0$ (*8*). Proc. Cambridge Phil. Soc. **50**, 26—32 (1954). — [6] Uniformly almost periodic solutions of non-linear differential equations of the second order (*8*). Proc. Cambridge Phil. Soc. **51**, 604—613 (1955). — [7] Periodic solutions of nonlinear differential equations of order $2n$ (*8*). J. London Math. Soc. **28**, 163—171 (1953).

OBMORŠEV, A. N.: [1] Investigation of phase trajectories at infinity (*9*). Prikl. Mat. Meh. SSSR, **14**, 383—390 (1950).

OLDENBURG, R.: [1] Frequency response (*2*). MacMillan 1955.

— R. C., and H. SARTORIUS: [1] Dynamik selbsttätiger Regelungen. 1. Band. Allgemeine und mathematische Grundlagen, stetige und unstetige Regelungen, Nichtlinearitäten, pp. 258, 2. Aufl. (*). München, Oldenbourg, 1951.

OLVER, F. W. J.: [1] The asymptotic solution of linear differential equations of the second order in a domain containing one transition point (*10*). Phil. Trans. Roy. Soc. London (A) **247**, 307—327, 328—368 (1954); **249**, 65—97 (1956); **250**, 479—517 (1958).

OPPELT, W.: [1] Theorie der Regelung und Steuerung (*). Naturforschung und Medizin in Deutschland, 1939—1946, Bd. 4, Teil 2, 127—135. Weinheim, Verlag Chemie, 1953.

OSGOOD, W. F.: [1] Beweis der Existenz einer Lösung einer Differentialgleichung (*1*). Monatshefte Math.-Phys. **9**, 331—345 (1898). — [2] On a theorem of oscillation (*5*). Bull. Amer. Math. Soc. **25**, 216—221 (1919). — [3] Mechanics (*). MacMillan 1937.

OTROKOV, N.: [1] Sur le nombre des cycles limites au voisinage d'un foyer (*9*). Doklady Akad. Nauk SSSR, (N. S.) **43**, 98—101 (1944). — [2] On the number of limit cycles of a differential equation in the neighborhood of a singular point (*9*). Mat. Sbornik, (N. S.) **34** (76), 127—144 (1954).

PAINLEVÉ, P.: [1] Sur les positions d'équilibre instable (*1, 8*). C. R. Acad. Sci. Paris **125**, 1021—1024 (1897).

PAPUS, P. N.: [1] On finding regular semi-stable limit cycles (*9*). Uspehi Mat. Nauk, (N. S.) **7**, no. 4 (50), 165—168 (1952).

PEANO, G.: [1] Sur le théorème général relatif a l'existence des intégrales des l'équations différentielles ordinaires (*1, 3*). Nouv. Ann. de Math. (3) **11**, 79—82 (1892).

PEIXOTO, M. M.: [1] On structural stability (*9*). Ann. of Math. **69**, 199—222 (1959).
—, and M. C. PEIXOTO: [1] Structural stability on the plane with enlarged boundary conditions (*9*). An. Acad. Brasil. Ci. **31**, 135—160 (1959).

PEKERIS, J.: [1] Asymptotic solutions for the normal modes in the theory of microwave propagation (*8*). J. Appl. Phys. **17**, 1108—1124 (1946).

PERLIS, S.: [1] Theory of matrices (*). Addison-Wesley 1952.

PERRON, O.: [1] Über die Poincarésche lineare Differentialgleichung (*3*). J. reine u. angew. Mat. **137**, 6—64 (1910). — [2] Über lineare Differentialgleichungen, bei denen die unabhängige Variable reell ist (*3*). Ibid. **142**, 254—270 (1913); **143**, 29—50 (1913). — [3] Über die Abhängigkeit der Integrale eines Systems linearer Differentialgleichungen von einem Parameter (*3*). Sitzgsber. Heidelberg. Akad. Wiss. **13**, 15 (1918); **1919**, 3. — [4] Über das Verhalten der Integrale einer linearen Differentialgleichung bei großen Werten der unabhängig Variablen (*3*). Math. Z. **1**, 27—43 (1918). — [5] Über nicht homogene lineare Differentialgleichungen (*3*). Ibid. **6**, 161—166 (1920). — [6] Über die Gestalt der Integralkurven einer Differentialgleichung erster Ordnung in der Umgebung eines singularen Punktes (*9*). Ibid. **15**, 121—146 (1922); **16**, 273—295 (1923). — [7] Über einen Grenzwertsatz (*3*). Ibid. **17**, 149—152 (1923). — [8] Über Stabilität und asymptotisches Verhalten der Integrale von Differentialgleichungssystemen (*3*). Ibid. **29**, 129—160 (1928). — [9] Über ein vermeintliches Stabilitätskriterium (*3*). Nachr. Math. Ges. Göttingen **1930**, 28—29. — [10] Die Stabilitätsfrage bei Differentialgleichungen (*3*). Math. Z. **32**, 703—728 (1930). — [11] Über eine Matrixtransformation (*3*). Ibid. **32**, 465—473 (1930). — [12] Über die Entwickelbarkeit der Integrale von Differentialgleichungen nach Potenzen eines Parameters und der Anfangswerte (*10*). Math. Ann. **113**, 292—303 (1936). — [13] Über Bruwiersche Reihen (*3*). Math. Z. **45**, 127—141 (1939). — [14] Die Ordnungszahlen linearer Differentialgleichungssysteme (*3*). Ibid. **31**, 748—766 (1930).

PERSEN, L.: [1] Über die Wronskische Determinante bei selbstadjungierten Differentialgleichungen (*5*). Norjke Vid. Selsk. Forh. Trondheim **24** (1951); **1952**, 12—15.

PERSIDSKIĬ, K.: [1] Über die Stabilität einer Bewegung nach der ersten Näherung (*7*). Math. Sbornik (1) **40**, 284—293 (1933). — [2] On an estimate for characteristic values (*7*). Izvestiya Akad. Nauk Kazah. SSSR, **2**, 36—45 (1948). — [3] On the stability of solutions of denumerable systems of differential equations (*7*). Ibid. **2**, 3—35 (1948). — [4] On the stability of the solution of an infinite system of equations (*7*). Prikl. Mat. Meh. SSSR, **12**, 597—612 (1948). — [5] On the characteristic numbers of the solution of an infinite system of linear differential equations (*7*). Doklady Akad. Nauk SSSR, (N. S.) **63**, 229—232 (1948). — [6] Uniform stability in the first approximation (*1, 3, 4, 7*). Prikl. Mat. Meh. SSSR, **13**, 229—240 (1949). — [7] On the stability of solutions of differential equations (*6, 7*). Izvestiya Akad. Nauk Kazah. SSSR, **4**, 3—18 (1950). —

[8] On the spectrum of characteristic values (*7*). Prikl. Mat. Meh. SSSR, **14**, 635—650 (1950). — [9] Some critical cases of denumerable systems (*6, 7*). Izvestiya Akad. Nauk Kazah. SSSR, **5**, 3—24 (1951). — [10] On characteristic numbers (*3*). Ibid. **6**, 64—76 (1952). — [11] On stability of solutions of differential equations (*7*). Ibid. **1938**, 29—45.

PETROV, V. N.: [1] The limits of applicability of S. TCHAPLYGIN's theorem on differential inequalities to linear equations with ordinary derivatives of the second order (*5*). Doklady Akad. Nauk SSSR, (N.S.) **51**, 255—258 (1946). — [2] Inapplicability of the theorem on the differential inequality of S. TCHAPLYGIN to certain nonlinear differential equations of the second order (*8*). Ibid. **51**, 497—499 (1946).

PETROVSKY, I. G.: [1] Über das Verhalten der Integralkurven eines Systems gewöhnlicher Differentialgleichungen in der Nähe eines Singularen Punktes (*9*). Mat. Sbornik **41**, 107—156 (1934).

—, and E. M. LANDIS: [1] On the number of limit cycles of the equation $dy/dx = M(x, y)/N(x, y)$, where M and N are polynomials of second degree (*9*). Dokl. Akad. Nauk SSSR, (N.S.) **102**, 29—32 (1955).

PÉYOVITCH, T.: [1] Sur une propriété asymptotique à zero des équations linéaire (*8*). C. R. 2. Congr. Math. Pays Slaves **64**, 158—159 (1935). — [2] Sur la valeur à l'infini des intégrales de certaines équations différentielles (*8*). Rev. Sci. **84**, 354—356 (1946). — [3] L'existence de solutions asymptotiques de certaines équations différentielles (*8*). Acad. Serbe Sci. Publ. Inst. Math. **1**, 88—92 (1947). — [4] Sur les solutions asymptotiques de certaines équations différentielles (*8*). Glas. Srpske. Akad. Nauka **191**, 189—196, 197—199 (1948). — [5] Sur les solutions asymptotiques des équations différentielles (*8*). Premier Congr. Math. Phys. Jugoslav, 1949. Naučna Knjiga, Belgrade **2**, 121—145 (1951). — [6] Sur les solutions asymptotiques des équations différentielles (*8*). Soc. Math. Phys. de Serbia, Naučna Knjiga, Belgrade, 1952, 52 pp.

PICARD, E.: [1] Traité d'Analyse (*), 3 vols. Paris, 1896.

PICONE, M.: [1] Sui valori eccezionali di un parametro da cui dipende un'equazione differenziale lineare del secondo ordine (*5*). Ann. Scuola Norm. Sup. Pisa (1) **11**, 1—141 (1910). — [2] Nuova analisi esistenziale e quantitativa delle soluzioni dei sistemi di equazioni differenziali ordinarie (*3*). Ibid. (2) **10**, 13—26 (1940). — [3] Ulteriore analisi quantitativa delle soluzioni di talune equazioni differenziali ordinarie (*8*). Ann. Mat. pura appl. (4) **28**, 195—203 (1949).

PINNEY, E.: [1] Nonlinear differential equations systems. Contrib. Theory nonlinear Oscillations **3**, 31—56 (1956). — [2] Nonlinear differential equations (*). Bull. Amer. Math. Soc **61**, 373—388 (1955). — [3] Ordinary differential-difference equations (*, *8*), pp. 262, Univ. of California Press 1958.

PIPES, L. A.: [1] The analysis of retarded control systems (*2*). J. Appl. Phys. **19**, 617—623 (1948). — [2] Matrix solution of equations of the Mathieu-Hill type (*4*). Ibid. **24**, 902—910 (1953).

PITT, H. R.: [1] The linear theory of neuron networks (*2*). Bull. Math. Biophs. **4**, 169—175 (1942); **5**, 23—31 (1943). — [2] On a class of integro-differential equations (*2*). Proc. Cambridge Phil. Soc. **40**, 199—211 (1944); **43**, 153—163 (1947).

PLATO, G.: [1] Über das Verhalten eines angefachten schwingungsfähigen Systems mit einem Freiheitsgrad, dessen Dämpfung dem Quadrat der Geschwindigkeit proportional ist (*8*). Z. angew. Math. Mech. **28**, 91—92 (1948). — [2] Über das Abklingen von Schwingungen mit schwacher in beliebiger Weise von der Geschwindigkeit abhängiger Dämpfung (*8*). Ibid. **25/27**, 93—94 (1947).

PLIS, A.: [1] On a topological method for studying the behavior of the integrals of ordinary differential equations (*9*). Bull. Acad. Polon. Sci. **2**, 415—418 (1954). —

[2] Remark sur le système dynamique dans le domaine doublement connexe (*9*). Ann. Polon. Math. **3**, 160—171 (1956).

Pliss, V. A.: [1] A qualitative picture of the integral curves in the large and the construction with arbitrary accuracy of the region of stability of a certain system of two differential equations (*7, 9*). Prikl. Mat. Meh. SSSR, **17**, 541—554 (1953). — [2] Necessary and sufficient conditions for stability in the large for a system of *n* differential equations (*7, 9*). Dokl. Akad. Nauk SSSR, (N.S.) **103**, 17—18 (1955).

Poincaré, H.: [1] Sur les proprietés des fonctions définies par les équations aux différences partielles (*1, 9*). Fac. Sci. Paris Thesis, 1879. — [2] Sur les équations linéaires aux différentielles ordinaires et aux différences finies (*9*). Amer. J. Math. **7**, 203—258 (1885). — [3] Sur les intégrales irrégulières des équations linéaires (*9*). Acta Math. **8**, 295—344 (1886). — [4] Sur le problème des trois corps et les équations de la dynamique (*8, 9*). Ibid. **1890**, 1—271. — [5] Sur les courbes définies par les equations différentielles (*9*). J. de math. (3) **7**, 375—422 (1881); **8**, 251—296 (1882); (4) **1**, 167—244 (1885); **2**, 151—217 (1886). — C. R. Acad. Sci. Paris **93**, 951—952 (1881); **98**, 287—289 (1884). — [6] Les méthodes nouvelles de la mécanique celeste (*). Paris, 3 vols., 1892, 1893, 1899. — [7] Leçons de mécanique celeste (*). Paris, 3 vols., 1905, 1907, 1910. — [8] Sur un théorème de Géométrie (*9*). Rend. Circ. Mat. Palermo **33**, 375—407 (1912).

Pol, B. van der: [1] Oscillation hysteresis in a triode generator (*8*). Phil. Mag. **43**, 700 (1922). — [2] Relaxation oscillations (*8*). Ibid. (7) **2**, 928 (1926). — [3] Forced oscillations in circuit with nonlinear resistance (*8*). Ibid. **3**, 65—80 (1927). — [4] (*8*). Proc. Inst. Radio Eng. **22**, 1051—1086 (1934). — [5] Note sur les propriétés des solutions d'une équation différentielle, que l'on peut déduire directement de l'équation différentielle elle-même (*8*). Actes du Colloque Internat. des Vibrations non linéaires, Ile de Porquerolles 1951, 159—167.

—, and J. van der Mark: [1] Le battement du coeur considéré comme oscillation de relaxation (*8*). Onde Électrique **7**, 365 (1928). — [2] Oscillations sinusoidales et de relaxation (*8*). Ibid. **9**, 293 (1930).

—, and M. J. O. Strutt: [1] Stability of solutions of Mathieu's equation (*4*). Phil. Mag. **5**, 18 (1928).

Pontryagin, L. S.: [1] On self-excited oscillations of systems close to Hamiltonians (*8*). Nat.-Fiz. Ž. SSSR, **6**, 25—28 (1934). — [2] On zeros of some transcendental functions (*2*). Izvetiya Akad. Nauk Kazah. SSSR, **6**, 115—134 (1942).

Popoff, K.: [1] Sugli integrali di alcune equazioni differenziali considerate come funzioni dei parametri che vi figurano, per grandi valori dei parametri (*5*) Rend. Accad. Italia (7) **3**, 524—531 (1942).

Popovici, C.: [1] Stabilité pondérée (*9*). Acad. Repub. Pop. Române. Bul. Sti., Sect. Sti. Mat. Fiz. **4**, 243—261 (1952).

Popovskiǐ, A. M.: [1] On the freedom of choice of the parameters of autonomic processes of regulation of several reciprocally related quantities (*2*). Avtom. i Telem. **10**, 401—423 (1949).

Porter, A.: [1] Introduction to Servomechanisms (*2*, *). Wiley 1950.

Potter, R. L.: [1] On self-adjoint differential equations of second order (*5*). Pacific J. Math. **3**, 467—491 (1953).

Predonzan, A.: [1] Sulle vibrazioni forzate di un sistema non dissipativo a due gradi di libertà (*8*). Ann. Scuola Norm. Super. Pisa (2) **12** (1943); **1947**, 173—183.

Prodi, G.: [1] Un'osservazione sugl'integrali dell'equazione $y'' + A(x)y = 0$ nel caso $A(x) \to +\infty$ per $x \to \infty$ (*5*). Rend. Accad. Lincei (8) **8**, 462—464 (1950). — [2] Nuovi criteri di stabilità per l'equazione $y'' + A(x)y = 0$ (*5*). Ibid. **10**,

447—451 (1951); **11**, 30—34 (1951). — [3] Intorno ad una formula asintotica di HARTMAN e WINTNER (*3*). Ann. Scuola Norm. Super. Pisa (3) **7**, 277—286 (1953).

PROSKURYAKOV, A. P.: [1] Characteristic numbers of the solutions of differential equations with periodic coefficients (*4*). Prikl. Mat. Meh. SSSR, **10**, 545—558 (1946).

PUTNAM, C. R.: [1] An application of spectral theory to a singular calculus of variations problem (*5*). Amer. J. Math. **70**, 780—803 (1948). — [2] An oscillation criterion involving a minimum principle (*5*). Duke Math. J. **16**, 633—636 (1949). — [3] On the spectra of certain boundary value problems (*5*). Amer. J. Math. **71**, 109—111 (1949). — [4] The cluster spectra of bounded potentials (*5*). Ibid. **71**, 612—620 (1949). — [5] On isolated eigenfunctions associated with bounded potentials (*5*). Ibid. **72**, 135—147 (1950). — [6] The comparison of spectra belonging to potentials with a bounded difference (*5*). Duke Math. J. **18**, 267—273 (1951). — [7] On the least eigenvalue of HILL's equation (*4, 5*). Quart. Appl. Math. **9**, 310—314 (1951). — [8] On the unboundedness of the essential spectrum (*5*). Amer. J. Math. **74**, 578—586 (1952). — [9] A sufficient condition for an infinite discrete spectrum (*5*). Quart. Appl. Math. **11**, 484—487 (1954). — [10] Note on a limit point criterion (*5*). J. London Math. Soc. **29**, 126—128 (1954). — [11] On the gaps in the spectrum of the Hill equation (*4, 5*). Quart. Appl. Math. **11**, 495—498 (1954). — [12] Stability and almost periodicity in dynamical systems (*6, 8*). Proc. Amer. Math. Soc. **5**, 352—356 (1954). — [13] A note on correlation functions and stability in dynamical systems (*8, 9*). ibid. **5**, 696—699 (1957). — [14] On the continuous spectra of singular boundary value problems (*5*). Canad. J. Math. **6**, 420—426 (1954). — [15] Integrable potentials and half-line spectra (*5*). Proc. Amer. Math. Soc. **6**, 243—246 (1955). [16] On dynamical systems with one degree of freedom (*5*). Canad. J. Math. **7**, 280—283 (1955). — [17] Necessary and sufficient conditions for the existence of negative spectra (*5*). Quart. Appl. Math. **13**, 335—337 (1955). — [18] Note on some oscillation criteria (*5*). Proc. Amer. Math. Soc. **6**, 950—952 (1955). [19] Note on a onedimensional nonconservative system (*5*). Amer. Math. Monthly **64**, 32—33 (1957). — [20] On future and past stability in incompressible systems (*6*). J. Math. Mech. **6**, 669—672 (1957).

—, and L. L. HELMS: [1] Stability in incompressible systems (*6*). J. Math. Mech. **7**, 901—904 (1958).

—, and A. WINTNER: [1] Linear differential equations with almost periodic or Laplace transform coefficients (*3, 4*). Amer. J. Math. **73**, 792—806 (1951).

RABINOVIC, YU. L.: [1] Estimate of the type and order of exponential growth of solutions of linear differential equations (*8*). Ucenye Zapiski Gos. Univ., Mat. **7**, 205—207 (1954).

RABINOVITCH, N. L.: [1] Sur les courbes définies par les équations différentielles (*9*). C. R. Acad. Sci. Paris **232**, 671—673 (1951).

RAPOPORT, I. M.: [1] On linear differential equations with periodic coefficients (*4*). Doklady Akad. Nauk SSSR, (N.S.) **76**, 793—795 (1951). — [2] On the stability of oscillations of material systems (*4*). Ibid. **77**, 25—28 (1951). — [3] On the asymptotic behavior of solutions of linear differential equations (*4*). Ibid. **78**, 1097—1100 (1951). — [4] On some asymptotic methods in the theory of differential equations (*8*, *), pp. 287. Kiev, 1954.

RAUCH, L. L.: [1] Oscillation of a third order nonlinear autonomous system (*9*). Contrib. to the Theory of nonlinear Oscillations. Ann. of Math. Studies 1950, no. 20, 39—88.

RAUSHER, M.: [1] Steady oscillations of systems with nonlinear and unsymmetrical elasticity (*8*). J. Appl. Mech. **5** (1938).

RAYLEIGH, Lord: [1] The theory of sound (*). Dover, 1945. — [2] Scientific papers (*1, 2, 8*). 1883, 1887. — [3] On maintained vibrations (*8*). Phil. Mag. **15**, 229 (1883).

RAYMOND, F. H.: [1] Sur la stabilité d'un assersissement linéaire multiple (*2*). C. R. Acad. Sci. Paris **235**, 508—510 (1952).

RAZMADÉ, A.: [1] Sur les solutions périodiques et les extrémales fermeés du calcul des variations (*8*). Math. Ann. **110**, 63—96 (1935).

REEB, G.: [1] Sur l'existence de solutions périodiques de certaines systèmes différentiels perturbés (*9*). Arch. Math. **2**, 205—206 (1950). — [2] Sur la stabilité et l'unicité des solutions périodiques de l'équation différentielle $X(x, y)\,dx + Y(x, y)\,dy = 0$ (*9*). Colloque de Topologie Strasbourg, 1951, 8 pp. — [3] Sur les solutions périodiques de certain systèmes differentiells perturbés (*8*). Canad. J. Math. **3**, 339—362 (1951). — [4] Über dynamische Systeme mit lauter periodischen Bewegungen (*9*). Abh. Math. Sem. Univ. Hamburg **17**, 98—103 (1951). — [5] Sur la stabilité des solutions périodiques de l'équation différentielle $X(x, y)\,dx + Y(x, y)\,dy = 0$ (*9*). Collectanea Math. **4**, no. 2, 51—56 (1951). — [6] Sur l'existence de mouvement périodiques de certain systèmes dynamiques (*9*). Arch. Math. **2**, 205—206 (1950); **3**, 76—78 (1952). — [7] Sur les mouvements périodiques de certain systèmes différential (*9*). C. R. Acad. Sci. Paris **227**, 1331—1332 (1948). — [8] Sur les trajectories fermées de certain champs de vecteurs (*9*). Ibid. **228**, 1097—1098 (1949). — [9] Sur les solutions périodiques de certain systèmes différentiels canoniques (*9*). Ibid. **228**, 1196—1198 (1949). — [10] Quelques proprietés globales des trajectoires de la dynamique dues à l'existence de l'invariant intégral de M. ELIE CARTAN. ibid. **229**, 969—971 (1949). — [11] Varietés de Rieman dont toutes le géodesiques sont fermées (*9*). Bull. Acad. Roy. Belg. Sci. (5) **36**, 324—329 (1950).

REISSIG, R.: [1] Erzwungene Schwingungen mit zäher Dämpfung und starker Gleitreibung (*8*). Math. Nachr. **11**, 231—238, 345—384 (1954); **12**, 119—128, 249—252, 283—300 (1954); **13**, 309—312 (1955). — [2] Über eine nichtlineare Differentialgleichung zweiter Ordnung (*8*). Ibid. **13**, 313—318 (1955). — [3] Über die Differentialgleichung $x'' + 2Dx' + \mu \operatorname{sgn} x' + x = \Phi(\eta t)$, wo $\Phi(\eta t + 2\pi) \equiv \Phi(\eta t)$ ist. Das Verhalten der Lösungen für $t \to \infty$ (*8, 9*). Abh Dtsch. Akad. Wiss. Berlin **1953**, 33 pp.

REIZIN, L. E.: [1] Behavior near to a singular point of integral curves of a system of three differential equations (*9*). Latvijas PSR Zinatnu Akad. Vestis (43), **1951**, no. 2, 333—346. Amer. Math. Soc. Translations (2) **1**, 239—252.

RELLICH, F.: [1] Über Lösungen nichtlinearer Differentialgleichungen (*9*). Festschrift Akad. Wiss. Göttingen, I. Math.-Phys. Kl., 168—174. Springer 1951.

RESETOV, M. R.: [1] On the boundedness of solutions and characteristic numbers of a denumerable system of linear differential equations of triangular form (*3*). Izvestiya Akad. Nauk Kazah. SSSR, **1950**, 109—114. — [2] On the stability of the solutions of a denumerable system of differential equations the linear parts of which have triangular form (*6, 7*). Ibid. **3**, 39—76 (1949).

REUTER, G. E. H.: [1] Subharmonics in a nonlinear system with unsymmetrical restoring force (*8*). Quart. J. Mech. Appl. Math. **2**, 198—207 (1949). — [2] A boundedness theorem for nonlinear differential equations of the second order (*8*). Proc. Cambridge Phil. Soc. **47**, 49—54 (1951). — [3] On certain nonlinear differential equations with almost periodic solutions (*8*). J. London Math. Soc. **26**, 215—221 (1951). — [4] Boundedness theorems for nonlinear differential equations of the second order (*8, 9*). J. London Math. Soc. **27**, 48—52 (1952).

RICCI, L.: [1] Sulle vibrazioni quasi armoniche di un sistema dissipativo (*4, 5*). Rend. Sem. Mat. Politecnico Torino **8**, 191—208 (1949).

RICHARD, U.: [1] Sulle successioni dei valori stazionari delle soluzioni delle equazioni differenziali lineari del 2° ordine (*5*). Rend. Sem. Mat. Politecnico Torino **9**, 309—324 (1950). — [2] Su un' equazione non lineare del secondo ordine (*5*). Ibid. **10**, 305—324 (1951). — [*3*] Sulla rappresentazione asintotica degli estremi delle soluzioni di equazioni differenziali lineari del 2° ordine (*5*). Rend. Accad. Lincei (8) **12**, 382—387 (1952) — [4] Su una classe di "funzioni ausiliarie" riguardanti le equazioni differenziali del 2° ordine (*5*). Atti Quarto Congr. Unione Mat. Ital., 2, 200—203. Taormina 1951.

ROBERSON, R.E.: [1] Synthesis of a nonlinear dynamic vibration absorber (*8*). J. Franklin Inst. **254**, 205—220 (1952). — [2] On the relationship between the Martenson and Duffing methods for nonlinear vibrations (*8*). Quart. Appl. Math. **10**, 270—272 (1952).

ROCARD, Y.: [1] Dynamique géneral des vibrations (*). Masson, 1949. — [2] Les oscillations de relaxation (*, *8*). Rev. Sci. **79**, 31—50 (1941). — [3] Attaque des systèmes vibrants par des moyens non linéaires (*8*). Ibid. **80**, 359—363 (1942). — [4] Les méfaits du roulement. Autooscillations et instabilités de route (*2*). Rev. Sci. **84**, 15—28 (1946). — [5] Étude de la stabilité des systèmes accessibles à des mesures (*2*). Ibid. **85**, 519—531 (1947). — [6] Sur les conditions d'auto-oscillation des systèmes vibrants (*2*). Proc. Phys. Soc. **61**, 393—402 (1948).

ROGERS, T.A., and W.C. HURTY: [1] Relay servomechanisms. The shunt-motor servo with inertia load (*2*). Trans. Amer. Soc. Mech. Engrs. **72**, 1163—1172 (1950).

ROSENBLATT, A.: [1] On the phenomenon of subresonance. Case of the generalized van der Pol equation with forced vibrations (*8*). Bol. Fac. Ing. Montevideo **3**, 116—126 (1945). — [2] On autoexcited oscillations. The galloping of electrical transmission lines (*8*). Rev. Ci. Lima **47**, 33—61 (1945). — [3] On the growth of the solutions of ordinary differential equations (*3*). Bull. Amer. Math. Soc. **51**, 723—727 (1945). — [4] On subharmonic resonance (*8*). Actas Acad. Ci. Lima **8**, 45—58 (1945). — [5] On the phenomenon of subresonance for the van der Pol equation (*8*). Bol. Fac. Ing. Montivideo **3**, 12 pp. (1949).

ROTHE, E.H.: [1] Asymptotic solution of a boundary value problem (*10*). Iowa State Coll. J. Sci. **13**, 369—372 (1939).

ROUDNEFF, G.V.: [1] Sur les équations de Sturm-Liouville ayant des singularités (*5*). Učenye Zapiski Moskov. Gos. Univ. **100**, 113—126 (1946).

ROUQUET LA GARRIGUE, V.: [1] Le sens de l'étude qualitative des équations différentielles (*, *9*). Trabajos Estadistica **2**, 273—289 (1951).

ROUTH, E.J.: [1] A treatise on the stability of a given state of motion (*), 1877. — [2] A treatise on the dynamics of a particle (*), 1898. Reprint G.E, Steckert 1954. — [3] A treatise of the dynamics of a system of rigid bodies (*2*). 1884. — [4] Die Dynamik der Systeme starrer Körper (*), 1898.

RUBERT, F.K.: [1] Erzwungene Pendelschwingungen endlicher Amplitude (*8*). Z. Physik **127**, 72—84 (1950).

RYABOV, B.A.: [1] Auto-oscillations in some servo-systems restrained by the presence of damping (Coulomb) friction (*8*). Doklady Akad. Nauk SSSR, (N.S.) **73**, 283—286 (1950).

RYABOV, YU.A.: [1] Generalization of a theorem of A. LYAPUNOV (*6*, *7*). Moskov. Gos. Univ. Uc. Zap. **1954**, 131—150.

RYŠKOV, S.S.: [1] On the regimes of operation of a vacuum-tube generator (*8*). Doklady Akad. Nauk SSSR, (N.S.) **96**, 921—924 (1954).

RYTOV, S.M.: [1] An extension of the limits of applicability of the small parameter method (*8*). Doklady Akad. Nauk SSSR, (N.S.) **47**, 181—184 (1945).

Rytov, S. M., and M. E. Zhabotinsky: [1] Application of the small parameter method to systems close to those of Sturm-Liouville (*8*). Izvestiya Akad. Nauk SSSR, **11**, 135—140 (1947).

Saharnikov, N. A.: [1] On conditions for the existence of a center and a focus (*9*). Prikl. Mat. Meh. SSSR, **14**, 513—526 (1950). — [2] Solution of the problem of the center and the focus in one case (*9*). Ibid. **14**, 651—658 (1950). — [3] A qualitative picture of the behavior of a trajectory near the boundary of a region of stability containing a singular point in the form of a center (*9*). Ibid. **15**, 349—354 (1951). — [4] On Frommer's conditions for the existence of a center (*9*). Ibid. **12**, 669—670 (1948).

San Juan, R.: [1] Le problème de Watson pour les solutions des équations différentielles linéaires homogènes (*10*). C. R. Acad. Sci. Paris **234**, 1338—1340 (1952).

Sansone, G.: [1] Studi asintotici sulle equazioni differenziali lineari nel campo reale (*5*). Atti 2. Congr. Unione Mat. Ital. Bologna 1940, 39—55. — [2] Sulle soluzioni di Emden dell'equazione di Fowler (*5*). Rend. Mat. Univ. Roma (5) **1**, 163—176 (1940). — [3] Studi asintotici sulle equazioni differenziali del secondo ordine (*5*). Rend. Sem. Mat.-Fis Milano **15**, 115—128 (1941). — [4] Problemi attuali sulla teoria delle equazioni differenziali ordinarie e su alcuni tipi di equazioni alle derivate parziali (*). Atti Convegno Mat. Roma **1942**; **1945**, 179—200. — [5] Le equazioni differenziali lineari, omogenee, del quarto ordine, nel campo reale (*5*). Ann. Scuola Norm. Sup. Pisa (2) **11**, 151—195 (1942). — [6] Studio degli integrali del sistema $y''+py=qz$, $z''+pz=ry+\omega y'$ (*5*) Annali Mat. pura appl. (4) **22**, 145—180 (1943). — [7] Su un criterio sufficiente di esistenza e di unicità per una classe di problemi ai limiti relativi alle equazioni differenziali lineari omogenee del quarto ordine (*5*). Boll Unione Mat. Ital. (2) **5**, 72—78 (1943). — [8] Su un problema ai limiti per l'equazione differenziale $y^{(n)}(x)+\lambda(n-1)!\,\omega(x)\,y(x)=0$ (*5*). Annali Mat. pura appl. (4) **24**, 209—236 (1945). — [9] Studi sulle equazioni differenziali lineari omogenee del terzo ordine nel campo reale (*5*). Revista Univ. Nac. Tucumán. A **6**, 195—253 (1948). — [10] Sopra l'equazione di A. Liénard delle oscillazioni di rilassamento (*8*). Annali Mat. pura appl. (4) **28**, 153—181 (1949). — [11] Valutazione asintotica degli integrali dell'equazione di Liénard che per $t\to-\infty$ tendono allo zero (*8*). Rend. Acad. Lincei (8) **6**, 13—18 (1949). — [12] Sopra una classe di equazioni di Liénard prive di integrali periodici (*8*). Ibid. (8) **6**, 156—160 (1949). — [13] Equazioni differenziali nel campo reale: comportamento asintotico delle soluzioni; punti singolari; soluzioni periodiche e valutazione del periodo (*). Rend. Mat. Univ. Roma (5) **10**, 265—289 (1951). — [14] Soluzioni periodiche dell'equazione di Liénard. Calcolo del periodo (*8*). Rend. Sem. Mat. Politecnico Torino **10**, 155—171 (1950). — [15] Le equazioni delle oscillazioni non lineari. Risultati analitici (*, *8*). Quarto Congr. Unione Mat. Ital., Taormina, 1951, vol. I, 186—217. — [16] Equazioni Differenziali nel Campo Reale (*), 2 vols., 2nd. edit. Zanichelli 1948/49, pp. 400, 475. — [17] Ordinary differential equations. (Russian Translation of [16], vol. 1.) Moscow 1953, 346 pp. — [18] Scritti matematici offerti a Luigi Berzolari, Pavia 1936, 385—403.

—, e R. Conti: [1] Sull'equazione di T. Uno ed R. Yokomi (*9*). Annali Mat. pura appl. (4) **37**, 37—59 (1954); **38**, 205—212 (1955). — [2] Equazioni differenziali non-lineari (*), pp. 647. Roma, Cremonese, 1956.

Sato, T.: [1] Über Stabilität einer linearen Differentialgleichung zweiter Ordnung (*5*). Jap. J. Math. **10**, 195—197 (1933). — [2] Sur l'équation différentielle contenant un paramètre (*10*). ibid. **17**, 299—305 (1941).

Savinov, G. V.: [1] Eigenschwingungssysteme mit stark ausgeprägter Nichtlinearität (*8*, *9*). Vestnik Moskov. Univ. **8**, no. 6, 77—83 (1953).

SCHÄFKE, F. W.: [1] Zur Parameterabhängigkeit beim Ausgangswertproblem für gewöhnliche lineare Differentialgleichungen (*10*). Math. Nachr. **3**, 20—39 (1949). — [2] Über die Stabilitätskarte der Mathieuschen Differentialgleichungen (*4*). Ibid. **4**, 175—183 (1951). — [3] Einige Stabilitätskriterien (*4*). Z. angew. Math. Mech. **33**, 283—285 (1953).

SCHEFFE, H.: [1] Asymptotic solutions of certain linear differential equations in which the coefficients of the parameter may have a zero. Trans. Amer. Math. Soc. **40**, 127—154 (1936).

SCHELKUNOFF, S. A.: [1] Solution of linear and slightly nonlinear differential equations (*5, 8*). Quart. Appl. Math. **3**, 348—355 (1945).

SCHWERDTFEGER, H.: [1] The eigenvalue problem of Hill's equation (*4*). J. Proc. Roy. Soc. New South Wales **79**, 176—189 (1946).

SEARS, D. B.: [1] Some properties of a differential equation (*5*). J. London Math. Soc. **27**, 180—188 (1952).

SEIFERT, G.: [1] On the existence of certain solutions of a nonlinear differential equation (*8*). Z. angew. Phys. **3**, 468—471 (1952). — [2] On certain solutions of a pendulum-type equation (*8*). Quart. Appl. Math. **11**, 127—131 (1953). — [3] A rotated vector approach to the problem of stability of solutions of pendulum-type equations. Contrib. Theory nonlinear Oscillations **3**, 1—16 (1956).

— H.: [1] Zur asymptotischen Integration von Differentialgleichungen (*9*). Math. Z. **48**, 173—192 (1942).

SESTAKOV, A. A.: [1] On the behavior of the integral curves of a system of ordinary differential equations in the neighborhood of a singular point (*9*). Doklady Akad. Nauk SSSR, **62**, 171—174, 591—594 (1948). — [2] On the behavior of the integral curves of a system of differential equations in the neighborhood of a singular point of higher order (*9*). Ibid. **65**, 139—142 (1949). — [3] Some theorems on stability in LYAPUNOV's sense (*9*). Ibid. **79**, 25—28 (1951). — [4] On the behavior of the integral curves of a system of n differential equations ($n \geq 3$) near to a singular point of higher order (*9*). Ibid. **79**, 205—208 (1951).

—, and A. U. PAIVIN: [1] On the asymptotic behavior of solutions of nonlinear systems of differential equations (*9*). Doklady Akad. Nauk SSSR, **62**, 495—498 (1948).

SERBIN, H.: [1] Periodic motions of a nonlinear dynamic system (*8, 9*). Quart. Appl. Math. **8**, 296—303 (1950).

SEROV, M. I.: [1] Remark on the number of zeros of the solution of a linear differential equation of the second order (*5*). Uspehi Mat. Nauk, (N. S.) **6**, no. 6 (46), 182—183 (1951).

SESTINI, G.: [1] Moto di un punto soggetto a resistenza e a forza di richiamo (*8*). Rend. Ist. Lombardo Sci. Lett. (3) **10**, (79), 117—134 (1946). — [2] Criterio di stabilità in un problema di meccanica non lineare (8). Riv. Mat. Univ. Parma **2**, 303—314 (1951). — [3] Criteri di stabilità per il moto di un punto soggetto a forza elastica, a resistenza e ad una forza disturbatrice (*8*). Atti IV. Congr. Unione Mat. Ital. Taormina 1951, 2, 559—564. — [4] Criterio di stabilità in un problema non lineare di meccanica dei sistemi a più gradi di libertà (*8*). Riv. Mat. Univ. Parma **5**, 227—232 (1954).

SEZAWA, K., and I. UTIDA: [1] On the phenomena of instability in undamped quasi-harmonic vibration (*4*). Proc. Imp. Acad. Tokyo **19**, 646—652 (1943); **20**, 128—132 (1944).

SHIMIZU, T.: [1] On the existence of limit cycles for some nonlinear differential equations (*9*). Math. Jap. **1**, 125—134 (1948). — [2] On differential equations for nonlinear oscillations, I (*9*). Ibid. **2**, 86—96 (1951).

SHOHAT, J.: [1] A new analytical method for solving VAN DER POL's and certain related types of nonlinear differential equations homogeneous and nonhomogeneous (*8*, *9*). J. Appl. Phys. **14**, 40—48 (1943). — [2] On van der Pol's and nonlinear differential equations (*8*, *9*). Ibid. **15**, 568—574 (1944).

Shtokalo, I. Z.: [1] Criteria for stability and instability of the solutions of linear differential equations with quasi-periodic coefficients (*4*). Akad. Nauk SSSR, Informaciinii Byuleten no. 1 (10—11), 1945, 38—39. — [2] Linear differential equations of the n-th order with quasiperiodic coefficients (*3*, *4*). Ibid. no. 1 (10—11), 1945, 40—42. — [3] Systems of linear differential equations with quasi-periodic coefficients (*3*,*4*). Ibid. no. 1 (10—11), 1945, 40—42. — [4] Generalized Gibbs formula for the case of linear differential equations with variable coefficients (*3*, *4*). Ibid. no. 1 (10—11), 1945, 42—45. — [5] Méthode asymptotique pour la solution de certaines classes d'équations différentielles linéaires à coefficients variables (*3*, *4*). Doklady Akad. Nauk SSSR, (N. S.) **46**, 51—52 (1945). — [6] Généralisation de la formule fondamentale de la méthode symbolique pour le cas des équations différentielles à coefficients variables (*4*). ,bid. **47**, 10—11 (1945). — [7] Linear differential equations of the n-th order with quasi-periodical coefficients (*3*, *4*). Dopovidi Akad. Nauk Ukrain. SSSR, **1946**, no. 3/4, 17—20. — [8] Systems of linear differential equations with quasi-periodical coefficients (*3*, *4*). Ibid. **1946**, no. 3/4, 21—24. — [9] Generalization of Heaviside's formula applied to linear differential equations with variable coefficients (*4*). Ibid. **1946**, no. 3/4, 25—29. — [10] A stability and instability criteria for solutions of linear differential equations with quasi-periodical cefficients (*3*, *4*). Mat. Sbornik, (N. S.) **19** (61), 263—283 (1946). — [11] Generalization of Heaviside's formula to a case of linear differential equations with variable coefficients (*3*, *4*). Doklady Akad. Nauk SSSR, (N. S.) **51**, 339—340 (1946). — [12] On the theory of linear differential equations with quasi-periodic coefficients (*3*, *4*). Akad. Nauk. Ukrain. SSSR., Zbirnik Prac Inst. Mat. 1946, **1947**, no. 8, 163—176. — [13] On a generalization of the fundamental formula of the symbolic method (*3*). Ukrain, Mat. Ž. no. 3, 51—59 (1949). — [14] On the form of solutions of certain classes of linear differential equations with variable coefficients (*3*, *4*). Ibid. **4**, 36—48 (1952).

Shwartzmann, A. P.: [1] On the question of boundedness of the solution of the differential equation $y''+p(x)\,y=0$ (*4*). Prikl. Mat. Meh. SSSR, **18**, 464—468 (1954).

Sibirskii, K. S.: [1] On conditions for the presence of a center and focus (*9*). Ucenge Zapiske Kisinevsk. Univ. **11**, 115—117 (1954).

Sibuya, Y.: [1] Sur un système des équations différentielles ordinaires linéaires à coefficients périodiques et contenant des paramètres (*4*). J. Fac. Sci. Univ. Tokyo, Sect. I **7**, 229—241 (1954). — [2] Sur les solutions périodiques d'un système des équations différentielles ordinaires non linéaires à coefficients périodiques (*8*). ibid. **7**, 243—254 (1954).

Siegel, C. L.: [1] Note on differential equations on the torus (*6*). Ann. of Math. (2) **46**, 423—428 (1945). — [2] Vorlesungen über Himmelmechanik (*), pp. 212, Springer 1956.

Signorini, A.: [1] Sul teorema di Whittaker (*6*). Rend. Accad. Lincei (5) **21**, 36—39 (1912). — [2] Esistenza di un'estremale chiusa dentro un contorno di Whittaker (*6*, *8*). Rend. Circ. Mat. Palermo **33**, 187—193 (1912).

Silov, G. E.: [1] Integral curves of a homogeneous equation of the first order (*9*). Uspehi Mat. Nauk, (N. S.) **5**, no. 5 (39), 193—203 (1950).

Simanov, S. N.: [1] On the theory of quasiharmonic oscillations (*4*, *8*, *9*). Prikl. Mat. Meh. SSSR, **16**, 129—146 (1952). — [2] On the stability of the solution of a nonlinear system of equations (*8*). Uspehi Mat. Nauk, (N. S.) **8**, no. 6 (58), 155—157 (1953). — [3] On the theory of oscillations of quasi-linear systems (*8*). Prikl. Mat. Meh. SSSR, **18**, 155—162 (1954). — [4] On a method of obtaining conditions for the existence of periodic solutions of nonlinear systems (*8*). Prikl. Mat. Meh. SSSR, **19**, 225—228 (1955).

Skalkina, M. A.: [1] On a connection between stability of solutions of differential and finite difference equations (*8*). Prikl. Mat. Meh. SSSR, **19**, 287—294 (1955).

Slezinger, I. N.: [1] Motion of a very simple mechanical system under the action of elastic forces and nonlinear friction (*8*). Akad. Nauk SSSR, Ž. Tehn. Fiz. **24**, 1660—1676 (1954).

Smirnov, V. I.: [1] A survey of the scientific work of Lyapunov (*). Prikl. Mat. Meh. SSSR, **12**, 479—560 (1948).

Smith, R. A.: [1] On an equation connected with the theory of triode oscillations (*9*). Proc. Cambridge Philos. Soc. **48**, 698—717 (1952). — [2] On the singularities in the complex plane of the solutions of $y'' + y'f(y) + g(y) = P(x)$, (*9*). Proc. London Math. Soc. (3) **3**, 498—512 (1953).

Snol, É.: [1] Behavior of eigenfunctions and the spectrum of Sturm-Liouville operators (*, *5*). Uspehi Mat. Nauk, (N. S.) **9**, no. 4 (62), 113—132 (1954).

Sobol, I. M.: [1] On the asymptotic behavior of the solutions of linear differential equations (*3*). Doklady Akad. Nauk SSSR, (N. S.) **61**, 219—222 (1948). — [2] On Riccati equations and the reduction to them of linear equations of the second order (*5*). Ibid. **65**, 275—278 (1949). — [3] Investigation with the aid of polar coordinates of the asymptotic behavior of solutions of a linear differential equation of the second order (*5*). Mat. Sbornik **28** (70), 707—714 (1951).

Sokolov, A. A.: [1] A criterion of stability for linear systems of regulation with distributed parameters (*2*). Akad. Nauk SSSR, Inzen. Sbornik **2**, no. 2, 3—26 (1946).

Solncev, Yu. K.: [1] On the asymptotic behavior of integral curves of a system of differential equations (*9*). Isvestiya Akad. Nauk SSSR, **9**, 233—240 (1945). — [2] Stability according to Lyapunov for solutions of systems of differential equations with discontinuous right sides (*9*). Uspehi Mat. Nauk, (N. S.) **5**, no. 4 (38), 140—141 (1950). — [3] On stability according to Lyapunov of the equilibrium position of a system of two differential equations in the case of discontinuous right hand side. Moskov. Gos. Univ. Uc. Zap. **148**, 4, 144—180 (1951).

Solodonikov, V. V.: [1] On an approximate method of investigation of the dynamics of a regulating system or a following system (*2*). Izvestiya Akad. Nauk SSSR, **1945**, 1179—1202. — [2] The frequency-response method in the theory of regulation (a survey) (*2*). Avtom. i Telem. **8**, 65—88 (1947).

Solutzev, G.: [1] On the asymptotic behavior of integral curves of a system of differential equations (*9*). Izvestiya Akad. Nauk SSSR, **9**, 233—240 (1945).

Soroka, W. W.: [1] Free periodic motions of an undamped two degrees of freedom oscillatory system with nonlinear unsymmetrical elasticity (*8*). J. Appl. Mech. **17**, 185—190 (1950).

Späth, H.: [1] Über das asymptotische Verhalten der Lösungen nichthomogener linearer Differentialgleichungen (*2*, *3*). Acta Math. **51**, 134—198 (1928). — [2] Über das asymptotische Verhalten der Lösungen nichthomogener linearer Differentialgleichungen (*2*, *3*). Math. Z. **30**, 487—513 (1929).

Spasskii, R. A.: [1] On a class of regulated systems (*7*). Prikl. Mat. Meh. SSSR, **18**, 329—344 (1954).

Starzinskii, V. M.: [1] On auto-oscillations of an electric governor (*8*). Prikl. Mat. Meh. SSSR, **13**, 41—50 (1949). — [2] Sufficient conditions for stability of a mechanical system with one degree of freedom (*7*). Ibid. **16**, 369—374 (1952). — [3] On the stability of unsteady motion in one case (*7*). Ibid. **16**, 500—504 (1952). — [4] On the stability of a mechanical system with one degree of freedom (*7*). Ibid. **17**, 117—122 (1953). — [5] Survey of works on conditions of stability of the trivial solution of a system of linear differential equations with periodic coefficients (*, *4*). Ibid. **18**, 469—510 (1954). Amer. Math. Soc. Translations (2) **1**, 189—238. — [6] On the stability of the trivial solution of differential equations of the second order with periodic coefficients (*4*). Eng. Sbornik **18**, 119—138 (1954). — [7] Remark on the investigation of stability of periodic motions (*4*). Prikl. Mat. Meh. SSSR, **19**, 119—120 (1955).

STEBAKOV, S. A.: [1] Qualitative investigation of the system $x' = P(x, y)$, $y' = Q(x, y)$ by mean of isoclines (*9*). Doklady Akad. Nauk SSSR, (N.S.) **82**, 677—680 (1952). — [2] Analysis of statically stable dynamical systems (*9*). Ibid. **95**, 455—458 (1954).

STEINBERG, T. S.: [1] On periodic solutions of a differential equation of nonlinear oscillations in the presence of "dry" and "viscous" friction (*9*). Izvestiya Akad. Nauk, no. 4, **1954**, 13—22.

STELLMACHER, K. L.: [1] Über erzwungene nicht-lineare Schwingungen hoher Erregerfrequenz und ihre Stabilität (*8*). Z. angew. Math. **34**, 105—119 (1954).

STEPANOV, V. V.: [1] On the solutions of a linear equation with periodic coefficients in the presence of periodic disturbing force (*4*). Prikl. Mat. Meh. SSSR, **14**, 311—312 (1950).

— W., and A. TIHONOV: [1] Über die Räume der fastperiodischen Funktionen (*6*). Math. Sbornik **41**, 166—178 (1934).

STERNBERG, R. L.: [1] Variational methods and nonoscillation theorems for systems of differential equations (*5*). Duke Math. J. **19**, 311—322 (1952).

—, W.: [1] Über die asymptotische Integration von Differentialgleichungen (*10*). Math. Ann. **81**, 119—180 (1920).

STOKER, J. J.: [1] Nonlinear Vibrations in Mechanical and Electrical Systems (*), pp. 273. Interscience, 1950. — [2] Periodic oscillations of nonlinear systems with infinitely many degrees of freedom (*8*). Coll. Internat. Vibrations nonlinéaires, Ile de Porquerolles **1951**, 61—74. — [3] Mathematical methods in nonlinear vibration theory (*, *8*). Proc. Symposium on nonlinear Circuit Analysis, 28—55. New York 1953. — [4] On the stability of mechanical system (*, *1*). Comm. Pure Appl. Math. **8**, 133—141 (1955).

—, and A. PETERS: [1] Nonlinear vibrations (*8*). Seminar notes New York Univ. 1943.

STOPPELLI, F.: [1] Su un'equazione differenziale della meccanica dei fili (*9*). Rend. Accad. Sci. Fiz. Mat. Napoli (4) **19** (1952); **1953**, 109—114.

STRELKOV, S. P.: [1] Introduction to the Theory of Vibration (*), pp. 344. Gos. Izdat. Moscow-Leningrad **1950**.

STRODT, W.: [1] Contributions to the asymptotic theory of ordinary differential equations in the complex domain (*10*). Mem. Amer. Math. Soc. **1954**, no. 13, pp. 81.

STRUTT, M. J. O.: [1] On HILL's problems with complex parameters and a real periodic function (*4*). Proc. Roy. Soc. Edinburgh, Sect. A **62**, 278—296 (1948).— [2] Zur Wellenmechanik des Atomgitters (*4*). Ann. Physik **86**, 819 (1928). — [3] Der charakteristische Exponent der Hillschen Differentialgleichung (*4*). Math. Ann. **101**, 559—569 (1929). — [4] Lamésche, Mathieusche und verwandte Funktionen (*, *4*). Ergebn. d. Math. **1**, 3, Springer 1932. — [5] Bounds for the characteristic values of Hill problems (*4*). Nederl. Akad. Wetensch. Versl. **52**, 83—90, 97—104 (1943). — [6] Characteristic curves of Hill problems (*4*). Ibid. **52**, 153—162, 212—222 (1943). — [7] Characteristic functions of Hill problems. I. Completeness of the sets of periodic and almost periodic characteristic functions (*4*). Ibid. **52**, 488—496 (1943). — [8] Characteristic functions of Hill problems. II. Expansion formulas in series of periodic and almost periodic characteristic functions (*4*). Ibid. **52**, 584—591 (1943).

Sturm, C.: [1] Sur les équations différentielles linéaires du second ordre (*5*). J. Math. Pure Appl. **1**, 106—186 (1836).

Sugiyama, S.: [1] Note on singularities of differential equations (*9*). Kodai Math. Sem. Rep. **1954**, 81—84. — [2] On the singularities of the differential equation $y''+f(x, y)y'+g(x, y)=P(x)$ (*9*). Ibid. **7**, 23—29 (1955).

Suyama, Y.: [1] On the zeros of solutions of second order linear differential equation (*5*). Mem. Fac. Sci. Kyusyu Univ., Ser. A **8**, 201—205 (1954). — [2] On the non-oscillatory solution of second-order linear differential equation (*5*). Ibid. **8**, 207—212 (1954).

Svarcman, A. P.: [1] On boundedness of solutions of the differential equation $y''+p(x)y=0$ (*5*). Prikl. Mat. Meh. SSSR, **18**, 464—468 (1954).

Synge, J. L.: [1] Hydrodynamical stability (*10*). Amer. Math. Soc. Semicentennial Publ. **2**, 227—269 (1938).

Szarski, J.: [1] Sur un système d'inégalités différentielles (*9*). Ann. Soc. Polon. Math. **20**, 126—134 (1947). — [2] Sur une propriété asymptotique des intégrales d'un système d'équations différentielles ordinaires (*9*). Ibid. **20**, 161—168 (1947). — [3] Sur certaines inégalités entre les intégrales des équations différentielles aux dérivées partielles du premier ordre (*9*). Ibid. **22**, 1—34 (1950). — [4] On an oscillatory property of successive approximations (*9*). Ibid. **22**, 201—206 (1949). — [5] Sur les systèmes majorants d'équations différentielles ordinaires (*9*). Ibid. **23**, 206—223 (1950).

Szmydt, Z.: [1] Sur la structure de l'ensemble engendré par les intégrales tendant vers le point singulier de système d'équations différentielles (*9*). Bull. Acad. Polon. Sci., Cl. III **1**, 223—227 (1953). — [2] Sur l'allure asymptotique des intégrales de certain systèmes d'équations différentielles nonlinéaires (*9*). Ann. Polon. Math. **1**, 253—276 (1955).

Szmydtowna, Z.: [1] Sur l'allure asymptotique des intégrales des équations différentielles ordinaires (*9*). Ann. Soc. Polon. Math. **24**, 17—34 (1951); **1954**, no. 2.

Ta, Li: [1] Die Stabilitätsfrage bei Differenzengleichungen (*2*). Acta Math. **63**, 99—141 (1934).

Taam, C.: [1] The boundedness of the solutions of a differential equation in the complex domain (*5*). Pacific. J. Math. **2**, 643—654 (1952). — [2] Non-oscillatory differential equations (*5*). Duke Math. J. **19**, 493—497 (1952). — [3] Linear differential equations with small perturbations (*5*). Ibid. **20**, 13—25 (1953). — [4] Nonoscillation and comparison theorems of linear differential equations with complex-valued coefficients (*5*). Portugaliae Math. **12**, 57—72 (1953). — [5] Oscillation theorems (*5*). Amer. J. Math. **74**, 317—324 (1952). — [6] On the complex zeros of Sturm-Liouville type (*5*). Pacific. J. Math. **3**, 837—843 (1953). — [7] The boundedness of the solutions of a nonlinear differential equation (*5*). Proc. Amer. Math. Soc. **1954**, 122—125. — [8] On the solutions of second order linear differential equations (*5*). Ibid. **4**, 876—879 (1953). — [9] An extension of Osgood's oscillation theorem for a nonlinear differential equation (*5*). Ibid. **5**, 705—714 (1954).

Tartakovskiĭ, V.: [1] Explicit formulas for local expansions about a nodal point (*8*). Doklady Akad. Nauk SSSR, (N. S.) **72**, 853—856 (1950). — [2] Explicit formulas for the local expansions of solutions of a system of ordinary differential equations (*8*). Ibid. **72**, 633—636 (1950). — [3] Explicit formulas for the solution of systems of ordinary differential equations (*8*). Ukrain Mat. Ž. **3**, 128—160 (1951).

Tatarkiewicz, K.: [1] Sur l'allure asymptotique des solutions de l'équation différentielle du second ordre (*9*). Ann. Univ. Mariae Curie-Sklodowska, Sect. A **7**, 19—81 (1953); **8**, 105—133 (1954).

TEOFILATO, P.: [1] Sopra alcuni sistemi differenziali a soluzioni sensibilmente costanti (*8*). Pont. Acad. Sci. Acta **8**, 112—130 (1944).

TEYANA, H.: [1] Some inequalities in the theory of linear differential equations (*3*). Tohoku Math. J. **47**, 210—216 (1940).

THEODORCHIK, K.: [1] Contribution to the theory of synchronization of relaxitory autooscillation (*8, 9*). Doklady Akad. Nauk SSSR, (N. S.) **40**, 54—57 (1943). — [2] Energy considerations for self-excited systems of THOMPSON's type with two degrees of freedom (*8, 9*). Ucenye Zapiski Moskov. Gos. Univ. Fizika **74**, 67—72 (1944). — [3] Energetics of asynchronous reactions (*8, 9*). Ibid. **77**, 112—116 (1945). — [4] Theory of synchronization of relaxation autooscillatory systems (*8, 9*). Acad. Sci. USSR. J. Phys. **9**, 139—143 (1945). — [5] On the theory of sinusoidal auto-oscillations in systems with many degrees of freedom (*8*). Doklady Acad. Nauk SSSR, (N. S.) **52**, 33—36 (1946). — [6] Limits of applicability of VAN DER POL's method (*8*). Ibid. **52**, 123—126 (1946). — [7] The laws of co-existence of frequencies in soft auto-oscillating systems (*8*). Izvestiya Akad. Nauk SSSR, **1946**, 385—388. — [8] The stability of harmonically autooscillating systems (*8*). Doklady Akad. Nauk SSSR, (N. S.) **56**, 367—369 (1947).

THOMPSON, E. W.: [1] On stability of periodic solutions of autonomous differential systems (*8*). (To appear).

THORNE, R. C.: [1] The asymptotic solution of linear second order differential equations in a domain containing a turning point and a regular singularity (*10*). Proc. Cambridge Phil. Soc. **53**, 382—398 (1957).

TIHONOV, A.: [1] On the dependence of the solutions of differential equations on a small parameter (*10*). Mat. Sbornik, (N. S.) **22** (64), 193—204 (1948). — [2] On systems of differential equations containing parameters (*10*). Ibid. **27** (60), 147—156 (1950). — [3] Systems of differential equations containing small parameters in the derivatives. Mat. Sbornik, (N. S.) **31**, 575—586 (1952).

TIMOSHENKO, S.: [1] Schwingungsprobleme der Technik (*). Springer 1932. — [2] Vibration problems in Engineering (*) 2nd edit. van Nostrand, 1941.

TITCHMARSH, E. C.: [1] Some theorems on perturbation theory (*10*). Proc. Roy. Soc. Lond., Ser. A **200**, 34—46 (1949). — [2] On the asymptotic distribution of eigenvalues (*5*). Quart. J. Math., Oxford Ser. **5**, 228—240 (1954).

TONELLI, L.: [1] Sulle orbite periodiche (*6, 8*). Rend. Accad. Lincei (5) **21**, sem. 1, 251—258, 332—334 (1912). — [2] Sulle soluzioni periodiche nel calcolo delle variazioni (*6, 8*). Ibid. (5) **24**, sem. 2, 317—324 (1915). — [3] Sulle orbite periodiche irreversibili (*8*). Mem. Accad. Sci. Bologna (6, 8) **1**, 21—25 (1924). — [4] Scritti matematici offerti a LUIGI BERZOLARI (*5*). **1936**, 404—405. — [5] Sull'unicità della soluzione di un'equazione differenziale ordinaria (*1*). Rend. Accad. Lincei (6) **1**, 272—277 (1925). — [6] Sulle estremali complete (*6*). Annali Scuola Normale Sup. Pisa (2) **5**, 159—168 (1936). — [7] Fondamenti di Calcolo delle Variazioni (*), 2 vols., Zanichelli, 1921, 1923.

TREVISAN, G.: [1] Sull'equazione differenziale $y'' + A(x)y = 0$ (*5*). Rend. Sem. Mat. Univ. Padova **23**, 340—342 (1954).

TRICOMI, F.: [1] Integrazione di una equazione differenziale presentatasi in elettrotecnica (*8*). Ann. Scuola Norm. Sup. Pisa (2) **2**, 1—20 (1933). — [2] Un nuovo metodo di studio delle equazioni differenziali lineari (*3*). Rend. Sem. Mat. Politecnico Torino **8**, 7—19 (1949). — [3] Equazioni differenziali (*), pp. 353, 2nd. edit., Einaudi, 1953. — [4] Equazioni differenziali con punti di transizione (turning points) (*10*). Rend. Accad. Lincei. (8) **17**, 137—141 (1954).

TRIMMER, J. D.: [1] Response of Physical Systems (*, *2*), pp. 268. Wiley, 1950.

TRJITZINSKY, W. J.: [1] Theory of linear differential equations containing a parameter (*3, 10*). Acta Math. **67**, 1—50 (1936). — [2] Theory of nonlinear singular differential systems (*8*). Trans. Amer. Math. Soc. **42**, 225—321 (1937). —

[3] Properties of growth for solutions of differential equations of dynamical type. Ibid. **50**, 252—294 (1941).

Troickiĭ, V. A.: [1] On the behavior of dynamical systems and systems of automatic regulation having several regulating organs near to the boundary of a region of stability (*6*, *7*). Prikl. Mat. Meh. SSSR, **17**, 673—684 (1953). — [2] On canonical transformations of the equations of the theory of automatic regulation (*7*, *8*). Ibid. **17**, 49—60 (1953).

Tschen, Y.: [1] Über das Verhalten der Lösungen einer Folge von Differentialgleichungen welche in Limes ausarten (*1*). Composito Math. **2**, 378—401 (1935).

Tulegenov, B.: [1] On stability of solutions of a system of differential equations of the second order (*6*). Izvestiya Akad. Nauk Kazah. SSSR, **4**, 57—72 (1950).

Tumura, M.: [1] The exact criterion for the stability of nonlinear vibrations (*8*). Tech. Rep. Osaka Univ. **1951**, 35—50; **1952**, 1—10.

Turbovič, I. T.: [1] On a question concerning nonlinear systems with variable parameters (*8*). Doklady Akad. Nauk SSSR, (N. S.) **57**, 351—352 (1947). — [2] Concerning nonlinear systems with variable parameters (*8*). Izvestiya Akad. Nauk Kazah. SSSR, **1948**, 203—208.

Turritin, H. L.: [1] Asymptotic solution of certain ordinary differential equations associated with multiple roots of the characteristic equation (*10*). Amer. J. Math. **58**, 364—376 (1936). — [2] Stokes multipliers for asymptotic solutions of a certain differential equation (*10*). Trans. Amer. Math. Soc. **68**, 304—329 (1950). — [3] Asymptotic expansions of solutions of systems of ordinary linear differential equations containing a parameter (*10*). Contrib. Theory nonlinear Oscillations **2**, 1952, 81—116. — [4] Convergent solutions of ordinary linear homogeneous differential equations in the neighborhood of an irregular singular point (*10*). Acta Math. **93**, 27—66 (1955).

Tuzov, A. P.: [1] Stability questions for a certain regulating system (*7*, *8*). Vestnik Leningrad. Univ. **1955**, no. 2, 43—70.

Ulanov, G. M.: [1] On the maximum deviation of the regulated quantity in a transient process (*2*). Avtom. i Telem. **9**, 168—175 (1948).

Uno, T.: [1] On the formation of limit cycles (*9*). Math. Japonicae **2**, 75—78 (1951). — [2] On the curves defined by some differential equations (*9*). Ibid. **2**, 119—126 (1952).

—, and R. Yokomi: [1] On some mode of appearance of limit cycles (*9*). Math. Japonicae **2**, 117—118 (1952).

Ura, T.: [1] Sur les courbes définies par les équations différentielles dans l'espace à *m* dimensions (*9*). Ann. École Norm. Sup. Paris (3) **70**, 287—360 (1953).

—, and Y. Hirasawa: [1] Sur les points singuliers des équations différentielles admettant un invariant intégral (*9*). Proc. Jap. Acad. **30**, 726—730 (1954).

Urabe, M.: [1] On the existence of periodic solutions for certain nonlinear differential equations (*8*). Math. Japonicae **2**, 23—26 (1950). — [2] Infinitesimal deformation of the periodic solution of the second kind and its application to the equation of a pendulum (*8*). J. Sci. Hiroshima Univ., Ser. A, **18**, 37—53, 183—219 (1954).

—, and S. Katsuma: [1] Generalization of Poincaré-Bendixson theorem (*9*). J. Sci. Hiroshima Univ., Ser. A, **17**, 365—370 (1954).

Urowski, A. V.: [1] On some criteria for stability of the integrals of systems of two linear differential equations with periodic coefficients (*4*). Doklady Akad. Nauk SSSR, **62**, 595—598 (1948).

Vasileva, A. B.: [1] On the differentiation of solutions of differential equations containing a small parameter (*8*, *10*). Doklady Akad. Nauk SSSR, **61**, 597 to 599 (1948). — [2] On the mathematical theory of catalysis (*8*, *10*). Vestnik

Moškov. Univ. Ser. Fiz.-Mat. **9**, no. 6, 39—46 (1954). — [3] On differentiation of solutions of systems of differential equations containing a small parameter (*8, 10*). Doklady Akad. Nauk SSSR, (N. S.) **75**, 483—486 (1950). — [4] On differentiation with respect to a small parameter of solutions of systems of differential equations (*8, 10*). Ibid. **78**, 845—848 (1951). — [5] On the differentiability of the solutions of differential equations containing a small parameter (*8, 10*). Mat. Sbornik, (N. S.) **28**, (70), 131—146 (1951). — [6] On differentiability of solutions of systems of differential equations containing small parameters with the derivatives (*8, 10*). Vestnik Moskov. Univ., Ser. Fiz.-Mat. **9**, no. 3, 29—40 (1954).

Veiga de Oliveira, F.: [1] Characteristic exponents. Application to stability (*). Univ. Lisboa Rev. Fac. Ci. (2) **2**, 201—288 (1952).

Villari, G.: [1] Un teorema di esistenza e di unicità per una classe di soluzioni dell'equazione $z''(t) + A(t) f(z) = 0$ (*1*). Riv. Mat. Univ. Parma **4**, 319—326 (1953). — [2] Sul comportamento asintotico degli integrali di una classe di equazioni differenziali non lineari (*8*). Ibid. **5**, 83—98 (1954).

Vinograd, R. É.: [1] On the limiting behavior of unbounded integral curves (*9*). [2] Sci. Rep. of Moscov State Univ. **135** (*3, 5*), 94—136 (1952). — [2] On a criterion of instability in the sense of Lyapunov of the solutions of a linear system of ordinary differential equations (*3*). Doklady Akad. Nauk SSSR, (N. S.) **84**, 201—204 (1952). — [3] Some criteria of boundedness of the solutions of a system of two linear differential equations (*3*). Ibid. **85**, 265—268 (1952). — [4] Instability of characteristic exponents of regular systems (*3*). Ibid. **91**, 999—1002 (1953). — [5] Negative solution of a question on stability of characteristic exponents of regular systems (*6*). Prikl. Mat. Meh. SSSR, **17**, 645 to 650 (1953). — [6] On boundedness of solutions of regular systems of differential equations with small added terms (*8*). Uspehi Mat. Nauk, (N. S.) **8**, no. 1 (53), 115—120 (1953). — [7] On an assertion of K. Persidskii (*6*). Ibid. **9**, no. 2 (60), 125—128 (1954). — [8] A new proof of Perron's theorem and certain properties of regulating systems (*3, 8*). Ibid. **9**, no. 2 (60), 129—136 (1954). — [9] Remark on the critical case of stability of a singular point in the plane (*9*). Doklady Akad. Nauk SSSR, (N. S.) **101**, 209—212 (1955). — [10] Instability of the smallest characteristic exponent of a control system (*9*). Ibid. **103**, 541—544 (1955). — [11] Inapplicability of the method of characteristic exponents to the study of nonlinear differential equations (*6*). Math. Sb. (N.S.) **41** (83), 431—438 (1957).

Viola, T.: [1] Sulle equazioni algebriche a coefficienti reali (*2*). Rend. Accad. Sci. Napoli (4) **8**, 76—83 (1937/38); **9**, 15—20 (1939). — [2] Sulla stabilità degli integrali delle equazioni differenziali lineari omogenee a coefficienti costanti (*2*). Rend. Accad. Italia (7) **1**, 238—244 (1940). — [3] Sul determinante di Hurwitz di una equazione algebrica (*2*). Boll. Unione Mat. Ital. (3) **4**, 40—45 (1949).

Viswanatham, B.: [1] The existence of harmonic vibrations (*9*). Proc. Amer. Math. Soc. **4**, 371—372 (1953).

Vogel, T.: [1] Les méthodes topologiques de discussion des problèmes aux oscillations non linéaires (*). Ann. Telecom. **6**, 2—10 (1951). — [2] Topologie des oscillations à déferlement (*, *8*). Coll. Internat. Vibrations non linéaires, Ile de Porquerolles 1951, 237—256. — [3] Sur les systèmes déferlants (*8*) Bull. Soc. Math. France **81**, 63—75 (1953).

Volk, I. M.: [1] A generalization of the method of small parameter in the theory of nonlinear oscillations of nonautonomous system (*8*). Doklady Akad. Nauk SSSR, (N. S.) **51**, 437—440 (1946). — [2] Elastic oscillations with damping proportional to a power of velocity (*8*). Prikl. Mat. Mech. SSSR, **10**, 125—134 (1946). — [3] Periodic solutions of nonautonomous systems depending upon the small parameter (*8*, *10*). Ibid. **10**, 559—574 (1946). — [4] Generalizations of the method of small parameters in the theory of periodic motions of nonautonomous systems (*8*, *10*). Ibid. **11**, 433—444 (1947). — [5] On periodic solutions of autonomous systems (*8*, *10*). Ibid. **12**, 29—38 (1948). — [6] On the stability of periodic motions when the equations and their periodic solutions are known only approximately (*8*). Ibid. **12**, 647—650 (1948). — [7] On a sufficient condition for the stability of a motion in the critical case of two roots with vanishing real parts (*8*). Ibid. **13**, 459—462 (1949).

Volosov, V. M.: [1] On differential equations with a small parameter in the highest derivative (*10*). Doklady Akad. Nauk SSSR, (N. S.) **73**, 873—876 (1950). — [2] Nonlinear differential equations of the second order with a small parameter with the highest derivative (*10*). Mat. Sbornik, (N. S.) **30** (72), 245—270 (1952). — [3] Quasihomogeneous differential equations of the second order having a small parameter (*10*). Mat. Sbornik, (N. S.) **36** (78), 501—554 (1955).

Volterra, V.: [1] Leçons sur les équations intégrales et les équations intégro-differentielles (*). Gauthier-Villars 1913. — [2] Leçons sur les fonctions lignes (*). Ibid. 1913. — [3] Variations and fluctuations of number of individuals in animal species living together (*8*). J. Conseil Int. Explor. de la mer **3**, 1 (1928). — [4] Leçons sur la théorie mathematique de la lutte pour la vie (*). Gauthier-Villars, 1931.

Voronov, A. A.: [1] Free oscillations of an oscillator with variable friction (*8*). Doklady Akad. Nauk SSSR, (N. S.) **81**, 517—520 (1951).

Voznink, L. L.: [1] Investigations sur la stabilité des solutions périodiques des equations de haut ordre. Akad. Nauk Ukrain. SSSR, Kiev **9**, no. 3 (1957).

Wagner, K.: [1] Einführung in die Lehre von den Schwingungen und Wellen (*). 2. Aufl. Dieterich 1947.

Wallach, S.: [1] The stability of differential equations with periodic coefficients (*4*, *5*). Proc. Nat. Acad. Sci. U.S.A. **34**, 203—204 (1948). — [2] The differential equation $y' = f(y)$ (*1*). Amer. J. Math. **70**, 345—350 (1948). — [3] On the location of spectra of differential equations (*5*). Ibid. **70**, 833—841 (1948). — [4] The spectra of periodic potentials (*5*). Ibid. **70**, 842—848 (1948).

Walther, A.: [1] Über nichthomogene lineare Differentialgleichungen (*3*). Göttinger Nachr. Math-Phys. Kl. **1926**, 103—118.

Wasow, W.: [1] On the asymptotic solution of boundary value problems for ordinary differential equations containing a parameter (*10*). J. Math. Phys. **23**, 173—193 (1944). — [2] (with Friedrichs) Singular perturbations of nonlinear oscillations (*10*). Duke Math. J. **13**, 367—381 (1946). — [3] On the construction of periodic solutions of singular perturbation problems (*10*). Contrib. Theory nonlinear Oscillations **1950**, 313—350. — [4] The complex asymptotic theory of a fourth order differential equation of hydrodynamics (*10*). Ann. of Math. **49**, 852—871 (1950). — [5] A study of the solutions of the differential equation $y^{(4)} + \lambda^2 (x y'' + y) = 0$ for large values of λ (*10*). Ann. of Math. **52**, 350—361 (1950). — [6] On singular perturbation problems in the theory of nonlinear vibrations. Actes Coll. Internat. Vibrations non linéaires. Ile de Porquerolles **1951**. — [7] Singular perturbation methodes for nonlinear oscillations (*10*). Proc. Symposium nonlinear circuit analysis. Poly. Inst. Brooklyn 1953, 75—98. — [8] Asymptotic solution of the differential equation of hydrodynamic stability in a domain containing a transition point (*10*). Ann. of Math. **58**, 222—252 (1953). — [9] On small disturbances of plane Couette flow (*10*). J. Res. Nat. Bur. Stand. **51**, 195—201 (1953). — [10] (with R. M. Redheffer) On the convergence of asymptotic solutions of linear differential equations (*10*). Pacific. J. Math. **5**, 817—834 (1955). — [11] The asymptotic theory of linear differential equations involving a parameter (*10*). Matematiche Catania **10**, 134—148 (1955). — [12] On the convergence of an approximation method of M. J. LIGHTHILL (*10*). J. Rat. Mech. Anal. **4**, 751—767 (1955). — [13] Singular perturbations of boundary value problems for nonlinear differential equations of the second order (*10*). Comm. Pure Appl. Math. **9**, 93—113 (1956). — [14] Solutions of certain nonlinear differential equations by series of exponential functions. Ill. J. Math. **2** (1958).

WATSON, A.C.D.: [1] The Sturmian theory of oscillations (*5*). Math. Gaz. **38**, 15—17 (1954).

WAX, N.: [1] On amplitude bounds for certain relaxation oscillations (*8*, *9*). J. Appl. Phys. **22**, 278—281 (1951).

WAŻEWSKI, T.: [1] Théorie des multiplicités régulières d'éléments de contact unis. Application aux transformations canoniques (*9*). Ann. Soc. Polon. Math. **18**, 55—112 (1945). — [2] Une méthode topologique de l'examen du phénomène asymptotique relativement aux équations différentielles ordinaires (*9*). Rend. Accad. Lincei (8) **3**, 210—215 (1947). — [3] Sur les intégrales asymptotiques des équations différentielles ordinaires (*9*). C. R. Soc. Sci. Lett. Varsovie **40** (1947); **1948**, 38—42. — [4] Sur les systèmes de deux équations différentielles linéaires dont les intégrales tendent asymptotiquement vers une ellipse (*9*). Ibid. **41** (1948); **1950**, 9—12. — [5] Sur la limitation des intégrales des systèmes d'équations différentielles linéaires ordinaires (*3*, *9*). Stud. Math. **10**, 48—59 (1948). — [6] Sur les intégrales d'un système d'équations différentielles tangentes aux hyperplans caractéristiques issue du point singulier (*9*). Ann. Soc. Polon. Math. **21** (1948); **1949**, 277—297. — [7] Sur l'allure asymptotique des intégrales d'une équation différentielle non linéaire (*9*). Bull. Acad. Polon. Sci. **1949**, 62—66. — [8] Sur certains lemmes relatifs au prolongement des intégrales des équations différentielles ordinaires (*9*). Ibid. **1949**, 73—74. — [9] Sur la coincidence asymptotique des intégrales de deux systèmes d'équations différentielles (*9*). Ibid. **1949/50**, 147—150. — [10] Systèmes des équations et des inégalités différentielles ordinaires aux deuxièmes membres monotones et leurs applications (*9*). Ann. Soc. Polon. Math. **23**, 112—166 (1950). — [11] Certaines propositions de caractère "epidermique" relatives aux inégalités différentielles (*9*). Ibid. **24** (1951); **1952**, 1—12. — [12] Sur certain conditions de coincidence asymptotique des intégrales des deux systèms d'équations différentielles (*9*). C. R. Soc. Sci. Lett. Varsovie **42** (1949); **1952**, 198—203. — [13] Sur l'évaluation du nombre des paramètres éssentiels dont dépend la famille des intégrales d'un système d'équations différentielles ayant une propriété asymptotique (*9*). Bull. Acad. Polon. Sci. **1**, 3—5 (1953). — [14] Sur la structure de l'ensemble engendré par les intégrales non asymptotiques des équations différentielles (*9*). Ibid. **3**, 143—148 (1955). — [15] Sur les intégrales de branchement des systèmes des équations différentielles ordinaires (*9*). Ann. Ann. Polon. Math. **19**, 338—345 (1955).

WEBER, E.: [1] Convex convolutions applied to nonlinear problems (*8*). Proc. Symp. Nonlinear Circuit Analys. P. I. Brooklyn **6**, 409—427 (1956).

—, H.E.: [1] Methodik der Berechnung von Regulierungen (*, *2*). Z. angew. Phys. **4**, 233—260 (1953).

WEGEL, R.L.: [1] Theory of Vibration of the Larynx (*8*). Bell Syst. Techn. J. **9**, 209 (1930).

WEIDENHAMMER, F.: [1] Resonanzlösungen in homogenen Mathieu Systemen (*4*). Z. angew. Math. Mech. **32**, 154—156 (1952).

WEINSTEIN, D.H.: [1] Characteristic values of MATHIEU's equation (*4*). Phil. Mag. **20**, 288 (1935).

WEISS, H. K.: [1] Analysis of relay servomechanisms (*2*). J. Aeronaut. Sci. **13**, 364—376 (1946).

WENDEL, J. G.: [1] On a van der Pol equation with odd coefficients (*8, 9*). J. London Math. Soc. **24**, 65—67 (1949). — [2] Singular perturbations of a van der Pol equation (*8, 10*). Contrib. to the Theory of nonlinear Oscillations. Ann. of Math. Studies 1950, no. 20, 243—290.

WENTZEL, G.: [1] Eine Verallgemeinerung der Quantenbedingungen für die Zwecke der Wellenmechanik (*10*). Z. Physik **38**, 518—529 (1926).

WEYL, H.: [1] Ramifications, old and new, of the eigenvalue problem (*5*). Bull. Amer. Math. Soc. **56**, 115—139 (1950). — [2] Über gewöhnliche Differentialgleichungen mit Singularitäten und die zugehörigen Entwicklungen willkürlicher Funktionen (*5*). Math. Ann. **68**, 220—269 (1910). — [3] Concerning classical problem in the theory of singular points of ordinary differential equations (*5*). Rev. Ciencias **46**, 23—112 (1944). — [4] Comment on the preceding paper (of N. LEVINSON [*6*]) (*3*). Amer. J. Math. **68**, 7—12 (1946).

WHITTAKER, E. T.: [1] On periodic orbits (*6, 8*). Monthly Notices Roy. Astr. Soc. London **62**, 186—193 (1902). — [2] Analytical Dynamics (*), Dover 1944.

WHYBURN, W. M.: [1] On self-adjoint ordinary differential equations of the fourth order (*5*). Amer. J. Math. **52**, 171—196 (1930). — [2] A nonlinear boundary value problem for second order differential systems (*9*). Pacific J. Math. **5**, 147—160 (1955).

WICHERT, A.: [1] Neuere Theorien der Schüttelerscheinungen elektrischer Lokomotiven mit Parallelkurbelgetrieben (*4*). Bol. ETZ **1921**, p. 42.

WIENER, N., and A. WINTNER: [1] On the ergodic dynamics of almost periodic systems (*9*). Amer. J. Math. **63**, 794—824 (1941).

WILKINS, J. E.: [1] On the growth of solutions of linear differential equations (*3*). Bull. Amer. Soc. **50**, 388—394 (1944). — [2] The converse of a theorem of Tchaplygin on differential inequalities (*5*). Ibid. **53**, 126—129 (1947).

WILLIAMS, J.: [1] Small oscillations with damping (*8*). Math. Gaz. **35**, 29—31 (1951).

WIMAN, A.: [1] Über die reellen Lösungen der linearen Differentialgleichungen zweiter Ordnung (*3, 5*). Ark. Mat. **22**, 1—22 (1919). — [2] Über eine Stabilitätsfrage in der Theorie der linearen Differentialgleichungen (*3, 5*). Acta Math. **66**, 121—145 (1936).

WINTNER, A.: [1] Small perturbations (*4*). Amer. J. Math. **67**, 417—430 (1945). — [2] Asymptotic equilibria (*3, 8*). Ibid. **68**, 125—132 (1946). — [3] Linear variations of constants (*3*). Ibid. **68**, 185—213 (1946). — [4] The nonlocal existence problem of ordinary differential equations (*1*). Ibid. **67**, 277—284 (1945); **68**, 173—178 (1946). — [5] Asymptotic integration constants in the singularity of BRIOT-BOUQUET (*8*). Ibid. **68**, 293—300 (1946). — [6] The adiabatic linear oscillator (*3, 5*). Ibid. **68**, 385—397 (1946). — [7] An Abelian lemma concerning asymptotic equilibria (*3, 8*). Ibid. **68**, 451—454 (1946). — [8] Asymptotic integration constants (*3*). Ibid. **68**, 553—559 (1946). — [9] (L^2)-connections between the potential and kinetic energies of linear systems (*5*). Ibid. **69**, 5—13 (1947). — [10] Asymptotic integrations of the adiabatic oscillator (*3, 5*). Ibid. **69**, 251—272 (1947). — [11] Vortices and nodes (*9*). Ibid. **69**, 815—824 (1947). — [12] Stability and high frequency (*5*). J. Appl. Phys. **18**, 941—942 (1947). — [13] Unrestricted Riccatian solution fields (*5*). Quart. J. Math., Oxford Ser. **18**, 65—71 (1947). — [14] A criterion for stable characteristic exponents (*4, 5*). Quart. Appl. Math. **5**, 232—236 (1947). — [15] A norm criterion for non-oscillatory differential equations (*5*). Ibid. **6**, 183—185 (1948). — [16] Asymptotic integrations of the adiabatic oscillator in its hyperbolic range (*5*). Duke Math. J. **15**, 55—67 (1948). — [17] The dissipation of internal energy in linear dynamical systems (*3*). Phil. Mag. (7) **39**, 722—728 (1948). — [18] On the location of continuous spectra (*5*). Amer. J. Math. **70**, 22—30 (1948). — [19] On the classical existence theorem of linear differential equations (*1*). Ibid. **71**, 331 to 338 (1949). — [20] On linear repulsive forces (*3*). Ibid. **71**, 362—366 (1949). — [21] A priori Laplace transformations of linear differential equations (*5*). Ibid. **71**, 587—594 (1949). — [22] On almost free linear motions (*3*, *5*). Ibid. **72**, 595—602 (1949). — [23] On linear asymptotic equilibria (*3*). Ibid. **71**, 853—858 (1949). — [24] On implicit analytic systems (*8*). Comm. Math. Helv. **23**, 294—302 (1949). [25] Linear differential equations and the oscillatory property of MACLAURIN's cosine series (*5*). Math. Gaz. **33**, 26—28 (1949). — [26] On free vibrations with amplitudinal limits (*3*). Quart. Appl. Math. **8**, 102—104 (1950). — [27] A criterion for the nonexistence of L^2-solutions of a nonoscillatory differential equation (*5*). J. London Math. Soc. **35**, 347—351 (1950). — [28] On the existence of Laplace solutions for linear differential equations of second order (*5*). Amer. J. Math. **72**, 442—450 (1950). — [29] On the nonexistence of conjugate points (*5*). Ibid. **73**, 368—380 (1951). — [30] On a theorem of BÔCHER in the theory of ordinary linear differential equations (*3*). Ibid. **76**, 183—190 (1954). — [31] Remarks to an earlier note (vol. **57**, 539—540) (*8*). Ibid. **76**, 717—720 (1954). — [32] On the bound of regularity of the solutions of analytic differential equations of first order (*8*). Quart. J. Math., Oxford Ser. (2) **5**, 145—149 (1954). — [33] On linear instability (*5*). Quart. Appl. Math. **13**, 192—195 (1955). — [34] Upon a theory of infinite systems of nonlinear implicit and differential equations (*8*). Amer. J. Math. **53**, 241—257 (1931). — [35] The analytical foundations of celestial mechanics (*). Princeton Univ. Press, 1947, pp. 448. — [36] On the Laplace-Fourier transcendents occurring in mathematical physics (*10*). Amer. J. Math. **69**, 87—98 (1947). — [37] A criterion for homogeneous differential equations with damped solutions (*5*). J. Math. a. Mech. **6**, 109—117 (1957). — [38] On criteria for linear stability (*3*). Ibid. **6**, 301—309 (1957). — [39] Stability and spectrum in the wave mechanics of lattices (*5*). Phys. Rev. (2) **72**, 81—82 (1947). — [40] Upon a theory of infinite systems of nonlinear implicit differential equations (*3*, *6*). Amer. J. of Math. **53**, 241—257 (1931). — [41] The infinities in the nonlocal existence problem of ordinary differential equations (*1*). Ibid. **68**, 173—178 (1946).

WITTICH, H.: [1] Ganze transzendente Lösungen algebraischer Differentialgleichungen (*8*). Nachr. Akad. Wiss. Göttingen **1946**, 71—73. — Math. Ann. **122**, 221—234 (1950). — [2] Über das Anwachsen der Lösungen linearer Differentialgleichungen (*8*). Ibid. **124**, 277—288 (1952).

WOLFSON, K. G.: [1] On the spectrum of a boundary value problem with two singular end points (*5*). Amer. J. Math. **72**, 713—719 (1950).

WOUK, A.: [1] Difference equations and *J*-matrices (*2*, *4*). Duke Math. J. **20**, 141—159 (1953).

WRIGHT, E. M.: [1] On the sequence defined by a nonlinear recurrence formala (*8*). J. London Math. Soc. **20**, 68—73 (1945). — [2] The nonlinear difference-differential equation (*8*). Quart. J. Math., Oxford Ser. (1) **17**, 245—252 (1946). — [3] Iteration of the exponential function (*2*). Ibid. **18**, 228—235 (1947). — [4] Linear difference-differential equations (*3*). Proc. Cambridge Phil. Soc. **44**, 179—185 (1948). — [5] The linear difference-differential equation with asymptotically constant coefficients (*3*). Amer. J. Math. **70**, 221—238 (1948). — [6] The linear difference-differential equation with constant coefficients (*2*). Proc. Roy. Soc. Edinburgh A **62**, 387—393 (1949). — [7] Perturbed functional equations (*8*). Quart. J. Math., Oxford Ser. **20**, 155—165 (1949). — [8] The stability of solutions of nonlinear difference-differential equations (*8*). Proc. Roy. Soc. Edinburgh A **63**, 18—26 (1950). — [9] The asymptotic expansion of integral functions defined by Taylor series (*10*). Philos. Trans. Roy. Soc. Lond. A **238**, 423—451 (1940); **239**, 217—232 (1941). — [10] The asymptotic expansion of integral functions and of the coefficients in their Taylor series (*10*). Trans. Amer. Math. Soc. **64**, 409—438 (1948).

WYLIE, C. R.: [1] On the forced vibrations of nonlinear springs (*8*). J. Franklin Inst. **236**, 273—284 (1943).

YAKUBOVIC, V. A.: [1] A certain criterion for reducibility of a system of differential equations (*3*). Doklady Akad. Nauk SSSR, (N.S.) **66**, 577—580 (1949). — [2] On the boundedness of the solutions of the equation $y'' + p(t)\,y = 0$, $p(t+\omega) = p(t)$ (*4*). Ibid. **74**, 901—903 (1950). — [3] Criteria of stability for systems of two canonical equations with periodic coefficients (*4*). Ibid. **78**, 221—224 (1951). — [4] Criteria for the stability of the solutions of systems of two linear differential equations with periodic coefficients (*4*). Uspehi Mat. Nauk **6**, 166—168 (1951). — [5] On the asymptotic behavior of the solutions of a system of differential equations (*8*). Mat. Sbornik, (N.S.) **28** (70), 217—240 (1951). — [6] An estimate of the characteristic exponents and criteria of stability for a linear differential equation of the second order with periodic coefficients (*4, 5*). Doklady Akad. Nauk SSSR, (N.S.) **87**, 345—384 (1952). — [7] Questions of stability of systems of two linear differential equations with periodic coefficients (*4*). Diss. Leningrad State Univ. **1953**, 1—13. — [8] Estimates of characteristic exponents of a system of linear differential equations with periodic coefficients (*4*). Prikl. Mat. Meh. SSSR. **18**, 533—546 (1954). — [9] Extension of LYAPUNOV's method of determining boundedness of solutions of the equation $y'' + p(t)\,y = 0$, $p(t+\omega) = p(t)$, to the case of a function $p(t)$ of variable sign (*4*). Ibid. **18**, 705—718 (1954).

YAMADA, H.: [1] On the calculation of free oscillations with intermediate nonlinearity (*8*). Rep. Res. Inst. Appl. Mech. Kyushu Univ. **1**, 11—21 (1952).

YAMAGUTI, M.: [1] On some properties of the non-linear differential (*8*) equations of the "parametric excitation". Mem. Coll. Sci. Univ. Kyoto **28**, 87—96 (1954).

YORISH, J. I.: [1] Constrained oscillations of systems in cases of broken characteristics of forces (*8*). Akad. Nauk SSSR Inzen. Sbornik **3**, no. 2, 126—136 (1947).

YOSHIZAWA, T.: [1] Note on the boundedness of solutions of a system of differential equations (*1, 6, 9*). Mem. Coll. Sci. Univ. Kyoto **28**, 27—32, 293—298 (1954). — [2] On the non-linear differential equation (*9*). Ibid. **28**, 133—141 (1954). — [3] On the convergence of solutions of the nonlinear differential equation (*9*). Ibid. **28**, 143—151 (1954). — [4] Note on the existence theorem of a periodic solution of the nonlinear differential equation (*9*). Ibid. **28**, 153—159 (1954). — [5] On the stability of solutions of a system of differential equations. Ibid. **29**, 27—33 (1955). — [6] Note on the equi-ultimate boundedness of solutions (*7*). Mem. Coll. Univ. Kyoto, (A) **31**, 211—217, 1958. — [7] Stability in the large (*7*). Ibid. **32**, 171—180, 1959. — [8] LIAPUNOV's functions (*7*). Funkcial. Ekvac. **2**, 95—142, 1959.

YOSIDA, K.: [1] On the asymptotic property of the differential equation $y'' + H(x)\,y = f(x, y, y')$ (*8*). Jap. J. Math. **9**, 145—152, 227—230 (1932).

YUROVSKII, A. V.: [1] On certain criteria for the stability of the integrals of a system of two linear differential equations with periodic coefficients (*4*). Doklady Akad. Nauk SSSR, (N.S.) **62**, 595—598 (1948).

ZADEH, L. A.: [1] The determination of the impulsive response of variable networks (*3*). J. Appl. Phys. **21**, 642—645 (1950).

ZAREMBA, S. K.: [1] Divergence of vector fields and differential equations (*9*). Amer. J. Math. **76**, 220—234 (1954).

ZATOPEK, A.: [1] Dynamical magnification of a seismograph excited by a shock of the form $\lambda^n e^{-\lambda\tau}\tau^n$ (*2*). Casopis Pest. Mat. Fys. **75**, 103—111 (1950).

ZEVAKIN, S.A.: [1] On finding the limit cycles of systems near to certain nonlinear ones (*9*). Prikl. Mat. Meh. SSSR, **15**, 237—244 (1951).

ZHABOTINSKII, M.E.: [1] Auto-oscillating systems with two degrees of freedom in the case of multiple frequencies (*8*). Akad. Nauk SSSR, Z. Eksper. Teoret. Fiz. **20**, 421—426 (1950). — [2] On periodic solutions of nonlinear partial differential equations (*8*). Doklady Akad. Nauk SSSR, (N.S.) **56**, 469—472 (1947). — [3] On a particular case of systems with two degrees of freedom (*8*). Akad. Nauk SSSR, Z. Eksper. Teoret. Fiz. **15**, 573—586 (1945).

ZHELEZCOV, N.A.: [1] Vysnegradskii diagrams for an isodromic regulator with non-linear action (*8*). Avtom. i Telem. **10**, 424—436 (1949). — [2] The method of point transformation and the problem of the forced vibrations of an oscillator with "combined friction" (*8*). Prikl. Mat. Meh. SSSR, **13**, 3—40 (1949). Amer. Math. Soc. Translation no. 57. — [3] On the theory of the symmetric multivibrator (*8*). Akad. Nauk SSSR., Z. Tehn. Fiz. **20**, 788—797 (1950).

ZIEGLER, H.: [1] Erzwungene Schwingungen mit konstanter Dämpfung (*2*). Ing.-Arch. **9** (1938).

ZLAMAL, M.: [1] Über ein Kriterium der Stabilität von LYAPUNOV (*4*). Cehoslovack. Mat. Z. **3** (78), 257—264 (1953). — [2] Oscillation criteria (*5*). Casopis Pest. Mat. Fys. **75**, 195, 213—218. — [3] Asymptotic properties of the solutions of the third order linear differential equations (*5*). Publ. Fac. Sci. Univ. Masaryk **1951**, 159—167. — [4] Nonlinear forced oscillations (*, *8*). Ibid. **77**, 53—64 (1952). — [5] Asymptotische Eigenschaften der Lösungen linearer Differentialgleichungen (*3*). Math. Nachr. **10**, 169—174 (1953).

ZUBOV, V.I.: [1] Some sufficient criteria for stability of a nonlinear system of differential equations (*3*). Prikl. Mat. Meh. SSSR, **17**, 506—508 (1953). — [2] On the theory of A. LYAPUNOV's second method (*7*). Doklady Akad. Nauk SSSR, (N.S.) **99**, 341—344 (1954); **100**, 857—859 (1955). — [3] Questions of the theory of LYAPUNOV's second method, construction of a general solution in the region of asymptotic stability (*7*). Prikl. Mat. Meh. SSSR, **19**, 179—210 (1955).

ZUKOVSKII, N.E.: [1] On the stability of motion (*1, 8*). Moscow Univ. Mem., Phys.-Math. **4**, 1—104 (1882). — [2] Conditions for finiteness of integrals of equations $y'' + py = 0$ (*4*). Mat. Sbornik **16**, 582—591 (1892).

ZWIRNER, G.: [1] Un criterio di confronto per equazioni differenziali del primo ordine (*6*). Ann. Univ. Ferrara (I) **7**, 213—216 (1948).

Index

Ergebnisse der Mathematik und ihrer Grenzgebiete

1. Bachmann: Transfinite Zahlen. DM 42,—
2. Miranda: Partial Differential Equations of Elliptic Type. DM 58,—
4. Samuel: Méthodes d'Algèbre Abstraite en Géométrie Algébrique. DM 29,—
5. Dieudonné: La Géométrie des Groupes Classiques. DM 38,—
6. Roth: Algebraic Threefolds with Special Regard to Problems of Rationality. DM 22,—
7. Ostmann: Additive Zahlentheorie. 1. Teil: Allgemeine Untersuchungen. DM 42,—
8. Wittich: Neuere Untersuchungen über eindeutige analytische Funktionen. DM 31,—
11. Ostmann: Additive Zahlentheorie. 2. Teil: Spezielle Zahlenmengen. DM 28,—
13. Segre: Some Properties of Differentiable Varieties and Transformations. DM 46,—
14. Coxeter/Moser: Generators and Relations for Discrete Groups. DM 35,—
15. Zeller/Beckmann: Theorie der Limitierungsverfahren. DM 64,—
16. Cesari: Asymptotic Behavior and Stability Problems in Ordinary Differential Equations. 3rd edition. DM 54,—
17. Severi: Il teorema di Riemann-Roch per curve-superficie e varietà questioni collegate. DM 26,—
18. Jenkins: Univalent Functions and Conformal Mapping. DM 37,—
19. Boas/Buck: Polynomial Expansions of Analytic Functions. DM 18,—
20. Bruck: A Survey of Binary Systems. DM 46,—
21. Day: Normed Linear Spaces. DM 17,80
23. Bergmann: Integral Operators in the Theory of Linear Partial Differential Equations. DM 40,—
25. Sikorski: Boolean Algebras. DM 42,—
26. Künzi: Quasikonforme Abbildungen. DM 43,—
27. Schatten: Norm Ideals of Completely Continuous Operators. DM 26,—
28. Noshiro: Cluster Sets. DM 40,—
29. Jacobs: Neuere Methoden und Ergebnisse der Ergodentheorie. Vergriffen
30. Beckenbach/Bellman: Inequalities. DM 38,—
31. Wolfowitz: Coding Theorems of Information Theory. DM 30,—
32. Constantinescu/Cornea: Ideale Ränder Riemannscher Flächen. DM 75,—
33. Conner/Floyd: Differentiable Periodic Maps. DM 29,—
34. Mumford: Geometric Invariant Theory. DM 24,—
35. Gabriel/Zisman: Calculus of Fractions and Homotopy Theory. DM 42,—
36. Putnam: Commutation Properties of Hilbert Space Operators and Related Topics. DM 31,—
37. Neumann: Varieties of Groups. DM 51,—
38. Boas: Integrability Theorems for Trigonometric Transforms. DM 20,—
39. Sz.-Nagy: Spektraldarstellung linearer Transformationen des Hilbertschen Raumes. DM 20,—
40. Seligman: Modular Lie Algebras. DM 43,—
41. Deuring: Algebren. DM 26,—
42. Schütte: Vollständige Systeme modaler und intuitionistischer Logik. DM 26,—
43. Smullyan: First-Order Logic. DM 36,—
44. Dembowski: Finite Geometries. DM 68,—
45. Linnik: Ergodic Properties of Algebraic Fields. DM 44,—
46. Krull: Idealtheorie. DM 28,—
47. Nachbin: Topology on Spaces of Holomorphic Mappings. DM 18,—
48. A. Ionescu Tulcea/C. Ionescu Tulcea: Topics in the Theory of Lifting. DM 36,—
49. Hayes/Pauc: Derivation and Martingales. DM 48,—
50. Kahane: Séries de Fourier Absolument Convergents. DM 44,—

51. Behnke/Thullen: Theorie der Funktionen mehrerer komplexer Veränderlichen. DM 48,—
52. Wilf: Finite Sections of Some Classical Inequalities. DM 28,—
53. Ramis: Sous-ensembles analytiques d'une variété banachique complexe. DM 36,—
54. Busemann: Recent Synthetic Differential Geometry. DM 32,—
55. Walter: Differential and Integral Inequalities. DM 74,—
56. Monna: Analyse non-archimédienne. DM 38,—
57. Alfsen: Compact Convex Sets and Boundary Integrals. DM 46,—
58. Greco/Salmon: Topics in m-Adic Topologies. DM 24,—
59. López de Medrano: Involutions on Manifolds. DM 36,—
60. Sakai: C*-Algebras and W*-Algebras. DM 66,—
61. Zariski: Algebraic Surfaces. DM 54,—